Novel Foods, Food-Grade Materials, and Environmentally Friendly Compounds from Food Waste

Novel Foods, Food-Grade Materials, and Environmentally Friendly Compounds from Food Waste

Guest Editors

Maria Martuscelli
Dino Mastrocola

MDPI

Basel • Beijing • Wuhan • Barcelona • Belgrade • Novi Sad • Cluj • Manchester

Guest Editors

Maria Martuscelli
Department of Bioscience and Technology for Food, Agriculture and Environment
University of the Studies of Teramo
Teramo
Italy

Dino Mastrocola
Department of Bioscience and Technology for Food, Agriculture and Environment
University of the Studies of Teramo
Teramo
Italy

Editorial Office

MDPI AG
Grosspeteranlage 5
4052 Basel, Switzerland

This is a reprint of the Special Issue, published open access by the journal *Foods* (ISSN 2304-8158), freely accessible at: https://www.mdpi.com/journal/foods/special_issues/novel_food_waste.

For citation purposes, cite each article independently as indicated on the article page online and as indicated below:

Lastname, A.A.; Lastname, B.B. Article Title. *Journal Name* **Year**, *Volume Number*, Page Range.

ISBN 978-3-7258-6598-7 (Hbk)
ISBN 978-3-7258-6599-4 (PDF)
https://doi.org/10.3390/books978-3-7258-6599-4

Cover image courtesy of Maria Martuscelli

Contents

About the Editors

Maria Martuscelli

Maria Martuscelli was born in Muro Lucano, Basilicata, Italy, in 1970. She currently works as an Associate Professor of Food Science and Technology (Department of Bioscience and Technology for Food, Agriculture, and Environment, University of Teramo, Italy). She is a Member of the Italian Society of Food Technology and is listed in the Register of Experts in Industrial Research. Many of her research outputs have involved scientific projects supported by public institutions or private companies. She has also contributed to foreign institutions such as the "Unité Mixte de Recherche et Flaveur Vision du consommateur comportement, of the Institut National de la Recherche Agronomique" in Dijon, France, and has been involved in data collection for the European Food Safety Authority, in relation to "biogenic amines in food". She is collaborating with Professor Dr. Meijin Guo's research group (State Key Laboratory of Bioreactor Engineering, East China University of Science and Technology, Shanghai, China). She is the author of 128 scientific publications, published in both international and national journals, as well as proceedings of national and international congresses and book chapters. Over her career, she was named a Top Scholar by ScholarGPS 2024 (in the top 0.5% of all scholars worldwide) in the biogenic amine area, ranking #25.

Dino Mastrocola

Dino Mastrocola was born in Guardiagrele (CH) in 1958. He currently works as a Full Professor of Food Science and Technology (Department of Bioscience and Technology for Food, Agriculture, and Environment) at the University of Teramo, Italy, and he was Rector of the same institution from November 2018 to October 2024. He is the President of the Italian Society of Food Sciences and Technologies, of which he is a founding member. He carries out research activities in the field of food science and technology and is involved in national and international research projects. He participated in the Food Improvement Project in China coordinated by prof. Antonino Zichichi at the International Centre for Scientific Culture—World Laboratory in Lausanne. He has conducted rigorous research in the R&D of Unilever Research in Colworth House (England) and at the Academy of Sciences in Beijing. From 1996 to 2003, he took part in the European food science dissemination project Flair Flow Europe. He is the author of 242 scientific publications, published in both international and national journals, as well proceedings of national and international congresses, including 20 essays.

foods MDPI

Editorial

New Perspectives to Enhance Wastes and By-Products from Agro-Food Processing

Maria Martuscelli [1,*], Luigi Esposito [1], Donatella Restuccia [2], Meijin Guo [3] and Dino Mastrocola [1]

1 Department of Bioscience and Technology for Food, Agriculture and Environment, University of Teramo, Via R. Balzarini 1, 64100 Teramo, Italy; lesposito2@unite.it (L.E.); dmastrocola@unite.it (D.M.)
2 Department of Pharmacy, Health and Nutritional Sciences, University of Calabria, 87036 Rende, Italy; donatella.restuccia@unical.it
3 State Key Laboratory of Bioreactor Engineering, East China University of Science and Technology, Shanghai 200237, China; guo_mj@ecust.edu.cn
* Correspondence: mmartuscelli@unite.it

The exploitation of by-products and waste from the agri-food industry represents a sustainable approach within the frame of the circular economy, the basis of the European Green Deal and ecological transition. The European Union recommends a reduction in food waste, but when this is not possible, the redirection of this waste for human consumption is recommended.

Bread and bakery products are among the most discarded food products in the world. They can be considered a high-glucose slurry; however, there is no research that considers the potential use of bread-waste hydrolysis, considering that this approach could also circumvent the costs of starch isolation, purification, and gelatinization. Sigüenza-Andrés et al. [1] studied a simultaneous hydrolysis of bread waste using α-amylase and glucoamylase; this research was carried out performing liquefaction and saccharification at the same time. This process was compared with a traditional sequential hydrolysis. The authors showed that the slurry produced with the simultaneous process had a yield higher than that obtained with sequential hydrolysis and could reduce both time and energy.

Specifically related to the biodegradation of waste cooking oils, studies on the development of large-scale bioremediation applications were carried out. Zahri et al. [2] studied the degradation of waste canola oil and pure canola oil using a consortium of native Antarctic soil bacteria. Zahri et al. [3] also confirmed that the Antarctic soil bacterial consortium (reference BS14) are able to biodegrade canola oil. In addition, kinetic studies were carried out to examine the ability of BS14 to produce biosurfactants via a biodegradation process. Secondary mathematical equations were chosen for kinetic analyses (Monod, Haldane, etc.). Biosurfactant production was confirmed through a preliminary screening test, with further optimisation by response surface methodology (RSM).

Recent studies reveal that by-products and waste are a significant source of bioactive compounds, which can be selectively extracted and reintroduced in the food chain to manufacture functional foods and/or be marketed as nutraceuticals.

Innovative techniques have been tested in comparison with traditional extraction systems. Ultrasonic processing has great potential to transform waste from the food and agriculture industries into value-added products. Wu et al. [4] discussed the use of ultrasound as a green technology, comparing it to conventional chemical extraction/processing methods. They also summarized the latest developments and their future potential as energy-efficient methods to convert food and agricultural waste into value-added products. Moreover, Mrkonjić et al. [5] demonstrated that the microwave-assisted extraction (MAE) technique is an efficient approach for the extraction of biologically active compounds from food waste. In particular, these authors applied this technique to enhance *Thymus serpyllum* herbal dust, which represents a high-value source of natural antioxidants with great potential for further use in various forms within different branches of the industry.

Aubourg et al. [6] showed the suitability of ethanol-containing green systems for the extraction of bioactive lipid compounds from squid (*Doryteuthis gahi*) by-products.

Research on food and related fields must encompass sustainability and encourage its introduction in food system production, aiming to obtain safe products with reduced preservative use. In this regard, promising results are reported by Panitsa et al. [7] regarding the use of encapsulated sodium benzoate in tubular cellulose derived from orange pulp to obtain orange juice using limited chemical preservatives. This system was capable of inhibiting the growth of spoilage microorganisms, also reducing their number (yeasts and lactic acid bacteria).

By-products from vegetable processing are produced in large amounts each year, by both vegetable production and transformation. A critical and decisive aspect regarding the use of food by-products as food or food ingredients is their approval by the responsible government bodies. In the European Union (EU), novel foods (i.e., foods not marketed before 1997) need approval (for safety and toxicological aspects) before they can be put on the market.

As an example, the coffee chain is a large producer of interesting by-products [8]. It has been suggested to increase the income of coffee farmers by valorising coffee leaf tea consumption. Coffee leaf tea is a traditional beverage in some coffee-producing countries and was authorized in 2020 within the EU according to novel food regulations [9]. Depending on selection and processing (age of the leaves, drying, fermentation, roasting, etc.), coffee leaf tea may exhibit a wide variety of flavours. Steger et al. [10] investigated the production of coffee leaf tea (*Coffea arabica*) in El Salvador and the influence of processing steps on non-volatile compounds and volatile aroma-active compounds.

Among other coffee by-products, coffee silver skin (CSS) is available in large amounts and is more stable due to its lower water content. Martuscelli et al. [11] carried out a study on CSS characterization and composition to verify its potentiality as a source of fibre, minerals, and bioactive molecules, while toxic minerals (e.g., nickel) were found at low levels. Martuscelli et al. [12] also investigated the use of coffee silver skin as an ingredient in meat products. In newly formulated chicken burgers, silver skin limited weight loss (after the cooking process) to 10% (1.5% addition) and 11% (3% addition), significantly lower ($p < 0.01$) than seen with the control (24%). Moreover, in cooked burgers, CSS reduced the occurrence of hexanal, and also showed good activity in the limitation of off-flavours, contrasting with the emergence of octanal, alcohols and other markers of lipid oxidation. The thiobarbituric acid reactive substances (TBARS) assay confirmed the protective effect of CSS on meat oxidation. Regarding novel food status assessment in the European Union (EU), CSS consumption is not recognized to a significant degree before 15 May 1997; therefore, consultation is currently in progress [13].

Other plant co-products such as cacao pod husks were also demonstrated to be valid new ingredients to improve the technological parameters, functional characteristics and stability of meat formulations. Delgado-Ospina et al. [14] added different cocoa pod husk flour levels as a starch replacement for reformulating frankfurters. Textural properties and sensory characteristics were affected, although these samples had higher water content, hardness, and adhesiveness, while their springiness decreased. Overall, newly formulated products demonstrated enriched dietary fibre and bioactive compounds, and had a high content of water-soluble pectins.

Bee pollen is a natural plant derivative that can be considered as a balanced, functional and health food. Only few studies exist on this natural product. Gonçalves et al. [15] expanded on the composition and phenolic profile of multifloral bee pollen from Abrantes (Portugal), as well as on its antioxidative and antidiabetic properties. These results contributed to the generation of deeper knowledge of potential health-promoting effects to develop novel pharmaceutical drugs, nutraceuticals, and dietary supplements.

The valorisation of by-products represents a sustainable method, both from an environmental and an economic point of view. In addition to this, these matrices are rich in biologically active compounds, with positive effects on the qualitative traits of products of animal origin, as well as on animal welfare. Grass-finished beef has demonstrated wide nutritional variations, with some grass-finished beef having a considerably higher *n*-6:*n*-3 ratio compared to grain-finished beef. Krusinski et al. [16] investigated the effects of commonly used supplemental feeds on the nutritional profile of grass-finished beef. There has also been increasing interest in using grass-fed beef by-products to augment the nutrient profile of eggs within local pasture-raising systems in the United States [17]. Another important aspect is the use of waste to improve the quality of milk and derived dairy products. The introduction of grape pomace in the diet of dairy ruminants has shown several advantages, particularly related to the fatty acid profiles of milk and cheese. Bennato et al. [18] demonstrated that dietary enrichment with grape pomace did not affect the yield and chemical composition of ewe's milk. In addition, total phenolic compounds and antioxidant activity were not affected by diet. The results of this study suggest that 10% of grape pomace can be included in the diet of lactating ewes without modifying milk gross composition but inducing significant changes in the fatty acid profile.

Finally, the prevalence of eco-friendly containers is increasing worldwide. Eco-friendly packaging materials include biodegradable and biomass products, and their continuous development is in progress. Lee et al. [19] carried out an interesting study in order to find out whether a biodegradable container made of pork gelatine can be used in industry. The containers were analysed with respect to morphological, mechanical and thermal properties, as well as biodegradability. Walnut shell powder increased the hardness, but slowed the biodegradation process. This study shows the high potentiality for applications of pork gelatine biodegradable containers for food packaging.

In conclusion, the collection of papers presented in this Editorial clearly shows that feasible and efficient solutions can be achieved by including food and agricultural wastes in a myriad of applications. These topics will be prevalent in future research globally. Finally, we cannot forget about the socio-economic benefits of the virtuous revalorisation of by-products. Due to their high quantities, they must be considered as independent production chains from which high value can be extracted.

Conflicts of Interest: The authors declare no conflict of interest.

References

1. Sigüenza-Andrés, T.; Pando, V.; Gómez, M.; Rodríguez-Nogales, J.M. Optimization of a Simultaneous Enzymatic Hydrolysis to Obtain a High-Glucose Slurry from Bread Waste. *Foods* **2022**, *11*, 1793. [CrossRef]
2. Zahri, K.N.M.; Zulkharnain, A.; Fuentes, C.; Sabri, S.; Khalil, K.A.; Convey, P.; Ahmad, S.A. The use of response surface methodology as a statistical tool for the optimisation of the waste and pure canola oil biodegradation by Antarctic soil bacteria. *Life* **2021**, *11*, 456. [CrossRef]
3. Zahri, K.N.M.; Khalil, K.A.; Gomez-Fuentes, C.; Zulkharnain, A.; Sabri, S.; Convey, P.; Lim, S.; Ahmad, S.A. Mathematical Modelling of Canola Oil Biodegradation and Optimisation of Biosurfactant Production by an Antarctic Bacterial Consortium Using Response Surface Methodology. *Foods* **2021**, *10*, 2801. [CrossRef]
4. Wu, Y.; Yao, S.; Narale, B.A.; Shanmugam, A.; Mettu, S.; Ashokkumar, M. Ultrasonic Processing of Food Waste to Generate Value-Added Products. *Foods* **2022**, *11*, 2035. [CrossRef]
5. Mrkonjić, Ž.; Rakić, D.; Takači, A.; Kaplan, M.; Teslić, N.; Zeković, Z.; Lazarević, I.; Pavlić, B. Polyphenols Recovery from *Thymus serpyllum* Industrial Waste Using Microwave-Assisted Extraction–Comparative RSM and ANN Approach for Process Optimization. *Foods* **2022**, *11*, 1184. [CrossRef]
6. Aubourg, S.P.; Trigo, M.; González, M.J.; Lois, S.; Medina, I. Comparative Study of Bioactive Lipid Extraction from Squid (*Doryteuthis gahi*) by-Products by Green Solvents. *Foods* **2022**, *11*, 2188. [CrossRef]
7. Panitsa, A.; Petsi, T.; Kandylis, P.; Kanellaki, M.; Koutinas, A.A. Tubular Cellulose from Orange Juice By-Products as Carrier of Chemical Preservatives; Delivery Kinetics and Microbial Stability of Orange Juice. *Foods* **2021**, *10*, 1882. [CrossRef]
8. Tritsch, N.; Steger, M.C.; Segatz, V.; Blumenthal, P.; Rigling, M.; Schwarz, S.; Zhang, Y.; Franke, H.; Lachenmeier, D.W. Risk assessment of caffeine and epigallocatechin gallate in coffee leaf tea. *Foods* **2022**, *11*, 263.
9. Lachenmeier, D.W.; de Rajcic, R.T.; Schwarz, S. An update on sustainable valorization of coffee by-products as novel foods within the European Union. *Biol. Life Sci. Forum* **2022**, *6*, 37.
10. Steger, M.C.; Rigling, M.; Blumenthal, P.; Segatz, V.; Quintanilla-Belucci, A.; Beisel, J.M.; Rieke-Zapp, J.; Schwarz, S.; Lachenmeier, D.W.; Zhang, Y. Coffee Leaf Tea from El Salvador: On-Site Production Considering Influences of Processing on Chemical Composition. *Foods* **2022**, *11*, 2553. [CrossRef]

11. Martuscelli, M.; Esposito, L.; Di Mattia, C.D.; Ricci, A.; Mastrocola, D. Characterization of Coffee Silver Skin as Potential Food-Safe Ingredient. *Foods* **2021**, *10*, 1367. [CrossRef]
12. Martuscelli, M.; Esposito, L.; Mastrocola, D. The Role of Coffee Silver Skin against Oxidative Phenomena in Newly Formulated Chicken Meat Burgers after Cooking. *Foods* **2021**, *10*, 1833. [CrossRef]
13. Lachenmeier, D.W.; Schwarz, S.; Rieke-Zapp, J.; Cantergiani, E.; Rawel, H.; Martín-Cabrejas, M.A.; Martuscelli, M.; Gottstein, V.; Angeloni, S. Coffee By-Products as Sustainable Novel Foods: Report of the 2nd International Electronic Conference on Foods—"Future Foods and Food Technologies for a Sustainable World". *Foods* **2022**, *11*, 3. [CrossRef]
14. Delgado-Ospina, J.; Martuscelli, M.; Grande-Tovar, C.D.; Lucas-González, R.; Molina-Hernandez, J.B.; Viuda-Martos, M.; Fernández-López, J.; Pérez-Álvarez, J.Á.; Chaves-López, C. Cacao Pod Husk Flour as an Ingredient for Reformulating Frankfurters: Effects on Quality Properties. *Foods* **2021**, *10*, 1243. [CrossRef]
15. Gonçalves, A.C.; Lahlou, R.A.; Alves, G.; Garcia-Viguera, C.; Moreno, D.A.; Silva, L.R. Potential Activity of Abrantes Pollen Extract: Biochemical and Cellular Model Studies. *Foods* **2021**, *10*, 2804. [CrossRef]
16. Krusinski, L.; Maciel, I.C.F.; Sergin, S.; Jambunathan, V.; Garg, E.; Garmyn, A.J.; Singh, S.; Bitler, C.A.; Rowntree, J.E.; Fenton, J.I. Effects of Hay, Baleage, and Soybean Hulls Waste Used as Supplemental Feeds on the Nutritional Profile of Grass-Finished Beef. *Foods* **2022**, *11*, 3856. [CrossRef] [PubMed]
17. Sergin, S.; Jambunathan, V.; Garg, E.; Rowntree, J.E.; Fenton, J.I. Fatty Acid and Antioxidant Profile of Eggs from Pasture-Raised Hens Fed a Corn- and Soy-Free Diet and Supplemented with Grass-Fed Beef Suet and Liver. *Foods* **2022**, *11*, 3404. [CrossRef]
18. Bennato, F.; Ianni, A.; Florio, M.; Grotta, L.; Pomilio, F.; Saletti, M.A.; Martino, G. Nutritional Properties of Milk from Dairy Ewes Fed with a Diet Containing Grape Pomace. *Foods* **2022**, *11*, 1878. [CrossRef]
19. Lee, S.-H.; Kim, H.-Y. Utilization of By-Products from Livestock: Study on the Mechanical and Thermal Properties of Biodegradable Containers Made with Pork Skin Gelatin Polymer. *Foods* **2022**, *11*, 2513. [CrossRef] [PubMed]

foods

MDPI

Review

Ultrasonic Processing of Food Waste to Generate Value-Added Products

Yue Wu [1,†], Shunyu Yao [1,†], Bhakti Anand Narale [2,†], Akalya Shanmugam [2,3,*], Srinivas Mettu [4,*] and Muthupandian Ashokkumar [1,*]

1 Sonochemistry Group, School of Chemistry, The University of Melbourne, Parkville, VIC 3010, Australia; wuyw7@student.unimelb.edu.au (Y.W.); shunyuy3@student.unimelb.edu.au (S.Y.)
2 Food Processing Business Incubation Centre, National Institute of Food Technology, Entrepreneurship and Management-Thanjavur, Thanjavur 613005, India; bhakti@iifpt.edu.in
3 Centre of Excellence in Non-Thermal Processing, National Institute of Food Technology, Entrepreneurship and Management-Thanjavur, Thanjavur 613005, India
4 Chemical and Environmental Engineering, School of Engineering, RMIT University, Melbourne, VIC 3000, Australia
* Correspondence: akalya@iifpt.edu.in (A.S.); srinivas.mettu@rmit.edu.au (S.M.); masho@unimelb.edu.au (M.A.)
† These authors contributed equally to this work.

Abstract: Ultrasonic processing has a great potential to transform waste from the food and agriculture industry into value-added products. In this review article, we discuss the use of ultrasound for the valorisation of food and agricultural waste. Ultrasonic processing is considered a green technology as compared to the conventional chemical extraction/processing methods. The influence of ultrasound pre-treatment on the soluble chemical oxygen demand (SCOD), particle size, and cell wall content of food waste is first discussed. The use of ultrasonic processing to produce/extract bioactives such as oil, polyphenolic, polysaccharides, fatty acids, organic acids, protein, lipids, and enzymes is highlighted. Moreover, ultrasonic processing in bioenergy production from food waste such as green methane, hydrogen, biodiesel, and ethanol through anaerobic digestion is also reviewed. The conversion of waste oils into biofuels with the use of ultrasound is presented. The latest developments and future prospective on the use of ultrasound in developing energy-efficient methods to convert food and agricultural waste into value-added products are summarised.

Keywords: ultrasound processing; food waste; extraction; bio-energy products; sustainable goals

1. Introduction

Food waste consists of amino acids, proteins, lipids, polysaccharide, fibre, and other compounds [1]. Generally, it is generated by fresh produce markets, the agricultural industry, commercial restaurants and canteens, food processing plants, and food supply chain industries. Nowadays, large amounts of food waste are generated, and the Food and Agriculture Organization has reported that over 1 billion tons of food from human consumption are wasted [2]. Therefore, food loss and food waste directly affect the economy, environment, and society, which may ultimately hinder the development and progress achieved through efficient agriculture production. In fact, around 931 million tons of edible food have been wasted at downstream stages, in which household consumption accounted for 61%, followed by food service (26%) and retail (13%) [3]. Usually, food waste is treated by landfilling, incineration, composting, and anaerobic digestion [4]. However, the organic matter content in food waste can reach up to 80% of the dry mass [5]. In addition, food waste has a high water content and abundant organic matter. These characteristics, which can be used to upgrade the food waste to value-added products, are wasted in the treatment of food waste using incineration, landfill, or anaerobic digestion technologies [6]. For example, anaerobic digestion has some limitations, such as long operation time, low efficiency, and environmental pollution. Similarly, landfills require enough space and a long degradation time to be able to treat food waste. Moreover, incineration is associated with high cost and energy consumption (fuel and oxygen). Therefore, novel technologies are emerging to improve the utilization of food waste in order to produce value-added products.

The chemical treatment of food waste to extract valuable compounds is not sustainable as it uses petrochemical-based solvents. For the sustainable valorisation of food waste, there is a need to develop clean technologies that do not use chemicals derived from fossil fuel. Ultrasound is one viable option emerging as a clean technique that can be utilised in food processing areas, for extraction [7–9], homogenisation [10], and physicochemical property modification [11]. When power ultrasound (20 kHz frequency) is applied to the materials, the cavitation bubbles are generated in a liquid medium. The cavitation bubbles oscillate and collapse, generating high shear forces (Figure 1). The tremendous energy is released during the cavitation [12], which then leads to increased mass transfer and cell disruption for extraction purposes [13,14]. Although high-intensity ultrasound is an effective method for cell rupture, specific kinds of spores and microorganisms [15] sometimes have hard cell walls that ultrasound cannot fully disrupt as a standalone method. Currently, ultrasound is a mature technology that can be utilised independently or associated with other technology. With other technique combinations, ultrasound can thus be highly effective. General ultrasonic combinations serve as a pre-treatment of food waste for enhanced bioenergy production, including physical, thermal [16], microwave [17–19], high hydrostatic pressure [7], and chemical approaches, which are used together with enzymes [20,21], ionic liquid [22], acid [23,24], and alkali [25,26]. For the purpose of creating improved properties, ultrasound can act as a modification method, directly influencing the physicochemical characteristics of food waste materials. Therefore, ultrasound can be utilized to generate valuable compounds from food waste.

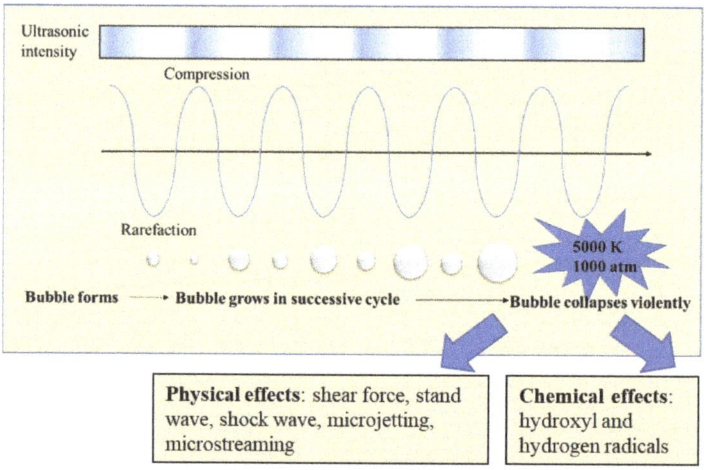

Figure 1. A schematic representation of the acoustic cavitation phenomenon.

This review focuses on the application of ultrasound and its effects on food waste. First, the influence of ultrasound pre-treatment on soluble chemical oxygen demand, particle size, and the cell wall content of food waste is presented. Then, the ultrasound-assisted extraction of oil, polyphenol, polysaccharides, and proteins from various food waste is summarised. Next, the ultrasound-promoted production of bioenergy, including bio-methane, bio-hydrogen, biodiesel, and bioethanol from food waste is discussed. Then, we discuss the current challenges and future developments in relation to the utilization of ultrasound on food waste.

2. The Influence of Ultrasound Pre-Treatment on Food Waste Materials

2.1. The Impact of Ultrasound on Organic Substance Solubilisation

Pre-treatment is a crucial step prior to enhanced bio-energy production. Anaerobic digestion yields can be accelerated using pre-treatment. The typical conversion process of food waste to bio-energy products includes sequential complex biochemical reactions during anaerobic digestion, which significantly impacts biogas production yield. It has been reported that organic substrate availability and mass transfer are the significant parameters that affect anaerobic digestion [27]. It is acknowledged that improving the organic matter solubilisation can effectively enhance the conversion rate, thus increasing the bio-energy production yields [28–30].

Ultrasound is an effective way to disrupt floc and cells due to the extreme pressures, high shear forces, violent turbulence, primary oxidative radicals, among other physical effects generated by acoustic cavitation [31–33]. Nevertheless, the efficiency of sonication was based on other parameters, such as ultrasound power, ultrasound time, and raw materials [34,35]. Figure 2 shows the morphology of the floc structure of the sludge before and after sonication pre-treatment using SEM. The sludge floc has an overall uneven structure at the start, and while sonication proceeds, a more consistent structure and smaller size can be observed, owing to floc disintegration [36]. The morphology shown in the figure indicates that with a longer sonication time, higher sonication power had more influence on the floc disintegration process.

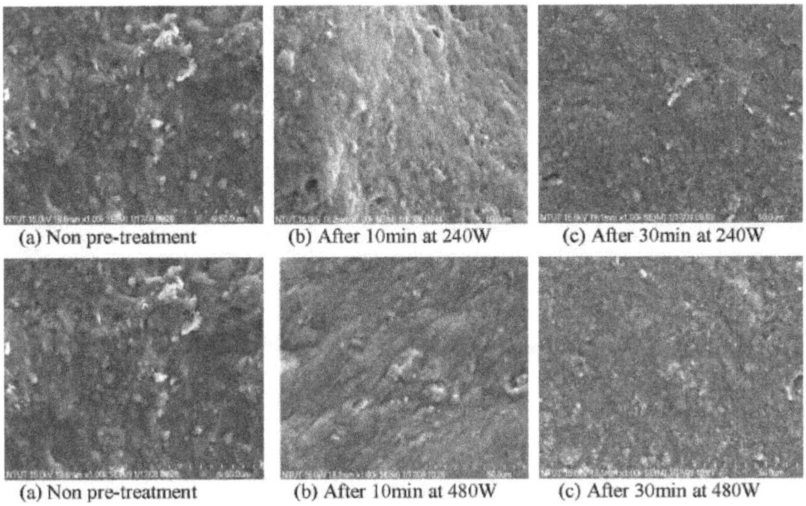

Figure 2. The morphology of the floc structure of sludge before and after ultrasonic pre-treatment under SEM [36]. Note: different lowercases represents food waste without ultrasound treatment (**a**), with sonication for 10 min (**b**) and with sonication for 30 min (**c**).

The chemical oxygen demand (COD) is normally used to determine the number of organic substances in the waste [37], and soluble chemical oxygen demand (SCOD) can be an indicative method of reflecting the influence of ultrasound on organic substance solubilisation, which has been listed in Table 1. It can be observed in the table that different parameters can not only affect the sonication efficiency, as mentioned, but they can also exert their influence on organic substance solubilisation. Abhijit [38] reported an increase of ~56% in SCOD release as compared to samples without sonication. Wang et al. [39] investigated the influence of sonication on food waste in relation to the production of volatile fatty acids. They first applied the same ultrasound conditions onto various total solid (TS) content food waste. Higher ΔSCOD was observed in lower TS content samples due to the intensified power density in the system. Joshi and Gogate [40] carried out a systematic study of different US parameters of food waste in order to improve the anaerobic digestion process for biogas production. The highest SCOD value of 18,500 ± 20 mg/L (with a 61.5% increase) was observed for the following conditions: 10 min sonication time, 0.4 W/mL power density, and 60% duty cycle (US pulse mode), which was treated as the optimum. Generally, a longer sonication time and higher power will give a higher SCOD value [41–44]. However, a decrease in SCOD value was observed when the US power density increased due to a higher power density, leading to the degradation of organic compounds and the cushioning effect [45]. In addition to the standalone utilisation of sonication, Elbeshbishy et al. [46] investigated the use of ultrasound with other techniques and their combined effects on organic matter solubilisation. They combined the thermal (heat) and chemical (acid and base) method with ultrasound, and the highest SCOD increase of 33% was obtained for the Ultrasound-base pre-treatment.

Table 1. The influence of ultrasound on SCOD changes.

Source/Composition of Food Waste	Ultrasonic Conditions	Outcomes	Reference
A mixture of rice, cabbage, pork, and tofu waste with different total solid (TS) content	20 kHz, 480 W/L, 15 min	TS = 40 g/L, ΔSCOD: +157% TS = 100 g/L, ΔSCOD: +172%	[39]
Sewage sludge	27 kHz, 200 W/L	2.5 min, ΔSCOD: +239% 10 min, ΔSCOD: +577%	[41]
Activated sludge	20 kHz, 21 kJ/g TS, 9 min	ΔSCOD: +246%	[47]
A mixture of wheat, gram flour, rice, fruit peel, and vegetable waste	20 kHz, 0.4 W/mL, 10 min	ΔSCOD: +61.5%	[40]
Food waste obtained from the Dufferin Organics Processing Facility	20 kHz, 79 kJ/g TS	ΔSCOD: +25%	[46]
Food waste from the Aurora treatment plant	20 kHz, 10,384, 15,577, 20,769 kJ/kg TS for 30, 45, 60 min US sample	30 min, ΔSCOD: +10.3% 45 min, ΔSCOD: +29.4% 60 min, ΔSCOD: +37%	[48]
Dairy waste	20 kHz, 0.6 W/mL, 25 min	ΔSCOD: +28.4%	[49]
Complex food waste	20 kHz, 16,875 kJ/kg TS, 15 min	ΔSCOD: +56.5%	[38]
Algae	20 kHz, 30 min, 36,000 KJ/Kg TS	Maximum ΔSCOD: +1950% was observed at 200 W US power	[50]
Pulp mill	20 kHz, 3.1 W/mL	15 min, ΔSCOD: +14.9% 60 min, ΔSCOD: +44.3%	[42]
Food waste	20 kHz, 5000 kJ/kg	ΔSCOD: +9.0%	[51]
Rice, noodles, vegetables, and meat waste obtained from a cafeteria at Harbin Institute	20 kHz, 1.25 W/mL, 30 min	ΔSCOD: +115%	[52]
Activated sludge	24 kHz, 1690 and 3380 kJ/kg TS for 5- and 10 min sonication sample	5 min, ΔSCOD: +17% 10 min, ΔSCOD: +21%	[43]
Activated sludge	20 kHz, 1.04 W/mL, 2.5 min	ΔSCOD: +12.6%	[53]
Food waste	20 kHz, 2 W/mL, 15 min	ΔSCOD: +71.5%	[54]
Algae sludge	40 kHz, 30 min, the power density was not mentioned	ΔSCOD: +520%	[55]
Food waste	20 kHz, 23 kJ/g TS, 30 min	ΔSCOD: +22.1%	[56]
Digestate	20 kHz, 3000 kJ/kg TS	ΔSCOD: +21%	[57]
Organic waste from the food industry	20 kHz, 50,000 kJ/kg TS	ΔSCOD: +20%	[58]
Dairy digestate waste	20 kHz, 15,000 kJ/kg TS	ΔSCOD: +15%	[59]
Food waste	20 kHz, 6946 kJ/kg TS, 30 min	ΔSCOD: +159%	[60]
Solid organic waste	20 kHz, 15,000 kJ/kg TS	ΔSCOD: +9.0%	[61]
Olive mill wastewater	20 kHz, 0.4 W/mL, 10 min	ΔSCOD: +23%	[62]
Solid waste	20 kHz, 0.2 W/mL, 60 min	ΔSCOD: +18.5%	[63]
Fermentation residues	20 kHz pulsed US (4 s on, 6 s off), 2 W/mL, 30 min	ΔSCOD: +39.5%	[25]
Seed sludge from food factory	20 kHz, 200 W/L, 45 min	ΔSCOD: +11%	[64]

2.2. The Influence of Ultrasound on Particle Size

As mentioned earlier, sonication is a practical approach for cell wall disruption [65–67]. The cell rupture occurs due to high shear forces and the generation of the extreme pressures and high local temperature conditions through acoustic cavitation. These conditions bring about characteristic modification via particle size alteration [68–70]. Therefore, ultrasound has been extensively applied in the biochemical processing area, such as in wastewater treatment [71], drug delivery [72], enhanced protein functionality [73], surface cleaning [74], among others. Today, ultrasound is also used in the pre-treatment of food waste to make it more viable for ensuing bio-energy production. Usually, the assistance of sonication can lead to particle size reduction, as depicted in Figure 3, hence expanding the surface area for the enzymatic attack [70]. Pejin et al. [75] implied that the ultrasound pre-treatment of triticale can reduce particle size, thereby enhancing the saccharification yield. Likewise, Nitayavardhana et al. [76] utilised ultrasound at 8 W/mL power density for 40 s to rupture the lignocellulosic cell structure in the cassava chip slurries; accordingly, particle size was considerably decreased from 600 to 15 μm. Similarly, another case of a significant reduction in particle size was reported by Khanal et al. [77]; the particle size of corn declined by nearly 20-fold with the application of ultrasonic pre-treatment under a high-power mode. An integrated investigation of the effectiveness of ultrasonic pre-treatment of food waste was carried out by Li et al. [70], who suggested that particle size reduction is a function of power and time of sonication. For instance, the particle size of the samples treated with 0.39 W/mL and 0.34 W/mL ultrasound was found to have decreased by 74% and 34%, respectively, when compared to the control sample. In terms of the waste mixture, the study of the sonication pre-treatment of a mixture of food waste and cardboard performed by Begum et al. [48] showed that ultrasound still had a beneficial effect on droplet size reduction, which decreased 10-fold for a 1:1 mix rate with ultrasonic pre-treatment as compared to the untreated samples.

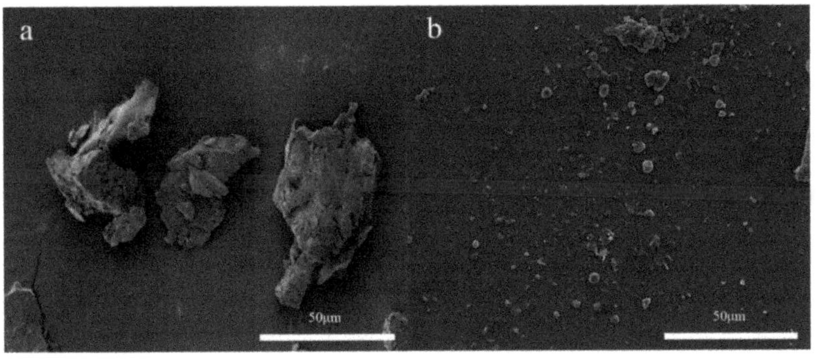

Figure 3. The morphology of food waste samples under SEM: (**a**) control (untreated) food waste sample, (**b**) sonicated food waste sample using a 35 mm horn (at 0.34 W/mL power). The SEM pictures were obtained at a magnification of 2000 [70]. Note: The scale bar is 50 μm.

2.3. The Effects of Ultrasound on Typical Polysaccharides

The composition of food waste varies between different species. Land plants (vegetables, fruits) are the most representative type of food waste (Table 2). The cell wall of these materials is ruptured as a result of ultrasound pre-treatment. Their fundamental components such as typical polymers (cutin, pectin) and a majority of polysaccharides (lignin, cellulose, hemicellulose) may be affected [78]. In other food wastes, the polysaccharides are primarily composed of lignin, cellulose, and hemicellulose, with lignin governing the interaction net between each content, thus leading to a complicated and obstinate structure [40]. This complex structure, typically with high lignocellulosic material content, makes further production less efficient. However, it is conceivable that after ultrasonic pre-treatment, these available polysaccharides are readily decomposed into monomeric sugars and further used for bio-energy production [79]. Moreover, it has been established that ultrasound is a proven predominant approach for lignin barrier disruption, hemicellulose removal, and cellulose crystallinity reduction. The removal of the lignin content is ascribed to bond oxidation during the ultrasound [80], and hemicellulose removal is attributed to its amorphous nature [81]. The reduction in lignin and hemicellulose content leads to an improvement in cellulose content [60]. Therefore, enhancing the accessibility to cellulose (disintegration of hydrogen bond in cellulose microfibrils) allows for further bio-energy production yield to be improved [42]. In previously reported studies, Ji et al. [82] investigated how different ultrasonic parameters affect these polysaccharides in vegetable wastes. It was discovered that in the single frequency sonication, the lignin removal rate decreased from 85.66 to 78.36%, when ultrasound frequency increased from 20 to 60 kHz, which might be ascribed to the decrease in ultrasonic cavitation. Perrone et. al. [83] noted that sugarcane bagasse samples pre-treated with sonication can effectively reduce hemicellulose and lignin content but enhance cellulose content from 41.3 to 51.8%. Likewise, Velmurugan and Muthukumar [84] also reported a dramatic reduction in hemicellulose and lignin percentage with alkaline-ultrasound pre-treatment, but a significant increase in the cellulose content of sugarcane bagasse was observed. On the contrary, high losses of cellulose content in spent coffee waste were reported by Ravindran [85], which might be attributed to a prolonged exposure to the ultrasonic environment. Comparably, the research carried out by Zhang et al. [86] found similar results in rice hull, which indicated that extrusion pre-treatment could lead to the degradation of cellulose, thereby reducing cellulose content even with the use of sonication. The delignification of biomass to hydrolyse lignin is a critical step in preparing the biomass for a further step in downstream processing such as fermentation [87]. Saratale et al. obtained a lignin removal rate of 25% in wheat waste with an individual ultrasonic pre-treatment [88]. Ong et al. [89] reported a delignification rate of 36.4% in oil palm fronds with an ultrasound-assisted deep eutectic solvent (DES) pre-treatment. However, Gaudino et al. [90] achieved the highest delignification rate of 45% in wheat straw, owing to the fact that higher lignin depolymerization was triggered by greater ultrasonic power. Mohapatra et al. [91] conducted a systematic study of the delignification efficiency on the Pennisetum sp. They reported that the highest delignification rates of 80.4% and 82.1% for Denanath grass (DG) and Hybrid Napier grass (HNG) were obtained with ultrasound-assisted pre-treatment. Furthermore, they also recognized that a more significant fraction of the sonication duty cycle, a higher temperature, and more intensive ultrasonic power have beneficial impacts on lignin elimination.

Table 2. The variation in the amounts of cellulose, hemicellulose, and lignin in an extract after ultrasonic treatment.

Source/Composition of Food Waste	Ultrasonic Conditions	Outcomes	Reference
Wheat waste	20 kHz, 30 min, NaOH (2% *w*/*v*) combination	Δcellulose: +13.2%; Δlignin: −10.1%	[88]
Vegetable waste	20 kHz, 20 min	Δcellulose: +23.1%; Δhemicellulose: −9.0%; Δlignin: −10.2%	[82]
Oil palm fronds	20 kHz, 50 min	Δlignin: −5.8%	[89]
Wheat straw	20 kHz, 120 min at 50 °C	Δlignin: −6.2%	[90]
Rice straw	30 kHz, water bath 90 °C, 4 h	Δlignin: −4.6%	[92]
Rice hull	40 kHz, 500 W, 1.5 h	Δcellulose: −2.8%; Δhemicellulose: −3.7%; Δlignin: +1.6%	[86]
Sugarcane bagasse	22 kHz, 50 W, 25 min, ozone/alkaline assisted	Δcellulose: +10.5%; Δhemicellulose: −8.1%; Δlignin: −15.6%	[83]
Sugarcane bagasse	24 kHz, 500 W, water bath 40 °C	Δcellulose: +21.4%; Δhemicellulose: −18.6%; Δlignin: −8.4%	[84]
Spent coffee waste	47 kHz, 310 W, 20 min	Δlignin: −8.8%	[93]

In addition to the polysaccharide content changes, the crystallinity of the amorphous component may also be altered after pre-treatment. The impacts of ultrasound are generated mainly by ultrasonic cavitation, which leads to violent shear forces, intense turbulence, high temperature, and pressure in a localized area [94]; this can in turn lead to the disruption of lignin content, thus increasing the accessibility of the cellulose component. Moreover, the increase in the crystallinity index after sonication is ascribed to the higher cellulose content, and the results can also be used to determine the extent of the delignification process [45]. Figure 4 demonstrates representative XRD patterns of the control and the pre-treated samples; XRD was utilized to examine the intensity of amorphous and crystalline areas at 18° and 22° (2θ), respectively [95]. It can be observed from the figure that at the 2θ = 22° crystalline region, a higher crystallinity was achieved after ultrasonic pre-treatment. The higher index is due to ultrasonic pre-treatment disrupting the lignin network and increasing the available surface area, as previously mentioned. Likewise, Velmurugan and Muthukumar [96] reported that in sugarcane bagasse, the crystallinity index of the pre-treated ultrasonic sample was found to be 16% more than the control sample. Sindhu et al. [97] observed that the crystallinity index in chilli post-harvest residue for the control and ultrasonic pre-treated samples were 35.43% and 52.32%, respectively. They also achieved the highest crystallinity index of 63.62% with the sono-assisted acid pre-treatment. Nonetheless, some investigations have also indicated that ultrasonic pre-treatment might decrease the crystallinity index. Hydroxyl groups in cellulose macromolecules form intramolecular and intermolecular hydrogen bonds, resulting in a variety of structured crystal lattices [98]. As mentioned, the extreme conditions caused by ultrasound can effectively disintegrate hydrogen bonds in the cellulose component and modify its crystalline arrangement, thus leading to the crystallinity reduction [99]. Jin et al. [100] investigated sweet potato and reported that ultrasound damaged its cellulose structure, thus resulting in a reduction in its crystallinity zone (2θ = 22°). Similarly, Ninomiya et al. [101] discovered that ultrasonic pre-treatment in bamboo powder was more effective in crystalline reduction than in the thermal method. It was found that sample products from the ultrasound treatment could be further reduced by 6.1% more than the thermal treatment.

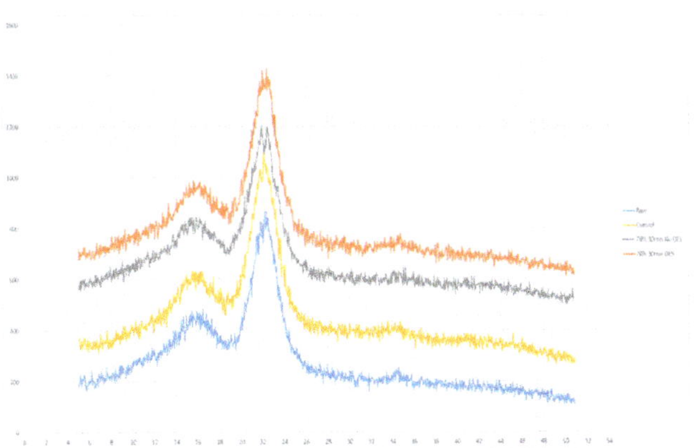

Figure 4. The X-ray diffraction (XRD) spectra of control and pre-treated samples [89].

3. Ultrasound-Assisted Extraction of Bioactives from Food Waste

3.1. Conventional Methods of Extraction

Extraction is a mass-transport process where a solute migrates from a matrix to a solvent [102]. Distillation by water, distillation by steam, distillation by a combination of water and steam, Soxhlet extraction, maceration, decoction, and percolation are the most common conventional processes used in the extraction of useful materials from waste [103]. Liquid hot water saccharification (particularly optimization at the lab scale) and steam explosion (especially suited for industry) are generally practiced amongst all the thermal treatments of extraction. Both treatments use a similar matrix destructuring mechanism, which is dependent on biomass treatment using hot water at high-pressure and its release, either slowly (in a liquid hot-water treatment) or abruptly (steam explosion) [104]. Recent research has discovered that a steam explosion-assisted extraction is highly effective, particularly for the extraction of phenolic acid, and the recovery yield also increases with a rise in residence time and temperature [105]. Conventional treatments are economical; however, there are multiple factors that bring out the complexity in the process, especially with the use of solvents, the longer processes, and the sequences in the processing method. The density, viscosity, reactivity, the volume of utilization of the medium employed in the process are some of the factors that complicate the processing. Even with the challenges above, Soxhlet extraction is very much used as a base technique for comparison against any new processes [106]. Modern tools of extraction include the use of ultrasound [107], pulse electric field [108], supercritical extraction and microwave [109], etc. Ultrasound is considered as one major emerging tool for the extraction of useful materials from food waste [107]. Ultrasound-based extraction has the following advantages in comparison to all emerging technologies: higher yield, lesser medium of extraction, lower capital, lesser time and energy, low operating temperature, and prevention of purity [107]. Figure 5 represents the schematic illustration of various conventional methods of extraction of bioactives or useful materials from different food wastes.

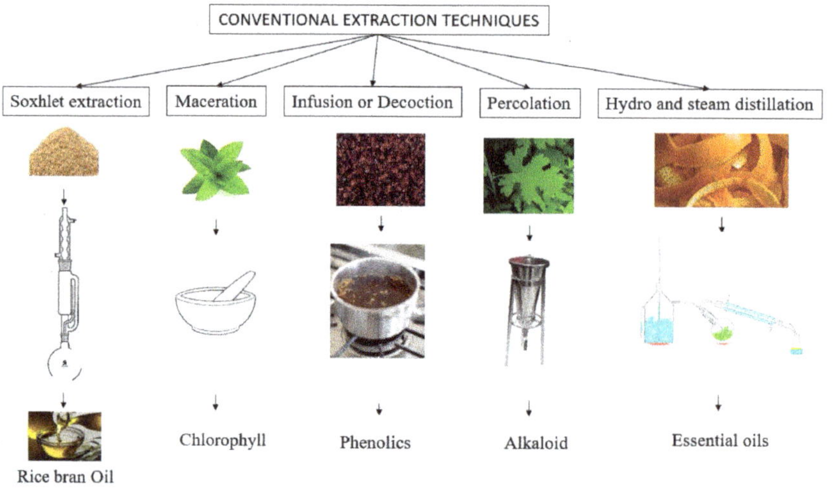

Figure 5. Schematic illustration of various conventional methods of extraction of bioactives or useful materials from different food wastes.

3.2. Extraction Using Ultrasound

The use of ultrasound in extraction processes has grown significantly in recent decades, especially in the food and natural product fields, for both traditional and environmentally novel extraction approaches. Ultrasound-assisted extraction (UAE) is an environmentally friendly, low-cost method that adheres to all green chemistry standards. When compared to conventional extraction processes, UAE reduces the extraction time, energy, and solvent consumption, while maintaining the high purity of the end product [110]. Shear forces, pressure changes, agitation, cavitation, microjets, radical generation as well as fragmentation, erosion, capillarity, and sonoporation all contribute to the extraction impact of ultrasound, which occurs both in the solvent and in the matrix, as pointed out by Chemat et al. [110]. The cavitation effects of ultrasound appear to be the main cause of this sonication effect. Microbubbles are formed when ultrasonic vibrations create a series of low-pressure and high-pressure zones in a medium such as solvent. Because of the rapid and continual pressure changes, these bubbles grow and eventually burst. When the bubbles implosively collapse, breaking points emerge on the surface of the cell, thus contributing to the increased transfer of mass in and out of cells. This approach is valuable for the extraction of useful materials such as polyphenols and many different molecules from a range of food wastes that employ different solvents [111].

3.3. Ultrasonic Extraction of Useful Materials from Cereal Processing Waste

With the significant production of husk, bran, and germ as the principal cereal by-products, the large yields are followed by post-harvest losses in all phases of product conversion. Plant by-products are prone to microbial breakdown due to their high water content and significant organic load, resulting in environmental difficulties when discarded and increasing waste treatment expenses for food makers. The recovery of biogenic substances from low-cost plant by-products could provide economic and environmental benefits as well as a cash return for enterprises rather than incurring disposal expenditures [112]. Food by-products such as wheat and rice brans, oat hulls, and wheat germ contain significant levels of phytosterols and polysaccharides. Phytosterols have been shown to have a positive impact on human health. Bioactive polysaccharides found mostly in cereals are known as β-glucans [113]. In comparison to rye (1–2%) and wheat (1%), barley (3–11%) and oat (3–7%) had higher levels of β-glucans.

Ultrasonication is a technique for destructing plant cell walls and releasing polysaccharides and other difficult-to-extract substances at lower temperatures and in less time. Small intervals of UAE were utilized for separating components of hemicellulose (mostly heteroxylans that are rich in polyphenols) from commercial wheat bran that gave sugar yields comparable to regular alkaline extraction. Furthermore, the procedure could be cut in half by using less sodium hydroxide, resulting in a 60% reduction in time [114]. UAE allows for a reduction in traditional solvent use or, alternatively, the use of water or greener solvents such as ethanol; hence, it is a reliable green technology. In one study, the response surface approach was used to optimise a UAE technique for the extraction of major phenolic acids (ferulic, salicylic, 2-hydroxycinnamic, p-coumaric, t-ferulic, caffeic, and vanillic acids) from the bran layers of certain key ancient wheat species. Of all the phenolic acids studied, ethanol had the highest extraction yield for any solvent type [115]. Another study predicted and optimised the UAE for monomeric anthocyanin and phenolic chemicals from black and purple rice bran. Extraction yield was studied in relation to pH, temperature, solvent concentration, and extraction time. Purple rice bran extracts had a greater total phenolic content of 2232 mg Gallic acid equivalent (GAE)/100 g and higher monomeric anthocyanin content of 34.86 mg C3G/L as compared to black rice bran extracts, which had values of 31.95 mg C3G/L and 1978.76 mg GAE/100 g under the optimum conditions. Furthermore, for black (753.89 mg GAE/100 g and 18.75 mg C3G/L) and purple rice bran (778.98 mg GAE/100 g and 21.82 mg C3G/L), the anthocyanin and total phenolic content yields using UAE were significantly higher than the usual extraction procedure [116]. Likewise, from corn cob waste, xylan was separated using UAE, with a reduction in the time of extraction, the amount of alkali usage, and the extraction temperature. In comparison with the xylan extracted without ultrasonic treatment in a 5% solution of NaOH, the composition of xylan and the physical structure remained unaltered but with improved bioactivity [114]. Moreover, polysaccharides extracted from the stem of sijiaoling with the use of UAE also showed better antioxidant capacity in comparison to the extraction process using hot water [117]. Kumar et al. [107] explored ultrasound in the production of oil from rice bran. Similarly, the oil from wheat germ was extracted by Teslic et al. [118] with a yield of 10.67%.

3.4. Ultrasonic Extraction of Useful Materials from Fruit and Vegetable Processing Waste

Fruit processing waste consists primarily of 40 to 50% cellulose, 10 to 25% lignin, and 20 to 30% hemicellulose and other polysaccharides [112]. Similarly, the wastes are also rich in polyphenol, pectins, and dietary fibre concentrations as compared to the edible components. The most common phenolic chemicals found in fruit and vegetable wastes are phenolic acids, flavonoids, and tannins [119]. Several research works have shown that using UAE increases pectin, a polysaccharide output, and reduces extraction time. Peels of pomegranate, orange, grapefruit, eggplant, grape pomace, and tomato waste have all resulted in pectin yields of above 25% in the UAE [107]. Wang et al. [120] examined the chemical and UAE extraction of pectin from grapefruit peel and found that UAE produced a greater yield (16.34%) and that the time of extraction was reduced by 37.78%. Likewise, de Oliveira et al. [121] found a 1.6-fold rise in the productivity of pectin on the UAE of peels obtained from passion fruit when compared to chemical extraction. Another study of UAE improved the extraction yield of pectin to 53% in the case of mango peel phenolic residue and to 31% in the case of rehydrated mango peel [122]. A substantial association between pectin emergence and tissue swelling was reported by Xu et al. [123], who came to the conclusion that the vegetal tissue disruption using ultrasonic treatment is the major mechanism for improving the extractability by UAE. Similarly, crude polysaccharide was successfully extracted from rambutan fruit peel using an efficient UAE approach. The ultrasonic process parameters of 110 W, 53 °C, and 41 min showed the best result. This could be used in any value addition of food products [124]. A different study on the dual frequency ultrasound treatment of ginger leaves and stem (DFGLS) and the triple frequency ultrasound treatment of ginger leaves and stem (TPGLS) for the extraction of polysaccharides resulted in a yield of 9.74% and 10.50%, respectively. The TFGLS polysaccharides had a larger total sugar, uronic acid, and sulphate radical concentrations than the DFGLS polysaccharides, but a lower molecular weight. The antioxidant capabilities of TFGLS polysaccharides were higher for scavenging DPPH, 2,2′-azinobis (3-ethylbenzothiazoline-6-sulphonic acid) (ABTS), superoxide radicals, and hydroxyl; they also had a high reducing power and chelating activity as well as better emulsifying property, hygroscopicity, and foamability. These findings suggest that dual- and triple-frequency ultrasound can enhance the functional properties and biological activities of polysaccharides obtained from leaves and stems of ginger [125]. A higher production of cavitation bubbles and the resulting physical phenomena in the case of multi-frequency ultrasound produces the above effects noted in TFGLS polysaccharides [126]. The polysaccharides obtained from various food wastes are used in the formulation of value-added food products such as texturizing, stabilizing, thickening, and emulsifying additives [127]. Likewise, the soluble dietary fibres are known to reduce cholesterol and the associated risk of cardiovascular diseases [128]. The findings of Zhang et al. [129] imply that papaya peel is a good source of soluble dietary fibre (SDF), and that ultrasound-assisted alkaline extraction yielded 36.99% of SDF. When comparing the SDF of the alkaline extraction method to the SDF of the ultrasound-assisted method (u-SDF), the u-SDF contains more essential amino acids and minerals. Furthermore, u-SDF has shown improvements in thermal stability and water holding, oil holding, and swelling capacity, implying that this process could significantly improve the properties of SDF, which could be a viable component in functional product formulations. Furthermore, the monosaccharide present is predominantly pectic saccharide, which belongs to the low-methoxy pectin family and forms stable food-grade gels with calcium ions [129]. Few studies have made a comparison between conventional solvent extraction and UAE solvent extraction for dietary fibres; those that did found that the UAE of the dietary fibre from citrus changshan-huyou peels [130] and apple pomace [131] gave higher yields, took less process time, and used lower temperatures. Similarly, UAE dietary fibre from culinary banana bract, peels of papaya, and residues of soyabean had superior yield, thermal stability, purity, water-holding, swelling, and oil-holding capabilities than alkaline-extracted dietary fibre [129,132]. Another comparative study between the UAE and the Soxhlet method on oil obtained from seeds of Moringa peregrina was carried out by Mohammadpour et al. [133]. The greatest oil yield with UAE was 53.101%, whereas it was 43% with the Soxhlet process after 11 h of extraction. The peroxide value (PV) of oil extracted by the Soxhlet method was higher than the UAE method, while other chemical properties such as iodine value, antioxidant activity (DPPH percent), and total phenolic content of oil extracted by the

Soxhlet method were lower than those obtained by the UAE method. As a result of several physiochemical investigations, it has been discovered that the UAE approach increases oil quality. Overall, it can be concluded that the UAE can be a cost-effective and economically viable technology. Some researchers have combined ultrasound with other processing tools such as microwave, pulse electric field to extract useful materials. In one such study [134], pectin was extracted from fig skin by ultrasound/microwave-aided extraction (UMAE). A combined sonication process with a time of 21.35 min, microwave power at 580.9 W, and an irradiation time of 11.67 min was found to provide the optimum conditions for the extraction yield (11.71%). The yield was higher for UAE and the hot water extraction method, which were 8.74% and 6.95%, respectively. A microstructure evaluation of the UMAE-treated sample revealed that there is enough disintegration of plant material, resulting in the increased release of pectic substance. The pectin obtained was studied for various bioactivities such as anti-radical, anti-oxidant, and anticancer activities. An improvement in functional property was observed due to uronic acids and many sugars such as glucose, fucose, arabinose, galactose, rhamnose, and mannose, with an average molecular weight of 6.89×103 kDa. Hence, this pectin from fig skin is a novel ingredient which could enable the formation of healthier function foods [134].

3.5. Ultrasonic Extraction of Useful Materials from Oilseeds and Nuts, Pulses and Legume Processing Waste

UAE technology was used to obtain the sunflower by-product pectin (SFBP). The UAE factors were successfully tuned, and the best result was 11.15% (irradiation period 30 min, temperature 33 °C, and ultrasound power 400 W). SFBP was found to have a low-esterified galacturonic acid content (72.94%). It was observed to have long side branches of galactan, arabinogalactan, and arabinan (with an average molecular weight of 175 kDa) under these sonication conditions. SFBP appeared to have strong thermal stability according to the thermal analysis. The results of functional qualities such as solubility, water-holding capacity, oil-holding capacity, emulsifying capacity, emulsion stability, foam capacity, foam stability, and antioxidant capabilities revealed that SFBP had a better value than commercial ones [135]. Similarly, UAE with parameters such as sonication time at 10 min, ultrasound power at 200 W, and a liquid-to-solvent ratio of 1.5 yielded 12.78% of pectin from walnut processing waste. In these conditions, the resultant pectin was rich in galacturonic acid (69.44%) and high in degree of esterification (59.21%), as determined by the NMR and FTIR spectra. Furthermore, the molecular weight distribution analysis revealed that the pectin obtained had low molecular weight (6.30–158.48 kDa) and was heterogeneous in nature. The walnut pectin XRD spectrum revealed an amorphous structure with few crystalline regions. Walnut pectin also demonstrated good water and oil retention capacities, radical scavenging activity, and emulsifying characteristics [136].

The goal of another study was to see how UAE affected the yield of polysaccharide gums (PSG) from flaxseed meal. Anti-nutritional components such as cyanide (HCN) and tannins were dramatically reduced with the use of ultrasound-assisted extraction. When the extraction temperature (°C) and amplitude level (%) were changed, and all other variables were kept constant, the PSG yield extracted from partially defatted flaxseed meal samples were in the range of 7.24–11.04%. Accompanied by chemical analysis, the viscosity, solution stability, and the foaming and emulsifying capabilities were investigated to see if they could be used as a new food additives [137]. In a different study, UAE was shown to be a superior approach for getting larger yields of proteases (up to 330 IU) and α-amylases (825 IU) while reducing extraction time and obtaining a more concentrated product [138]. Likewise, Chanioti and Tzia [139] studied the goodness of UAE in oil recovery from olive pomaces. The yield of oil extraction and its unsaponifiable matter, total phenol content, and antioxidant activity were all assessed. The extraction yield was found to be 2.5–4.4% of unsaponifiable matter, 10.9% oil, 0.16–0.21 mg Trolox g-1 oil, and 0.14–0.26 mg GA g-1 oil for the total phenolic content. Hence, the emerging extraction technology of ultrasound in combination with inexpensive and natural raw materials such as olive pomace confirmed it to be an economical alternative to conventional methods of extraction, in accordance with the demands of the food industry and sustainable development. A recent work developed and validated an ultrasound-assisted extraction technique for extracting phenolic compounds from Moroccan almond cold-pressed oil residue. The ideal extraction conditions were determined using response surface methods, which included using an ultrasound frequency of 27.0 kHz and 53.0% (v/v) aqueous ethanol as a green solvent for a 29.4 min extraction time. Upon comparison with the conventional heat reflux extraction, the ultrasound-assisted extraction allowed for a significant rise in extraction efficiency. Under ideal conditions for ultrasound-assisted extraction, total phenolic content was found to be 13.86 mg/g dry weight. The main phenolic chemicals found in the valuable waste were identified by doing an HPLC analysis, which used chlorogenic acid, protocatechuic acid, followed by p-hydroxybenzoic acid, and p-coumaric acid [140].

Another study optimised canola oil extraction with the help of 35 kHz, 800 W ultrasonic waves using solvents such as hexane-isopropanol and hexane at the ratio of 2:3. With the help of the Box–Behnken design, the effects of the ultrasound treatment time, solvent-to-canola (sample) ratio, and extraction temperature on the oxidative stability and yield of extracted oil were investigated. The optimum conditions obtained for canola oil extraction with an efficiency of 22.39% using hexane were: ultrasonic treatment for 87 min at 55 °C and a solvent-to-canola proportion of 6.39 (% v/w), according to the data. When compared to Soxhlet extraction, the results showed that ultrasound waves greatly improved extraction efficiency. When a hexane–isopropanol combination was used, the extraction efficiency increased. An increase in ultrasonic treatment resulted in a significant improvement in oxidative stability. In comparison to the extracted oil using hexane as a solvent, the oxidative stability of the extracted oil with a mixed solvent (hexane–isopropanol) was higher. The fatty acid compositions of the ultrasound-assisted and the Soxhlet-extracted oils were not significantly different when compared using gas chromatography [141]. In a combination process, an ultrasound-assisted aqueous enzymatic extraction method was employed in the cooking pre-treatment to extract date seed oil. The cooking temperature and solvent ratio can affect the oil quality because a prolonged heating process will cause the thermal degradation of oil, resulting in the formation of hydroperoxides and other products resulting from oil degradation and oxidation [142]. The results showed that a sample-to-solvent ratio of 1: 2 and a cooking period of 40 min were the best conditions for extracting date seed oil. Overall, the findings of this study show that it was critical to choose the right sample-to-solvent ratio and to use a cooking pre-treatment in order to extract oil with high yield and quality. This study found that by using the cooking pre-treatment, the innovative, ultrasound-assisted, aqueous enzymatic extraction procedure improved the extraction of oil from the sample [142].

In a recent study, soluble dietary fibres (SDFs) were extracted from black soybean hulls using ultrasound–microwave co-modification and enzyme modification (cellulase and hemicellulase). SDF structural, physical, and chemical characteristics as well as their binding capacity with cholesterol were investigated previous to the modification and also after the modification. SDFs obtained using the ultrasound–microwave co-modified hulls and the enzyme-modified hulls had a molecular weight reduction of 33.21% and 45.29%, respectively, from the raw black soybean hulls. The extracted SDFs modified by the ultrasound–microwave method had a water-holding capacity of 3.79 g/g, a water-swelling capacity of 1.39 mL/g, and an oil-holding capacity of 1.14 g/g, and an increase of 9.54%, 23.01%, and 17.53%, respectively, over the raw SDF values. When compared to raw SDFs, the enzyme-modified SDFs had a water-holding capacity of 3.59 g/g, a water-selling capacity of 1.25 mL/g, and an oil-holding capacity of 1.03 g/g, signifying an increase of 3.76%, 10.62%, and 6.19%, respectively. On comparing the values of the cholesterol-binding capacity of SDFs modified by the ultrasound–microwave and enzyme methods to raw SDFs, the values were 13.82 and 12.34 mg/g, respectively, indicating a 47.98% and a 32.20% increase [143]. This can help in lowering cholesterol and in the improvement of cardiovascular health. Overall, the UAE process helps to further improve the extraction of useful components such as oil, polysaccharides, phenolic compounds, dietary fibres, pectins, etc., through the imposed physical effects of acoustic cavitation and the implied mass transfer. A combination with other techniques has always proved to be beneficial, as confirmed by the above discussion.

3.6. Ultrasonic Extraction Design and Optimization Methods

Ultrasonic extraction is a multi-variable system that depends on power, frequency of ultrasound, sonication time as well the viscosity of the medium being sonicated. The design of experiments (DOE) as well as the optimization of experimental parameters are crucial for improving the extraction and energy efficiency of sonication systems. In this regard, several ultrasonic extraction design and optimization methods have been explored [144–146]. Typical optimization methods are the Response Surface Methodology (including the Box–Behnken design), the Plackett–Burman design, and the orthogonal experimental design. Typically, for high-intensity ultrasonic extraction, frequency is fixed at around 20 kHz. Then, the optimization variables left are the power of ultrasound, sonication time, and the viscosity of the medium.

The viscosity of the medium is generally dependent on the ratio of raw materials to water. The higher the concentration of raw materials, the higher the viscosity is. Recently, we have shown that the efficient extraction of lipids by the rupture of microalgae cells required around 20% solids [14]. However, such an optimization needs to be carried out for individual systems since the viscosity dependence on solid concentration is material-dependent. Additionally, the rheological behaviour of the medium needs to be characterized. The rheological behaviour of solid slurry could be Newtonian, shear-thinning, or shear-thickening. The shear-thinning materials would perform better when ultrasonicated as compared to the shear-thickening material. The decrease in viscosity with shear (with applied ultrasound) would aid in the extraction, whereas the increase in viscosity for the shear-thickening material would hinder the extraction due to energy dissipation and an insufficient penetration of ultrasonic waves.

One of the well-used process optimization methods is the RSM (Response Surface Methodology) [146]. RSM optimises the process of selecting the factorial variable so that the desired response from the combination of variables is either maximized or minimized. The RSM can be used to screen for factorial variables first, followed by a three-level factorial response study. The response contour plots as a function of the system variables are used to locate the variable combination, either for a minimized or a maximized response. In a process system involving multiple parameters and their interaction, the use of RSM reduces the number of experimental trials needed to optimise the system, hence making the optimization less laborious and saving costs.

Generally, the first-order polynomial response (Y) as a function of independent system variables (n) is obtained, as shown below from the Plackett–Burman design method [147]:

$$Y = \beta_o + \sum_{i=1}^{n} \beta_i X_i \quad (1)$$

where Y is the predicted response, (β_o) is the intercept, β_i is the linear regression coefficient, and X_i is the coded independent variable. This model assumes no interactions between the independent variables. The PB design method is used for the screening of important independent variables that can have a significant effect on the response.

Once the screening has been completed, three important independent variables are chosen, to which the Box–Behnken Design is applied, as shown below:

$$Y = \beta_o + \sum_{i=1}^{3} \beta_i X_i + \sum_{i=1}^{3} \beta_{ii} X_i^2 + \sum_{i=1}^{2} \sum_{j=i+1}^{3} \beta_{ij} X_i X_j \quad (2)$$

where Y is the predicted response; (β_o) is the intercept; β_i is the linear, β_{ii} is the quadratic, and β_{ij} is the interaction regression coefficient; X_i and X_j are the coded independent variables. Generally, the Design-Expert Software (Stat-Ease Inc., Minneapolis, MN, USA) was used for the regression analysis of the experimental design, in data analysis, and therefore to obtain the quadratic polynomial response function.

3.7. Alterations after Exposure to Ultrasound and the Possible Effects and Consequences for Product Quality

Solvents are typically used for plant extraction in the food industry, which is polluting, labour-intensive, and expensive. Soxhlet extraction is generally used with the most popular oil extraction method. The benefits of employing low-frequency ultrasound have been reported, which shows a rise in the recovery of the compounds under softer extraction conditions within a shorter period of time, with the help of more environmentally friendly organic solvents such as ethanol or by using other green solvents. Although extraction yields are higher when compared to conventional extraction procedures, food items with high fat content exposed to ultrasonic-for-extraction reasons have also shown alterations in the organoleptic and/or chemical properties [148].

The samples had different fatty acid compositions, and the presence of the lipid-degrading compounds (2E, 4E)-deca-2,4-dienal and limonene (Z)-hept-2-enal indicated that the kiwi seed oil had undergone oxidation. This oil has a high amount of polyunsaturated fatty acids (PUFAs), 57% of which are linolenic acid (C18:3), and a low amount of tocols, i.e., tocopherol/tocotrienol (35 mg kg^{-1}), which can affect the stability it has against oxidation [149]. However, the value of peroxides rose during the lipid extraction of flaxseed with ultrasonic assistance, which led the scientists to speculate that free radicals may possibly have been produced [150]. Despite the two soybean varieties improving the yield of oil with the UAE (11.2%), a general drop in unsaturated fatty acids and a rise in saturated ones, at a 3.4% oxidation rate, were also noted [151]. Studies using ultrasound as a pre-treatment for the aqueous enzymatic oil extraction of thyme leaves (*Thymus vulgaris* L.) showed an increase in carvacrol, thymol, and p-cymene, while a significant decrease was observed for γ-terpinene [152]. Compounds such as E(Z)hexenal and nonanal were found in the volatiles from grapes (hybrid cultivar *Othello Vitis* sp.) extracted using ultrasound, and this was attributed to an enzymatic breakdown of unsaturated fatty acids during the injury-induced stress response to sample preparation [153].

4. Ultrasound-Assisted Production of Bio-Energy Products from Food Waste

4.1. Biodiesel Production

Biodiesel is sulphur-free, biodegradable, less toxic, and can be used in a compression ignition engine [154]. Biodiesel fuel is the alkyl esters of long-chain fatty acids. Biodiesel from food waste is obtained by the transesterification and alcoholysis of triacylglycerols such as waste fats, waste greases, and waste cooking oils [155]. There are two types of transesterifications: indirect and direct transesterification [156]. Normally, microorganisms can metabolize the carbon source into microbial oils for the indirect transesterification, whereas the alkaline or acidic catalysts in nature can be utilized for the direct transesterification (Figure 6). Ultrasound has been successfully applied in the transesterification reactions of waste cooking oils and the side-streaming products from oil production. Ultrasonic probe (horn) and bath have been employed in indirect/direct transesterification [155,157]. It has been recognized that the formation of emulsion between immiscible fluids generated by the mechanical effects was the major factor for increasing the direct/indirect transesterification reaction rate [158]. However, hydrogen and hydroxyl radicals have minor effects on the acceleration of this reaction [159]. Compared with high-frequency ultrasound, the mechanical effects are stronger in low-frequency ultrasound [160]. Therefore, low-frequency ultrasound is normally used for enhancing the biodiesel production from food waste.

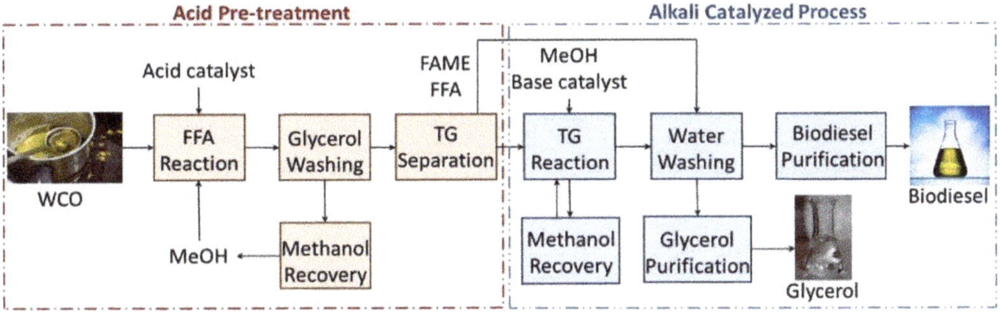

Figure 6. The direct transesterification of waste cooking oil (WCO) using acidic and alkaline catalysts [155]. WCO: waste cooking oil; FFA: free fatty acid; MeOH: methanol; TG: triglycerides; FAME: fatty acid methyl esters.

The ultrasound-assisted direct/indirect transesterification of food waste is summarised in Table 3. Carmona-Cabello et al. [161] reported that low-frequency ultrasound-assisted transesterification can significantly save both energy and reaction time to produce high-quality biodiesel from solid food waste. They also recommended the addition of an antioxidant in biodiesel to improve its storage stability. Oza et al. [162] optimised the methyl alcohol/oil molar ratio (6:1), temperature (50 °C), and the concentration of KOH (0.5%) for the ultrasound-assisted transesterification of waste cottonseed cooking oil; they found that the concentration of KOH was the major factor that influenced the yield of biodiesel. Similar results were found by Sharma et al. [163], who identified that the optimised methanol/oil molar proportion of ultrasound-assisted CaO and the KOH-catalysed transesterification of waste cottonseed cooking oil were 6.1:1 and 10.9:1, and that the biodiesel yield could reach 97.76% and 96.16%. In addition, the ultrasound-assisted conversion follows a pseudo-first-order reaction kinetics, and continuous sonication has higher effectiveness as compared with pulse sonication [163]. Yasvanthrajan et al. [164] found that the ultrasound-assisted transesterification could significantly shorten the reaction time (8 h reduction) and improve the biodiesel yield as compared with conventional processes. Meanwhile, the conversion of waste bio-oil was found to increase first, followed by a decrease with an increase in ultrasonic amplitude (from 20% to 100%) [164]. Apart from the above-mentioned catalysts, ultrasound has also been successfully employed in the biodiesel production from other catalysts, including modified coal fly ash [165], calcium diglyceroxide [166], and hydrotalcite [167].

Table 3. Ultrasound-assisted transesterification of food waste to produce biodiesel.

Materials	Ultrasonic Frequency	Ultrasonic Power	Temperature	Biodiesel Yield	Reference
Solid food waste oil	20 kHz	50% amplitude	52.5 °C	93.23%	[161]
Waste cottonseed cooking oil	20 kHz	500 W	50 °C	98%	[162]
Waste cottonseed cooking oil	20 kHz	500 W	40–60 °C	70.21~97.76%	[163]
Waste bio-oil	20 kHz	20~100% amplitude	45 °C	98.7%	[164]
Waste cooking oil	20 kHz	108 W	70 °C	95.57%	[165]
Waste cooking oil	22 kHz	120 W	60 °C	93.5%	[166]
Waste cooking oil	20 kHz	55% amplitude	57 °C	76.45%	[167]

4.2. Bio-Methane Production

Anaerobic digestion is the main pathway for producing biogas such as bio-methane and carbon dioxide from food waste [168]. Generally, the organic materials can be degraded by microorganisms in an anaerobic environment in order to generate bio-methane [169]. Due to the low yield of bio-methane during anaerobic digestion, ultrasound has been applied as a pre-treatment to solubilize the organic materials, disintegrate cell structure, and break the structure of cellulose crystalline so that the yield of bio-methane could be enhanced [170]. Some researchers have explored the effect of ultrasound on bio-methane production during anaerobic digestion (Table 4). Mirko et al. [171] compared the effect of five different pre-treatments (alkaline, thermo-alkaline, low temperature thermal, microwave, and ultrasound) on the yield of bio-methane during a 30-day anaerobic digestion. Compared with other pre-treatments, ultrasound has a minor influence on the biomethane yields. On the contrary, Yue et al. [17] reported that the bio-methane yield increased from 726.85 mL/g total volatile solids (TVS) to 927.97 mL/g TVS, with increased ultrasonic energy levels (from 1000 kJ/kg to 50,000 kJ/kg) during the 30-day anaerobic digestion of food waste. Shanthi et al. [172] found that the combination of surfactant addition and sonication pre-treatment could enhance the bio-methane during the 30-day anaerobic digestion of fruit and vegetable residue. Apart from the direct anaerobic digestion of food waste, it can also be co-digested with other materials such as manure, sludge, and crude glycerine [33]. Quiroga et al. [33] found that the daily production of bio-methane was increased by 31% for sonicated samples during the anaerobic co-digestion of food waste, cattle manure, and sludge. Similar results were also reported by Ormaechea et al. [173], in which the yield of bio-methane was increased by around two times in sonication pre-treated materials as compared with untreated materials after anaerobic co-digestion. Recently, researchers have tried to apply ultrasound in both the pre-treatment and the anaerobic processes. It was found that such a combination has a satisfactory influence on bio-methane production [40]. However, ultrasonic power, time, and duty cycle could affect such digestion progress and bio-methane production [40]. Therefore, the condition of ultrasound as a pre-treatment should be optimised according to the different substrates.

Table 4. Effect of ultrasound on bio-methane production from food waste during anaerobic digestion.

Materials	Ultrasonic Frequency	Ultrasonic Power	Temperature	Bio-Methane Yield	Reference
Buckwheat hull	40 kHz	110 W	25 °C	141.9 NL Kg VS-1	[171]
Restaurant waste	20 kHz	0–500 W	Room temperature	647.49 mL/g TVS~927.97 mL/g	[17]
Fruit and vegetable residue	25 kHz	90 W	-	0.61 g/g COD	[172]
Mixture of food waste, cattle manure, and sludge	24 kHz	400 W	-	0.85 L/L day	[33]
Mixture of food waste, cattle manure, and crude glycerine	24 kHz	400 W	55 °C	520 L/kg VS	[173]
Food waste	20 kHz and 25 kHz	120 W and 200 W	35 °C	~0.26 L/day	[40]

4.3. Bio-Hydrogen Production

Generally, food waste can be degraded into bio-methane and some other biogas such as bio-hydrogen from fermentation. Dark fermentation has been proven to be an effective pathway for producing bio-hydrogen [174]. Due to its mechanical effects, ultrasound can promote the production of bio-hydrogen by enhancing the solubilization of substrates. Up to now, ultrasound has been applied before and during the fermentation of food wastes (Table 5). Bundhoo et al. [60] compared the effects of ultrasound and microwave on dark fermentation from food and yard waste; they found that ultrasound was more effective at the solubilization of organic matter and contributed to a decrease in the production of bio-hydrogen. Elbeshbishy et al. [175] applied ultrasound to pre-treat food waste at 20 kHz and found that the longer pre-treatment time could significantly increase the yield of hydrogen.

Later, their team compared the effects of four different pre-treatment methods, including ultrasound, heat shock, base, and acid, on the hydrogen yield [46]. In fact, ultrasonication pre-treatment reached the maximum production of hydrogen (894 mL), and such benefits could be inhibited by the combination of ultrasound with other processes, such as heat or base treatment [46]. Elbeshbishy et al. [56] investigated the effect of sonication on anaerobic hydrogen production from pulp waste and explained that sonication pre-treatment may promote the release of proteins and carbohydrates into the liquid phase to improve the yield of hydrogen (88% as compared to the control group). Similar results were also reported by Gadhe et al. [38], who found that the optimised ultrasonication conditions were 12 times and 8% solid content, and that this condition could enhance the yield of hydrogen by 75%. Additionally, Emmanuel et al. [176] also found that the hydrogen production of fermented palm oil mill effluent was increased by 38% after sonication pre-treatment, and that *Clostridium* spp. and *Thermoanaerobacterium* spp. were the dominant microflora in the fermented substrate. Arun and Sivashanmugam [49] also confirmed the positive effects of ultrasonication pre-treatment on dark fermentation. A 10 mL/g VS increase in the yield of hydrogen was observed in ultrasound pre-treated dairy waste as compared with untreated waste. However, apart from the enhancement of solubilization, other pathways still need to be explored.

Table 5. Effect of ultrasound on bio-hydrogen production from food waste during fermentation.

Materials	Ultrasonic Frequency	Ultrasonic Power	Temperature	Bio-Hydrogen Yield	Reference
Mixture of food and yard waste	20 kHz	0~500 W	-	7.87 mL/g VS, decreased by 11%	[60]
Food waste	20 kHz	500 W	<30 °C	1.55 mol/mol VS, increased by 120%	[175]
Food waste	20 kHz	500 W	<30 °C	97 mL/g VS, increased by 131%	[46]
Food waste	20 kHz	500 W	<30 °C	80 mL/g VS, increased by 88%	[56]
Food waste	20 kHz	1200 W	<30 °C	0.62~5.23 mL/h, increased by 75%	[38]
Food waste	20 kHz	130 W	-	0.7 mmol/g COD, increased by 38%	[176]
Dairy waste	20 kHz	0.3~1.1 W/mL	25 °C	15.51 mL/g VS, increased by 3 times	[49]

4.4. Bio-Ethanol Production

Bio-ethanol is another product of food waste, and fermentation is the most common process to acquire this biogas. However, there is scant literature on the ultrasound-assisted production of bioethanol from food waste. Suresh et al. [177] reported that ultrasound pre-treatment at 340 W could improve the yield of bioethanol from potato waste after fermentation to 54.1 g/L. In addition, ultrasound-combined HCl was more efficient in the enhancement of bio-ethanol as compared with ultrasound-combined enzyme pre-treatment. Sindhu et al. [178] applied ultrasound-assisted alkali as a pre-treatment to ferment chili residue and obtained 1.94% bio-ethanol after fermentation. Battista et al. [179] compared different pre-treatment methods to improve bioethanol production from olive oil waste and found that ultrasound pre-treatment at 1800 W for 30 min could increase the yield of bio-ethanol by 50%. In addition, ultrasound was applied to hydrolyse sweet lime peel in a study on bio-ethanol production by John et al. [180], who confirmed that the maximum bio-ethanol yield after fermentation could reach 64%. Although the positive effects of ultrasound pre-treatment on bio-ethanol production have been proven, the mechanism, the effects of ultrasonic parameters, and the change in other properties during the fermentation of food waste are not clear and should be further investigated.

5. Conclusions and Perspectives

As discussed throughout the review, ultrasound has shown great potential as a green technology that can valorise various types of food waste materials in order to extract value-added products. Ultrasound accelerates the disintegration of large chunks of food waste into fine particles, thereby increasing the mass transfer and hence the diffusion of bioactives from food waste into the extraction medium. Ultrasound has been shown to rupture the cell walls of not only soft cells such as microalgae, but also hard cell walls such as plant cells. Ultrasound has been shown to be effective in extracting value-added materials from wheat waste, vegetable waste, oil palm fronds, wheat straw, rice straw, rice hull, sugarcane bagasse, and spent coffee waste. Ultrasonication can also be used to modify the crystal structure of the extracted materials. For plant cell walls, ultrasound has been proven to be a predominant approach for lignin barrier disruption, hemicellulose removal, and cellulose crystallinity reduction. Ultrasonication has proven to be an efficient method for extracting valuable materials from fruit and vegetable waste, oilseeds and nuts, pulse and legume processing waste, and cereal processing waste. From cereal processing waste, bioactive compounds such as phytosterols and β-glucans can be extracted by using ultrasonication.

From fruit and vegetable waste, ultrasonication increases the pectin output and reduces the extraction time. When ultrasonication was used, the pectin yields increased by 25% from the peels of pomegranate, orange, grapefruit, eggplant, grape pomace, and tomato waste. Ultrasonication can also reduce the concentration of unwanted components such as cyanide (HCN) and tannins, which are released into extraction medium.

In the algae biofuel industry, ultrasonication is proving to be an efficient method, not only for cell rupture to extract TAGs, but also for further transesterification reactions to convert the TAGs into biodiesel. In the case of enzymatic reactions, ultrasound enhances the reaction kinetics, thereby decreasing the time required and saving on costs. In the case of bio-methane production from anaerobic digestion, ultrasound has proven to be an efficient pre-treatment method for solubilizing the organic materials, disintegrating the cell structure, and breaking the structure of cellulose crystalline so that the yield of bio-methane could be enhanced. In the case of bio-hydrogen production from the fermentation of food waste, ultrasound can promote the production of bio-hydrogen by enhancing the solubilization of substrates. Ultrasound was shown to be effective in producing bio-hydrogen before and during the fermentation of food wastes. In the case of bioethanol production from the fermentation of potato waste, ultrasonication has been shown to increase the bioethanol yield.

Although ultrasonication has proven to be effective in extracting valuable materials from food and agriculture wastes, more research needs to be performed on the cost-effectiveness of the treatment when applied in a large scale. There have been a lot of developments in the scaling up of ultrasonic extraction technologies. A number of companies are producing flow-through sonication systems to reduce the footprint of ultrasonic reactors. The flow-through systems work well when the viscosities of the sonication medium are low to medium. However, when the viscosities increase during the sonication, the effectiveness may decrease with time. There are few studies in the literature that have studied the optimum viscosity levels of algae slurry so that cells could be ruptured well and oils could be extracted efficiently [14,181]. The energy input into the system for the unit extraction of value-added materials was calculated and compared with existing technologies. These kinds of in-depth studies are needed so that the energy efficiency of ultrasonication can be proved.

Author Contributions: Conceptualization, S.M. and M.A.; investigation, Y.W. and S.Y.; resources, M.A.; writing—original draft preparation, Y.W., S.Y., B.A.N., A.S., and S.M.; supervision, S.M., A.S., and M.A.; project administration, M.A. All authors have read and agreed to the published version of the manuscript.

Funding: This research received no external funding.

Institutional Review Board Statement: Not applicable.

Informed Consent Statement: Not applicable.

Data Availability Statement: Not applicable.

Acknowledgments: This work was supported by the China Scholarship Council and the University of Melbourne Research Scholarship.

Conflicts of Interest: The authors declare no conflict of interest.

References

1. Li, H.; Xu, J.; Nyambura, S.M.; Wang, J.; Li, C.; Zhu, X.; Feng, X.; Wang, Y. Food waste pyrolysis by traditional heating and microwave heating: A review. *Fuel* **2022**, *324*, 124574. [CrossRef]
2. Gustavsson, J.; Cederberg, C.; Sonesson, U.; Van Otterdijk, R.; Meybeck, A. *Global Food Losses and Food Waste*; FAO: Rome, Italy, 2011.
3. Amicarelli, V.; Lagioia, G.; Bux, C. Global warming potential of food waste through the life cycle assessment: An analytical review. *Environ. Impact Assess. Rev.* **2021**, *91*, 106677. [CrossRef]
4. Xu, Z.; Qi, H.; Yao, D.; Zhang, J.; Zhu, Z.; Wang, Y.; Cui, P. Modeling and comprehensive analysis of food waste gasification process for hydrogen production. *Energy Convers. Manag.* **2022**, *258*, 115509. [CrossRef]
5. Singh, D.; Yadav, S.; Bharadwaj, N.; Verma, R. Low temperature steam gasification to produce hydrogen rich gas from kitchen food waste: Influence of steam flow rate and temperature. *Int. J. Hydrog. Energy* **2020**, *45*, 20843–20850. [CrossRef]
6. Xiong, S.; He, J.; Yang, Z.; Guo, M.; Yan, Y.; Ran, J. Thermodynamic analysis of CaO enhanced steam gasification process of food waste with high moisture and low moisture. *Energy* **2019**, *194*, 116831. [CrossRef]
7. Grassino, A.N.; Ostoji, J.; Mileti, V.; Djakovi, S.; Bosiljkov, T.; Zori, Z.; Jeek, D.; Brni, S.R.; Brni, M. Application of high hydrostatic pressure and ultrasound-assisted extractions as a novel approach for pectin and polyphenols recovery from tomato peel waste —ScienceDirect. *Innov. Food Sci. Emerg. Technol.* **2020**, *64*, 102424. [CrossRef]
8. Mouratoglou, E.; Malliou, V.; Makris, D.P. Novel Glycerol-Based Natural Eutectic Mixtures and Their Efficiency in the Ultrasound-Assisted Extraction of Antioxidant Polyphenols from Agri-Food Waste Biomass. *Waste Biomass-Valoriz.* **2016**, *7*, 1377–1387. [CrossRef]
9. Liu, B.; Cai, J.; Huai, X.; Li, F. Cavitation Bubble Collapse Near a Heated Wall and Its Effect on the Heat Transfer. *J. Heat Transf.* **2013**, *136*, 022901. [CrossRef]
10. Ozdemir, D.S.; Baspinar, N.; Akalin, P.P. Effects of ultrasound homogenisation on the activities of superoxide dismutase, glutathione peroxidase, catalase and levels of lipid peroxide in liver homogenates. *Eurasian J. Vet. Sci.* **2015**, *31*, 16. [CrossRef]
11. Ashokkumar, M.; Sunartio, D.; Kentish, S.; Mawson, R.; Simons, L.; Vilkhu, K.; Versteeg, C. Modification of food ingredients by ultrasound to improve functionality: A preliminary study on a model system. *Innov. Food Sci. Emerg. Technol.* **2008**, *9*, 155–160. [CrossRef]
12. Alpert, A.J. Hydrophilic-interaction chromatography for the separation of peptides, nucleic acids and other polar compounds. *J. Chromatogr. A* **1990**, *499*, 177–196. [CrossRef]
13. Tutunchi, P.; Roufegarinejad, L.; Hamishehkar, H.; Alizadeh, A. Extraction of red beet extract with β-cyclodextrin-enhanced ultrasound assisted extraction: A strategy for enhancing the extraction efficacy of bioactive compounds and their stability in food models. *Food Chem.* **2019**, *297*, 124991–124994. [CrossRef] [PubMed]
14. Yao, S.; Mettu, S.; Law, S.Q.; Ashokkumar, M.; Martin, G.J. The effect of high-intensity ultrasound on cell disruption and lipid extraction from high-solids viscous slurries of *Nannochloropsis* sp. biomass. *Algal Res.* **2018**, *35*, 341–348. [CrossRef]
15. Silva, F.V. Evelyn Resistant moulds as pasteurization target for cold distributed high pressure and heat assisted high pressure processed fruit products. *J. Food Eng.* **2020**, *282*, 109998. [CrossRef]
16. Dhar, B.R.; Nakhla, G.; Ray, M.B. Techno-economic evaluation of ultrasound and thermal pretreatments for enhanced anaerobic digestion of municipal waste activated sludge. *Waste Manag.* **2012**, *32*, 542–549. [CrossRef]
17. Yue, L.; Cheng, J.; Tang, S.; An, X.; Hua, J.; Dong, H.; Zhou, J. Ultrasound and microwave pretreatments promote methane production potential and energy conversion during anaerobic digestion of lipid and food wastes. *Energy* **2021**, *228*, 120525. [CrossRef]
18. Pollini, L.; Rocchi, R.; Cossignani, L.; Maes, J.; Campagnone, D.; Blasi, F. Phenol Profiling and Nutraceutical Potential of Lycium spp. Leaf Extracts Obtained with Ultrasound and Microwave Assisted Techniques. *Antioxidants* **2019**, *8*, 260. [CrossRef]
19. Guo, Z.; Zhao, B.; Li, H.; Miao, S.; Zheng, B. Optimization of ultrasound-microwave synergistic extraction of prebiotic oligosaccharides from sweet potatoes (*Ipomoea batatas* L.). *Innov. Food Sci. Emerg. Technol.* **2019**, *54*, 51–63. [CrossRef]
20. Görgüç, A.; Bircan, C.; Yılmaz, F.M. Sesame bran as an unexploited by-product: Effect of enzyme and ultrasound-assisted extraction on the recovery of protein and antioxidant compounds. *Food Chem.* **2019**, *283*, 637–645. [CrossRef]
21. Wang, Z.; Wang, C.; Zhang, C.; Li, W. Ultrasound-assisted enzyme catalyzed hydrolysis of olive waste and recovery of antioxidant phenolic compounds. *Innov. Food Sci. Emerg. Technol.* **2017**, *44*, 224–234. [CrossRef]
22. Ji, Q.; Yu, X.; Yagoub, A.E.-G.A.; Li, M.; Fakayode, O.A.; Yan, D.; Zhou, C.; Chen, L. Ultrasound-Ionic Liquid Pretreatment Enhanced Conversion of the Sugary Food Waste to 5-Hydroxymethylfurfural in Ionic Liquid/Solid Acid Catalyst System. *Catal. Lett.* **2019**, *150*, 1373–1388. [CrossRef]

23. Karanicola, P.; Patsalou, M.; Stergiou, P.-Y.; Kavallieratou, A.; Evripidou, N.; Christou, P.; Panagiotou, G.; Damianou, C.; Papamichael, E.M.; Koutinas, M. Ultrasound-assisted dilute acid hydrolysis for production of essential oils, pectin and bacterial cellulose via a citrus processing waste biorefinery. *Biresour. Technol.* **2021**, *342*, 126010. [CrossRef] [PubMed]
24. Germec, M.; Tarhan, K.; Yatmaz, E.; Tetik, N.; Karhan, M.; Demirci, A.; Turhan, I. Ultrasound-assisted dilute acid hydrolysis of tea processing waste for production of fermentable sugar. *Biotechnol. Prog.* **2016**, *32*, 393–403. [CrossRef]
25. Hu, Y.; Shen, Y.; Wang, J. Pretreatment of antibiotic fermentation residues by combined ultrasound and alkali for enhancing biohydrogen production. *J. Clean. Prod.* **2020**, *268*, 122190. [CrossRef]
26. Qiu-Ying, W.U.; Chen, F.S.; Shi, D.M.; Gong, B.W.; Zhang, P.L. Optimization of ultrasound-assisted alkali extraction of xylan from peanut shell according by response surface methodology. *Sci. Technol. Food Ind.* **2012**.
27. Li, Y.Y.; Noike, T. Upgrading of anaerobic digestion of waste activated sludge by thermal pretreatment. *Water Sci. Technol.* **1992**, *26*, 857–866. [CrossRef]
28. Lafitte-Trouqué, S.; Forster, C. The use of ultrasound and γ-irradiation as pre-treatments for the anaerobic digestion of waste activated sludge at mesophilic and thermophilic temperatures. *Bioresour. Technol.* **2002**, *84*, 113–118. [CrossRef]
29. Dewil, B. Principles and potential of the anaerobic digestion of waste-activated sludge. *Prog. Energy Combust. Sci.* **2008**, *34*, 755–781.
30. Cesaro, A.; Belgiorno, V. Pretreatment methods to improve anaerobic biodegradability of organic municipal solid waste fractions. *Chem. Eng. J.* **2014**, *240*, 24–37. [CrossRef]
31. Li, W.; Ashokkumar, M. Introduction to Ultrasound and Sonochemistry. *Electrochem. Soc. Interface* **2018**, *27*, 43–46. [CrossRef]
32. Tao, Y.; Wu, P.; Dai, Y.; Luo, X.; Manickam, S.; Li, D.; Han, Y.; Show, P.L. Bridge between mass transfer behavior and properties of bubbles under two-stage ultrasound-assisted physisorption of polyphenols using macroporous resin. *Chem. Eng. J.* **2022**, *436*, 135158. [CrossRef]
33. Quiroga, G.; Castrillón, L.; Fernández-Nava, Y.; Marañón, E.; Negral, L.; Rodríguez-Iglesias, J.; Ormaecha, P. Effect of ultrasound pre-treatment in the anaerobic co-digestion of cattle manure with food waste and sludge. *Bioresour. Technol.* **2014**, *154*, 74–79. [CrossRef] [PubMed]
34. Show, K.-Y.; Mao, T.; Lee, D.-J. Optimisation of sludge disruption by sonication. *Water Res.* **2007**, *41*, 4741–4747. [CrossRef] [PubMed]
35. Hua, I.; Hoffmann, M.R. Optimization of Ultrasonic Irradiation as an Advanced Oxidation Technology. *Environ. Sci. Technol.* **1997**, *31*, 2237–2243. [CrossRef]
36. Chang, T.-C.; You, S.-J.; Damodar, R.A.; Chen, Y.-Y. Ultrasound pre-treatment step for performance enhancement in an aerobic sludge digestion process. *J. Taiwan Inst. Chem. Eng.* **2011**, *42*, 801–808. [CrossRef]
37. Kosseva, M.R. *Sources, Characterization, and Composition of Food Industry Wastes*; Academic Press: Cambridge, MA, USA, 2013; pp. 37–60. [CrossRef]
38. Gadhe, A.; Sonawane, S.S.; Varma, M.N. Ultrasonic pretreatment for an enhancement of biohydrogen production from complex food waste. *Int. J. Hydrog. Energy* **2014**, *39*, 7721–7729. [CrossRef]
39. Jiang, J.; Gong, C.; Wang, J.; Tian, S.; Zhang, Y. Effects of ultrasound pre-treatment on the amount of dissolved organic matter extracted from food waste. *Bioresour. Technol.* **2014**, *155*, 266–271. [CrossRef]
40. Joshi, S.M.; Gogate, P.R. Intensifying the biogas production from food waste using ultrasound: Understanding into effect of operating parameters. *Ultrason. Sonochem.* **2019**, *59*, 104755. [CrossRef]
41. Grönroos, A.; Kyllönen, H.; Korpijärvi, K.; Pirkonen, P.; Paavola, T.; Jokela, J.; Rintala, J. Ultrasound assisted method to increase soluble chemical oxygen demand (SCOD) of sewage sludge for digestion. *Ultrason. Sonochem.* **2005**, *12*, 115–120. [CrossRef]
42. Hay, J.X.W.; Wu, T.Y.; Juan, J.C.; Jahim, J.M. Improved biohydrogen production and treatment of pulp and paper mill effluent through ultrasonication pretreatment of wastewater. *Energy Convers. Manag.* **2015**, *106*, 576–583. [CrossRef]
43. Riau, V.; De la Rubia, M.A.; Pérez, M. Upgrading the temperature-phased anaerobic digestion of waste activated sludge by ultrasonic pretreatment. *Chem. Eng. J.* **2015**, *259*, 672–681. [CrossRef]
44. Zorba, G.T.; Sanin, F.D. Disintegration of Sludge by Sonication and Improvement of Methane Production Rates in Batch Anaerobic Digesters. *CLEAN—Soil Air Water* **2012**, *41*, 396–402. [CrossRef]
45. Subhedar, P.B.; Gogate, P.R. Alkaline and ultrasound assisted alkaline pretreatment for intensification of delignification process from sustainable raw-material. *Ultrason. Sonochem.* **2014**, *21*, 216–225. [CrossRef] [PubMed]
46. Elbeshbishy, E.; Hafez, H.; Dhar, B.R.; Nakhla, G. Single and combined effect of various pretreatment methods for biohydrogen production from food waste. *Int. J. Hydrog. Energy* **2011**, *36*, 11379–11387. [CrossRef]
47. Tian, X.; Ng, W.J.; Trzcinski, A.P. Optimizing the synergistic effect of sodium hydroxide/ultrasound pre-treatment of sludge. *Ultrason. Sonochem.* **2018**, *48*, 432–440. [CrossRef]
48. Begum, S.; Anupoju, G.R.; Eshtiaghi, N. Anaerobic co-digestion of food waste and cardboard in different mixing ratios: Impact of ultrasound pre-treatment on soluble organic matter and biogas generation potential at varying food to inoculum ratios. *Biochem. Eng. J.* **2020**, *166*, 107853. [CrossRef]
49. Arun, C.; Sivashanmugam, P. Enhanced production of biohydrogen from dairy waste activated sludge pre-treated using multi hydrolytic garbage enzyme complex and ultrasound-optimization. *Energy Convers. Manag.* **2018**, *164*, 277–287. [CrossRef]

50. Snehya, A.; Sundaramahalingam, M.; Rajeshbanu, J.; Anandan, S.; Sivashanmugam, P. Studies on evaluation of surfactant coupled sonication pretreatment on Ulva fasciata (marine macroalgae) for enhanced biohydrogen production. *Ultrason. Sonochem.* **2021**, *81*, 105853. [CrossRef] [PubMed]
51. Elbeshbishy, E.; Nakhla, G. Comparative study of the effect of ultrasonication on the anaerobic biodegradability of food waste in single and two-stage systems. *Bioresour. Technol.* **2011**, *102*, 6449–6457. [CrossRef]
52. Guo, W.; Wu, Q.; Yang, S.; Luo, H.; Peng, S.; Ren, N. Optimization of ultrasonic pretreatment and substrate/inoculum ratio to enhance hydrolysis and volatile fatty acid production from food waste. *RSC Adv.* **2014**, *4*, 53321–53326. [CrossRef]
53. Li, X.; Peng, Y.; He, Y.; Jia, F.; Wang, S.; Guo, S. Applying low frequency ultrasound on different biological nitrogen activated sludge types: An analysis of particle size reduction, soluble chemical oxygen demand (SCOD) and ammonia release. *Int. Biodeterior. Biodegrad.* **2016**, *112*, 42–50. [CrossRef]
54. Liu, N.; Jiang, J.; Yan, F.; Gao, Y.; Meng, Y.; Aihemaiti, A.; Ju, T. Enhancement of volatile fatty acid production and biogas yield from food waste following sonication pretreatment. *J. Environ. Manag.* **2018**, *217*, 797–804. [CrossRef] [PubMed]
55. Li, L.; Li, Z.; Song, K.; Gu, Y.; Gao, X. Improving methane production from algal sludge based anaerobic digestion by co-pretreatment with ultrasound and zero-valent iron. *J. Clean. Prod.* **2020**, *255*, 120214. [CrossRef]
56. Elbeshbishy, E.; Hafez, H.; Nakhla, G. Viability of ultrasonication of food waste for hydrogen production. *Int. J. Hydrog. Energy* **2011**, *37*, 2960–2964. [CrossRef]
57. Azman, S.; Milh, H.; Somers, M.H.; Zhang, H.; Huybrechts, I.; Meers, E.; Meesschaert, B.; Dewil, R.; Appels, L. Ultrasound-assisted digestate treatment of manure digestate for increased biogas production in small pilot scale anaerobic digesters. *Renew. Energy* **2020**, *152*, 664–673. [CrossRef]
58. Boni, M.R.; D'Amato, E.; Polettini, A.; Pomi, R.; Rossi, A. Effect of ultrasonication on anaerobic degradability of solid waste digestate. *Waste Manag.* **2016**, *48*, 209–217. [CrossRef]
59. Somers, M.H.; Azman, S.; Sigurnjak, I.; Ghyselbrecht, K.; Meers, E.; Meesschaert, B.; Appels, L. Effect of digestate disintegration on anaerobic digestion of organic waste. *Bioresour. Technol.* **2018**, *268*, 568–576. [CrossRef]
60. Bundhoo, Z.M. Effects of microwave and ultrasound irradiations on dark fermentative bio-hydrogen production from food and yard wastes. *Int. J. Hydrog. Energy* **2017**, *42*, 4040–4050. [CrossRef]
61. Cesaro, A.; Belgiorno, V. Sonolysis and ozonation as pretreatment for anaerobic digestion of solid organic waste. *Ultrason. Sonochem.* **2013**, *20*, 931–936. [CrossRef]
62. Oz, N.A.; Uzun, A.C. Ultrasound pretreatment for enhanced biogas production from olive mill wastewater. *Ultrason. Sonochem.* **2015**, *22*, 565–572. [CrossRef]
63. Cesaro, A.; Naddeo, V.; Amodio, V.; Belgiorno, V. Enhanced biogas production from anaerobic codigestion of solid waste by sonolysis. *Ultrason. Sonochem.* **2012**, *19*, 596–600. [CrossRef] [PubMed]
64. Oz, N.A.; Yarimtepe, C.C. Ultrasound assisted biogas production from landfill leachate. *Waste Manag.* **2014**, *34*, 1165–1170. [CrossRef] [PubMed]
65. Peng, Y.; Zhang, Z.; Wang, M.; Shi, X.; Zhou, Y.; Zhou, Y.; Kong, Y. Inactivation of harmful Anabaena flos-aquae by ultrasound irradiation: Cell disruption mechanism and enhanced coagulation. *Ultrason. Sonochem.* **2020**, *69*, 105254. [CrossRef] [PubMed]
66. Krzendrfer, A. Vibrations and ultrasound in food processing—Sources of vibrations, adverse effects, and beneficial applications—An overview. *J. Food. Eng.* **2021**, *324*, 110875. [CrossRef]
67. Martín, J.F.G.; Guillemet, L.; Feng, C.-H.; Sun, D.-W. Cell viability and proteins release during ultrasound-assisted yeast lysis of light lees in model wine. *Food Chem.* **2013**, *141*, 934–939. [CrossRef] [PubMed]
68. Saha, M.; Eskicioglu, C.; Marin, J. Microwave, ultrasonic and chemo-mechanical pretreatments for enhancing methane potential of pulp mill wastewater treatment sludge. *Bioresour. Technol.* **2011**, *102*, 7815–7826. [CrossRef] [PubMed]
69. Bhatia, S.; Sharma, K.; Dahiya, R.; Bera, T. *Modern Applications of Plant Biotechnology in Pharmaceutical Sciences*; Academic Press: Cambridge, MA, USA, 2015; pp. 121–126. [CrossRef]
70. Li, X.; Mettu, S.; Martin, G.J.; Ashokkumar, M.; Lin, C.S.K. Ultrasonic pretreatment of food waste to accelerate enzymatic hydrolysis for glucose production. *Ultrason. Sonochem.* **2018**, *53*, 77–82. [CrossRef]
71. Romero-Pareja, P.; Aragon, C.A.; Quiroga, J.M.; Coello, M.D. Evaluation of a biological wastewater treatment system combining an osa process with ultrasound for sludge reduction. *Ultrason. Sonochem.* **2017**, *36*, 336–342. [CrossRef]
72. Deprez, J.; Lajoinie, G.; Engelen, Y.; De Smedt, S.; Lentacker, I. Opening doors with ultrasound and microbubbles: Beating biological barriers to promote drug delivery. *Adv. Drug Deliv. Rev.* **2021**, *172*, 9–36. [CrossRef]
73. Juodeikiene, G.; Zadeike, D.; Trakselyte-Rupsiene, K.; Gasauskaite, K.; Bartkiene, E.; Lele, V.; Viskelis, P.; Bernatoniene, J.; Ivanauskas, L.; Jakstas, V. Functionalisation of flaxseed proteins assisted by ultrasonication to produce coatings enriched with raspberries phytochemicals. *LWT* **2020**, *124*, 109180. [CrossRef]
74. Maskooki, A.; Mortazavi, S.A.; Maskooki, A. Cleaning of spiralwound ultrafiltration membranes using ultrasound and alkaline solution of EDTA. *Desalination* **2010**, *264*, 63–69. [CrossRef]
75. Pejin, D.J.; Mojović, L.V.; Pejin, J.D.; Grujić, O.S.; Markov, S.L.; Nikolić, S.B.; Marković, M.N. Increase in bioethanol production yield from triticale by simultaneous saccharification and fermentation with application of ultrasound. *J. Chem. Technol. Biotechnol.* **2011**, *87*, 170–176. [CrossRef]
76. Nitayavardhana, S.; Rakshit, S.K.; Grewell, D.; van Leeuwen, J.; Khanal, S.K. Ultrasound pretreatment of cassava chip slurry to enhance sugar release for subsequent ethanol production. *Biotechnol. Bioeng.* **2008**, *101*, 487–496. [CrossRef] [PubMed]

77. Khanal, S.K.; Montalbo, M.; van Leeuwen, J.; Srinivasan, G.; Grewell, D. Ultrasound enhanced glucose release from corn in ethanol plants. *Biotechnol. Bioeng.* **2007**, *98*, 978–985. [CrossRef]
78. Walton, J.D. Deconstructing the Cell Wall. *Plant Physiol.* **1994**, *104*, 1113–1118. [CrossRef] [PubMed]
79. Joshi, S.M.; Gogate, P.R. Intensified Synthesis of Bioethanol from Sustainable Biomass. In *Waste Biomass Management—A Holistic Approach*; Springer: Cham, Switzerland, 2017; Volume 58, pp. 251–287. [CrossRef]
80. García, A.; Erdocia, X.; Alriols, M.G.; Labidi, J. Effect of ultrasound treatment on the physicochemical properties of alkaline lignin. *Chem. Eng. Process. Process Intensif.* **2012**, *62*, 150–158. [CrossRef]
81. Leite, A.L.M.P.; Zanon, C.D.; Menegalli, F.C. Isolation and characterization of cellulose nanofibers from cassava root bagasse and peelings. *Carbohydr. Polym.* **2017**, *157*, 962–970. [CrossRef]
82. Ji, Q.; Yu, X.; Yagoub, A.E.-G.A.; Chen, L.; Zhou, C. Efficient removal of lignin from vegetable wastes by ultrasonic and microwave-assisted treatment with ternary deep eutectic solvent. *Ind. Crop. Prod.* **2020**, *149*, 112357. [CrossRef]
83. Perrone, O.M.; Colombari, F.M.; Rossi, J.S.; Moretti, M.M.S.; Bordignon, S.E.; Nunes, C.d.C.C.; Gomes, E.; Boscolo, M.; Da-Silva, R. Ozonolysis combined with ultrasound as a pretreatment of sugarcane bagasse: Effect on the enzymatic saccharification and the physical and chemical characteristics of the substrate. *Bioresour. Technol.* **2016**, *218*, 69–76. [CrossRef]
84. Velmurugan, R.; Muthukumar, K. Sono-assisted enzymatic saccharification of sugarcane bagasse for bioethanol production. *Biochem. Eng. J.* **2012**, *63*, 1–9. [CrossRef]
85. Velmurugan, R.; Muthukumar, K. Ultrasound-assisted alkaline pretreatment of sugarcane bagasse for fermentable sugar production: Optimization through response surface methodology. *Bioresour. Technol.* **2012**, *112*, 293–299. [CrossRef] [PubMed]
86. Zhang, Y.; Li, T.; Shen, Y.; Wang, L.; Zhang, H.; Qian, H.; Qi, X. Extrusion followed by ultrasound as a chemical-free pretreatment method to enhance enzymatic hydrolysis of rice hull for fermentable sugars production. *Ind. Crop. Prod.* **2020**, *149*, 112356. [CrossRef]
87. Loow, Y.-L.; Wu, T.Y.; Jahim, J.; Mohammad, A.W.; Teoh, W.H. Typical conversion of lignocellulosic biomass into reducing sugars using dilute acid hydrolysis and alkaline pretreatment. *Cellulose* **2016**, *23*, 1491–1520. [CrossRef]
88. Saratale, G.D.; Saratale, R.G.; Varjani, S.; Cho, S.-K.; Ghodake, G.S.; Kadam, A.; Mulla, S.I.; Bharagava, R.N.; Kim, D.-S.; Shin, H.S. Development of ultrasound aided chemical pretreatment methods to enrich saccharification of wheat waste biomass for polyhydroxybutyrate production and its characterization. *Ind. Crop. Prod.* **2020**, *150*, 112425. [CrossRef]
89. Ong, V.Z.; Wu, T.Y.; Lee, C.B.T.L.; Cheong, N.W.R.; Shak, K.P.Y. Sequential ultrasonication and deep eutectic solvent pretreatment to remove lignin and recover xylose from oil palm fronds. *Ultrason. Sonochem.* **2019**, *58*, 104598. [CrossRef] [PubMed]
90. Gaudino, E.C.; Tabasso, S.; Grillo, G.; Cravotto, G.; Dreyer, T.; Schories, G.; Altenberg, S.; Jashina, L.; Telysheva, G. Wheat straw lignin extraction with bio-based solvents using enabling technologies. *Comptes Rendus. Chim.* **2018**, *21*, 563–571. [CrossRef]
91. Mohapatra, S.; Dandapat, S.J.; Thatoi, H. Physicochemical characterization, modelling and optimization of ultrasono-assisted acid pretreatment of two Pennisetum sp. using Taguchi and artificial neural networking for enhanced delignification. *J. Environ. Manag.* **2017**, *187*, 537–549. [CrossRef] [PubMed]
92. Sakdaronnarong, C.; Srimarut, N.; Lucknakhul, N.; Na-Songkla, N.; Jonglertjunya, W. Two-step acid and alkaline ethanolysis/alkaline peroxide fractionation of sugarcane bagasse and rice straw for production of polylactic acid precursor. *Biochem. Eng. J.* **2014**, *85*, 49–62. [CrossRef]
93. Ravindran, R.; Jaiswal, S.; Abu-Ghannam, N.; Jaiswal, A.K. Evaluation of ultrasound assisted potassium permanganate pretreatment of spent coffee waste. *Bioresour. Technol.* **2017**, *224*, 680–687. [CrossRef]
94. Yusof, N.S.M.; Babgi, B.; Alghamdi, Y.; Aksu, M.; Madhavan, J.; Ashokkumar, M. Physical and chemical effects of acoustic cavitation in selected ultrasonic cleaning applications. *Ultrason. Sonochem.* **2016**, *29*, 568–576. [CrossRef]
95. New, E.K.; Wu, T.Y.; Lee, C.B.T.L.; Poon, Z.Y.; Loow, Y.-L.; Foo, L.Y.W.; Procentese, A.; Siow, L.F.; Teoh, W.H.; Daud, N.N.N.; et al. Potential use of pure and diluted choline chloride-based deep eutectic solvent in delignification of oil palm fronds. *Process Saf. Environ. Prot.* **2018**, *123*, 190–198. [CrossRef]
96. Velmurugan, R.; Muthukumar, K. Utilization of sugarcane bagasse for bioethanol production: Sono-assisted acid hydrolysis approach. *Bioresour. Technol.* **2011**, *102*, 7119–7123. [CrossRef] [PubMed]
97. Sindhu, R.; Binod, P.; Pandey, A. A novel sono-assisted acid pretreatment of chili post harvest residue for bioethanol production. *Bioresour. Technol.* **2016**, *213*, 58–63. [CrossRef]
98. Park, S.; Baker, J.O.; Himmel, M.E.; Parilla, P.A.; Johnson, D.K. Cellulose crystallinity index: Measurement techniques and their impact on interpreting cellulase performance. *Biotechnol. Biofuels* **2010**, *3*, 1–10. [CrossRef]
99. Alemdar, A.; Sain, M. Biocomposites from wheat straw nanofibers: Morphology, thermal and mechanical properties. *Compos. Sci. Technol.* **2008**, *68*, 557–565. [CrossRef]
100. Jin, J.; Lin, H.; Yagoub, A.E.A.; Xiong, S.; Xu, L.; Udenigwe, C. Effects of high power ultrasound on the enzymolysis and structures of sweet potato starch. *J. Sci. Food Agric.* **2020**, *100*, 3498–3506. [CrossRef] [PubMed]
101. Ninomiya, K.; Ohta, A.; Omote, S.; Ogino, C.; Takahashi, K.; Shimizu, N. Combined use of completely bio-derived cholinium ionic liquids and ultrasound irradiation for the pretreatment of lignocellulosic material to enhance enzymatic saccharification. *Chem. Eng. J.* **2012**, *215*, 811–818. [CrossRef]
102. Galanakis, C.M.J.F.; Processing, B. Emerging technologies for the production of nutraceuticals from agricultural by-products: A viewpoint of opportunities and challenges. *Food Bioprod. Process.* **2013**, *91*, 575–579. [CrossRef]

103. Giacometti, J.; Kovačević, D.B.; Putnik, P.; Gabrić, D.; Bilušić, T.; Krešić, G.; Stulić, V.; Barba, F.J.; Chemat, F.; Barbosa-Cánovas, G.; et al. Extraction of bioactive compounds and essential oils from mediterranean herbs by conventional and green innovative techniques: A review. *Food Res. Int.* **2018**, *113*, 245–262. [CrossRef]
104. Wood, I.P.; Cook, N.M.; Wilson, D.R.; Ryden, P.; Robertson, J.A.; Waldron, K.W. Ethanol from a biorefinery waste stream: Saccharification of amylase, protease and xylanase treated wheat bran. *Food Chem.* **2016**, *198*, 125–131. [CrossRef]
105. Liu, L.; Zhao, M.; Liu, X.; Zhong, K.; Tong, L.; Zhou, X.; Zhou, S. Effect of steam explosion-assisted extraction on phenolic acid profiles and antioxidant properties of wheat bran. *J. Sci. Food Agric.* **2016**, *96*, 3484–3491. [CrossRef]
106. Azmir, J.; Zaidul, I.S.M.; Rahman, M.M.; Sharif, K.M.; Mohamed, A.; Sahena, F.; Jahurul, M.H.A.; Ghafoor, K.; Norulaini, N.A.N.; Omar, A.K.M. Techniques for extraction of bioactive compounds from plant materials: A review. *J. Food Eng.* **2013**, *117*, 426–436. [CrossRef]
107. Kumar, K.; Srivastav, S.; Sharanagat, V.S. Ultrasound assisted extraction (UAE) of bioactive compounds from fruit and vegetable processing by-products: A review. *Ultrason. Sonochem.* **2020**, *70*, 105325. [CrossRef] [PubMed]
108. Ranjha, M.M.A.N.; Kanwal, R.; Shafique, B.; Arshad, R.N.; Irfan, S.; Kieliszek, M.; Kowalczewski, P.; Irfan, M.; Khalid, M.Z.; Roobab, U.; et al. A Critical Review on Pulsed Electric Field: A Novel Technology for the Extraction of Phytoconstituents. *Molecules* **2021**, *26*, 4893. [CrossRef]
109. Pattnaik, M.; Pandey, P.; Martin, G.J.O.; Mishra, H.N.; Ashokkumar, M. Innovative Technologies for Extraction and Microencapsulation of Bioactives from Plant-Based Food Waste and Their Applications in Functional Food Development. *Foods* **2021**, *10*, 279. [CrossRef]
110. Chemat, F.; Rombaut, N.; Sicaire, A.-G.; Meullemiestre, A.; Fabiano-Tixier, A.-S.; Abert-Vian, M. Ultrasound assisted extraction of food and natural products. Mechanisms, techniques, combinations, protocols and applications. A review. *Ultrason. Sonochem.* **2017**, *34*, 540–560. [CrossRef]
111. Gallo, M.; Ferrara, L.; Naviglio, D. Application of Ultrasound in Food Science and Technology: A Perspective. *Foods* **2018**, *7*, 164. [CrossRef] [PubMed]
112. Tlais, A.Z.A.; Fiorino, G.M.; Polo, A.; Filannino, P.; Di Cagno, R. High-Value Compounds in Fruit, Vegetable and Cereal Byproducts: An Overview of Potential Sustainable Reuse and Exploitation. *Molecules* **2020**, *25*, 2987. [CrossRef]
113. Golisch, B.; Lei, Z.; Tamura, K.; Brumer, H. Configured for the Human Gut Microbiota: Molecular Mechanisms of Dietary β-Glucan Utilization. *ACS Chem. Biol.* **2021**, *16*, 2087–2102. [CrossRef]
114. Hromádková, Z.; Košt'álová, Z.; Ebringerová, A. Comparison of conventional and ultrasound-assisted extraction of phenolics-rich heteroxylans from wheat bran. *Ultrason. Sonochem.* **2008**, *15*, 1062–1068. [CrossRef]
115. Cengiz, M.F.; Babacan, Ü.; Akinci, E.; Kesci, S.T.; Kaba, A. Extraction of phenolic acids from ancient wheat bran samples by ultrasound application. *J. Chem. Technol. Biotechnol.* **2020**, *96*, 134–141. [CrossRef]
116. Das, A.B.; Goud, V.V.; Das, C. Extraction of phenolic compounds and anthocyanin from black and purple rice bran (*Oryza sativa* L.) using ultrasound: A comparative analysis and phytochemical profiling. *Ind. Crop. Prod.* **2017**, *95*, 332–341. [CrossRef]
117. Raza, A.; Li, F.; Xu, X.; Tang, J. Optimization of ultrasonic-assisted extraction of antioxidant polysaccharides from the stem of Trapa quadrispinosa using response surface methodology. *Int. J. Biol. Macromol.* **2016**, *94*, 335–344. [CrossRef] [PubMed]
118. Teslić, N.; Bojanić, N.; Čolović, D.; Fišteš, A.; Rakić, D.; Solarov, M.B.; Zeković, Z.; Pavlić, B. Technology, Conventional versus novel extraction techniques for wheat germ oil recovery: Multi-response optimization of supercritical fluid extraction. *Sep. Sci. Technol.* **2021**, *56*, 1546–1561. [CrossRef]
119. Trigo, J.P.; Alexandre, E.M.; Saraiva, J.A.; Pintado, M.E. High value-added compounds from fruit and vegetable by-products–Characterization, bioactivities, and application in the development of novel food products. *Crit. Rev. Food Sci. Nutr.* **2022**, *60*, 1388–1416. [CrossRef]
120. Wang, W.; Ma, X.; Xu, Y.; Cao, Y.; Jiang, Z.; Ding, T.; Ye, X.; Liu, D. Ultrasound-assisted heating extraction of pectin from grapefruit peel: Optimization and comparison with the conventional method. *Food Chem.* **2015**, *178*, 106–114. [CrossRef]
121. De Oliveira, C.F.; Giordani, D.; Lutckemier, R.; Gurak, P.D.; Cladera-Olivera, F.; Marczak, L.D.F. Extraction of pectin from passion fruit peel assisted by ultrasound. *LWT* **2016**, *71*, 110–115. [CrossRef]
122. Guandalini, B.B.V.; Rodrigues, N.P.; Marczak, L.D.F. Sequential extraction of phenolics and pectin from mango peel assisted by ultrasound. *Food Res. Int.* **2018**, *119*, 455–461. [CrossRef] [PubMed]
123. Xu, Y.; Zhang, L.; Bailina, Y.; Ge, Z.; Ding, T.; Ye, X.; Liu, D. Effects of ultrasound and/or heating on the extraction of pectin from grapefruit peel. *J. Food Eng.* **2014**, *126*, 72–81. [CrossRef]
124. Maran, J.P.; Priya, B. Ultrasound-assisted extraction of polysaccharide from *Nephelium lappaceum* L. fruit peel. *Int. J. Biol. Macromol.* **2014**, *70*, 530–536. [CrossRef] [PubMed]
125. Chen, X.; Wang, Z.; Kan, J. Polysaccharides from ginger stems and leaves: Effects of dual and triple frequency ultrasound assisted extraction on structural characteristics and biological activities. *Food Biosci.* **2021**, *42*, 101166. [CrossRef]
126. Feng, R.; Zhao, Y.; Zhu, C.; Mason, T. Enhancement of ultrasonic cavitation yield by multi-frequency sonication. *Ultrason. Sonochem.* **2002**, *9*, 231–236. [CrossRef]
127. De Souza, M.A.; Vilas-Boas, I.T.; Leite-da-Silva, J.M.; Abrahão, P.d.N.; Teixeira-Costa, B.E.; Veiga-Junior, V.F.J.P. Polysaccharides in Agro-Industrial Biomass Residues. *Polysaccharides* **2022**, *3*, 95–120. [CrossRef]
128. Dhingra, D.; Michael, M.; Rajput, H.; Patil, R.T. Dietary fibre in foods: A review. *J. Food Sci. Technol.* **2012**, *49*, 255–266. [CrossRef] [PubMed]

129. Zhang, W.; Zeng, G.; Pan, Y.; Chen, W.; Huang, W.; Chen, H.; Li, Y. Properties of soluble dietary fiber-polysaccharide from papaya peel obtained through alkaline or ultrasound-assisted alkaline extraction. *Carbohydr. Polym.* **2017**, *172*, 102–112. [CrossRef]
130. Chen, K.; Yan, W.; Zhang, X.; Tang, X.; Han, X.; Li, J. Ultrasound-assisted extraction of dietary fibre from Citrus Changshan-huyou peels. *J. Chem. Pharm. Res.* **2014**, *6*, 312–318.
131. Li, X.; He, X.; Lv, Y.; He, Q. Extraction and Functional Properties of Water-Soluble Dietary Fiber from Apple Pomace. *J. Food Process Eng.* **2014**, *37*, 293–298. [CrossRef]
132. Begum, Y.A.; Deka, S.C. Effect of processing on structural, thermal, and physicochemical properties of dietary fiber of culinary banana bracts. *J. Food Process. Preserv.* **2019**, *43*, e14256. [CrossRef]
133. Mohammadpour, H.; Sadrameli, S.M.; Eslami, F.; Asoodeh, A. Optimization of ultrasound-assisted extraction of Moringa peregrina oil with response surface methodology and comparison with Soxhlet method. *Ind. Crop. Prod.* **2019**, *131*, 106–116. [CrossRef]
134. Gharibzahedi, S.M.T.; Smith, B.; Guo, Y. Ultrasound-microwave assisted extraction of pectin from fig (Ficus carica L.) skin: Optimization, characterization and bioactivity. *Carbohydr. Polym.* **2019**, *222*, 114992. [CrossRef]
135. Ezzati, S.; Ayaseh, A.; Ghanbarzadeh, B.; Heshmati, M.K. Pectin from sunflower by-product: Optimization of ultrasound-assisted extraction, characterization, and functional analysis. *Int. J. Biol. Macromol.* **2020**, *165*, 776–786. [CrossRef]
136. Asgari, K.; Labbafi, M.; Khodaiyan, F.; Kazemi, M.; Hosseini, S.S. High-methylated pectin from walnut processing wastes as a potential resource: Ultrasound assisted extraction and physicochemical, structural and functional analysis. *Int. J. Biol. Macromol.* **2019**, *152*, 1274–1282. [CrossRef] [PubMed]
137. Akhtar, M.N.; Mushtaq, Z.; Ahmad, N.; Khan, M.K.; Ahmad, M.H.; Hussain, A.I.; Imran, M. Optimal Ultrasound-Assisted Process Extraction, Characterization, and Functional Product Development from Flaxseed Meal Derived Polysaccharide Gum. *Processes* **2019**, *7*, 189. [CrossRef]
138. Sun, J.; Zhang, Z.; Xiao, F.; Wei, Q.; Jing, Z. Ultrasound-assisted alkali extraction of insoluble dietary fiber from soybean residues. *IOP Conf. Ser. Mater. Sci. Eng.* **2018**, *392*, 052005. [CrossRef]
139. Chanioti, S.; Tzia, C. Optimization of ultrasound-assisted extraction of oil from olive pomace using response surface technology: Oil recovery, unsaponifiable matter, total phenol content and antioxidant activity. *LWT* **2017**, *79*, 178–189. [CrossRef]
140. Tungmunnithum, D.; Elamrani, A.; Abid, M.; Drouet, S.; Kiani, R.; Garros, L.; Kabra, A.; Addi, M.; Hano, C. A Quick, Green and Simple Ultrasound-Assisted Extraction for the Valorization of Antioxidant Phenolic Acids from Moroccan Almond Cold-Pressed Oil Residues. *Appl. Sci.* **2020**, *10*, 3313. [CrossRef]
141. Jalili, F.; Jafari, S.M.; Emam-Djomeh, Z.; Malekjani, N.; Farzaneh, V. Optimization of Ultrasound-Assisted Extraction of Oil from Canola Seeds with the Use of Response Surface Methodology. *Food Anal. Methods* **2017**, *11*, 598–612. [CrossRef]
142. Amigh, S.; Dinani, S.T. Combination of ultrasound-assisted aqueous enzymatic extraction and cooking pretreatment for date seed oil recovery. *Heat Mass Transf.* **2020**, *56*, 2345–2354. [CrossRef]
143. Shen, M.; Weihao, W.; Cao, L. Soluble dietary fibers from black soybean hulls: Physical and enzymatic modification, structure, physical properties, and cholesterol binding capacity. *J. Food Sci.* **2020**, *85*, 1668–1674. [CrossRef]
144. Cacace, J.; Mazza, G. Optimization of Extraction of Anthocyanins from Black Currants with Aqueous Ethanol. *J. Food Sci.* **2003**, *68*, 240–248. [CrossRef]
145. Liyana-Pathirana, C.; Shahidi, F. Optimization of extraction of phenolics compounds from wheat using response surface methodology. *Food. Chem.* **2005**, *93*, 45–46.
146. Zhong, K.; Wang, Q. Optimization of ultrasonic extraction of polysaccharides from dried longan pulp using response surface methodology. *Carbohydr. Polym.* **2010**, *80*, 19–25. [CrossRef]
147. Chen, F.; Zhang, Q.; Fei, S.; Gu, H.; Yang, L. Optimization of ultrasonic circulating extraction of samara oil from Acer saccharum using combination of Plackett–Burman design and Box–Behnken design. *Ultrason. Sonochem.* **2017**, *35*, 161–175. [CrossRef]
148. Ahmad, T.; Aadil, R.M.; Ahmed, H. Treatment and utilization of dairy industrial waste: A review. *Trends Food Sci. Technol.* **2019**, *88*, 361–372. [CrossRef]
149. Cravotto, G.; Bicchi, C.; Mantegna, S.; Binello, A.; Tomao, V.; Chemat, F. Extraction of kiwi seed oil: Soxhlet versus four different non-conventional techniques. *Nat. Prod. Res.* **2011**, *25*, 974–981. [CrossRef]
150. Metherel, A.H.; Taha, A.Y.; Izadi, H.; Stark, K.D. The application of ultrasound energy to increase lipid extraction throughput of solid matrix samples (flaxseed). *Prostaglandins Leukot. Essent. Fat. Acids* **2009**, *81*, 417–423. [CrossRef] [PubMed]
151. Li, H.; Pordesimo, L.; Weiss, J. High intensity ultrasound-assisted extraction of oil from soybeans. *Food Res. Int.* **2004**, *37*, 731–738. [CrossRef]
152. Kowalski, R.; Wawrzykowski, J. Effect of ultrasound-assisted maceration on the quality of oil from the leaves of thyme *Thymus vulgaris* L. *Flavour. Fragr. J.* **2009**, *24*, 69–74. [CrossRef]
153. Radulović, N.; Blagojević, P.; Palić, R. Volatiles of the grape hybrid cultivar Othello (*Vitis vinifera* x (*Vitis labrusca* x *Vitis riparia*)) cultivated in Serbia. *J. Essent. Oil. Res.* **2010**, *22*, 616–619. [CrossRef]
154. Živković, S.; Veljković, M. Environmental impacts the of production and use of biodiesel. *Environ. Sci. Pollut. Res.* **2017**, *25*, 191–199. [CrossRef] [PubMed]
155. Goh, B.H.H.; Chong, C.T.; Ge, Y.; Ong, H.C.; Ng, J.-H.; Tian, B.; Ashokkumar, V.; Lim, S.; Seljak, T.; Józsa, V. Progress in utilisation of waste cooking oil for sustainable biodiesel and biojet fuel production. *Energy Convers. Manag.* **2020**, *223*, 113296. [CrossRef]
156. Ma, F.; Hanna, M.A. Biodiesel production: A review. *Bioresour. Technol.* **1999**, *70*, 1–15. [CrossRef]

157. Veljković, V.B.; Avramović, J.M.; Stamenković, O.S. Biodiesel production by ultrasound-assisted transesterification: State of the art and the perspectives. *Renew. Sustain. Energy Rev.* **2011**, *16*, 1193–1209. [CrossRef]
158. Singh, A.K.; Fernando, S.D.; Hernandez, R. Base-Catalyzed Fast Transesterification of Soybean Oil Using Ultrasonication. In Proceedings of the 2006 ASAE Annual Meeting, American Society of Agricultural and Biological Engineers, St. Joseph, MI, USA, 9–12 July 2006.
159. Kalva, A.; Sivasankar, T.; Moholkar, V. Physical Mechanism of Ultrasound-Assisted Synthesis of Biodiesel. *Ind. Eng. Chem. Res.* **2009**, *48*, 534–544. [CrossRef]
160. Wu, Y.; Li, W.; Martin, G.J.; Ashokkumar, M. Mechanism of low-frequency and high-frequency ultrasound-induced inactivation of soy trypsin inhibitors. *Food Chem.* **2021**, *360*, 130057. [CrossRef]
161. Carmona-Cabello, M.; Sáez-Bastante, J.; Pinzi, S.; Dorado, M. Optimization of solid food waste oil biodiesel by ultrasound-assisted transesterification. *Fuel* **2019**, *255*, 115817. [CrossRef]
162. Oza, S.; Prajapati, N.; Kodgire, P.; Kachhwaha, S.S. An ultrasound-assisted process for the optimization of biodiesel production from waste cottonseed cooking oil using response surface methodology. *Water-Energy Nexus* **2021**, *4*, 187–198. [CrossRef]
163. Sharma, A.; Kodgire, P.; Kachhwaha, S.S. Investigation of ultrasound-assisted KOH and CaO catalyzed transesterification for biodiesel production from waste cotton-seed cooking oil: Process optimization and conversion rate evaluation. *J. Clean. Prod.* **2020**, *259*, 120982. [CrossRef]
164. Yasvanthrajan, N.; Sivakumar, P.; Muthukumar, K.; Murugesan, T.; Arunagiri, A. Production of biodiesel from waste bio-oil through ultrasound assisted transesterification using immobilized lipase. *Environ. Technol. Innov.* **2020**, *21*, 101199. [CrossRef]
165. Xiang, Y.; Wang, L.; Jiao, Y. Ultrasound strengthened biodiesel production from waste cooking oil using modified coal fly ash as catalyst. *J. Environ. Chem. Eng.* **2016**, *4*, 818–824. [CrossRef]
166. Gupta, A.R.; Yadav, S.V.; Rathod, V.K. Enhancement in biodiesel production using waste cooking oil and calcium diglyceroxide as a heterogeneous catalyst in presence of ultrasound. *Fuel* **2015**, *158*, 800–806. [CrossRef]
167. Anuar, M.R.; Abdullah, A.Z. Ultrasound-assisted biodiesel production from waste cooking oil using hydrotalcite prepared by combustion method as catalyst. *Appl. Catal. A Gen.* **2016**, *514*, 214–223. [CrossRef]
168. Khan, M.U.; Lee, J.T.E.; Bashir, M.A.; Dissanayake, P.D.; Ok, Y.S.; Tong, Y.W.; Shariati, M.A.; Wu, S.; Ahring, B.K. Current status of biogas upgrading for direct biomethane use: A review. *Renew. Sustain. Energy Rev.* **2021**, *149*, 111343. [CrossRef]
169. Rahman, K.M.; Woodard, R.; Manzanares, E.; Harder, M.K. An assessment of anaerobic digestion capacity in Bangladesh. *Renew. Sustain. Energy Rev.* **2014**, *32*, 762–769. [CrossRef]
170. Erden, G.; Filibeli, A. Biotechnology, Ultrasonic pre-treatment of biological sludge: Consequences for disintegration, anaerobic biodegradability, and filterability. *J. Chem. Technol. Biotechnol.* **2010**, *85*, 145–150. [CrossRef]
171. Mirko, C.; Daniela, P.; Chiara, T.; Giovanni, G. Pretreatments for enhanced biomethane production from buckwheat hull: Effects on organic matter degradation and process sustainability. *J. Environ. Manag.* **2021**, *285*, 112098. [CrossRef]
172. Shanthi, M.; Banu, J.R.; Sivashanmugam, P. Effect of surfactant assisted sonic pretreatment on liquefaction of fruits and vegetable residue: Characterization, acidogenesis, biomethane yield and energy ratio. *Bioresour. Technol.* **2018**, *264*, 35–41. [CrossRef]
173. Ormaechea, P.; Castrillón, L.; Marañón, E.; Fernández-Nava, Y.; Negral, L.; Megido, L. Influence of the ultrasound pretreatment on anaerobic digestion of cattle manure, food waste and crude glycerine. *Environ. Technol.* **2017**, *38*, 682–686. [CrossRef]
174. Levin, D.B.; Pitt, L.; Love, M. Biohydrogen production: Prospects and limitations to practical application. *Int. J. Hydrog. Energy* **2004**, *29*, 173–185. [CrossRef]
175. Elbeshbishy, E.; Hafez, H.; Nakhla, G. Enhancement of biohydrogen producing using ultrasonication. *Int. J. Hydrog. Energy* **2010**, *35*, 6184–6193. [CrossRef]
176. Leaño, E.P.; Anceno, A.J.; Babel, S. Ultrasonic pretreatment of palm oil mill effluent: Impact on biohydrogen production, bioelectricity generation, and underlying microbial communities. *Int. J. Hydrog. Energy* **2012**, *37*, 12241–12249. [CrossRef]
177. Suresh, T.; Sivarajasekar, N.; Balasubramani, K.; Ahamad, T.; Alam, M.; Naushad, M. Process intensification and comparison of bioethanol production from food industry waste (potatoes) by ultrasonic assisted acid hydrolysis and enzymatic hydrolysis: Statistical modelling and optimization. *Biomass-Bioenergy* **2020**, *142*, 105752. [CrossRef]
178. Sindhu, R.; Binod, P.; Mathew, A.K.; Abraham, A.; Gnansounou, E.; Ummalyma, S.B.; Thomas, L.; Pandey, A. Development of a novel ultrasound-assisted alkali pretreatment strategy for the production of bioethanol and xylanases from chili post harvest residue. *Bioresour. Technol.* **2017**, *242*, 146–151. [CrossRef] [PubMed]
179. Battista, F.; Mancini, G.; Ruggeri, B.; Fino, D. Selection of the best pretreatment for hydrogen and bioethanol production from olive oil waste products. *Renew. Energy* **2016**, *88*, 401–407. [CrossRef]
180. John, I.; Pola, J.; Appusamy, A. Optimization of Ultrasonic Assisted Saccharification of Sweet Lime Peel for Bioethanol Production Using Box–Behnken Method. *Waste Biomass Valoriz.* **2017**, *10*, 441–453. [CrossRef]
181. Mozhiarasi, V. Overview of pretreatment technologies on vegetable, fruit and flower market wastes disintegration and bioenergy potential: Indian scenario. *Chemosphere* **2021**, *288*, 132604. [CrossRef] [PubMed]

foods

MDPI

Article

Polyphenols Recovery from *Thymus serpyllum* Industrial Waste Using Microwave-Assisted Extraction–Comparative RSM and ANN Approach for Process Optimization

Živan Mrkonjić [1], Dušan Rakić [1], Aleksandar Takači [1], Muammer Kaplan [2], Nemanja Teslić [3], Zoran Zeković [1], Ivana Lazarević [1] and Branimir Pavlić [1,*]

[1] Faculty of Technology, University of Novi Sad, Bulevar Cara Lazara 1, 21000 Novi Sad, Serbia; zivan_mrkonjic@hotmail.com (Ž.M.); dusan.rakic@uns.ac.rs (D.R.); stakaci@tf.uns.ac.rs (A.T.); zzekovic@tf.uns.ac.rs (Z.Z.); ivana.lazarevic702@gmail.com (I.L.)

[2] TUBITAK Marmara Research Centre, Institute of Chemical Technology, P.O. Box 21, Gebze 41470, Kocaeli, Turkey; muammer.kaplan@tubitak.gov.tr

[3] Institute of Food Technology, University of Novi Sad, Bulevar Cara Lazara 1, 21000 Novi Sad, Serbia; nemanja.teslic@fins.uns.ac.rs

* Correspondence: bpavlic@uns.ac.rs; Tel.: +381-63-874-34-20

Abstract: The aim of this study was to valorize *Thymus serpyllum* L. herbal dust, the particular fraction distinguished as industrial waste from filter-tea production. This work demonstrated comparable analysis considering model fitting, influence analysis and optimization of microwave-assisted extraction (MAE) of bioactive compounds from the aforementioned herbal dust using face-centered central composite experimental design within the response surface methodology (RSM), as well as artificial neural networks (ANN). In order to increase yield and amount of compounds of interest and minimize solvent, time and energy consumption, the ethanol concentration (45, 60 and 75%), extraction time (5, 12.5 and 20 min), liquid–solid ratio (10, 20 and 30 mL/g) and irradiation power (400, 600 and 800 W) were used as independent variables. Total extraction yield (Y), total phenols yield (TP), as well as antioxidant activity parameters obtained by DPPH and ABTS assays, were selected as responses. It could be concluded that the MAE technique is an efficient approach for the extraction of biologically active compounds from *T. serpyllum* herbal dust, which represents a high-value source of natural antioxidants with great potential for further use in various forms within different branches of industry.

Keywords: wild thyme; MAE; antioxidant activity; multi-response optimization; artificial neural network; by-product; Lamiaceae

1. Introduction

Thymus serpyllum L., known as a highly potent source of bioactive compounds, is widely spread across northern and central Europe. As a perennial shrub, whose herb has been collected and used in many cultures as a healing agent, nowadays mostly as a tea product, it has shown great potential for further research. The first written traces showed that even in ancient Greece, wild thyme was used to treat respiratory and digestive tract [1]. Since then, many more characteristics such as antibacterial, antiviral, astringent, anti-inflammatory, sedative, anticonvulsant, expectorant, antispasmodic, choleretic, analgesic, diuretic, diaphoretic, anticholesterolemic, immunostimulant and anthelmintic have been identified [1–3].

It was noted that during the process of production and packing different types of various teas a fraction remained unused. The reason lies in the particle size diameter that is smaller than the pores of filter-tea bags. Although presented as industrial waste, the filter-tea fraction presents a valuable source of polyphenols [4–7]. Considering the above, wild thyme herbal dust has been chosen as a starting point for the investigation of microwave-assisted extraction (MAE). MAE belongs to green extraction techniques characterized by short extraction time and low energy and solvent consumption [8]. External energy in form of microwaves generates heat energy which enhances the cellular disruption of plant material and accelerates the mass transfer from the matrix into the liquid phase [9]. The heat energy occurs due to dipole rotation and ionic conduction, which is caused by the friction between molecules that rotate and by resistance and friction that appear during ions migration throughout the solution [10].

Parameters that can significantly affect MAE efficiency are the type of solvent, extraction time, liquid–solid ratio, irradiation power, number of cycles, temperature and type of raw material. Polar solvents, i.e., solvents with a high dielectric constant, have an advantage over non-polar ones in terms of better microwave energy adsorption [11]. It must be taken into account that highly polar solvents such as water, due to low microwave dissipation factor, might prevent heating inside the plant matrix and reduce the release efficiency of compounds of interest [12]. For that reason, the mixture of ethanol and water, generally recognized as safe, is considered the first choice for the isolation of compounds of interest using microwave energy [8]. The liquid–solid ratio also plays a significant role in the optimization process of MAE and it should be adjusted in order to entirely soak the plant matrix without leading to a reduction of polyphenols recovery [13]. Temperature has a close connection with irradiation power and extraction time, so an adequate setting of temperature will minimize the risk of thermal degradation and provide extraction of compounds of interest [14]. So far, many studies have successfully conducted microwave-assisted extraction (MAE) in order to maximize the total phenols yield (TP) and antioxidant activity of extracts obtained from various plants from the Lamiaceae family [15–18].

The main objective of our study was to valorize *T. serpyllum* herbal dust by MAE and to evaluate extraction parameters (ethanol concentration, extraction time, liquid–solid ratio and irradiation power) affecting total extraction yield (Y), total phenols yield (TP) and antioxidant properties of obtained extracts. To our best knowledge, there are no papers investigating the application of response surface methodology (RSM) and artificial neural networks (ANN) on the MAE of *T. serpyllum*. Therefore, RSM and ANN approaches were applied to optimize antioxidants recovery from *T. serpyllum* herbal dust and to conduct comparative analysis in terms of influence analysis, model fitting and optimization accuracy. Finally, validation was performed and qualitative polyphenols profiles of extracts obtained at the central point and under optimal conditions by RSM were determined by HPLC-MS/MS analysis.

2. Materials and Methods

2.1. Sample

The sample was kindly donated by Macval D.O.O. Novi Sad (Serbia) and it is considered a representative sample of the population that unites plants grown and collected in the territory of the Republic of Serbia. Due to its mean particle size of less than 0.315 mm and the impracticality of having this fraction packed in the filter-tea bag, it is characterized as a by-product. The *T. serpyllum* herbal dust was stored in a paper bag at room temperature prior to extractions.

2.2. Chemicals

Folin–Ciocalteu reagent, (±)-6-hydroxy-2,5,7,8-tetramethylchromane-2-carboxylic acid (Trolox) (≥98%), 2,2-diphenyl-1-picrylhydrazyl (DPPH) (≥90%), 2,4,6-tris(2-pyridyl)-s-triazine (≥99.0%) and gallic acid (≥98.0%) were supplied from Sigma-Aldrich (Steinheim, Germany). 2,2′-azino-bis(3-ethylbenzothiazoline-6-sulfonic acid) diammonium salt (ABTS) (98%) was purchased from J&K, Scientific Ltd., Beijing, China. Additionally, ferric chloride hexahydrate (>99%) and sodium carbonate anhydrous (>99%) were supplied from Centrohem, Stara Pazova, Serbia, while potassium peroxydisulfate (>99.6%) and acetic acid (99.8%) were purchased from Lach-Ner, Neratovice, the Czech Republic. Sodium acetate anhydrous (>99%) was purchased from Kemika, Zagreb, Croatia. The ultra-pure water was obtained by a Milli-Q Plus system (EMD Millipore, Billerica, MA, USA).

2.3. Microwave-Assisted Extraction (MAE)

MAE was conducted in a homemade open microwave system consisting of a microwave oven (MM817ASM, Bosch, Gerlingen, Germany), a round glass flask (500 mL) and a condenser. The MAE experiments were performed in a single mode at a fixed frequency of 2.45 GHz, where the temperature varied depending on the boiling point of the mixture of raw material and applied solvent. The glass flask was fixed at the same position as the microwave system and no additional agitation was applied. Obtained crude extracts were cooled down, filtered through filter paper under vacuum (V-700, Büchi, Flawil, Switzerland), collected into plastic vials and stored at 4 °C prior to analysis.

2.4. Determination of Y and TP

Using the rotary vacuum evaporator (RV 05, IKA, Staufen im Breisgau, Germany) 10 mL of crude extract was evaporated under 45 °C and further dried in the oven at 105 °C until constant mass. Results of Y were presented as a mass of total extractable solids per 100 g of dry plant material (%; *w/w*). Determination of TP of obtained extracts was performed by Folin–Ciocalteu assay [19]. Absorbances were recorded at 750 nm using a spectrophotometer (Jenway, model 6300, Stone, UK). All experiments were performed in triplicate and mean values of TP of obtained extracts were presented as grams of gallic acid equivalent (GAE) per 100 g of sample dry weight (g GAE/100 g).

2.5. Antioxidant Activity of Extracts

Antioxidant activity of extracts was determined by spectrophotometric methods towards DPPH [20] and ABTS radicals [21], as well as by the ferric ion reducing antioxidant power (FRAP) assay [22]. DPPH assay was performed by mixing 100 µL of extract with 2900 µL of DPPH solution, which was previously prepared at a concentration of 26 mg/L of methanol. After 1 h in the dark at room temperature, the absorbances were recorded at a wavelength of 517 nm. ABTS assay was performed by mixing 100 µL of extract with 2900 µL of ABTS reagent. After 5 h in the dark at room temperature, the absorbances were recorded at a wavelength of 734 nm. The FRAP assay was performed by mixing 100 µL of extract with 2900 µL of FRAP reagent. After incubation in the dark at 37 °C for 10 min, the absorbances were recorded at a wavelength of 593 nm. All experiments were performed in triplicate and mean values were presented in the case of DPPH and ABTS assays as mM of Trolox equivalent (TE) per gram of sample dry weight (mM TE/g) and in the case of FRAP assay mM of Fe^{2+} per gram of sample dry weight (mM Fe^{2+}/g).

2.6. Q Exactive Hybrid Quadrupole-Orbitrap LC-MS/MS Analysis

With the aim of testing free, ester-linked and glycoside-linked phenolic compounds in samples, the Q Exactive LC-MS/MS–Orbitrap (Thermo Scientific, Hemel Hempstead, UK) was employed. The same method was applied and described in detail in the study by Pavlić et al. [17]. Chromatographic separation of compounds was achieved on a Poroshell 120 EC-C18 column (3.0 × 100 mm, 2.7 µm, Agilent) setting gradient flow at 0.6 mL/min (mobile phase A: 0.1% formic acid–water, mobile phase B: methanol; 0–5 min, mobile phase B concentration changed as 0–9% B; 5–9 min, 9–2% B; 16–35 min, 2–18% B; 35–50 min, 18–20% B; 50–65 min, 20–30% B and 65–80 min, 30% B). The injection volume was 10 µL. A Q Exactive hybrid quadrupole-Orbitrap mass spectrometer equipped with an ESI source working in both negative and positive ionization mode was used for accurate mass measurements. The following parameters were set: ion spray voltage, 2.8 kV; capillary temperature, 300 °C; capillary voltage, 35 V and tube lens voltage, 95 V; sheath gas, 19 (arbitrary units); auxiliary gas, 7 (arbitrary units). Mass spectra were recorded covering the m/z range of 55–1000 Da. Default values were used for most other acquisition parameters (Automatic gain control (AGC) target 3×10^6 ions). The data processing was achieved using XCalibur 2.2 software (Thermo Fisher Scientific, Waltham, MA, USA). An external calibration for mass accuracy was performed before the analysis.

2.7. Design of Experiments and Statistical Methods

With the aim of investigating the impact of the MAE parameters on target responses and optimizing the extraction process, response surface methodology (RSM) and artificial neural networks (ANNs) were used. The impacts of ethanol concentration (45, 60 and 75%), extraction time (5, 12.5 and 20 min), liquid–solid ratio (10, 20 and 30 mL/g) and irradiation power (400, 600 and 800 W) were used as input parameters in both methods. A study of Mrkonjić et al. [23], that investigated the impact of different concentrations of ethanol on Y, TP and DPPH using conventional solid–liquid extraction of *T. serpyllum* herbal dust, made it easier to define an appropriate domain of concentration of ethanol. Extraction time and liquid–solid ratio were set by the pattern of similar studies, while irradiation power was based on the capabilities of the microwave oven [17,18,24].

2.7.1. Response Surface Methodology (RSM)

Under the RSM, the face-centered central composite experimental design (CCD), which included four input parameters arranged at three levels, was used. Taking into account five replicates at the central point, it consisted of a total of 29 runs. Multiple regression was applied and the results were fitted to a second-order polynomial model (Equation (1)).

$$Y = \beta_0 + \sum_{i=1}^{4} \beta_i X_i + \sum_{i=1}^{4} \beta_{ii} X_i^2 + \sum_{i=1}^{3} \sum_{j=i+1}^{4} \beta_{ij} X_i X_j \quad (1)$$

Y represents the response variable, X_i and X_j are the independent variables affecting the response, and β_0, β_i, β_{ii}, and β_{ij} are the regression coefficients for intercept, linear, quadratic and cross-product terms. The goodness of fit was determined by analysis of variance (ANOVA), while model adequacy was evaluated by the coefficient of determination (R^2), coefficient of variance (CV) and *p*-values for the model and lack of fit. Optimal extraction conditions were determined considering desired Y and TP, as well as antioxidant activity parameters obtained by DPPH and ABTS assays. The selection of optimal conditions was based on the desirability function (*D*) [25]. In order to verify obtained empirical models, validation was performed by using the extracts prepared at optimal conditions determined by the selected statistical methodologies. For the listed analysis, Design-Expert v.11 software (Stat-Ease, Minneapolis, MN, USA) was used with a significance level of 0.05.

2.7.2. Artificial Neural Network (ANN)

Constructed ANN models consisted of four input parameters and one output, which means that each model individually for each investigated response was developed. The input layer consisted of 4 nodes representing the input parameters, 1 hidden layer with 5–10 nodes and the output layer consisted of 1 node representing the response. The number of hidden nodes differs in each network as the node number is a subject of optimization. For training, 70% of data was used, 15% for testing and 15% for validation. The training process consisted of three steps, where in the first step feedforward of the input training pattern was performed. In the second step, calculation and backpropagation of the associated error were conducted and finally, the gradient-dependent adjustment of the calculated weights was conducted. ANN calculations were performed using Statistica 10 software package.

Error estimation was achieved by the Symbiotic organisms search (SOS) method and the training algorithm was achieved by the Broyden–Fletcher–Goldfarb–Shanno (BFGS) algorithm. For the error measurement, the R^2 quotient was used. Out of 20 networks for each output, the best were chosen based on the R values and a global sensitivity analysis was performed for each network. The optimization was based according to validation error minimization. In order to find the optimal values of input parameters, a new neural network was constructed with 4 input and 4 output nodes and one hidden layer. The inputs and outputs were normalized. The sum of normalized outputs was the value that was maximized in the optimization process.

3. Results and Discussion

3.1. Total Extraction Yield (Y) and Total Phenols Yield (TP)

Experimentally observed values of Y and TP according to face-centered CCD, including four input parameters (ethanol concentration, extraction time, liquid–solid ratio and irradiation power) on three levels, are presented in Table 1. In order to normalize independent variables, their actual values are presented in the form of coded values that were set by the range from −1 to 1.

Table 1. Face-centered CCD with coded and actual values of input parameters and experimentally observed values of investigated responses.

Run	Input Parameters								Responses				
	A: Ethanol Concentration (%)		B: Extraction Time (min)		C: Liquid–Solid Ratio (mL/g)		D: Irradiation Power (W)		Y (%)	TP (g GAE/100 g DW)	DPPH (mM TE/g)	FRAP (mM Fe^{2+}/g)	ABTS (mM TE/g)
1	45	−1	5	−1	10	−1	400	−1	15.21	3.8200	0.2027	0.5945	0.4570
2	75	1	5	−1	10	−1	400	−1	10.69	3.9155	0.1861	0.5763	0.4373
3	45	−1	20	1	10	−1	400	−1	15.00	4.2637	0.2145	0.6395	0.5018
4	75	1	20	1	10	−1	400	−1	13.25	3.2721	0.1836	0.3590	0.3857
5	45	−1	5	−1	30	1	400	−1	14.06	3.2983	0.1734	0.5445	0.3789
6	75	1	5	−1	30	1	400	−1	14.50	4.1299	0.2057	0.6792	0.4208
7	45	−1	20	1	30	1	400	−1	11.54	2.8903	0.1567	0.4678	0.3377
8	75	1	20	1	30	1	400	−1	15.66	4.0896	0.2033	0.6032	0.4970
9	45	−1	5	−1	10	−1	800	1	15.54	4.1036	0.2150	0.6932	0.4163
10	75	1	5	−1	10	−1	800	1	14.86	4.0438	0.2208	0.7071	0.5023
11	45	−1	20	1	10	−1	800	1	11.72	3.5056	0.1666	0.5372	0.4318
12	75	1	20	1	10	−1	800	1	15.10	4.0288	0.2148	0.6319	0.4942
13	45	−1	5	−1	30	1	800	1	12.85	3.0775	0.1613	0.4854	0.3707
14	75	1	5	−1	30	1	800	1	15.62	3.9141	0.2091	0.5634	0.3918
15	45	−1	20	1	30	1	800	1	11.90	3.6965	0.1806	0.5713	0.4237
16	75	1	20	1	30	1	800	1	12.20	3.8593	0.1845	0.5768	0.4393
17	45	−1	12.5	0	20	0	600	0	13.64	3.5796	0.1874	0.7182	0.3438
18	75	1	12.5	0	20	0	600	0	14.82	4.2085	0.2382	0.4889	0.4634
19	60	0	5	−1	20	0	600	0	15.39	4.1064	0.2208	0.7109	0.5120
20	60	0	20	1	20	0	600	0	15.06	3.6245	0.2199	0.6465	0.4009
21	60	0	12.5	0	10	−1	600	0	15.58	3.8154	0.2197	0.4470	0.3991
22	60	0	12.5	0	30	1	600	0	16.84	3.9876	0.2125	0.4525	0.4960
23	60	0	12.5	0	20	0	400	−1	14.86	4.0288	0.2208	0.4416	0.4453
24	60	0	12.5	0	20	0	800	1	16.90	3.6371	0.2218	0.5167	0.3284
25	60	0	12.5	0	20	0	600	0	17.60	3.6558	0.2346	0.5725	0.3855
26	60	0	12.5	0	20	0	600	0	15.21	3.8706	0.2253	0.5196	0.4193
27	60	0	12.5	0	20	0	600	0	15.15	3.5922	0.1936	0.5549	0.3429
28	60	0	12.5	0	20	0	600	0	15.64	3.7930	0.2280	0.4882	0.4570
29	60	0	12.5	0	20	0	600	0	17.19	3.8580	0.2348	0.5786	0.4000

According to Table 1, Y was in the range 10.69–17.60%, where the lowest value of Y was observed using 75% of ethanol, liquid–solid ratio of 10 mL/g at 400 W for 5 min (run 2). On the other hand, using 60% of ethanol, liquid–solid ratio of 20 mL/g at 600 W for 12.5 min led to highest value of Y (run 25). Recovery of polyphenols was the lowest (2.8903 g GAE/100 g DW) by using 45% of ethanol, liquid–solid ratio of 30 mL/g at 400 W for 20 min (run 7). Even though the recovery of polyphenols was the highest using 45% of ethanol, but with a different liquid–solid ratio (10 mL/g) at 400 W for 20 min (4.2637 g GAE/100 g DW; run 3), the lower liquid–solid ratio is desirable in the sense of preventing excessive dilution of liquid extracts and decreasing cost of their further processing [17]. In addition, an observation that the highest values of TP were obtained using 50% of ethanol concentration was noticed by Jovanović et al. [26], who investigated maceration, heat-assisted and ultrasound-assisted extractions (UAE) of *T. serpyllum*. In comparison with ultrasound-assisted and pressurized-liquid extraction (PLE), it could be seen that MAE provided particularly lower Y values, values that are close to those obtained by conventional extraction [23,27]. On the other hand, observed high values of TP and lower values of Y indicate that MAE could be considered a good approach for the selective isolation of polyphenols from *T. serpyllum*, facilitating the downstream process.

The significance of the influence of input parameters and their interactions on target responses are presented in Table 2.

Table 2. Significance of linear, cross-product and quadratic terms on Y, TP, DPPH, FRAP and ABTS.

Coefficients	Y	TP	DPPH	FRAP	ABTS
			p-Value		
Linear					
X_1-Ethanol concentration	<0.0001 *	0.0001 *	<0.0001 *	0.3235	0.0616
X_2-Extraction time	0.0448 **	0.1554	0.0014 *	0.6279	0.2024
X_3-Liquid–solid ratio	0.0004 *	<0.0001 *	0.6736	0.0339	<0.0001 *
X_4-Irradiation power	0.2989	0.8638	0.5279	0.0545	0.5151
Cross-product					
X_1X_2	-	0.1157	0.6323	-	-
X_1X_3	-	0.0321 **	0.0161 **	-	-
X_1X_4	-	-	0.5840	-	0.3010
X_2X_3	0.0191 **	-	0.6873	-	-
X_2X_4	0.2665	0.8809	-	0.2744	-
X_3X_4	-	0.6695	0.2751	-	-
Quadratic					
X_1^2	0.0377 **	0.0090 *	0.1226	-	0.0216 **
X_2^2	0.0729	0.7068	0.6754	-	-
X_3^2	0.2946	0.4842	0.0311 **	-	-
X_4^2	0.1989	0.8243	-	0.0098 *	0.2836

* $p < 0.01$—significant, ** $0.01 < p < 0.05$—moderately significant.

According to *p*-values obtained by ANOVA analysis, linear terms of ethanol concentration and liquid–solid ratio exhibited a highly significant effect ($p < 0.01$) on Y and TP, while the linear term of extraction time had a moderately significant effect ($0.01 < p < 0.05$) only on Y. Ethanol concentration had a negative influence on Y (Figure S1a), but in the case of TP, it was in the range of experimental data and its maximum value was obtained at an ethanol concentration of 55.22%, after which it had more intensive negative influence (Figure S2a). The liquid–solid ratio had a negative influence on Y (Figure S1b), but positive on TP (Figure S2b). The same trend with ethanol concentration was reported by Zeković et al. [17], whose investigation was based on UAE and MAE optimization of *Salvia officinalis* L. herbal dust. Linear term of irradiation power exhibited almost negligible influence (Figure S2d), but extraction time provided a slightly positive one on TP (Figure S2c). By increasing the irradiation power, the Y value at one point reaches a maximum after which it begins slightly to decline (Figure S1d). On the other hand, increasing the extraction time does not necessarily mean that the extraction of polyphenols will be efficient, due to the possibility of potential extraction of non-target compounds. In addition, prolonged time in combination with microwave energy could enhance the denaturation of the plant cell walls and accelerate the diffusion of the aforementioned compounds to the liquid phase [28,29]. The influence of interactions was noticed as moderately significant only between extraction time and liquid–solid ratio on Y, as well as between ethanol concentration and liquid–solid ratio on TP. The influence of ethanol concentration quadratic term was pronounced in the case of TP as highly significant and moderately significant in the case of Y. Although the influence of ethanol concentration was not in the range of experimental data, the explanation for this lies in the occurrence of potential non-selective extraction, as well as in the technical difficulties during the filtration process of the obtained crude extract [23].

3.2. Antioxidant Activity of *T. serpyllum* Extracts

The antioxidant activity of *T. serpyllum* extracts was assessed through DPPH, FRAP and ABTS tests. Polyphenols free radicals, which arise by hydrogen or electron transfer reactions, were measured using DPPH and ABTS antioxidant capacity. On the other hand, the reducing power of extracts was measured using the FRAP test, due to the capability of polyphenols to prevent oxidation by chelation and reduction of prooxidant metals and inhibition of oxidizing enzymes [30,31]. The experimental values, obtained by the already mentioned tests, are presented in Table 1. The lowest scavenging capacity of DPPH (0.1567 mM TE/g) was observed in run 7 (45% ethanol, 20 min, 30 mL/g and 400 W), while the highest one (0.2382 mM TE/g) was observed in run 18 using the following values of input parameters, 75% ethanol, 12.5 min, 20 mL/g and 600 W. In the case of FRAP, the lowest value was observed in run 4 (0.3590 mM Fe^{2+}/g) using 75% of ethanol, 20 min extraction time, 10 mL/g liquid–solid ratio and irradiation power of 400 W. Using 45% of ethanol, 12.5 min extraction time, 20 mL/g liquid–solid ratio and irradiation power of 600 W the highest value of FRAP (0.7182 mM Fe^{2+}/g) was accomplished (run 17). Extract obtained in run 24 (60% ethanol, 12.5 min, 20 mL/g and 800 W) exhibited the lowest scavenging capacity of free ABTS radicals (0.3284 mM TE/g), while the highest one (0.5120 mM TE/g) was exhibited in run 19 using 60% of ethanol, liquid–solid ratio of 20 mL/g at 600 W for 5 min. Obtained values were in accordance with the results provided by Mrkonjić et al. [23], who investigated the antioxidant activity of UAE wild thyme herbal dust extracts using the same tests. Đukić et al. investigated the antioxidant activity of *T. serpyllum* extracts obtained by MAE (96% ethanol, 30 min, 20 mL/g, 600 W). However, obtained results were expressed as IC_{50} (29.60 mg/mL), which could not be compared with results obtained by the present study [32].

Linear terms of ethanol concentration and extraction time exhibited a highly significant influence on DPPH, while the linear term of the liquid–solid ratio showed a significant impact on ABTS (Table 2). Only a combination of ethanol concentration and liquid–solid ratio showed moderate significance on DPPH, while all other interaction terms did not show a significant impact on other responses. The quadratic term of ethanol concentration exhibited a moderately significant impact on ABTS, the liquid–solid ratio the same on DPPH, while irradiation power showed a significant impact on FRAP. Taking into account the aforementioned quadratic terms and responses, ABTS, DPPH and FRAP values were in the range of experimental data. Positive or negative influence of input parameters on DPPH, FRAP and ABTS was presented in Figures S3, S4 and S5, respectively. Due to the significant impact of the quadratic term (Table 2), it could be observed that ethanol concentration had a positive influence on ABTS until reaching the ethanol concentration of 57.41% (Figure S5a), after which the impact started to decline. Considering that antioxidant activity and TP are closely related, it could be noticed that the same trend with ethanol concentration was observed by the finding of Simić et al. [24], whose goal was to maximize TP of *Aronia melanocarpa* MAE extracts. This could be explained by increasing the dipole moment of the solvent throughout higher absorption of microwaves. Moreover, such a kind of solvent shows a higher affinity towards moderately polar polyphenols in terms of enhanced diffusion of the solvent into the plant cells, which leads to better heating and facilitates the mass transfer from solid to liquid phase [29,33].

3.3. *RSM versus ANN Model Fitting*

Adequacy and statistical significance of the applied quadratic model obtained by RSM were checked by analysis of variance (ANOVA) (Table 3). Based on values of R^2 obtained for Y, TP, DPPH, FRAP and ABTS (0.9242, 0.8487, 0.9216, 0.4733 and 0.6661, respectively), it could be concluded that there is a good fit between experimentally observed and predicted values, except in the case of FRAP and ABTS.

The values of coefficient of variance (CV), that describe the dispersion degree of the data, for Y, TP, DPPH, FRAP and ABTS, were 4.11, 4.55, 4.14, 12.57 and 8.44%, respectively. Those values lead to the conclusion that there is a good fitness of the applied model, but again except in the case of FRAP and ABTS. The lack of fit for all responses excluding FRAP was not statistically significant ($p > 0.05$). Considering the above, only FRAP was excluded in the process of both RSM and ANN optimization.

Table 3. ANOVA and descriptive statistics parameters for applied quadratic models.

Response	Source	Sum of Squares	df	Mean Square	*F*-Value	*p*-Value
Y	Model	79.21	10	7.92	21.94	<0.0001
	Residual	6.5	18	0.361		
	Lack of Fit	5.86	14	0.4187	2.64	0.1804
	Pure Error	0.6355	4	0.1589		
	Cor Total	85.71	28			
	R^2 = 0.9242					
	CV = 4.11%					
TP	Model	2.66	12	0.2218	7.48	0.0002
	Residual	0.4744	16	0.0296		
	Lack of Fit	0.2434	12	0.0203	0.3513	0.9284
	Pure Error	0.231	4	0.0577		
	Cor Total	3.14	28			
	R^2 = 0.8487					
	CV = 4.55%					
DPPH	Model	0.0135	12	0.0011	15.68	<0.0001
	Residual	0.0011	16	0.0001		
	Lack of Fit	0.0006	12	0.0001	0.383	0.9114
	Pure Error	0.0005	4	0.0001		
	Cor Total	0.0146	28			
	R^2 = 0.9216					
	CV = 4.14%					
FRAP	Model	0.1038	6	0.0173	3.29	0.0182
	Residual	0.1155	22	0.0052		
	Lack of Fit	0.1123	18	0.0062	7.8	0.0298
	Pure Error	0.0032	4	0.0008		
	Cor Total	0.2193	28			
	R^2 = 0.4733					
	CV = 12.57%					
ABTS	Model	0.0535	7	0.0076	5.99	0.0006
	Residual	0.0268	21	0.0013		
	Lack of Fit	0.0188	17	0.0011	0.5563	0.8241
	Pure Error	0.008	4	0.002		
	Cor Total	0.0803	28			
	R^2 = 0.6661					
	CV = 8.44%					

df—degrees of freedom.

Recently, several studies have focused on the comparative analysis of RSM and ANN modeling applied to the optimization of polyphenols recovery from various plant resources such as blackcurrants [34], Meghalayan cherry fruit [35] and kidney beans [36]. However, Teslić et al. [30] recently proposed a comprehensive comparison of these statistical tools applied to the optimization of polyphenols isolation from defatted wheat germ. Therefore, the comparison between RSM and ANN approaches in this work was performed on three levels, where in the first one the goodness of fit was evaluated. Considering obtained regression plots and their coefficients, it has been noticed that corresponding coefficients for ANN were slightly higher than those obtained for RSM (Figure 1). This is in accordance with results published by Teslić et al. [30].

Fitting of experimental results by RSM was done using second-order polynomial model. On the other hand, fitting of experimental results by ANN was done using nonlinear models. Due to potential "over-fitting", ANN simulation was realized with 5 to 10 neurons in the hidden layer. Summary of active networks and their training and test performances were presented in Table S1. Considering particularly high values of R^2 for Y, TP, DPPH and ABTS (0.9570, 0.8960, 0.9740 and 0.8020, respectively), it could be concluded that ANN provided adequate fit of experimental data for aforementioned responses (Figure 1b).

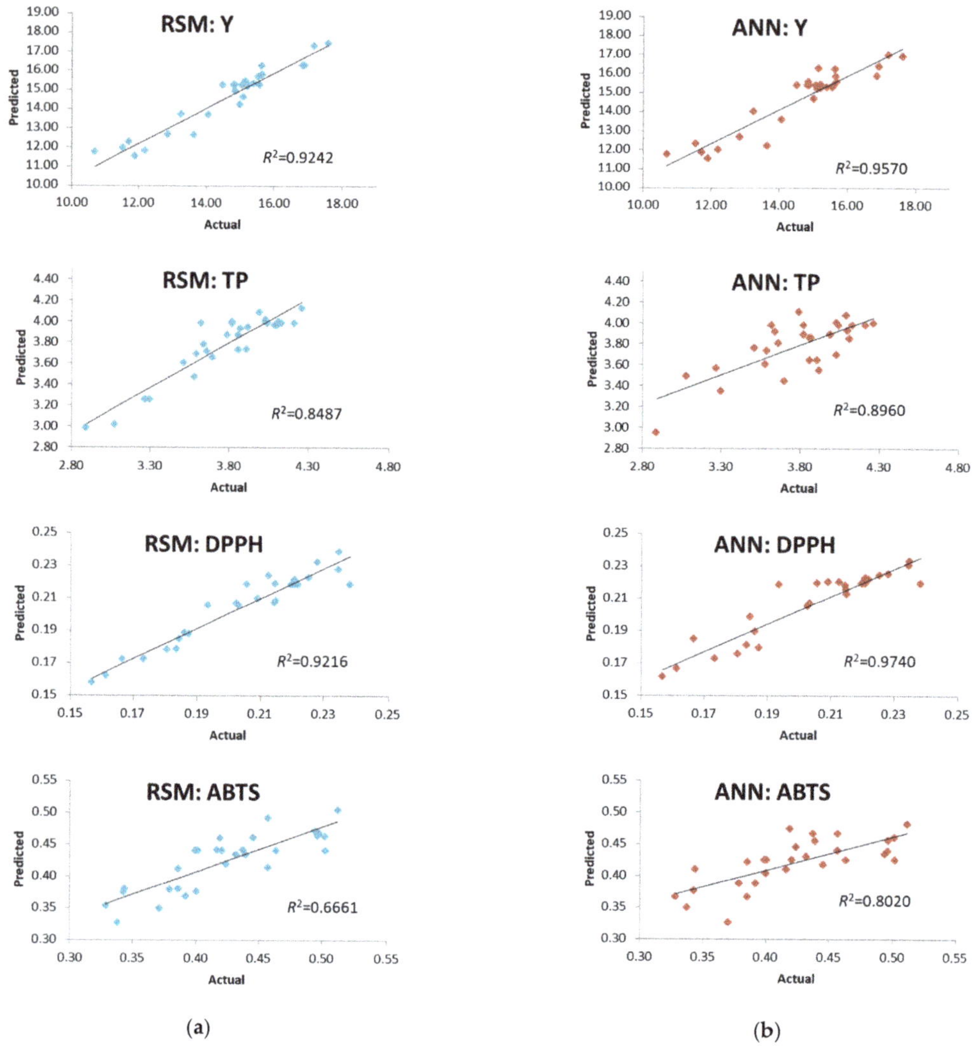

Figure 1. Comparison of experimentally observed and predicted values determined by (**a**) RSM and (**b**) ANN.

The second level of comparison between RSM and ANN was evaluated through contributions of input parameters (Figure 2).

In consideration of contributions of linear, interaction and quadratic terms obtained by RSM and contributions obtained by sensitivity analysis throughout the ANN approach, it could be concluded that ethanol concentration was the most significant parameter that influenced Y, TP and DPPH, except in the case of ABTS, where the liquid–solid ratio was treated as the most significant parameter (Figure 2a). On the other hand, ANN showed that ethanol concentration was the most significant parameter in the case of all responses (Figure 2b). This is in agreement with the finding by Mrkonjić et al. [23], who investigated the significance of UAE parameters on the same target responses. However, another study by Mrkonjić et al. [27], which covered the PLE of *T. serpyllum* herbal dust, singled out ethanol concentration as the third most significant parameter, after temperature and extraction time. According to Veggi et al. [33], the most important factor that affects MAE efficiency is solvent choice. Furthermore, solvents used in conventional extractions could differ in terms of the capacity of the solvent to absorb the microwave energy and heat up. Besides solvent, the liquid–solid ratio is considered the second most important parameter affecting MAE efficiency. The reason for that lies in more energy consumption when a high liquid–solid ratio is used. In addition, the inadequate setting of this parameter could require additional time downstream of the process and cause nonuniform distribution and exposure to microwaves [13,33].

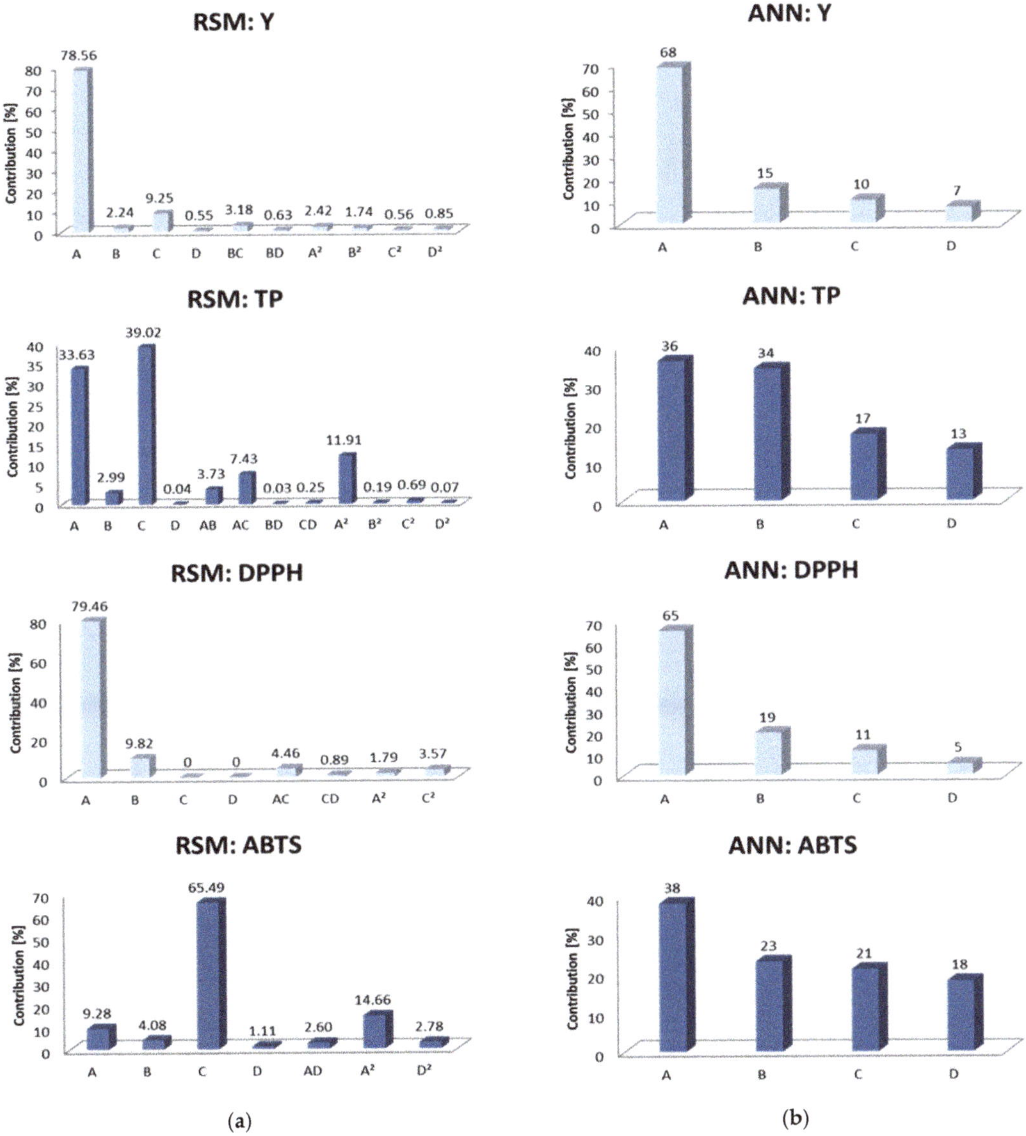

Figure 2. Contributions of ethanol concentration (A), extraction time (B), liquid–solid ratio (C) and irradiation power (D) observed by (**a**) RSM and (**b**) ANN.

For all responses, ANN has identified extraction time as the second, liquid–solid ratio as the third, and irradiation power as the fourth most significant parameter. RSM provided the same influence as ANN on DPPH, while the liquid–solid ratio was the second, extraction time the third and irradiation power the fourth most significant parameter that affected Y and TP. In the case of RSM and contributions obtained for ABTS, the significance of parameters was as follows: liquid–solid ratio > ethanol concentration > extraction time > irradiation power. However, the significance of irradiation power was almost negligible, which was not the case with ANN, where the contribution was in the range of 5 to 18% (Figure 2).

The third level of comparison between RSM and ANN was evaluated considering experimentally observed and predicted values obtained after the optimization process by RSM and ANN, which was explained in Section 3.4. However, given the accuracy of the methodology, the ANN approach has a slight advantage over RSM. The same trend was noticed in the finding of Simić et al. [24], who used RSM and ANN for modeling and optimizing the MAE of TP from *Aronia melanocarpa*.

According to the significance of linear, cross-product and quadratic terms on all responses excluding FRAP, the predictive model equations are presented in Table 4.

Table 4. Predictive model equations in coded values for total extraction and phenols yield and antioxidant activity obtained by RSM.

Response	Model Equation
Y	$Y = 15.2400 - 1.8100X_1 + 0.3053X_2 - 0.6210X_3 + 0.1515X_4 - 0.3865X_2X_3 - 0.1722X_2X_4 - 0.8383X_1^2 + 0.7117X_2^2 - 0.4033X_3^2 - 0.4983X_4^2$
TP	$TP = 3.9900 - 0.2028X_1 + 0.0605X_2 + 0.2184X_3 - 0.0071X_4 + 0.0716X_1X_2 + 0.1011X_1X_3 - 0.0066X_2X_4 - 0.0187X_3X_4 - 0.3184X_1^2 + 0.0410X_2^2 - 0.0767X_3^2 + 0.0242X_4^2$
DPPH	$DPPH = 0.2185 - 0.0222X_1 + 0.0077X_2 + 0.0009X_3 + 0.0013X_4 + 0.0057X_1X_3 - 0.0024X_3X_4 - 0.0082X_1^2 - 0.0119X_3^2$
FRAP	$FRAP = 0.6253 - 0.0172X_1 - 0.0084X_2 + 0.0386X_3 - 0.0347X_4 - 0.0203X_2X_4 - 0.0785X_4^2$
ABTS	$ABTS = 0.4400 - 0.0166X_1 + 0.0111X_2 + 0.0443X_3 - 0.0056X_4 + 0.0095X_1X_4 - 0.0480X_1^2 + 0.0213X_4^2$

X_1—Ethanol concentration; X_2—Extraction time; X_3—Liquid–solid ratio; X_4—Irradiation power.

3.4. Process Optimization and Experimental Verification

The planned experiment allowed us to realize the process optimization and to find the optimal working conditions using both RSM and ANN. All responses except FRAP were taken into account in the optimization process, where the goal of optimization was to maximize all of them. Optimized conditions obtained by RSM were the ethanol concentration of 52.19%, extraction time of 20 min, liquid–solid ratio of 23.64 mL/g and irradiation power of 400 W, with a desirability function of 0.88. On the other hand, optimized conditions obtained by ANN were the ethanol concentration of 45.00%, extraction time of 5 min, a liquid–solid ratio of 30.00 mL/g and the same value of irradiation power, as in the case of RSM. The optimization process of ANN was performed in C++ programming language by the brute force method, which ran all input values with the increment of 0.01. Predicted values for ANN were calculated using developed mathematical models by Statistica software, while in the case of RSM predicted values were obtained based on quadratic model equations given in Table 4. The same methodology was applied in our previous research focused on optimization of the extraction process by RSM and ANN approaches [30]. According to obtained optimal conditions, the experimental verification was conducted (Table 5).

Table 5. Predicted and experimental values of investigated responses obtained at optimal conditions according to RSM and ANN optimization.

Input Parameters	RSM		ANN	
	Optimal Parameters			
Ethanol concentration [%]	52.19		45	
Extraction time [min]	20		5	
Liquid–solid ratio [mL/g]	23.64		30	
Irradiation power [W]	400		400	
Output Parameters	**Predicted Values**	**Experimental Values**	**Predicted Values**	**Experimental Values**
Y [%]	16.07	16.89	15.68	17.46
TP [g GAE/100 g]	4.16	4.86 ± 0.1830	4.54	4.36 ± 0.1556
DPPH [mM TE/g]	0.23	0.26 ± 0.0039	0.23	0.20 ± 0.0020
ABTS [mM TE/g]	0.49	0.53 ± 0.0142	0.58	0.57 ± 0.0185

It could be concluded that both RSM and ANN approaches showed satisfactory correlation between predicted and experimental values. Despite the different conditions under which both verifications were performed, the experimentally obtained values were high and relatively close to each other. Furthermore, both approaches could be recognized as an infallible choice for MAE optimization of *T. serpyllum* herbal dust. According to the techno-economical aspect, a slight preference is given to ANN, where a high-quality extract would be produced in the shortest possible time.

3.5. Polyphenols Profile

The qualification of polyphenols in extracts obtained on the central point (Sample MAE-CP) and under the optimal conditions (Sample MAE-OPT) was carried out using Q Exactive hybrid quadrupole-orbitrap LC-MS/MS analysis (Table 6). A total of 35 different compounds were detected.

Detected phenolic acids in the extract obtained on the central point were gallic, vanillic, protocatechuic, 3-*p*-coumaroylquinic, 4-*p*-coumaroylquinic, *p*-coumaric acid, coumaric acid hexoside isomer-1, coumaric acid hexoside isomer-2, coumaric acid hexoside isomer-3, *cis*- and *trans*-coutaric, caffeic and ellagic acid. In extract obtained under optimal conditions all aforementioned phenolic acids were identified except for *cis*- and *trans*-coutaric acid. In comparison with the present study, Boros et al. [37] have additionally identified chlorogenic, ferulic and rosmarinic acids in methanol extracts of *T. serpyllum*, whereas the raw material originated from Hungarian and Romanian habitats.

Compounds that were identified in both extracts were dihydroxycoumarin, quercetin in form of hexoside isomer-1, hexoside isomer-2, 3-*O*-galactoside, 3-*O*-glucoside, 3-*O*-rutinoside and glucuronide, kaempferol, kaempferol-3-rutinoside, 3-galactoside and 3-glucoside, naringenin-7-*O*-glucoside, isorhamnetin-3-*O*-galactoside, eriodictyol and luteolin. Furthermore, monogalloyl-glucose, a compound that belongs to the tannin group, was identified in both extracts as well. However, (+)-catechin and (-)-epicatechin together with naringenin were identified only in extract MAE-OPT. On the other hand, quercetin in a free form and its derivatives, quercetin pentoside isomer-1 and quercetin pentoside isomer-2, were identified only in extract MAE-CP.

In the study of Ivasenko et al. [3], although based on UAE of bioactive compounds from the population of *T. serpyllum* from Central Kazahstan, phenolic acids and flavonoids prevailed in the extracts obtained with 70% of ethanol, a liquid–solid ratio of 1:20 at room temperature for 30 min. Among the contents of extracts, the following bioactive compounds stood out: luteolin-7-*O*-glucoside, rosmarinic acid, naringenin, epicatechin, catechin, myricetin, quercetin, apigenin, kaempferol, caffeic, gallic, chlorogenic, ferulic acid, rutin and luteolin. According to Fecka and Turek [38], in methanolic extract of *T. serpyllum* caffeic acid and its derivatives, eriocitrin, luteolin, luteolin-7-*O*-rutinoside, luteolin-7-*O*-glucuronide, eriodictyol, lithospermic and rosmarinic acid were detected. Considering the influence of the locality of origin, environmental and other conditions, as well as extraction process and all factors included, a certain similarity among differently obtained extracts could be recognized. It may be concluded that all the aforementioned could greatly affect the chemical composition and yield of the *T. serpyllum*. For that reason, the qualitative profile of the extract is of great importance, with the purpose of its further use in the pharmaceutical, cosmetic and food industries.

Table 6. Polyphenols profile of extracts obtained on the central point (Sample MAE-CP) and under the optimal conditions (Sample MAE-OPT).

Retention Time [min]	Compound	Sample MAE-CP	Sample MAE-OPT
		Measured Mass [m/z]/Error [mDa]	
12.67	Monogalloyl-glucose	331.07/0.48	331.07/0.58
14.14	Gallic acid	169.01/0.01	169.01/0.46
16.15	Vanillic acid	167.03/0.14	167.03/−0.04
19.28	Protocatechuic acid	153.02/0.12	153.02/0.32
37.61	3-*p*-Coumaroylquinic acid	337.09/0.92	337.09/0.92
37.61	4-*p*-Coumaroylquinic acid	337.09/0.92	337.09/0.92
39.33	(+)-Catechin	ND	289.07/0.69
39.33	(−)-Epicatechin	ND	289.07/0.69
39.53	Coumaric acid hexoside isomer-1	325.09/0.30	325.09/0.30
39.53	Coumaric acid hexoside isomer-2	325.09/0.30	325.09/0.30
39.53	Coumaric acid hexoside isomer-3	325.09/0.30	325.09/0.30
39.53	*p*-Coumaric acid	163.04/−0.19	163.04/−0.19
43.32	Dihydroxycoumarin	177.02/0.36	177.02/−0.07
44.46	Caffeic acid	179.03/0.13	179.03/0.05
52.08	*cis*-Coutaric acid	295.05/0.17	ND
52.08	*trans*-Coutaric acid	295.05/0.17	ND
66.28	Quercetin hexoside isomer-1	463.09/−0.14	463.09/0.98
66.28	Quercetin hexoside isomer-2	463.09/−0.14	463.09/0.98
66.28	Quercetin-3-*O*-galactoside	463.09/−0.14	463.09/0.98
66.28	Quercetin-3-*O*-glucoside	463.09/−0.14	463.09/0.98
66.61	Kaempferol-3-rutinoside	593.15/−0.11	593.15/−0.22
72.90	Naringenin-7-*O*-glucoside	433.11/−0.01	433.11/0.34
76.06	Quercetin-3-*O*-rutinoside	609.15/−0.09	609.15/0.01
76.37	Quercetin glucuronide	477.07/1.37	477.07/1.50
76.45	Kaempferol-3-galactoside	447.09/−0.59	447.09/0.03
76.45	Kaempferol-3-glucoside	447.09/−0.59	447.09/0.03
76.52	Isorhamnetin-3-*O*-galactoside	477.10/−0.64	477.10/−0.13
76.83	Eriodictyol	287.06/0.57	287.06/0.68
77.27	Naringenin	ND	271.06/1.42
78.10	Kaempferol	285.04/0.24	285.04/0.45
78.10	Luteolin	285.04/0.24	285.04/0.45
78.74	Quercetin pentoside isomer-1	433.08/0.30	ND
78.74	Quercetin pentoside isomer-2	433.08/0.30	ND
78.84	Quercetin	301.03/−1.45	ND
78.90	Ellagic acid	301.00/−0.80	301.00/−0.70

ND = not detected.

4. Conclusions

In comparison with ANN, RSM could provide further insights into the influence analysis of cross-product and quadratic terms. Moreover, RSM could also provide a more detailed graphical visualization of the influence of input parameters on responses. Both techniques suggested that ethanol concentration was the most significant factor affecting all responses, except in the case of RSM on ABTS, where the liquid–solid ratio was the most significant parameter. Optimal conditions obtained by RSM were the ethanol concentration of 52.19%, extraction time of 20 min, a liquid–solid ratio of 23.64 mL/g and irradiation power of 400 W, while optimal conditions obtained by ANN were an ethanol concentration of 45.00%, extraction time of 5 min, liquid–solid ratio of 30.00 mL/g and irradiation power of 400 W. Verification demonstrated that both RSM and ANN approaches showed good correlation between predicted and experimental values. Considering the fitting quality, influence analysis and optimization, both mathematical procedures RSM and ANN could be successfully used in order to describe experimental data. According to the techno-economical aspect, a slight preference is given to ANN, where, by reducing extraction time, the quality of polyphenol-enriched wild thyme MAE extract may remain unchanged. It could be also concluded that MAE represents a cost-effective approach for the isolation of natural bioactive compounds from *T. serpyllum* herbal dust. Furthermore, it could be used in a form of liquid or powder extract in the agri-food sector, pharmaceutical and cosmetic industry as a substitution for synthetic products. Liquid extracts obtained by MAE will be processed into powder form by spray drying technology. Further research on wild thyme herbal dust extracts will include supercritical fluid extraction of lipophilic bioactive compounds and evaluation of their chemical profile and bioactivity. The perspective of the application of wild thyme powder and lipophilic extracts as natural additives in food products such as meat and dairy products will be determined in our following studies.

Supplementary Materials: The following supporting information can be downloaded at: https://www.mdpi.com/article/10.3390/foods11091184/s1, Figure S1: Influence of (a) ethanol concentration, (b) extraction time, (c) liquid–solid ratio and (d) irradiation power on Y; Figure S2: Influence of (a) ethanol concentration, (b) extraction time, (c) liquid–solid ratio and (d) irradiation power on TP; Figure S3: Influence of (a) ethanol concentration, (b) extraction time, (c) liquid–solid ratio and (d) irradiation power on DPPH; Figure S4: Influence of (a) ethanol concentration, (b) extraction time, (c) liquid–solid ratio and (d) irradiation power on FRAP; Figure S5: Influence of (a) ethanol concentration, (b) extraction time, (c) liquid–solid ratio and (d) irradiation power on ABTS; Table S1: Summary of active networks and their training and test performances title.

Author Contributions: Conceptualization, B.P. and D.R.; methodology, Ž.M., M.K., N.T. and I.L.; software, A.T. and D.R.; validation, B.P. and D.R.; formal analysis, Ž.M. and M.K.; investigation, Ž.M., I.L. and N.T.; resources, B.P.; data curation, Ž.M., B.P., D.R. and A.T.; writing—original draft preparation, Ž.M. and I.L.; writing—review and editing, D.R., B.P., N.T., Z.Z., A.T. and M.K.; visualization, B.P.; supervision, B.P. and D.R.; project administration, B.P. and Z.Z.; funding acquisition, Z.Z. and B.P. All authors have read and agreed to the published version of the manuscript.

Funding: This research was supported by the Science Fund of the Republic of Serbia, 7750168, Novel extracts and bioactive compounds from under-utilized resources for high-value applications–BioUtilize.

Institutional Review Board Statement: Not applicable.

Informed Consent Statement: Not applicable.

Data Availability Statement: Data are contained within the article or Supplementary Materials.

Acknowledgments: The authors would like to thank the Ministry of Education, Science and Technological Development, Republic of Serbia, for financial support (Project No. 451-03-68/2022-14/200134).

Conflicts of Interest: The authors declare no conflict of interest.

References

1. Jarić, S.; Mitrović, M.; Pavlović, P. Review of ethnobotanical, phytochemical, and pharmacological study of *Thymus serpyllum* L. *Evid.-Based Complement. Altern. Med.* **2015**, *2015*, 101978. [CrossRef] [PubMed]
2. Carpentieri, S.; Soltanipour, F.; Ferrari, G.; Pataro, G.; Donsì, F. Emerging Green Techniques for the Extraction of Antioxidants from Agri-Food By-Products as Promising Ingredients for the Food Industry. *Antioxidants* **2021**, *10*, 1417. [CrossRef] [PubMed]
3. Ivasenko, S.; Orazbayeva, P.; Skalicka–wozniak, K.; Ludwiczuk, A.; Marchenko, A.; Ishmuratova, M.; Poleszak, E.; Korona-Glowniak, I.; Akhmetova, S.; Karilkhan, I.; et al. Antimicrobial Activity of Ultrasonic Extracts of Two Chemotypes of *Thymus serpyllum* L. of Central Kazakhstan and their Polyphenolic Profiles. *Open Access Maced. J. Med. Sci.* **2021**, *9*, 61–67. [CrossRef]
4. Ramić, M.; Vidović, S.; Zeković, Z.; Vladić, J.; Cvejin, A.; Pavlić, B. Modeling and optimization of ultrasound-assisted extraction of polyphenolic compounds from *Aronia melanocarpa* by-products from filter-tea factory. *Ultrason. Sonochem.* **2015**, *23*, 360–368. [CrossRef] [PubMed]
5. Naffati, A.; Vladić, J.; Pavlić, B.; Radosavljević, R.; Gavarić, A.; Vidović, S. Recycling of filter tea industry by-products: Application of subcritical water extraction for recovery of bioactive compounds from *A. uva-ursi* herbal dust. *J. Supercrit. Fluids* **2017**, *121*, 197–206. [CrossRef]
6. Pavlić, B.; Naffati, A.; Hojan, T.; Vladić, J.; Zeković, Z.; Vidović, S. Microwave-assisted extraction of wild apple fruit dust—production of polyphenol-rich extracts from filter tea factory by-products. *J. Food Process Eng.* **2017**, *40*, e12508. [CrossRef]
7. Pavlić, B.; Teslić, N.; Vidaković, A.; Vidović, S.; Velićanski, A.; Versari, A.; Radosavljević, R.; Zeković, Z. Sage processing from by-product to high quality powder: I. Bioactive potential. *Ind. Crops Prod.* **2017**, *107*, 81–89. [CrossRef]
8. Osorio-Tobón, J.F. Recent advances and comparisons of conventional and alternative extraction techniques of phenolic compounds. *J. Food Sci. Technol.* **2020**, *57*, 4299–4315. [CrossRef]
9. Yahya, N.A.; Attan, N.; Wahab, R.A. An overview of cosmeceutically relevant plant extracts and strategies for extraction of plant-based bioactive compounds. *Food Bioprod. Process.* **2018**, *112*, 69–85. [CrossRef]
10. Sparr Eskilsson, C.; Björklund, E. Analytical-scale microwave-assisted extraction. *J. Chromatogr. A* **2000**, *902*, 227–250. [CrossRef]
11. Wang, L.; Weller, C.L. Recent advances in extraction of nutraceuticals from plants. *Trends Food Sci. Technol.* **2006**, *17*, 300–312. [CrossRef]
12. Ajila, C.M.; Brar, S.K.; Verma, M.; Tyagi, R.D.; Godbout, S.; Valéro, J.R. Extraction and Analysis of Polyphenols: Recent trends. *Crit. Rev. Biotechnol.* **2011**, *31*, 227–249. [CrossRef] [PubMed]
13. Eskilsson, C.S.; Björklund, E.; Mathiasson, L.; Karlsson, L.; Torstensson, A. Microwave-assisted extraction of felodipine tablets. *J. Chromatogr. A* **1999**, *840*, 59–70. [CrossRef]
14. Akhtar, I.; Javad, S.; Yousaf, Z.; Iqbal, S.; Jabeen, K. Microwave assisted extraction of phytochemicals an efficient and modern approach for botanicals and pharmaceuticals. *Pak. J. Pharm. Sci.* **2019**, *32*, 223–230.
15. Jacotet-Navarro, M.; Rombaut, N.; Fabiano-Tixier, A.S.; Danguien, M.; Bily, A.; Chemat, F. Ultrasound versus microwave as green processes for extraction of rosmarinic, carnosic and ursolic acids from rosemary. *Ultrason. Sonochem.* **2015**, *27*, 102–109. [CrossRef] [PubMed]
16. Nabet, N.; Gilbert-López, B.; Madani, K.; Herrero, M.; Ibáñez, E.; Mendiola, J.A. Optimization of microwave-assisted extraction recovery of bioactive compounds from *Origanum glandulosum* and *Thymus fontanesii*. *Ind. Crop. Prod.* **2019**, *129*, 395–404. [CrossRef]
17. Pavlić, B.; Kaplan, M.; Bera, O.; Oktem Olgun, E.; Canli, O.; Milosavljević, N.; Antić, B.; Zeković, Z. Microwave-assisted extraction of peppermint polyphenols—Artificial neural networks approach. *Food Bioprod. Process.* **2019**, *118*, 258–269. [CrossRef]
18. Zeković, Z.; Pintać, D.; Majkić, T.; Vidović, S.; Mimica-Dukić, N.; Teslić, N.; Versari, A.; Pavlić, B. Utilization of sage by-products as raw material for antioxidants recovery—Ultrasound versus microwave-assisted extraction. *Ind. Crops Prod.* **2017**, *99*, 49–59. [CrossRef]
19. Singleton, V.L.; Rossi, J.A.J. Colorimetry of Total Phenolics with Phosphomolybdic- Phosphotungstic Acid Reagents. *Am. J. Enol. Vitic.* **1965**, *16*, 144–158. [CrossRef]
20. Brand-Williams, W.; Cuvelier, M.E.; Berset, C. Use of a free radical method to evaluate antioxidant activity. *LWT Food Sci. Technol.* **1995**, *28*, 25–30. [CrossRef]
21. Re, R.; Pellegrini, N.; Proteggente, A.; Pannala, A.; Yang, M.; Rice-Evans, C. Antioxidant activity applying an improved ABTS radical cation decolorization assay. *Free Radic. Biol. Med.* **1999**, *26*, 1231–1237. [CrossRef]
22. Benzie, I.F.; Strain, J.J. The ferric reducing ability of plasma (FRAP) as a measure of "antioxidant power": The FRAP assay. *Anal. Biochem.* **1996**, *239*, 70–76. [CrossRef] [PubMed]
23. Mrkonjić, Ž.; Rakić, D.; Olgun, E.O.; Canli, O.; Kaplan, M.; Teslić, N.; Zeković, Z.; Pavlić, B. Optimization of antioxidants recovery from wild thyme (*Thymus serpyllum* L.) by ultrasound-assisted extraction: Multi-response approach. *J. Appl. Res. Med. Aromat. Plants* **2021**, *24*, 100333. [CrossRef]
24. Simic, V.M.; Rajkovic, K.M.; Stojic, S.S.; Velic, D.T.; Nada, C.; Simic, V.M.; Karabegovic, I.T.; Lazic, M.L.; Simić, V.M.; Rajković, K.M.; et al. Optimization of microwave-assisted extraction of total polyphenolic compounds from chokeberries by response surface methodology and artificial neural network. *Sep. Purif. Technol.* **2016**, *160*, 89–97. [CrossRef]
25. Derringer, G.; Suich, R. Simultaneous Optimization of Several Response Variables. *J. Qual. Technol.* **1980**, *12*, 214–219. [CrossRef]

26. Jovanović, A.A.; Đorđević, V.B.; Zdunić, G.M.; Pljevljakušić, D.S.; Šavikin, K.P.; Gođevac, D.M.; Bugarski, B.M. Optimization of the extraction process of polyphenols from *Thymus serpyllum* L. herb using maceration, heat- and ultrasound-assisted techniques. *Sep. Purif. Technol.* **2017**, *179*, 369–380. [CrossRef]
27. Mrkonjić, Ž.; Rakić, D.; Kaplan, M.; Teslić, N.; Zeković, Z.; Pavlić, B. Pressurized-liquid extraction as an efficient method for valorization of *Thymus serpyllum* herbal dust towards sustainable production of antioxidants. *Molecules* **2021**, *26*, 2548. [CrossRef]
28. Dragović-Uzelac, V.; Garofulić, I.E.; Jukić, M.; Penić, M.; Dent, M. The Influence of Microwave-Assisted Extraction on the Isolation of Sage (*Salvia officinalis* L.) Polyphenols. *Food Technol. Biotechnol.* **2012**, *50*, 377–383.
29. Ranic, M.; Nikolic, M.; Pavlovic, M.; Buntic, A.; Siler-Marinkovic, S.; Dimitrijevic-Brankovic, S. Optimization of microwave-assisted extraction of natural antioxidants from spent espresso coffee grounds by response surface methodology. *J. Clean. Prod.* **2014**, *80*, 69–79. [CrossRef]
30. Teslić, N.; Bojanić, N.; Rakić, D.; Takači, A.; Zeković, Z.; Fišteš, A.; Bodroža-Solarov, M.; Pavlić, B. Defatted wheat germ as source of polyphenols—Optimization of microwave-assisted extraction by RSM and ANN approach. *Chem. Eng. Process. Process Intensif.* **2019**, *143*, 107634. [CrossRef]
31. Sharma, R. Polyphenols in health and disease: Practice and mechanisms of benefits. In *Polyphenols in Human Health and Disease*; Academic Press: Cambridge, MA, USA, 2017; pp. 757–778.
32. Đukić, D.; Mašković, P.; Moračanin, S.V.; Kurćubić, V.; Milijašević, M.; Babić, J. Conventional and unconventional extraction methods applied to the plant, *Thymus serpyllum* L. In Proceedings of the 59th International Meat Industry Conference, Zlatibor, Serbia, 1–4 October 2017; Volume 85, p. 12064. [CrossRef]
33. Veggi, P.C.; Martinez, J.; Meireles, M.A.A. Fundamentals of Microwave Extraction. In *Microwave-Assisted Extraction for Bioactive Compounds: Theory and Practice*; Chemat, F., Cravotto, G., Eds.; Springer Science+ Business Media: New York, NY, USA, 2012; pp. 15–52.
34. Bu, X.; Xu, Y.; Zhao, M.; Li, D.; Zou, J.; Wang, L.; Bai, J.; Yang, Y. Simultaneous extraction of polysaccharides and polyphenols from blackcurrant fruits: Comparison between response surface methodology and artificial neural networks. *Ind. Crops Prod.* **2021**, *170*, 113682. [CrossRef]
35. Kashyap, P.; Riar, C.S.; Jindal, N. Optimization of ultrasound assisted extraction of polyphenols from Meghalayan cherry fruit (*Prunus nepalensis*) using response surface methodology (RSM) and artificial neural network (ANN) approach. *J. Food Meas. Charact.* **2021**, *15*, 119–133. [CrossRef]
36. Yang, Q.Q.; Gan, R.Y.; Zhang, D.; Ge, Y.Y.; Cheng, L.Z.; Corke, H. Optimization of kidney bean antioxidants using RSM & ANN and characterization of antioxidant profile by UPLC-QTOF-MS. *LWT* **2019**, *114*, 108321. [CrossRef]
37. Boros, B.; Jakabová, S.; Dörnyei, Á.; Horváth, G.; Pluhár, Z.; Kilár, F.; Felinger, A. Determination of polyphenolic compounds by liquid chromatography-mass spectrometry in *Thymus* species. *J. Chromatogr. A* **2010**, *1217*, 7972–7980. [CrossRef]
38. Fecka, I.; Turek, S. Determination of polyphenolic compounds in commercial herbal drugs and spices from Lamiaceae: Thyme, wild thyme and sweet marjoram by chromatographic techniques. *Food Chem.* **2008**, *108*, 1039–1053. [CrossRef]

foods

MDPI

Article

Comparative Study of Bioactive Lipid Extraction from Squid (*Doryteuthis gahi*) by-Products by Green Solvents

Santiago P. Aubourg *, Marcos Trigo, María Jesús González, Salomé Lois and Isabel Medina

Department of Food Technology, Marine Research Institute, CSIC, C/E. Cabello, 6, 36208 Vigo, Spain; mtrigo@iim.csic.es (M.T.); mjgp@iim.csic.es (M.J.G.); salomel@iim.csic.es (S.L.); medina@iim.csic.es (I.M.)
* Correspondence: saubourg@iim.csic.es

Abstract: A novel approach of bioactive lipid extraction by different green solvents was carried out on squid (*Doryteuthis gahi*) by-products. By-products (viscera, heads, skin, tails, etc.), considered as a single product, were subjected to the following solvent systems: ethanol, acetone, ethyl acetate, 1/1 ethanol/acetone, 1/1 ethanol/ethyl acetate, and 1/1 acetone/ethyl acetate. Analyses carried out included lipid yield, lipid class content, and fatty acid (FA) composition. Results were compared to the lipid extract obtained by the traditional procedure (1/1 chloroform/methanol). Lipid yields obtained by green solvents led to a 33.4–73.2% recovery compared to traditional extraction; the highest values ($p < 0.05$) were obtained by ethanol-containing systems. Compared to the traditional procedure, ethanol systems showed an 85.8–90.3% recovery of phospholipid compounds and no differences ($p > 0.05$) in the ω3/ω6 ratio. Green-extracting systems led to higher average values for eicosapentaenoic acid content (15.66–18.56 g·100 g^{-1} total FAs) and polyene index (1.93–3.29) than chloroform/methanol extraction; differences were significant ($p < 0.05$) for systems including acetone and ethyl acetate. No differences ($p > 0.05$) were detected for docosahexaenoic acid content between the traditional procedure and green systems, with all values being included in the 31.12–32.61 g·100 g^{-1} total FA range. The suitability of EtOH-containing green systems for extraction of bioactive lipid compounds from squid by-products was concluded.

Keywords: Patagonian squid; by-products; bioactive lipids; green extraction; ethanol; acetone; ethyl acetate; phospholipids; PUFA; ω3/ω6 ratio

1. Introduction

Chemical diversity observed in marine species makes them an extraordinary source of highly valuable constituents such as unsaturated fatty acids, essential minerals, lipid-soluble vitamins, and nutritional and digestible proteins [1]. Marine lipids differ from lipids from other plant and animal sources in that they contain a wider range of fatty acids (FAs), longer-chain FAs, and a larger proportion of highly unsaturated FAs, particularly ω3 FAs such as docosahexaenoic (DHA) and eicosapentaenoic (EPA) acids [2]. Interestingly, the positive role of ω3 FA compounds in preventing certain human diseases (i.e., cardiovascular, neurodegenerative, etc.) has provoked a great deal of attention on marine lipids [3,4].

Fishing activities are reported to generate a high amount of waste. Marine by-products (blood, viscera, heads, bellies, bones, skin, etc.) are a main concern for current fishery management policies and legislations [5]. Remarkably, marine by-products constitute a relevant source of main constituents such as lipids, proteins, minerals, and vitamins, in addition to minor components such as enzymes, amino acids, pigments, chitin, collagen, and other bioactive compounds [6,7]. Consequently, employment of such compounds has been found useful for the development of food, pharmaceutical, nutraceutical, and cosmeceutical industries [8,9].

Concerning the lipid fraction of seafood by-products, great efforts have been carried out on the extraction of valuable constituents with highly nutritional and functional properties for human health such as flavonoids, vitamins A and D, ω3 FA concentrates, phospholipids or polyphenols in general [10–12]. Among them, physical fractionation [13], wet pressing [14,15], supercritical fluid extraction [16], urea concentration [17], and enzymatic hydrolysis [18,19] have been tested satisfactorily. Concerning the procedure of solvent extraction, great attention has been paid in recent decades to replace organic solvents with eco-friendly solvents. Remarkably, such efforts have been especially focused on agricultural by-products and have employed different kinds of green solvents such as ethanol, acetone, glycerol, and ethyl acetate [20–22]. Concerning green solvent employment on marine products, previous research can be considered scarce. Thus, acetone/ethanol mixtures were tested for lipid class extraction from Antarctic krill (*Euphausia superba*) [23] and ethanol/hexane was checked as an extracting system for obtaining antioxidants (i.e., vitamin E) from microalga (*Scenedesmus dimorphus*) [24].

The present study focused on a novel and alternative extraction of bioactive marine lipids by green solvents. As a marine substrate, by-products obtained after commercialisation of Patagonian squid (*Doryteuthis gahi*) were employed. This species was chosen as being an important fishery resource widely distributed along the Pacific and Atlantic coasts of South America [25]. By-products (viscera, tails, heads, tentacles, skin, etc.), considered as a single product, were lyophilised and then subjected to the following solvent systems: ethanol (EtOH), acetone (AcMe), ethyl acetate (AcOEt), 1/1 EtOH/AcMe (*v/v*), 1/1 EtOH/AcOEt (*v/v*), and 1/1 AcMe/AcOEt (*v/v*). Analyses carried out included lipid yield, lipid class content (phospholipids, PLs; free fatty acids, FFAs; sterols, STs; triacylglycerols, TAGs) and FA content (saturated FAs, STFAs; monounsaturated FAs, MUFAs; polyunsaturated FAs, PUFAs; ω3/ω6 ratio; polyene index, PI). Results were compared to those corresponding to the lipid extract obtained by the traditional procedure (i.e., 1/1 $CHCl_3$/MeOH, *v/v*) [26].

2. Materials and Methods

2.1. Solvents, Chemicals, and Standars

Solvents and chemical reagents used were of reagent grade and purchased from Merck (Darmstadt, Germany). The following solvents were employed: ethanol, acetone, ethyl acetate, chloroform, methanol, hexane, and toluene. The following reagents were used: ammonium molybdate, cupric acetate, pyridine, acetic acid, acetic anhydride, acetyl chloride, NaCl, anhydride sodium sulphate, sulphuric acid, ferric trichloride, perchloric acid, nitric acid, and hydroxylamine.

Quantitative standards (1,2-dipalmitoyl-rac-glycero-3-phosphocholine, oleic acid, cholesterol, methyl stearate, nonadecanoic acid) were purchased from Sigma-Aldrich (St. Louis, MO, USA). Qualitative (Qualmix Fish) and quantitative (FAME Mix) FAME standards were obtained from Supelco, Inc. (Bellefonte, PA, USA) and Larodan (Malmo, Sweden), respectively.

2.2. Initial Squid, Lyophilisation, and Extracting Systems

Patagonian squid (*Doryteuthis gahi*) by-products (viscera, heads, skin, tails, etc.) were provided by SERPESBA S. L. U. (Vigo, Spain). The squid were obtained near the Argentinean coast from the southwest Atlantic Ocean. Squid samples were frozen (−40 °C) and transported to the factory in Vigo (Spain) where samples were thawed (overnight storage at 4 °C), the mantel being taken for commercialisation and the resulting by-products pooled together and transported (for 15 min) in refrigerated conditions (4 °C) to our laboratory placed in the same town.

In order to analyse the initial by-products, three × 10 g portions of by-products were separated and subjected to moisture assessment and lipid extraction by the traditional procedure according to the methodology described later on. Lipid extracts were stored at −40 °C before subsequent lipid composition analysis.

On the same day, two × 1 kg of by-products were subjected to the lyophilisation process (−70 °C, 72 h, 0.05 mTorr) (Model FD8515-C60, Ilshin Biobase Europe, Ede, The Netherlands). After this time, lyophilised samples were pooled together, minced, and employed for lipid extraction by different extracting mixtures. For it, the following green solvent systems were applied: EtOH, AcMe, AcOEt, 1/1 EtOH/AcMe (*v/v*), 1/1 EtOH/AcOEt (*v/v*), and 1/1 AcMe/AcOEt (*v/v*). Additionally, lipid extraction by the traditional procedure (1/1 $CHCl_3$/MeOH, *v/v*) was also carried out.

According to preliminary trials carried out in our laboratory, the lyophilisation step was found necessary in order to substantially increase the lipid and PL yields by using the green solvent mixtures. This procedure, including moisture elimination in general, has already been described as necessary in previous related research including green solvents such as acetone and ethanol [23,27].

2.3. Moisture Assessment, Lipid Extraction, and Lipid Yield

Moisture was determined in initial by-products and in lyophilised by-products as the weight difference (1–2 g) before and after 4 h at 105 °C according to official method 950.46 B [28]. Results were calculated as g·kg^{-1} squid by-products.

Traditional lipid extraction of initial by-products and lyophilised by-products was carried out by the Bligh and Dyer [26] method; this procedure employs a single-phase solubilisation of the lipids using a $CHCl_3$/MeOH (1/1) mixture.

Lipid extraction of lyophilised by-products by employment of the above-mentioned green solvent systems was carried out as follows: 3.5 g of lyophilised by-products were mixed with 10 mL of the extracting system, stirred for 1 min at 4 °C, centrifuged at 3500× *g* for 10 min at 4 °C, and the supernatant was collected. The procedure was repeated two more times, with all supernatants being collected together. The resulting extracts were subjected to partial evaporation of the solvent (rotary evaporator; 10 min at 30 °C), and were brought up to a 15 mL volume and stored at −40 °C before analysis of lipid composition.

Each of the extracting systems (traditional and green) was carried out in triplicate (n = 3). In all kinds of lipid extracts, quantification of lipid extracts was carried out according to Herbes and Allen [29]. Results were calculated as g·kg^{-1} by-products.

2.4. Lipid Class Analysis

The PL content of lipid extracts was spectrophotometrically (710 nm; Beckman Coulter DU 640 spectrophotometer, Brea, CA, USA) measured according to the method of Raheja et al. [30], which is based on the formation of a coloured complex with ammonium molybdate. For quantification purposes, different quantities (0, 5, 10, 20, 40, 60, 80, 100, 130, and 150 µL) of a 1,2-dipalmitoyl-rac-glycero-3-phosphocholine (DPPC) solution in chloroform (15.3 mg/5 mL) were employed. The validity range was 16.1–483.0 µg DPPC and the R^2 value of the analytical procedure was 0.9995. Results were calculated as g DPPC·kg^{-1} lipids and g DPPC·kg^{-1} by-products.

The FFA content of lipid extracts was spectrophotometrically (715 nm) determined following the method of Lowry and Tinsley [31], which is based on the formation of a complex with cupric acetate-pyridine. For quantification purposes, different quantities (0, 5, 10, 20, 40, 60, 80, 100, 130, and 150 µL) of an oleic acid solution in toluene (705.3 mg/25 mL) were employed. The validity range was 0.5–15.0 µmol oleic acid and the R^2 value of the analytical procedure was 0.9998. Results were calculated as g oleic acid·kg^{-1} lipids and g oleic acid·kg^{-1} by-products.

The ST content in lipid extracts was assessed spectrophotometrically (615 nm) by the method of Huang et al. [32], based on the reaction with acetic anhydride in acetic acid (Liebermann–Buchard reaction). For quantification purposes, different quantities (0, 5, 10, 20, 40, 60, 80, 100, 130, and 150 µL) of a cholesterol solution in acetic acid (12.2 mg/5 mL) were employed. The validity range was 11.5–345.0 µg cholesterol and the R^2 value of the analytical procedure was 0.9998. Results were calculated as g choleterol·kg^{-1} lipids and g choleterol·kg^{-1} by-products.

To evaluate the TAG presence, lipid extracts were first purified on 20 × 20 cm thin-layer chromatography plates coated with a 0.5 mm-layer of silica gel G from Merck using a mixture of hexane-ethyl ether-acetic acid (90/10/1, *v/v/v*; two times) as eluent [33]. Once the TAG fraction was purified, the spectrophotometric (520 nm) method of Vioque and Holman [34] was used to measure the ester linkage content according to the conversion of the esters into hydroxamic acids and subsequent complexion with Fe (III). For quantification purposes, different quantities (0, 2, 5, 10, 20, and 40 µL) of a methyl stearate solution in toluene (41.0 mg/5 mL) were employed. The validity range was 16.4–328.0 µg methyl stearate and the R^2 value of the analytical procedure was 0.9995. Results were calculated as g tristearine·kg^{-1} lipids and g tristearine·kg^{-1} by-products.

2.5. Analysis of the FA Composition

Lipid extracts were converted into fatty acid methyl esters (FAMEs) by using acetyl chloride in methanol and then analysed by gas–liquid chromatography (GLC; PerkinElmer 8700 chromatograph, Madrid, Spain) [33]. The quantitative response of the equipment was checked with a GLC quantitative standard (FAME Mix, Supelco, Inc., Bellefonte, PA, USA). Peaks corresponding to FAMEs were identified by comparison of their retention times with those of a standard mixture (Qualmix Fish, Larodan, Malmo, Sweden). Peak areas were automatically integrated. C19:0 FA was used as internal standard for quantitative purposes; for it, 100 µL (i.e., 40 µg C19:0) of a 0.4 mg·mL^{-1} solution in toluene were added to each sample before the methylation reaction with acetyl chloride. Limits of detection and quantification were 500 and 1500 area units, respectively. The content of each FA was calculated as g·100 g^{-1} total FAs.

Results concerning FA groups (STFAs, MUFAs, PUFAs, ω3 FAs, and ω6 FAs) and FA ratios (ω3/ω6 and EPA/DHA) were calculated according to quantification of individual FA compounds. Additionally, the PI was calculated on the basis of the following FA content ratio [33]: (EPA + DHA)/C16:0.

2.6. Statistical Analysis

Data ($n = 3$) obtained from the different lipid analyses (yield, lipid classes, and FA profile and ratios) were subjected to a one-way ANOVA ($p < 0.05$) to investigate differences among the different kinds of lipid extracts (traditional and green systems) (Statistica version 6.0, 2001; Statsoft Inc., Tulsa, OK, USA). Comparison of means was performed using a least-squares difference (LSD) method.

3. Results and Discussion

3.1. Determination of Moisture and Lipid Contents

Initial by-products employed in the current study showed a moisture value of 842.6 ± 2.2 g·kg^{-1}. Concerning the lipid content of the initial sample, extraction carried out by the traditional procedure led to a level of 19.0 ± 0.5 g·kg^{-1} by-products (Table 1). Values for both constituents agree with those obtained recently during a seasonal study of by-product composition of the same squid species [35]; at that time, value ranges detected for moisture and lipids were 829.0–842.8 and 17.5–21.8 g·kg^{-1} by-products, respectively. Previous research related to moisture and lipid content of cephalopod by-products shows varying values according to the species and the particular body tissue taken into account. Thus, similar values as found in the current study were obtained by Kacem et al. [36] for moisture (75–85%), and lipids (0.58–4.02%) in viscera fractions (stomach, intestines, and pyloric caeca) from *Sepia officinalis* captured off the Tunisian coasts at different seasons. However, squid (*Loligo formosana*) ovary showed lower levels for both constituents, with values obtained being 72.1% and 0.5% for moisture and lipids, respectively [37]. Contrary to these results, a higher lipid content (199.8 g·kg^{-1} viscera) than in the present study was detected by Toyes-Vargas et al. [38] in giant squid (*Dosidicus gigas*) viscera. Notably, a marked difference in lipid content was detected by Saito et al. [39] when analysing the liver (15.7–17.9%) and gonad (1.0–1.4%) of Humboldt squid (*Dosidicus gigas*).

Table 1. Determination * of lipid yield and lipid class content of initial by-products and lyophilised by-products obtained by different lipid-extracting systems **.

Substrate	Extracting System	Lipid Determination			
		Lipid Yield ($g \cdot kg^{-1}$ by-products)	FFAs ($g \cdot kg^{-1}$ lipids)	STs ($g \cdot kg^{-1}$ lipids)	TAGs ($g \cdot kg^{-1}$ lipids)
Initial by-products	$CHCl_3$/MeOH	19.0 ± 0.5 a	228.2 ± 6.1 c (4.34)	103.6 ± 2.6 a (0.20)	8.5 ± 0.1 c (0.02)
Lyophilised by-products	$CHCl_3$/MeOH	103.8 ± 1.8 e	194.4 ± 6.7 a (2.02)	102.6 ± 4.8 a (1.06)	8.4 ± 0.7 c (0.09)
	EtOH	65.6 ± 4.2 d	210.1 ± 4.5 b,c (1.38)	138.5 ± 2.6 b (0.91)	1.2 ± 0.7 a (0.01)
	AcMe	36.3 ± 0.4 b	366.3 ± 14.8 f (1.33)	260.0 ± 9.5 e (0.94)	7.0 ± 0.7 b,c (0.03)
	AcOEt	46.8 ± 1.6 c	256.7 ± 8.0 d (1.20)	188.3 ± 6.9 c (0.88)	6.1 ± 1.1 b (0.03)
	EtOH/AcMe	73.6 ± 2.7 d	197.72 ± 2.2 a (1.46)	127.8 ± 6.7 b (0.94)	1.6 ± 0.3 a (0.01)
	EtOH/AcOEt	71.1 ± 1.2 d	201.3 ± 2.9 a,b (1.43)	132.5 ± 1.1 b (0.94)	2.2 ± 0.9 a (0.02)
	AcMe/AcOEt	41.6 ± 2.1 c	307.6 ± 2.4 e (1.28)	232.0 ± 14.8 d (0.97)	5.4 ± 1.0 b (0.02)

* Results expressed as average values of three independent determinations ($n = 3$) ± standard deviations. Data included in brackets correspond to average contents of lipid classes expressed as $g \cdot kg^{-1}$ by-products. In each column, different lowercase letters (a–f) indicate significant differences ($p < 0.05$). ** Extracting systems: 1/1 chloroform/methanol ($CHCl_3$/MeOH), ethanol (EtOH), acetone (AcMe), ethyl acetate (AcOEt), 1/1 ethanol/acetone (EtOH/AcMe), 1/1 ethanol/ethyl acetate (EtOH/AcOEt), and 1/1 acetone/ethyl acetate (AcMe/AcOEt). Other abbreviations employed: FFAs (free fatty acids), STs (sterols), and TAGs (triacylglycerols).

The current lyophilisation process of squid by-products led to a substantial moisture loss, so that a 52.1 ± 0.9 $g \cdot kg^{-1}$ value was detected. Therefore, lipid extraction carried out by all extracting systems tested led to a substantial content increase in this constituent (Table 1) when compared to initial by-products; remarkably, the highest values ($p < 0.05$) were obtained by applying the traditional procedure. Concerning the green solvent extraction, values detected were included in the 36.3–73.6 $g \cdot kg^{-1}$ by-products range (Table 1); this corresponded to a 33.4–73.2% recovery when compared to the lipid yield obtained in lyophilised samples by the traditional procedure. This different recovery can be explained on the basis that green solvents tested are more polar than the chloroform/methanol mixture and, therefore, would not be likely to extract entirely non-polar lipid classes such as TAGs, waxes, cholesterol esters, etc. On the other side, great differences were found among the different extracting systems tested. Thus, those including EtOH led to higher ($p < 0.05$) lipid recoveries than in the case of employing extracting systems not including this solvent. Remarkably, the EtOH/AcMe-extracting system reached the highest average recovery (i.e., 73.2%). In previous research, Gigliotti et al. [23] tested the lipid extraction from Antarctic krill (*Euphausia superba*) with different AcMe/EtOH ratios; as in the present research, an increasing proportion of ethanol led to a lipid yield increase so that higher lipid levels were obtained by using solvent ratios (*v/v*) of 1/30 (ca. 13.5%) and 1/12 (ca. 12.0%) than with 1/9 (ca. 9.5%) and 1/6 (ca. 7.5%) ratios.

3.2. Determination of Lipid Classes

Initial by-products showed PL values included in the 450.8 ± 9.9 $g \cdot kg^{-1}$ lipid range (0.86 ± 0.03 $g \cdot kg^{-1}$ by-products) (Figure 1), according to a previous report on by-products from the same squid species measured during a seasonal study (359.2–463.5 g·kg^{-1} lipids) [34]. This PL presence in the lipid fraction can be considered similar to the one reported in edible parts of lean fish species such as rainbow trout (*Oncorhynchus mykiss*) [40] and megrim (*Lepidorhombus whiffiagonis*) [41].

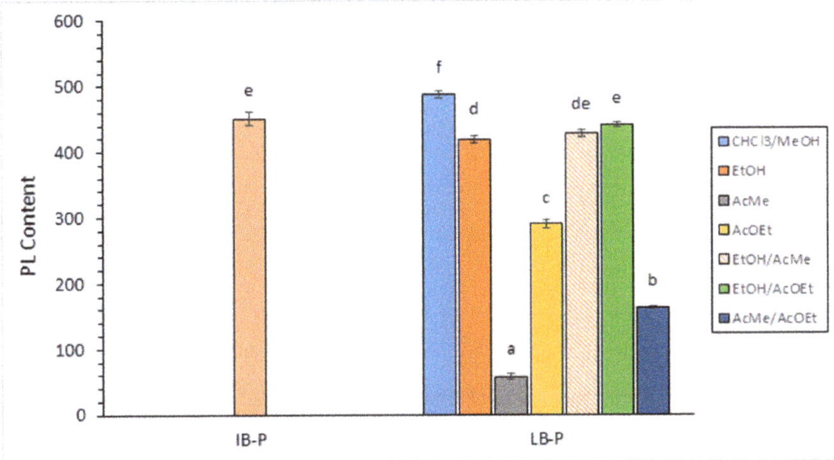

Figure 1. Determination of phospholipid (PL) content (g·kg^{-1} lipids) in initial by-products (IB-P) and lyophilised by-products (LB-P) obtained by different lipid-extracting systems. Average values of three independent determinations ($n = 3$); standard deviations are indicated by bars. Values accompanied by different lowercase letters (a–f) indicate significant differences ($p < 0.05$). Extracting systems as expressed in Table 1.

Comparison between extracts obtained by the traditional procedure on initial by-products and lyophilised by-products revealed a slight increase ($p < 0.05$) as a result of the lyophilisation process (Figure 1). Concerning green solvent extraction, varying PL values were detected, all of them lower than that obtained by the traditional procedure (Figure 1). Among green systems, the highest levels ($p < 0.05$) were obtained by applying systems including EtOH (418.5–440.4 g·kg^{-1} lipids; 2.75–3.15 g·kg^{-1} by-products). On the contrary, the lowest levels ($p < 0.05$) were detected for the AcMe solvent (57.6 g·kg^{-1} lipids; 0.21 g·kg^{-1} by-products). It is worth pointing out that systems including EtOH provided recovery values accounting for 85.8–90.3% when compared to $CHCl_3$/MeOH extraction. Therefore, EtOH-containing systems can be considered an interesting green procedure in order to obtain this highly valuable lipid class. A greater recovery with EtOH-containing systems can be explained on the basis of the higher polarity of EtOH than that of AcMe and AcOEt, so that polar lipids such as PL classes would be extracted more abundantly. Contrary to these extractability results, a higher PL yield (ca. 30%) was detected by applying a 1/6 ratio (AcMe/EtOH) than in the case of using a higher presence of ethanol (1/9, 1/12, and 1/30, AcMe/EtOH; ca. 21–22%) during PL extraction from krill (*E. superba*) [23].

PL compounds have been described as being important constituents of cell membranes and having an important structural role in living bodies in general. Furthermore, and on the basis of their amphiphilic character, PLs have recently attracted great attention for serving as drug delivery systems and having a high bioavailability and protecting effect on different kinds of diseases [42,43]. Thus, profitable functions related to pharmaceutical and food production industries have recently been developed for marine PL compounds [12]. Therefore, the present study has shown that EtOH-including systems can provide an accurate way of extracting valuable PL compounds by a green procedure.

Results obtained for FFA and ST classes are presented in Table 1. Average values obtained in initial by-products for both lipid classes (228.2 g FFA·kg^{-1} lipids and

4.34 g FFA·kg^{-1} by-products; 103.6 g ST·kg^{-1} lipids and 0.20 g ST·kg^{-1} by-products), agree with the previous research obtained during a seasonal study on by-products corresponding to the same squid species [35]; in that study, values obtained for FFA and ST classes were included in the ranges 156.6–282.0 and 115.0–132.1 g·kg^{-1} lipids, respectively. Compared to values detected in edible parts of cephalopod species and marine species in general [33,44,45], FFA levels of current by-products can be considered relatively high. Such high values can be explained on the basis of the great presence of lipases and phospholipases in visceral tissues [7,19]. Thus, higher FFA levels (g·100 g^{-1} tissue) were detected in ovary (0.03–0.10) and hepatopancreas (0.18–0.59) than in arm (0.01) and mantle (0.01) during a seasonal study on common octopus (*O. vulgaris*) [46]; when compared to present by-products, ovary FFA values were found to be lower, but FFA presence in hepatopancreas was found to be similar. Higher FFA levels than in the present study were detected in *Loligo duvauceli* liver (10.67 g·kg^{-1} tissue) by Vairamani et al. [47]. Concerning previous research on the ST level, higher values were also obtained in ovary (0.16–0.31 g·100 g^{-1} tissue) and hepatopancreas (0.28–0.69 g·100 g^{-1} tissue) than in edible tissues such as arm (0.03–0.05 g·100 g^{-1} tissue) and mantle (0.03–0.06 g·100 g^{-1} tissue) during a seasonal study on octopus (*O. vulgaris*) [46]; a comparison to present by-products showed that hepatopancreas values were found to be higher, but ovary presence was found to be similar.

Comparison between lipid extracts obtained by the traditional procedure on initial by-products and lyophilised by-products revealed an average decrease in FFA and ST contents as a result of lyophilisation (Table 1); this decrease was found significant ($p < 0.05$) in the case of the FFA value. Compared to the traditional extraction on lyophilised by-products, all green-extracting systems led to an increased average value of both lipid classes. Among the green-extracting systems, the following increasing ($p < 0.05$) sequence for solvent extractability was observed: EtOH < AcOEt < AcMe; remarkably, the highest ($p < 0.05$) proportions of both lipid classes were detected in the extract corresponding to the AcMe system. Results obtained for the comparative extractability of the three green solvents can be explained on the basis of their relative polarities; thus, a lower polarity of AcMe than that of EtOH and AcOEt has shown a higher ability for extracting lipid classes such as FFAs and STs. According to this result, Gigliotti et al. [23] observed an increasing cholesterol yield by increasing the AcMe presence in an AcMe/EtOH-extracting system when applied to Antarctic krill (*E. superba*).

A very low content of TAG compounds was detected in initial by-products (Table 1). Such samples showed slightly lower values than those reported for by-products from the same squid species (9.5–13.1 g·kg^{-1} lipids) [35]. Additionally, present TAG levels (0.02 ± 0.00 g·kg^{-1} by-products) were found lower than those detected by Sieiro et al. [46] in a seasonal study carried out on the ovary (0.32–0.52 g·100 g^{-1} tissue) and the digestive gland (2.51–3.71 g·100 g^{-1} tissue) of octopus (*O. vulgaris*).

This low TAG presence in the lipid fraction was maintained in the lipid extract of lyophilised by-products obtained by the traditional procedure (Table 1). Remarkably, average values obtained by all kinds of green extraction solvents were found lower than those obtained by the traditional procedure; differences were found significant ($p < 0.05$) in all cases except for the comparison to the AcMe system ($p > 0.05$). Comparison among extracting systems led to a similar trend in the case of the FFA and ST classes. Thus, extracting systems including EtOH led to lower ($p < 0.05$) TAG levels than systems not including this solvent. Furthermore, the AcMe solvent provided the highest average levels of this lipid class. As for FFA and ST classes, TAG results obtained concerning the comparative ability of the extracting solvents tested can be explained on the basis of their different polarities. As being a non-polar lipid class, TAGs have been extracted more extensively by the less polar solvent tested (i.e., AcMe). Contrary to the present results, Gigliotti et al. [23] showed that TAG extraction in lipids from Antarctic krill (*E. superba*) increased with the ethanol presence in the AcMe/EtOH-extracting system; thus, the 1/12 ratio showed higher TAG yields (ca. 2%) than 1/6 and 1/9 (ca. 1%) ratios.

3.3. Fatty Acid (FA) Analysis

The FA profile of initial by-products is shown in Figure 2. Composition (g·100 g^{-1} total FAs) was as follows: 3.19 ± 0.12 (C14:0), 0.50 ± 0.03 (C15:0), 25.45 ± 0.87 (C16:0), 1.51 ± 0.10 (C16:1ω7), 1.18 ± 0.04 (C17:0), 4.88 ± 0.05 (C18:0), 3.68 ± 0.05 (C18:1ω9), 2.25 ± 0.06 (C18:1ω7), 0.50 ± 0.01 (C18:2ω6), 5.66 ± 0.05 (C20:1ω9), 0.48 ± 0.04 (C20:2ω6), 2.53 ± 0.09 (C20:4ω6), 0.78 ± 0.04 (C22:1ω9), 15.88 ± 0.04 (C20:5ω3, EPA), 0.21 ± 0.02 (C22:4ω6), 0.89 ± 0.03 (C24:1ω9), 0.48 ± 0.01 (C22:5ω3), and 29.95 ± 0.01 (C22:6ω3, DHA).

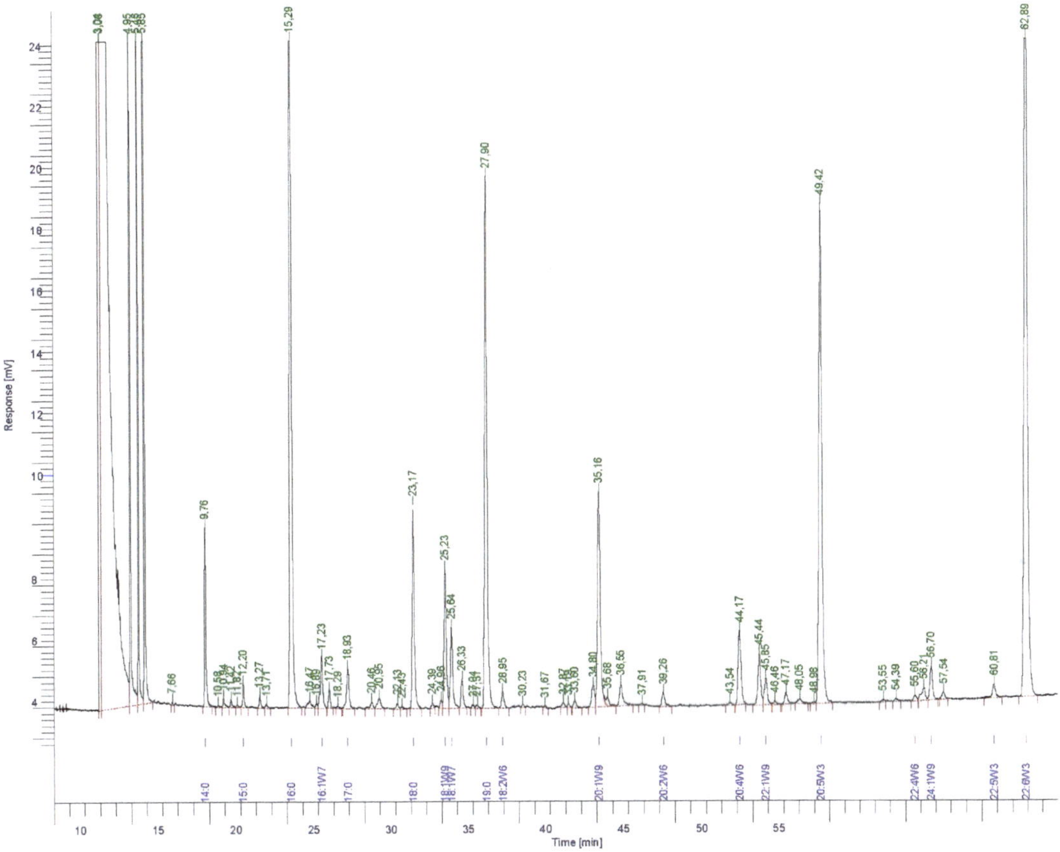

Figure 2. Fatty acid (FA) profile of initial by-products. Retention time (min) of individual FA is indicated in green. Assignation of peaks is expressed in blue.

According to such profile, DHA, C16:0, and EPA showed to be the most abundant FA compounds present in the squid by-products. Other relatively abundant FA components were C20:1ω9, C18:0, C18:1ω9, C14:0, and C20:4ω6. This FA composition agrees with previous results obtained on by-products corresponding to the same squid species [35]. A similar distribution of the main FA constituents (i.e., DHA, EPA, C16:0, C18:0, C20:4ω6, C18:1ω9, and 20:1ω9) was detected for ovary and digestive gland from octopus (*O. vulgaris*) [46].

Among the ω3 PUFAs, great attention has been accorded to DHA and EPA values in view of their beneficial health effects. Thus, clinical and epidemiological studies have associated EPA consumption with low prevalence of coronary, circulatory, and inflammatory diseases [48], whereas DHA has been related to foetal development, prevention of neurodegenerative diseases, and correct functioning of the nervous system and visual organs in the foetus [49]. In the present study, a comparison between traditional lipid extracts of initial by-products and lyophilised by-products showed that the lyophilisation process led to an average value increase for both PUFA compounds; such increase was found significant ($p < 0.05$) for DHA (Table 2). Increased average DHA values were also detected in all green-extracting systems when compared to the traditional procedure, although differences were not found significant ($p > 0.05$). In the case of EPA, lower average values were detected in samples corresponding to EtOH-extracting systems when compared to the traditional procedure; on the contrary, higher levels ($p < 0.05$) were obtained in AcMe-, AcOEt-, and AcMe/AcOEt-extracting systems when compared to the $CHCl_3$/MeOH system.

Table 2. Fatty acid (FA) analysis (g·100 g^{-1} total FAs) * in initial by-products and lyophilised by-products obtained by different lipid-extracting systems **.

Substrate	Extracting System	FA Determination					
		STFAs	MUFAs	PUFAs	DHA	EPA	PI
Initial by-products	$CHCl_3$/MeOH	35.20 ± 1.10 d	14.47 ± 0.06 b	50.03 ± 1.13 a	29.95 ± 0.17 a	15.88 ± 0.04 a,b	1.80 ± 0.10 a
Lyophilised by-products	$CHCl_3$/MeOH	35.10 ± 0.94 d	13.73 ± 0.07 a	51.17 ± 0.89 a,b	31.12 ± 0.72 b	16.05 ± 0.16 b	1.83 ± 0.08 a
	EtOH	32.58 ± 0.96 c	15.40 ± 0.48 c	52.02 ± 0.50 b	32.26 ± 0.44 b	15.72 ± 0.08 a	1.98 ± 0.09 a
	AcMe	23.35 ± 0.31 a	21.35 ± 0.41 f	55.30 ± 0.66 d	31.47 ± 0.51 b	18.48 ± 0.25 c	3.29 ± 0.09 b
	AcOEt	26.12 ± 1.18 b	18.03 ± 0.18 d	55.85 ± 1.35 c,d	32.61 ± 1.04 b	18.34 ± 0.27 c	2.82 ± 0.21 b
	EtOH/AcMe	33.22 ± 0.53 c	14.90 ± 0.20 b,c	51.88 ± 0.33 a,b	31.92 ± 0.20 b	15.82 ± 0.16 a,b	1.93 ± 0.04 a
	EtOH/AcOEt	33.04 ± 1.61 c,d	14.59 ± 0.06 b	52.37 ± 1.59 a,b,c	32.58 ± 1.25 b	15.66 ± 0.27 a,b	1.98 ± 0.17 a
	AcMe/AcOEt	24.49 ± 1.54 a,b	19.60 ± 0.23 e	55.91 ± 1.31 c,d	32.28 ± 1.04 b	18.56 ± 0.17 c	3.09 ± 0.28 b

* Results expressed as average values of three independent determinations ($n = 3$) ± standard deviations. In each column, different lowercase letters (a–f) indicate significant differences ($p < 0.05$). ** Abbreviations employed: STFAs (saturated fatty acids), MUFAs (monounsaturated fatty acids), PUFAs (polyunsaturated fatty acids), DHA (docosahexaenoic acid), EPA (eicosapentaenoic acid), and PI (polyene index). Extracting systems as expressed in Table 1.

Previous research on cephalopod by-products has shown varying values for DHA and EPA content, according to the species and the body tissue taken into account. Thus, Kacem et al. [36] found lower values than in the current study for DHA (9.1 g·100 g^{-1} total FAs) and EPA (11.6 g·100 g^{-1} total FAs) during a seasonal study in cuttlefish (*S. officinalis*) viscera. Toyes-Vargas et al. [38] denoted levels of 19.4 and 16.9 g·kg^{-1} viscera in giant squid (*D. gigas*) viscera for DHA and EPA, respectively, which can be considered similar for EPA but lower for DHA when compared to the present values. Recently, Šimat et al. [50] analysed the presence of both PUFA constituents in different kinds of fish by-products; thus, the lowest values (g·100^{-1} g total FAs) were detected in EPA (ca. 3.3) and DHA (ca. 8.3) for sea bass (*Sparus aurata*) and seabream (*Dicentrarchus labrax*) by-products, whereas the highest levels were found in by-products from tuna (*Thunnus thynnus*) and sardine (*Sardina pilchardus*) (9–14 and 13–21, for EPA and DHA, respectively). Therefore, present values obtained for DHA and EPA can be considered as relatively high for a by-product and be valuable as a source of both essential FA compounds.

Comparison among FA group contents showed the following decreasing ($p < 0.05$) tendency in initial by-products (Table 2): PUFA > STFA > MUFA; notably, PUFA presence reached ca. 50% value. Such marked differences among FA groups agree with previous related research on cephalopod by-products [35,46] and were maintained in all kinds of lipid extracts corresponding to lyophilised samples. Comparison of lipid fractions obtained by a traditional extracting procedure on initial by-products and lyophilised by-products showed that the lyophilisation process led to average decreases in STFA and MUFA contents and an average increase in PUFA group value; differences were only found significant ($p < 0.05$) in the case of the MUFA group. Compared to the traditional procedure, green solvent systems provided lower average STFA values and higher average MUFA and PUFA levels; for the MUFA group, such higher values were found significant ($p < 0.05$) for all green systems.

Comparison among green-extracting systems showed that the highest average values for STFAs were found in samples corresponding to extracting systems including EtOH, whereas the lowest average value was obtained for the AcMe system (Table 2). For the MUFA group, the highest value ($p < 0.05$) was obtained in samples corresponding to AcMe extraction, with the lowest average values obtained in lipid extracts corresponding to the extracting systems including EtOH. Concerning the PUFA group, the highest average value was found in samples corresponding to the AcMe/AcOEt-extracting system; on the contrary, the lowest average values were obtained for systems including EtOH. Remarkably, all green-extracting systems tested led to values included in the 51.8–55.9% range for the PUFA group.

Previous research has paid great attention to the ω3/ω6 ratio on the basis that most Western countries do not consume adequate levels of ω3 FA compounds [51,52]. In order to prevent inflammatory, cardiovascular, and neurological disorders, the World Health Organization (WHO) currently recommends that this ratio should not be below 1/10 in the human diet [53]; additionally, the European Nutritional Society reported that a human diet with an ω3/ω6 ratio of 1/5 or higher would have health benefits [54].

Values obtained for this FA ratio are presented in Figure 3 and agree with previous research on by-products corresponding to the same squid species [35]. Comparison between lipid fractions obtained by the traditional procedure on initial by-products and lyophilised samples showed a slight increase ($p < 0.05$) of this ratio as a result of the lyophilisation process. Additionally, lower average values were detected in lipid extracts corresponding to systems including AcMe, AcOEt, or both solvents when compared to the traditional procedure. Remarkably, extracting systems including EtOH did not provide differences ($p > 0.05$) with extracts corresponding to the traditional procedure, with all values being included in the 13.18–13.57 range. In spite of differences found for some green extraction systems, and according to the above-mentioned recommended levels for the ω3/ω6 ratio, all values obtained by the green extraction procedures can be considered as highly valuable for the human diet.

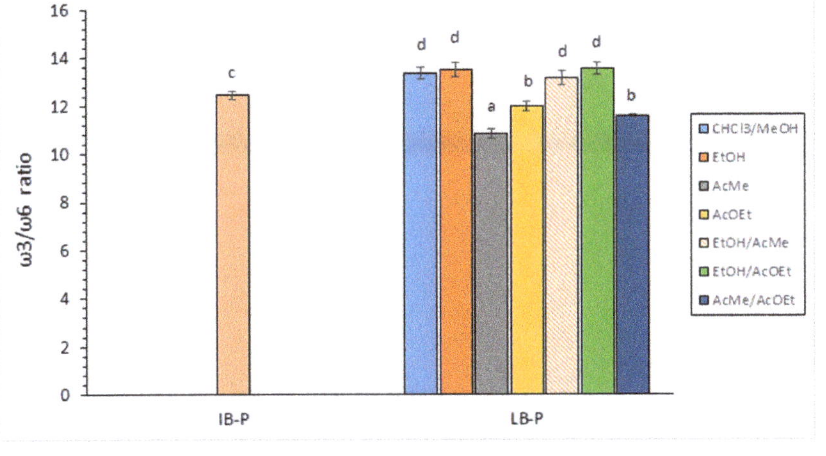

Figure 3. Determination of the ω3/ω6 ratio in initial by-products (IB-P) and lyophilised by-products (LB-P) obtained by different lipid-extracting systems. Average values of three independent determinations ($n = 3$); standard deviations are indicated by bars. Values accompanied by different lowercase letters (a–d) indicate significant differences ($p < 0.05$). Extracting systems as expressed in Table 1.

Present ω3/ω6 ratio values can be considered similar to those obtained for edible tissues of wild fish species such as blackspot seabream (*P. bogaraveo*) [33] and megrim (*L. whiffiagonis*) [41]. On the contrary, previous studies on seafood by-products have shown lower ω3/ω6 values than in the present research. Thus, Šimat et al. [50] obtained values included in the 6–10 range for tuna (*T. thynnus*) and sardine (*S. pilchardus*) by-products, whereas markedly lower ratios (0–2 range) were obtained for tuna (*T. thynnus*) liver and sea bass (*S. aurata*) and seabream (*D. labrax*) by-products. Lower ω3/ω6 ratio values (5.07 ± 0.02) than in the current study were also obtained by Chakraborty and Joseph [27] when extracting edible tissues from Indian sardine (*S. longiceps*) by applying the cooking and wet-pressing procedure. Recently, Rodríguez et al. [15] obtained ω3/ω6 ratio values included in the 7.6–8.0 range for squid (*Illex argentinus*) viscera.

The assessment of the PI has recently attracted great attention as a way of measuring the possible increase or decrease in the PUFA presence in the lipid fraction during seafood processing in general, and is directly related to the nutritional value of the corresponding lipid food [3,4]. Results obtained in the present study for this index are depicted in Table 2, with all values included in the 1.80–3.29 range. Such values can be considered similar to those obtained in previous research for squid *Illex argentinus* viscera (1.65–2.10 range) [15] and for squid (*D. gahi*) by-products [35].

Comparison between lipid fractions of initial by-products and lyophilised by-products extracted with the traditional procedure showed that the lyophilisation process did not produce changes ($p > 0.05$) in the PI. Furthermore, lipid fraction obtained from by-products by employing extracting systems including EtOH did not reveal differences ($p > 0.05$) with samples extracted with $CHCl_3$/MeOH. On the contrary, higher ($p < 0.05$) PI scores were detected in lipid fractions obtained by applying extracting systems including AcMe and AcOEt when compared to the traditional procedure. Remarkably, values obtained for all green-extracting systems were included in the 1.93–3.29 range, which can be considered as highly valuable levels and very similar to values reported for edible tissues obtained from marine species such as megrim (*L. whiffiagonis*) [41], rainbow trout (*O. mykiss*) [40], or blackspot seabream (*P. bogaraveo*) [33].

4. Conclusions

A novel approach of bioactive lipid extraction by different green solvent systems was carried out on Patagonian squid (*Doryteuthis gahi*) by-products. As a result, lipid yields obtained by green solvent systems led to a 33.4–73.2% recovery when compared to traditional extraction; remarkably, the highest values ($p < 0.05$) were obtained by EtOH-containing systems. Compared to the traditional procedure, EtOH systems showed an 85.8–90.3% recovery of PL compounds and no differences ($p > 0.05$) in the ω3/ω6 ratio. Green-extracting systems led to higher average values for EPA content (15.66–18.56 $g \cdot 100\ g^{-1}$ total FAs) and PI (1.93–3.29) than $CHCl_3$/MeOH extraction; differences were found significant ($p < 0.05$) for systems including AcMe and AcOEt. No differences ($p > 0.05$) were detected for DHA content between the traditional procedure and green systems, with all values being included in the 31.12–32.61 $g \cdot 100\ g^{-1}$ total FA range.

The suitability of green solvent systems tested is concluded. Their employment as an alternative method for lipid extraction could be recommended with the aim of reducing the risk of chemical exposure to humans and the environment. Additionally, by-products resulting from Patagonian squid commercialisation are considered in the current work as a single product, implying a simplified handling procedure for subsequent industrial use. Current results revealed that lipid extraction from the present by-products by green processing can be an interesting option for recovering high-value compounds such as PLs, DHA, EPA, and ω3 PUFAs in general, as well as a lipid extract including valuable ω3/ω6 and PI ratios. Remarkably, EtOH-containing systems have shown the highest levels of lipid yield, PL recovery, and ω3/ω6 ratio. On the basis of the high relevance of the present results, further research is envisaged to optimise and scale-up the extraction conditions (i.e., green solvent mixtures) of bioactive lipid compounds for the utilisation of the present squid by-products as a source of valuable marine constituents susceptible to be used in food, pharmaceutical, and nutraceutical industries.

Author Contributions: Conceptualisation, S.P.A. and I.M.; methodology, M.T., M.J.G. and S.L.; data curation, M.T., M.J.G. and S.L.; writing—original draft preparation, S.P.A.; writing—review and editing, S.P.A. and I.M. All authors have read and agreed to the published version of the manuscript.

Funding: This research was funded by the SPANISH MINISTRY OF RESEARCH AND INNOVATION, grant number RTI2018-095659-B-I00 (2019–2022), and the APC was partially funded by the Spanish National Research Council (CSIC).

Institutional Review Board Statement: Not applicable.

Data Availability Statement: All related data and methods are presented in this paper. Additional inquiries should be addressed to the corresponding author.

Acknowledgments: SERPESBA S. L. U. (Vigo, Spain) is greatly acknowledged for kindly providing the squid by-product samples.

Conflicts of Interest: The authors declare no conflict of interest.

References

1. Tilami, S.K.; Sampels, S. Nutritional Value of Fish: Lipids, Proteins, Vitamins, and Minerals. *Rev. Fish. Sci.* **2018**, *26*, 242–253.
2. Aubourg, S.P. Lipid compounds. In *Handbook of Seafood and Seafood Products Analysis*; Nollet, L., Toldrá, F., Eds.; CRC Press: Boca Raton, FL, USA; Francis and Taylor Group: Boca Raton, FL, USA, 2010; pp. 69–86.
3. Minihane, A.; Armah, C.; Miles, E.; Madden, J.; Clark, A.; Caslake, M.; Calder, P. Consumption of fish oil providing amounts of eicosapentaenoic acid and docosahexaenoic acid that can be obtained from the diet reduces blood pressure in adults with systolic hypertension: A retrospective analysis. *J. Nutr.* **2016**, *146*, 516–523. [CrossRef] [PubMed]
4. Schunck, W.; Konkel, A.; Fischer, R.; Weylandt, K. Therapeutic potential of omega-3 fatty acid-derived epoxy eicosanoids in cardiovascular and inflammatory diseases. *Pharmacol. Ther.* **2018**, *183*, 177–204. [CrossRef]
5. Ferraro, V.; Cruz, I.B.; Jorge, R.F.; Malcata, F.X.; Pintado, M.E.; Castro, P.M.L. Valorisation of natural extracts from marine source focused on marine by-products: A review. *Food Res. Int.* **2010**, *43*, 2221–2223. [CrossRef]
6. Rubio-Rodríguez, N.; Beltrán, S.; Jaime, I.; de Diego, S.M.; Sanz, M.T.; Carballido, J.R. Production of omega-3 polyunsaturated fatty acid concentrates: A review. *Innov. Food Sci. Emerg. Technol.* **2010**, *11*, 1–12. [CrossRef]
7. Atef, M.; Ojagh, M. Health benefits and food applications of bioactive compounds from fish byproducts: A review. *J. Funct. Foods* **2017**, *35*, 673–681. [CrossRef]
8. Shahidi, F. *Maximising the Value of Marine By-Products*; CRC Press: Boca Raton, FL, USA, 2007.
9. Vázquez, J.A.; Meduiña, A.; Durán, A.I.; Nogueira, M.; Fernández-Compás, A.; Pérez-Martín, R.I.; Rodríguez-Amado, I. Production of valuable compounds and bioactive metabolites from by-products of fish discards using chemical processing, enzymatic hydrolysis, and bacterial fermentation. *Mar. Drugs* **2019**, *17*, 139. [CrossRef]
10. Falch, E.; Rustad, T.; Aursand, M. By-products from gadiform species as raw material for production of marine lipids as ingredients in food or feed. *Process Biochem.* **2006**, *41*, 666–674. [CrossRef]
11. Blanco, M.; Sotelo, C.G.; Chapela, M.J.; Pérez-Martín, R. Towards sustainable and efficient use of fishery resources: Present and future trends. *Trends Food Sci. Technol.* **2007**, *18*, 29–36. [CrossRef]
12. Takahashi, K.; Inoue, Y. Marine by-product phospholipids as booster of medicinal compounds. *Adv. Food Nutr. Res.* **2012**, *65*, 31–46.
13. Głowacz-Rozynska, A.; Tynek, M.; Malinowska-Panczyk, E.; Martysiak-Zurowska, D.; Pawłowicz, R.; Kołodziejsk, I. Comparison of oil yield and quality obtained by different extraction procedures from salmon (*Salmo salar*) processing byproducts. *Eur. J. Lipid Sci. Technol.* **2016**, *118*, 1759–1767. [CrossRef]
14. Pudtikajorn, K.; Benjakul, S. Simple wet rendering method for extraction of prime quality oil from skipjack tuna eyeballs. *Eur. J. Lipid Sci. Technol.* **2020**, *122*, 2000077. [CrossRef]
15. Rodríguez, A.; Trigo, M.; Aubourg, S.P.; Medina, I. Optimisation of healthy-lipid content and oxidative stability during oil extraction from squid (*Illex argentinus*) viscera by green processing. *Mar. Drugs* **2021**, *19*, 616. [CrossRef] [PubMed]
16. Rubio-Rodríguez, N.; de Diego, S.M.; Beltrán, S.; Jaime, I.; Sanz, M.T.; Rovira, J. Supercritical fluid extraction of fish oil from fish by-products: A comparison with other extraction methods. *J. Food Eng.* **2012**, *109*, 238–248. [CrossRef]
17. Pando, M.E.; Rodríguez, A.; Galdames, A.; Berríos, M.M.; Rivera, M.; Romero, N.; Valenzuela, M.A.; Ortiz, J.A.; Aubourg, S.P. Maximization of the docosahexaenoic and eicosapentaenoic acids content in concentrates obtained from a by-product of rainbow trout (*Oncorhynchus mykiss*) processing. *Eur. Food Res. Technol.* **2018**, *244*, 937–948. [CrossRef]
18. Gbogouri, G.; Linder, M.; Fanni, J.; Parmentier, M. Analysis of lipids extracted from salmon (*Salmo salar*) heads by commercial proteolytic enzymes. *Eur. J. Lipid Sci. Technol.* **2006**, *108*, 766–775. [CrossRef]
19. Rustad, T.; Storro, I.; Slizyte, R. Possibilities for the utilisation of marine by-products. *Int. J. Food Sci. Technol.* **2011**, *46*, 2001–2014. [CrossRef]
20. Pateiro, M.; Gómez-Salazar, J.A.; Jaime-Patlán, M.; Sosa-Morales, M.E.; Lorenzo, J.M. Plant extracts obtained with green solvents as natural antioxidants in fresh meat products. *Antioxidants* **2021**, *10*, 181. [CrossRef]

21. Silva Pinho, L.; Palazzolli da Silva, M.; Thomazini, M.; Cooperstone, J.L.; Campanella, O.H.; da Costa Rodrigues, C.E.; Favaro-Trindade, C.S. Guaraná (*Paullinia cupana*) by-product as a source of bioactive compounds and as a natural antioxidant for food applications. *J. Food Proc. Preserv.* **2021**, *45*, e15854.
22. Gil-Martín, E.; Forbes-Hernández, T.; Romero, A.; Cianciosi, D.; Giampieri, F.; Battino, M. Influence of the extraction method on the recovery of bioactive phenolic compounds from food industry by-products. *Food Chem.* **2022**, *378*, 131918. [CrossRef]
23. Gigliotti, J.C.; Davenport, M.P.; Beamer, S.K.; Tou, J.C.; Jaczynski, J. Extraction and characterisation of lipids from Antarctic krill (*Euphausia superba*). *Food Chem.* **2011**, *125*, 1028–1036. [CrossRef]
24. Li, C.J.; Xin, M.R.; Sun, Z.L. Selection of extraction solvents for edible oils from microalgae and improvement of the oxidative stability. *J. Biosci. Bioeng.* **2021**, *132*, 365–371. [CrossRef] [PubMed]
25. FAO. Fishery Division. *Species Fact Sheets. Loligo gahi (Orbigny, 1835)*. 2021. Available online: www.fao.org/fishery/species/2713/en (accessed on 9 March 2022).
26. Bligh, E.; Dyer, W. A rapid method of total extraction and purification. *Can. J. Biochem. Physiol.* **1959**, *37*, 911–917. [CrossRef] [PubMed]
27. Chakraborty, K.; Joseph, D. Cooking and pressing is an effective and eco-friendly technique for obtaining high quality oil from *Sardinella longiceps*. *Eur. J. Lipid Sci. Technol.* **2015**, *117*, 837–850. [CrossRef]
28. AOAC. *Official Methods for Analysis of the Association of Analytical Chemistry*, 15th ed.; Association of Official Chemists, Inc.: Arlington, VA, USA, 1990; pp. 931–937.
29. Herbes, S.E.; Allen, C.P. Lipid quantification of freshwater invertebrates: Method modification for microquantitation. *Can. J. Fish. Aquat. Sci.* **1983**, *40*, 1315–1317. [CrossRef]
30. Raheja, R.; Kaur, C.; Singh, A.; Bhatia, A. New colorimetric method for the quantitative determination of phospholipids without acid digestion. *J. Lipid Res.* **1973**, *14*, 695–697. [CrossRef]
31. Lowry, R.; Tinsley, I. Rapid colorimetric determination of free fatty acids. *J. Am. Oil Chem. Soc.* **1976**, *53*, 470–472. [CrossRef]
32. Huang, T.; Chen, C.; Wefler, V.; Raftery, A. A stable reagent for the Liebermann-Buchardt reaction. *Anal. Chem.* **1961**, *33*, 1405–1407. [CrossRef]
33. Álvarez, V.; Medina, I.; Prego, R.; Aubourg, S.P. Lipid and mineral distribution in different zones of farmed and wild blackspot seabream (*Pagellus bogaraveo*). *Eur. J. Lipid Sci. Technol.* **2009**, *111*, 957–966. [CrossRef]
34. Vioque, E.; Holman, R. Quantitative estimation of esters by thin-layer chromatography. *J. Am. Oil Chem. Soc.* **1962**, *39*, 63–66. [CrossRef]
35. Aubourg, S.P.; Trigo, M.; Prego, R.; Cobelo-García, A.; Medina, I. Nutritional and healthy value of chemical constituents obtained from Patagonian squid (*Doryteuthis gahi*) by-products captured at different seasons. *Foods* **2021**, *10*, 2144. [CrossRef] [PubMed]
36. Kacem, M.; Sellami, M.; Kammoun, W.; Frikh, F.; Miled, N.; Rebah, F.B. Seasonal variations in proximate and fatty acid composition of viscera of *Sardinella aurita*, *Sarpa salpa*, and *Sepia officinalis* from Tunisia. *J. Aquat. Food Prod. Technol.* **2011**, *20*, 233–246. [CrossRef]
37. Singh, A.; Benjakul, S.; Kishimura, H. Characteristics and functional properties of ovary from squid *Loligo formosana*. *J. Aquat. Food Prod. Technol.* **2017**, *26*, 1083–1092. [CrossRef]
38. Toyes-Vargas, E.; Robles-Romo, A.; Méndez, L.; Palacios, E.; Civera, R. Changes in fatty acids, sterols, pigments, lipid classes, and heavy metals of cooked or dried meals, compared to fresh marine by-products. *Anim. Feed Sci. Technol.* **2016**, *221*, 195–205. [CrossRef]
39. Saito, H.; Sakai, M.; Wakabayashi, T. Characteristics of the lipid and fatty acid compositions of the Humboldt squid, *Dosidicus gigas*: The trophic relationship between the squid and its prey. *Eur. J. Lipid Sci. Technol.* **2014**, *116*, 360–366. [CrossRef]
40. Ortiz, J.; Palma, Ó.; González, N.; Aubourg, S.P. Lipid damage in farmed rainbow trout (*Oncorhynchus mikiss*) after slaughtering and chilling storage. *Eur. J. Lipid Sci. Technol.* **2008**, *110*, 1127–1135. [CrossRef]
41. Barbosa, R.G.; Trigo, M.; Prego, R.; Fett, R.; Aubourg, S.P. The chemical composition of different edible locations (central and edge muscles) of flat fish (*Lepidorhombus whiffiagonis*). *Int. J. Food Sci. Technol.* **2018**, *53*, 271–281. [CrossRef]
42. Küllenberg, D.; Taylor, L.A.; Schneider, M.; Massing, U. Health effects of dietary phospholipids. *Lipids Health Dis.* **2012**, *11*, 3. [CrossRef]
43. Li, J.; Wang, X.; Zhang, T.; Huang, Z.; Luo, X.; Deng, Y. A review on phospholipids and their main applications in drug delivery systems. *Asian J. Pharm. Sci.* **2015**, *10*, 81–98. [CrossRef]
44. Piclet, G. Le poisson aliment. Composition-Intérêt nutritionnel. *Cah. Nutr. Diét.* **1987**, *XXII*, 317–335.
45. Testi, S.; Bonaldo, A.; Gatta, P.; Badiani, A. Nutritional traits of dorsal and ventral fillets from three farmed fish species. *Food Chem.* **2006**, *98*, 104–111. [CrossRef]
46. Sieiro, M.P.; Aubourg, S.P.; Rocha, F. Seasonal study of the lipid composition in different tissues of the common octopus (*Octopus vulgaris*). *Eur. J. Lipid Sci. Technol.* **2006**, *108*, 479–487. [CrossRef]
47. Vairamani, S.; Sofia, V.; Sudharsan, S.; Vasanthkumar, S.; Ramasubrmanian, V.; Madeswaran, P.; Srinivasan, A.; Shanmugam, A. Reclamation of *Loligo duvauceli* (Orbigny, 1848) digestive gland (liver) waster for the extraction of oil and its lipid composition. *J. Biol. Sci.* **2013**, *13*, 634–639. [CrossRef]
48. Swanson, S.; Block, R.; Mousa, S. Omega-3 fatty acids EPA and DHA: Health benefits throughout life. *Adv. Nutr.* **2012**, *3*, 1–7. [CrossRef]

49. Ofosu, F.K.; Daliri, E.B.M.; Lee, B.H.; Yu, X. Current trends and future perspectives on omega-3 fatty acids. *Res. J. Biol.* **2017**, *5*, 11–20.
50. Šimat, V.; Vlahović, J.; Soldo, B.; Mekinić, I.G.; Čagalj, M.; Hamed, I.; Skroza, D. Production and characterization of crude oils from seafood processing by-products. *Food Biosci.* **2020**, *33*, 100484. [CrossRef]
51. Uauy, R.; Valenzuela, A. Marine oils: The health benefits of n-3 fatty acids. *Nutrition* **2000**, *16*, 680–684. [CrossRef]
52. Komprda, T. Eicosapentaenoic and docosahexaenoic acids as inflammation-modulating and lipid homeostasis influencing nutraceuticals: A review. *J. Funct. Foods* **2012**, *4*, 25–38. [CrossRef]
53. Kumari, P.; Kumar, M.; Reddy, C.R.; Jha, B. Algal lipids, fatty acids and sterols. In *Functional Ingredients from Algae for Foods and Nutraceuticals*; Domínguez, H., Ed.; Woodhead Publishing: Cambridge, UK, 2013; pp. 87–134.
54. Simopoulos, A.P. The importance of the ratio of omega-6/omega-3 essential fatty acids. *Biomed. Pharmacother.* **2002**, *56*, 365–379. [CrossRef]

foods

Article

Mathematical Modelling of Canola Oil Biodegradation and Optimisation of Biosurfactant Production by an Antarctic Bacterial Consortium Using Response Surface Methodology

Khadijah Nabilah Mohd Zahri [1], Khalilah Abdul Khalil [2], Claudio Gomez-Fuentes [3,4], Azham Zulkharnain [5], Suriana Sabri [6], Peter Convey [7,8], Sooa Lim [9] and Siti Aqlima Ahmad [1,4,*]

1 Department of Biochemistry, Faculty of Biotechnology and Biomolecular Sciences, Universiti Putra Malaysia, Serdang 43400, Selangor, Malaysia; khadijahnabilah95@gmail.com
2 School of Biology, Faculty of Applied Sciences, Universiti Teknologi MARA, Section 2, Shah Alam 45000, Selangor, Malaysia; khali552@uitm.edu.my
3 Department of Chemical Engineering, Universidad de Magallanes, Avda. Bulnes 01855, Punta Arenas, Chile; claudio.gomez@umag.cl
4 Center for Research and Antarctic Environmental Monitoring (CIMAA), Universidad de Magallanes, Avda. Bulnes 01855, Punta Arenas, Chile
5 Department of Bioscience and Engineering, College of Systems Engineering and Science, Shibaura Institute of Technology, 307 Fukasaku, Minuma-ku, Saitama 337-8570, Japan; azham@shibaura-it.ac.jp
6 Department of Microbiology, Faculty of Biotechnology and Biomolecular Sciences, Universiti Putra Malaysia, Serdang 43400, Selangor, Malaysia; suriana@upm.edu.my
7 British Antarctic Survey, NERC, High Cross, Madingley Road, Cambridge CB3 0ET, UK; pcon@bas.ac.uk
8 Department of Zoology, University of Johannesburg, P.O. Box 524, Auckland Park 2006, South Africa
9 Department of Pharmaceutical Engineering, Hoseo University, Asan-si 31499, Chungnam, Korea; salim0609@hoseo.edu
* Correspondence: aqlima@upm.edu.my

Abstract: An Antarctic soil bacterial consortium (reference BS14) was confirmed to biodegrade canola oil, and kinetic studies on this biodegradation were carried out. The purpose of this study was to examine the ability of BS14 to produce biosurfactants during the biodegradation of canola oil. Secondary mathematical equations were chosen for kinetic analyses (Monod, Haldane, Teissier–Edwards, Aiba and Yano models). At the same time, biosurfactant production was confirmed through a preliminary screening test and further optimised using response surface methodology (RSM). Mathematical modelling demonstrated that the best-fitting model was the Haldane model for both waste (WCO) and pure canola oil (PCO) degradation. Kinetic parameters including the maximum degradation rate (μ_{max}) and maximum concentration of substrate tolerated (S_m) were obtained. For WCO degradation these were 0.365 min^{-1} and 0.308%, respectively, while for PCO they were 0.307 min^{-1} and 0.591%, respectively. The results of all preliminary screenings for biosurfactants were positive. BS14 was able to produce biosurfactant concentrations of up to 13.44 and 14.06 mg/mL in the presence of WCO and PCO, respectively, after optimisation. The optimum values for each factor were determined using a three-dimensional contour plot generated in a central composite design, where a combination of 0.06% salinity, pH 7.30 and 1.55% initial substrate concentration led to the highest biosurfactant production when using WCO. Using PCO, the highest biosurfactant yield was obtained at 0.13% salinity, pH 7.30 and 1.25% initial substrate concentration. This study could help inform the development of large-scale bioremediation applications, not only for the degradation of canola oil but also of other hydrocarbons in the Antarctic by utilising the biosurfactants produced by BS14.

Keywords: kinetic modelling; canola oil; degradation; biosurfactant; Antarctic bacteria; central composite design

1. Introduction

Antarctica is commonly considered the most pristine landmass on Earth, making it the perfect place for monitoring the spread of global pollutants as well as being a sensitive indicator of global climate change [1]. Nevertheless, pollution events such as hydrocarbon oil spills (diesel, petroleum and engine as well as waste oil) have taken place in many parts of the Antarctic, affecting terrestrial, freshwater, marine and ice environments [2]. Shipping has a high potential to pollute seawater through waste oil generated from grey water and food waste, which includes waste cooking oil; for instance, a larger cruise ship carrying 2700 passengers can generate over one tonne per day [3]. Anthropogenic activities are clearly responsible for the release of oil into the environment, causing chronic toxicity and sub-lethal effects in natural ecosystems, often affecting localised areas [4–8].

The potential of microorganisms has been recognized in the development of strategies to remediate oil spills. Specifically relating to the biodegradation of waste cooking oils, Zahri et al. [9] studied the degradation of waste canola oil (WCO) and pure canola oil (PCO) using a consortium of native Antarctic soil bacteria. Microorganisms are increasingly used to treat and transform waste products, and are considered as an eco-friendly technology for the remediation of contaminated environments. Detailed mathematical regression approaches have been applied to model the kinetics of hydrocarbon bioremediation and identify the optimal environmental conditions required [10]. Such kinetic studies allow the determination of the time needed for a contaminant to be degraded to a target concentration and underpin the design of biodegradation kinetics models.

Biosurfactants are a useful tool in bioremediation that can act as emulsion stabilisers (either as an emulsifier or de-emulsifier) and anti-adhesive agents through mobilisation and solubilisation mechanisms [11]. The biosurfactants' function is derived from their amphiphilic composition, via the water-soluble and water-insoluble elements included in their molecular structure [12]. Biosurfactants have the capability to increase the surface area of hydrocarbons by lowering the interfacial tension between the hydrophilic and hydrophobic parts and, thus, can play a key role in the bioremediation process. The efficiency of bacterial production of biosurfactants can be affected by various nutritional and environmental factors. Factors such as salinity, temperature, pH, type and concentration of substrate can determine the stability of microbial biosurfactants [13]. Thus, biosurfactant production factors must be monitored and maintained within a specific range of operating conditions to achieve the optimum production of biosurfactants.

A native Antarctic soil bacterial consortium, designated BS14, has been previously confirmed to degrade canola oil [9,14]. Further studies on the potential of this consortium are described here, involving degradation kinetics modelling using secondary non-linear regression. The ability of the consortium to produce biosurfactants during the biodegradation of canola oil was also analysed and optimised through the application of response surface methodology (RSM). This research will contribute to the development of bioremediation approaches involving the use of biosurfactants and offering efficient and cost-effective means to treat contaminated areas, thereby improving environmental protection.

2. Materials and Methods

Consortium BS14 was originally obtained from soil collected in the vicinity of the Chilean General Bernardo O'Higgins Riquelme research station (Trinity Peninsula, northwest Antarctic Peninsula; 63°19′20.6″ S 57°53′53.6″ W). The isolation and preparation of the consortium are described by Zahri et al. [9]. Both waste canola oil (WCO) and pure canola oil (vegetal PCO) (Belmont, Chile) were acquired from the research station's kitchen. In Antarctica, WCO is stored in steel oil drums at room temperature. There was no specific information available on the number of times the canola oil was repeatedly heated. This is because the uses of canola oil in Antarctic include various type of cooking (frying, grilling and boiling) and types of food prepared.

2.1. Shake Flask Culture Conditions

A modified medium from Zahri et al. [9] was used for the biodegradation of canola oil: minimal salt medium (MSM). For study of the degradation of WCO the medium consisted (per L) of 8.34 g K_2HPO_4, 2.61 g KH_2PO_4, 0.13% w/v NaCl, 1 g $(NH_4)_2SO_4$, 1.13 g yeast extract and 1% v/v WCO with the pH adjusted to 7.13 using HCl. The MSM for PCO consisted (per L) of 9.52 g K_2HPO_4, 2.00 g KH_2PO_4, 0.75 g $(NH_4)_2SO_4$, 1.00 g yeast extract and 1% PCO, with the pH adjusted to pH 7.25 using HCl. The bacterial consortium culture (1 mL) was inoculated into 50 mL of both MSM media and incubation took place on an orbital shaker (150 rpm) at 10 °C.

2.2. Biodegradation Kinetics Modelling

The specific degradation rate with different initial substrate concentrations of WCO and PCO was obtained through assessment of the percentage of degradation each day over the 7 days incubation period. The fit of the data obtained to different widely used non-linear regression models was then assessed, including Monod, Haldane, Teissier–Edwards, Aiba and Yano models [15–19]. Varying environmental conditions have impacts on the performance of each kinetic model. Important kinetic parameters can be obtained through these established regression models, including the maximum degradation rate (μ_{max}), half-saturation constant (K_s), inhibition constant (K_i) and maximum initial concentration of substrate tolerated (S_m). The data were fitted using CurveExpert Professional software (version 2.6, Hyams Development, Huntsville, AL, USA).

Percentage Degradation of WCO and PCO

The amount of residual canola oil after biodegradation was determined gravimetrically after day 7 [20]. The residual oil was extracted using n-hexane organic solvent in a 1:1 ratio with the media in a 100 mL separating funnel. The mixture was vigorously shaken and then left to stand for 15 min to allow separation to occur. The aqueous layer was removed, and the organic layer was transferred into a Petri dish and concentrated by evaporation in a fume hood for 24 h. The mass of residual oil was determined and corrected by taking into consideration the residual oil from the abiotic control. The percentage biodegradation of WCO and PCO was calculated using Equation (1) [21]:

$$\text{Biodegradation } (\%) = \left(\frac{\begin{array}{c} Weight\ of\ residual\ canola\ oil\ (abiotic\ control) - \\ weight\ of\ residual\ canola\ oil\ (sample) \end{array}}{Original\ weight\ of\ introduced\ canola\ oil} \right) \times 100 \quad (1)$$

2.3. Analysis of Biosurfactant Production

The ability of BS14 to produce biosurfactants was screened by assessing haemolytic activity, microbial adhesion to hydrocarbons (MATH) assay, drop-collapse test, oil displacement test and emulsification index (E_{24}). A separate Antarctic bacterial consortium (BS5) known to not produce biosurfactants was used as the negative control in each assay. BS5 was also obtained from the vicinity of the research station.

2.3.1. Haemolytic Activity

Blood agar plates were used to screen for biosurfactant activity. BS14 cultured on MSM media (incubated for 2 days) were inoculated onto blood agar plates containing Columbia agar with 5% sheep blood and incubated for 5–6 days at 10 °C. The presence of a clear zone around the colonies confirmed the presence of biosurfactant synthesis by the bacteria [22].

2.3.2. Microbial Adhesion to Hydrocarbons (MATH) Assay

One-day-old BS14 cultures grown on both MSM media were harvested by centrifugation at 5000 rpm for 20 min at 4 °C, washed and rinsed twice with 1× phosphate-buffered saline (PBS). The cells were suspended for standardisation at 1 ± 0.01 $OD_{600\ nm}$ using PBS. Then, 300 µL of the organic solvents hexadecane and tetrahexadecane was added to 5 mL of bacterial cell suspension in separate test tubes. The mixtures in the test tubes were vortexed for 2 min and then left to stand for 15 min to allow for the separation of the solvent and medium [23]. The aqueous phase was carefully removed from the tubes, and its absorbance was measured at 600 nm using a spectrophotometer. Microbial adhesion to hydrocarbons was calculated using Equation (2):

$$\text{MATH}\ (\%) = \left[1 - \left(\frac{\text{Final absorbance reading (nm)}}{\text{Initial absorbance reading (nm)}}\right)\right] \times 100 \quad (2)$$

2.3.3. Drop-Collapse Test and Oil Displacement

This method was modified from Rani et al. [24]. MSM BS14 grown cultures (7 days) containing WCO or PCO were centrifuged at 5000 rpm for 20 min at 4 °C to obtain cell-free supernatants. Then, about 20 µL of used engine oil (obtained from a motorcycle repair shop in Sri Serdang, Selangor, Malaysia) was set on a grease-free glass slide followed by the addition of 50 µL of the cell-free supernatant onto the centre of the engine oil. Flattening of the oil drop is considered a positive indication of biosurfactant production.

Oil displacement activity was assessed using 2 mL of used engine oil added to a Petri dish containing 50 mL of sterile H_2O. Then, 100 µL of cell-free WCO or PCO MSM culture supernatants was carefully added to the engine oil surface and the diameter of the clear zone on the oil surface was measured [25].

2.3.4. Emulsification Index (E24)

Three millilitres of cell-free WCO or PCO MSM culture supernatant (grown for 7 days) was added to an equal amount of different types of hydrocarbons (commercial canola oil and toluene) in separate test tubes. This included WCO and PCO that consisted of frying canola oil, 100% canola oil and canola vegetable oil (Belmont, Chile). The tubes were then thoroughly mixed by vortexing for 5 min. The height of the emulsion layer was measured after 24 h. The emulsification activity index E_{24} was calculated using Equation (3) [26]:

$$\text{E24}\ (\%) = \left(\frac{\text{Height of emulsification (cm)}}{\text{Total height of the mixture (cm)}}\right) \times 100 \quad (3)$$

All commercialised PCO oil consisted of different compositions of oil, where (1) frying canola oil was composed of high oleic canola oil, high oleic wonder oil, corn oil, canola oil, vitamin E, citric acid and dimethylpolysiloxane, (2) 100% canola oil was composed of canola oil, vitamin E, citric acid as well as dimethylpolysiloxane and (3) canola vegetable oil was composed of soybean oil (90%), canola oil (10%), citric acid and dimethylpolysiloxane.

2.4. Extraction of Biosurfactants

To extract the biosurfactants produced, the cell-free WCO and PCO MSM culture supernatants were first centrifuged at 5000 rpm for 20 min at 4 °C and then filtered through a Whatman No.1 filter paper. Then, any remaining WCO or PCO in the supernatant was removed before the extraction process, and the pH was adjusted to pH 2 using 2N HCl. Subsequently, a 2:1 ratio of chloroform:methanol solution was added to the test tubes. The mixtures were vigorously vortexed for 15 min to allow a separation process to occur [27]. The organic phase (chloroform + biosurfactant) was placed in glass Petri dishes and dried in an oven at 80 °C for 1 day to completely evaporate the solvent. After drying, the glass Petri dishes were weighed, and the dry mass of the biosurfactants was calculated using Equation (4) [28]:

$$\text{Biosurfactant mass (mg)} = \text{Mass of plate after drying} - \text{Mass of the empty plate} \quad (4)$$

2.5. Experimental Design and Optimisation

The production of biosurfactant was screened daily for 10 days to determine the maximum biosurfactant production period. Then, the BS14 consortium was cultured under different conditions of salinity, pH, temperature and initial substrate concentration for the optimisation process with the incubation time at the highest production of biosurfactant during the screening process.

The Plackett–Burman design (PBD) was used to screen for the critical factors and eliminate all non-significant factors for further optimisation in the central composite design (CCD). The design did not designate the interaction between the factors [29]. Twelve experimental runs were generated where each experiment contained two factorials: low (1) and high (+) levels for each factor (Table 1). The range of low and high levels was based on our previous study, optimising bacterial growth and biodegradation of WCO and PCO by BS14 [9].

Table 1. Plackett–Burman experimental design for biosurfactant production by WCO and PCO using BS14 bacterial consortium.

Run	Standard Deviation	A	B	C	D
1	3	0.00	7.50	15.00	0.50
2	6	0.25	7.50	15.00	0.50
3	5	0.25	7.50	10.00	2.00
4	2	0.25	7.50	10.00	2.00
5	10	0.25	7.00	10.00	0.50
6	4	0.25	7.00	15.00	2.00
7	8	0.00	7.00	15.00	2.00
8	11	0.00	7.50	10.00	0.50
9	7	0.00	7.50	15.00	2.00
10	9	0.00	7.00	10.00	2.00
11	12	0.00	7.00	10.00	0.50
12	1	0.25	7.00	15.00	0.50

WCO: waste canola oil, PCO: pure canola oil, BS14: coded Antarctic bacterial consortium, A: salinity (% *w*/*v*); B: pH; C: temperature (°C); and D: WCO/PCO initial concentration (% *v*/*v*).

Statistical optimisation of biosurfactant production was carried out using a CCD, which is useful in identifying significant pairwise interactions between factors which affect the optimisation process. CCDs require only a small number of experiments and are used to create statistically designed experiments, estimate coefficients, predict the experimental response and validate the model [30]. The second-order quadratic model was chosen for the process in the optimisation, as given by Equation (5), to identify the presence of parameter interactions:

$$Y = \beta_0 + \sum \beta_i x_i + \left(\sum \beta_{ii} x_i\right)^2 + \sum \beta_{ij} x_i x_j \quad (5)$$

where β_0 is the regression coefficient term, β_i is the linear coefficient term, β_{ii} is the quadratic coefficient term, β_{ij} is the interaction coefficient term and x_i as well as x_j are the levels of the independent factors.

The use of a CCD in RSM generates an assessment of the overall significance of the model (*p*-value), the coefficient of determination, adequate precision and the coefficient of variance through the use of ANOVA. Although a key element of a CCD is in its determination of significant interactions between factors, it also identifies the significance of each individual factor [31].

Model verification was performed based on the point prediction in the post-analysis of the process, by comparing the predicted and experimentally determined values. Table 2 gives the experimental design for the validation of biosurfactant production by WCO and PCO cultures.

Table 2. Experimental design for the validation of biosurfactant production by BS14 bacterial consortium.

Substrate	A	B	C	Predicted (mg/mL)
WCO	0.00	7.35	2.0	12.65
PCO	0.05	7.27	1.76	10.94

A: salinity (% *w*/*v*), B: pH, C: WCO/PCO initial concentration (% *v*/*v*).

2.6. Statistical Analyses

All experiments were carried out in triplicate and variation was indicated by the standard error of the mean (SEM). One-way analysis of variance (ANOVA) was used to analyse the data, and where significant was followed by post hoc pairwise Tukey's tests, carried out using GraphPad InStat (version 3.1). $p < 0.05$ was accepted as statistically significant.

3. Results

3.1. Degradation Kinetics Modelling

WCO and PCO degradation was assessed over 7 days in cultures with different initial substrate concentrations (0.5% to 3.0%). For both WCO and PCO, the lowest initial concentration of the substrate yielded high degradation percentages over the 168 h culture period. The highest levels of degradation achieved were about 93% and 96%, respectively, at 0.5% initial substrate concentration. The percentage of WCO and PCO degradation reduced rapidly at initial concentrations above 1.5% (Figure 1), although the absolute amount of WCO and PCO degraded at 3% initial concentration was still approximately three times greater than that at 0.5% initial concentration.

The relationship between the specific degradation rate and the initial substrate concentration is shown in Figure 2. As noted above, the highest degradation rates of WCO and PCO were achieved at the lowest initial substrate concentration (0.5%), with the rate then decreasing as the initial substrate concentration increased. The WCO degradation rate decreased more rapidly with an increasing substrate concentration than did that of PCO.

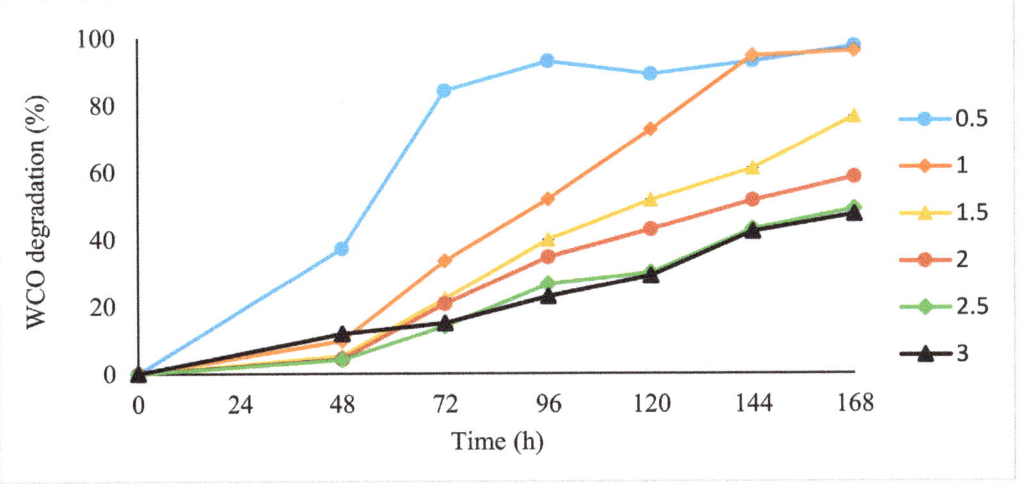

(a)

Figure 1. *Cont.*

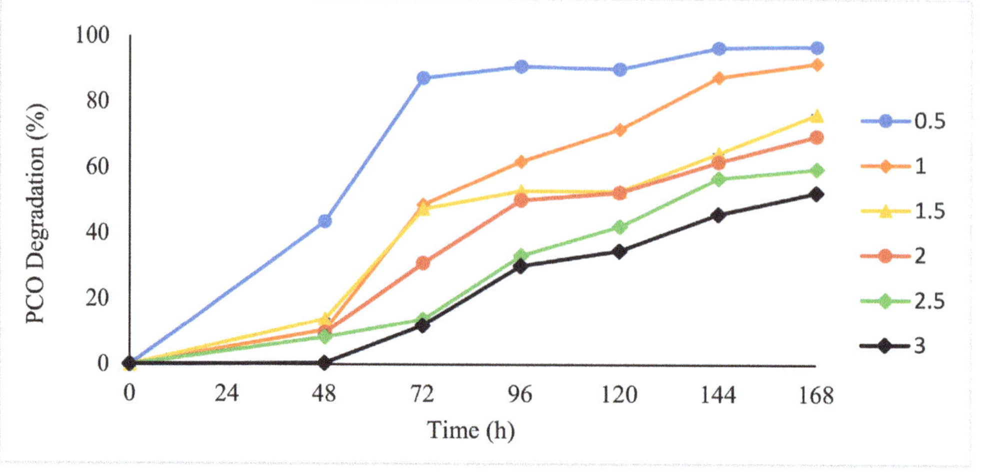

(**b**)

Figure 1. Effect of initial substrate concentration on degradation of (**a**) WCO and (**b**) PCO by BS14 bacterial consortium. Error bars represent mean ± standard error (SEM) and are contained within the points.

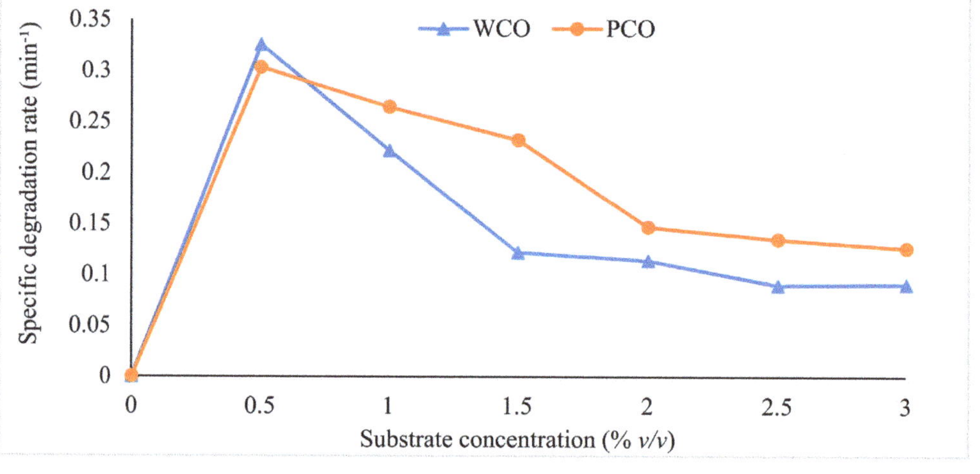

Figure 2. Specific degradation rates of WCO and PCO with increasing initial substrate concentration.

The relationship between degradation rate and initial substrate concentration was modelled using a range of standard non-linear models (Figure 3 and Supplementary Materials Table S1). The Haldane and Yano models provided the mathematical best fit to the experimental data; both models assume substrate inhibition at higher concentrations. Table 3 shows the statistical analysis across all the models tested, again identifying that the Haldane model provided the mathematical best fit for both WCO and PCO degradation. However, the fits of the Haldane and Yano models were very similar.

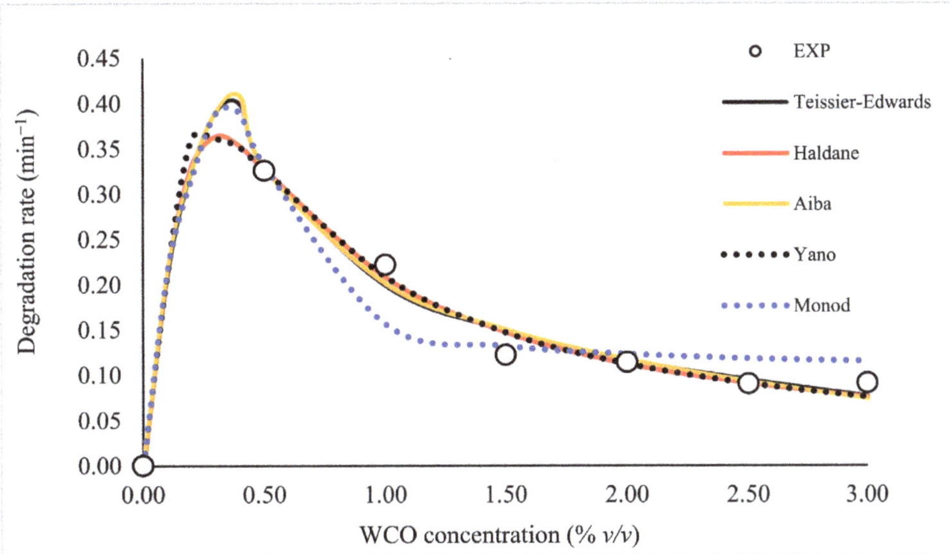

(a)

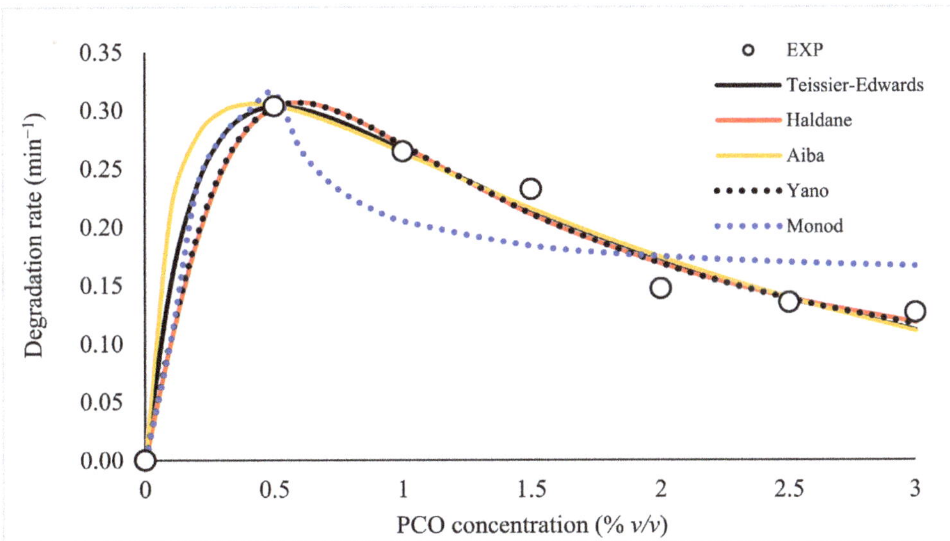

(b)

Figure 3. Kinetic modelling of (**a**) WCO and (**b**) PCO degradation using a range of standard non-linear models.

The half-saturation constant (K_s) and inhibition constant (K_i) in the Haldane model for WCO biodegradation were 5.14% and 0.019%, respectively. Meanwhile, in the biodegradation of PCO, the K_s value was 16.65% while the K_i value was 0.021%. As these values were obtained from the regressed equation, the maximum concentration of substrate tolerated (S_m) and the actual maximum degradation rate (μ_{max}) could be calculated. The S_m value for WCO (0.31%) was somewhat lower than that for PCO (0.59%), while the μ_{max} value for WCO was slightly higher (0.37 min^{-1}) than that for PCO (0.31 min^{-1}).

Table 3. Statistical analysis of different standard models applied to WCO and PCO degradation kinetics.

Model	R^2	Adj R^2	DF	RMSE	AICc	SSE	AF	BF
WCO								
Haldane	0.985	0.992	4	0.017	−54.814	0.001	0.999	0.999
Yano	0.985	0.992	3	0.019	−47.758	0.001	0.999	0.999
Teissier–Edwards	0.978	0.989	4	0.030	−45.251	0.004	1.000	1.000
Aiba	0.977	0.988	4	0.020	−51.869	0.002	0.999	0.999
Monod	0.910	0.954	5	0.035	−46.654	0.006	1.000	1.000
PCO								
Haldane	0.983	0.991	4	0.016	−54.821	0.001	0.999	0.999
Yano	0.984	0.992	3	0.019	−47.821	0.001	0.999	0.999
Teissier–Edwards	0.981	0.991	4	0.017	−53.802	0.001	0.999	0.999
Aiba	0.981	0.990	4	0.016	−53.553	0.001	0.999	0.999

R^2: coefficient determination, Adj R^2: Adjacent coefficient, DF: degree of freedom, RMSE: root mean square error, AICc: Akaike information criterion, SSE: sum of squared estimate of errors, AF: accuracy factor, BF: bias factor.

3.2. Preliminary Screening

The ability of bacterial consortium BS14 to produce biosurfactants during WCO and PCO biodegradation was next assessed. Preliminary screening for biosurfactant production utilised the haemolytic test, a microbial adhesion to hydrocarbons assay, the oil-spreading test, the drop-collapse test and emulsification index (E_{24}).

3.2.1. Haemolytic Test

The ability of the BS14 consortium to produce biosurfactants was tested on blood agar. Inocula from both WCO and PCO media showed positive results (Figure 4). Clear zones indicating β-haemolytic activity can be seen surrounding the bacterial growth, indicating the presence of biosurfactant-producing bacteria.

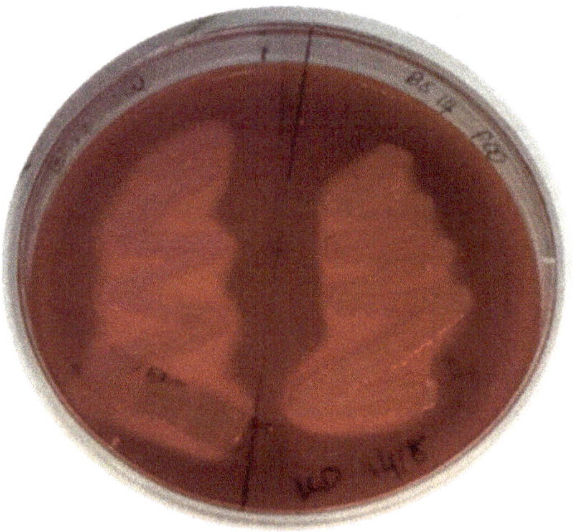

Figure 4. BS14 consortium inoculated from WCO (left) and PCO (right) media demonstrating β-haemolytic activity on a blood agar plate after a 2 days incubation.

3.2.2. Microbial Adhesion to Hydrocarbons (MATH) Assay

Organic solvents were used in this assay as the hydrocarbon substrates. The assay indicated that WCO and PCO samples were characterised by >50% microbial adhesion to hexadecane and tetrahexadecane (Figure 5).

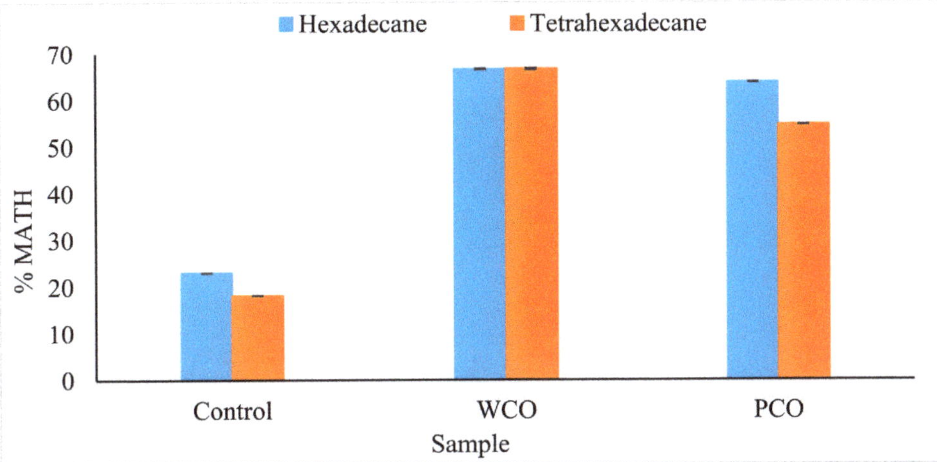

Figure 5. Microbial adhesion of BS14 bacterial consortium to two different hydrocarbon substrates. Error bars represent mean ± standard error (SEM).

3.2.3. Oil-Spreading and Drop-Collapse Test

Both these assays are qualitative tests that determine the presence of biosurfactants in the media. Oil displacement was clear in the oil-spreading test for both WCO and PCO media, with magnitudes of 23.00 and 31.17 mm diameter, respectively. The supernatant's ability to collapse a hydrocarbon droplet also showed positive results (Table 4).

Table 4. Oil spreading and drop collapse test results achieved by BS14 bacterial consortium in WCO and PCO media.

Sample	Diameter (mm)	
	Oil Spreading Test	Drop Collapse Test
Control	5.67	6.67
WCO	23.00	9.33
PCO	31.17	9.00

3.2.4. Emulsification Index (E_{24})

The emulsification test is an indirect method used to screen for biosurfactant production. Figure 6 shows the ability of cell-free WCO and PCO MSM culture supernatants to emulsify different types of cooking oil as hydrocarbon substrates. The biosurfactants in the supernatants emulsified >40% of the cooking oils and >10% of the toluene organic solvent. Even though the E_{24} for toluene was lower than those of other hydrocarbons, the capability of the biosurfactant to emulsify this organic solvent was greater than that of the negative control.

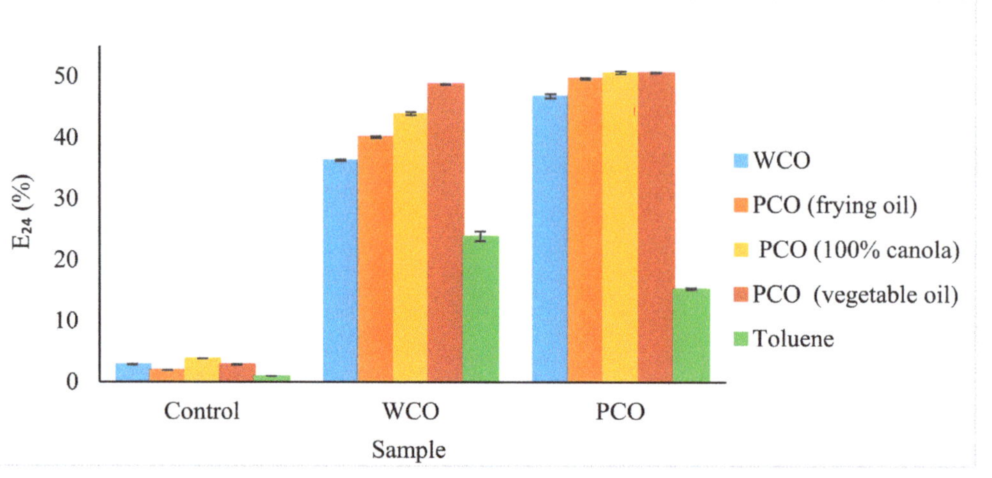

Figure 6. Emulsification index after a 24 h incubation (E_{24}) for both cell-free WCO and PCO MSM culture supernatants. Error bars represent mean ± standard error (SEM).

3.3. Measurement of Biosurfactant Production

The production of biosurfactants in WCO and PCO media was measured over 10 days of incubation. The highest production of biosurfactant was measured on day seven, with 7 and 8 mg/mL being produced in WCO and PCO media, respectively (Figure 7). The amount of biosurfactant produced in both cultures dropped sharply after day seven.

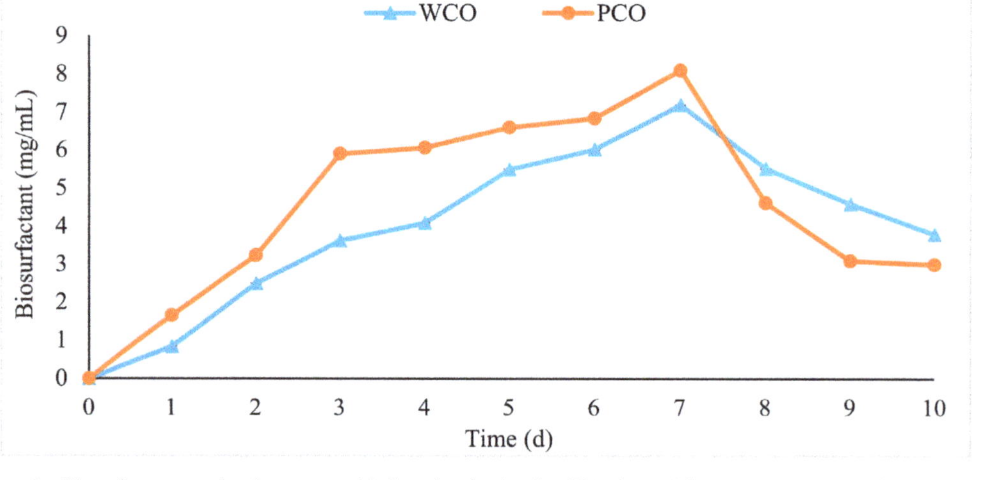

Figure 7. Biosurfactant production over a 10 days incubation by BS14 bacterial consortium. Error bars represent mean ± standard error (SEM) and are contained within the points.

3.4. Optimisation of Biosurfactant Production

Tables 5 and 6 display the analysis of variance (ANOVA) for PBD for biosurfactant production using WCO or PCO as a substrate, respectively. The factors with a significant influence on the degradation response were salinity (A), pH (B) and substrate concentration (C) for both WCO and PCO media.

Table 5. Analysis of variance (ANOVA) for Plackett-Burman design for biosurfactant production using WCO as a substrate.

Source	Sum of Squares	DF	*F* Value	Prob > *F*
Model	82.83	4	17.06	0.0078 **
A	9.82	1	14.16	0.0197 *
B	8.43	1	12.16	0.0252 *
C	2.21	1	3.18	0.1491
D	28.13	1	40.56	0.0031 **
Residual	2.77	7		
Cor Total	85.61	11		
R-squared	0.968		Pred *R*-squared	0.745
Adj *R*-squared	0.911		Adeq Precision	11.58

A: salinity (% *w*/*v*), B: pH, C: temperature (°C), D: WCO concentration (% *v*/*v*). * $p < 0.05$, ** $p < 0.01$.

Table 6. Analysis of variance (ANOVA) for Plackett-Burman design for biosurfactant production using PCO as a substrate.

Source	Sum of Squares	DF	*F* Value	Prob > *F*
Model	54.83	4	32.72	0.0023 **
A	11.33	1	47.34	0.0023 **
B	1.96	1	8.17	0.0460 *
C	0.54	1	2.25	0.2082
D	7.68	1	32.07	0.0048 **
Residual	0.96	7		
Cor Total	55.79	11		
R-squared	0.993		Pred *R*-squared	0.861
Adj *R*-squared	0.953		Adeq Precision	18.02

A: salinity (% *w*/*v*), B: pH, C: temperature (°C), D: PCO concentration (% *v*/*v*). * $p < 0.05$, ** $p < 0.01$.

All significant factors (A, B and D) were then further optimised using a CCD (Table 7). Twenty runs were generated from the software with the predicted and actual values for the biosurfactant production using either WCO or PCO as a substrate.

Table 7. Central composite experimental design for biosurfactant production by BS14 bacterial consortium using either WCO or PCO as substrate.

Run	Standard Deviation	A	B	C	WCO		PCO	
					Actual	Predicted	Actual	Predicted
1	17	0.13	7.25	1.25	13.80	12.95	6.73	5.29
2	8	0.25	7.50	2.00	3.02	4.06	13.50	13.80
3	13	0.13	7.25	−0.01	1.31	1.37	12.46	13.98
4	18	0.13	7.25	1.25	10.36	12.99	1.71	4.55
5	20	0.13	7.25	1.25	11.73	12.93	3.50	4.17
6	2	0.25	7.00	0.50	2.43	4.04	7.27	6.81
7	3	0.00	7.50	0.50	3.15	4.04	4.40	4.98
8	12	0.13	7.67	1.25	5.44	6.19	5.40	6.31
9	11	0.13	6.83	1.25	3.27	2.24	13.50	13.90
10	4	0.25	7.50	0.50	5.27	3.69	14.30	14.00
11	14	0.13	7.25	2.51	6.88	6.53	3.95	4.73
12	15	0.13	7.25	1.25	13.87	12.90	4.23	4.15
13	19	0.13	7.25	1.25	13.96	13.00	14.76	14.99
14	7	0.00	7.50	2.00	12.97	11.57	3.83	2.92

Table 7. *Cont.*

Run	Standard Deviation	A	B	C	WCO		PCO	
					Actual	Predicted	Actual	Predicted
15	10	0.34	7.25	1.25	5.64	5.51	4.32	4.31
16	6	0.25	7.00	2.00	3.35	2.67	3.77	1.99
17	9	−0.09	7.25	1.25	9.23	9.06	4.69	5.43
18	1	0.00	7.00	0.50	1.6	0.77	10.69	8.78
19	16	0.13	7.25	1.25	13.91	13.50	6.20	4.71
20	5	0.00	7.00	2.00	4.75	6.54	15.63	14.99

A: salinity (% *w*/*v*), B: pH, C: WCO/PCO concentration (% *v*/*v*).

The second-order fitting equation's coefficients were calculated by the software and the model acceptability was verified using ANOVA. The second-order polynomial coded equations are given as Equations (6) and (7), for WCO and PCO, respectively:

$$Y = -2634.60 + 47.40A + 721.46B + 18.58C - 19.07AC - 128.07A^2 - 49.43B^2 - 5.66C^2 \quad (6)$$

$$Y = -1.94 + 1.68BC - 49.54A^2 + 0.098B^2 - 4.87C^2 \quad (7)$$

The significant coefficients identified in WCO media included A, B, C, AC, A^2, B^2 and C^2, while in PCO media these included BC, A^2, B^2 and C^2.

In CCDs, different types of adequacy of the model were tested, including linear, 2FI, quadratic and cubic. For both WCO and PCO, the statistical model showed a high value of R^2 (>0.9) in the quadratic (Tables 8 and 9) and cubic models. The R^2 values generated in the cubic model were somewhat higher than the quadratic model: 0.971 (WCO) and 0.973 (PCO). Nevertheless, the model suggested by the software was the quadratic model for both WCO and PCO.

Table 8. Analysis of variance (ANOVA) for central composite design (CCD) for biosurfactant production using WCO as substrate.

Source	Sum of Squares	DF	*F* Value	Prob > *F*
Model	54.83	9	16.08	<0.0001 ***
A	11.33	1	5.69	0.0383 *
B	1.96	1	6.94	0.0249 *
C	0.54	1	12.05	0.0060 **
AB	7.68	1	2.46	0.1481
AC	12.47	1	9.54	0.0115 *
BC	25.50	1	0.57	0.4670
A^2	9.45	1	21.54	0.0009 ***
B^2	0.96	1	51.34	<0.0001 ***
C^2	0.55	1	54.45	<0.0001 ***
Residual	0.41	10		
Lack of Fit	55.79	5	1.28	0.3962
Pure Error	54.83	5		
Cor Total	11.33	19		
R-squared	0.935		Pred *R*-squared	0.638
Adj *R*-squared	0.877		Adeq Precision	10.521

A: salinity (% *w*/*v*), B: pH, C: WCO concentration (% *v*/*v*). * $p < 0.05$, ** $p < 0.01$, *** $p < 0.001$.

ANOVA for both CCDs in biosurfactant production using WCO and PCO showed that the model designs were highly significant (Tables 8 and 9). The importance of the single factors in CCDs can be inferred from the F values calculated in the design. Table 8 shows that $F_{\text{WCO concentration}} > F_{\text{pH}} > F_{\text{NaCl}}$, suggesting that WCO concentration was the most influential factor towards biosurfactant production when using WCO as a substrate.

When PCO was used as substrate, pH was the most influential factor, followed by NaCl and then PCO concentration (Table 9). Significant interactions between any two factors during the optimisation process are also identified through CCDs. Significant interactions were identified between salinity (A) and WCO concentration (C) in biosurfactant production using WCO as a substrate, and between pH (B) and PCO concentration (C) using PCO as a substrate.

Table 9. Analysis of variance (ANOVA) for central composite data design (CCD) for biosurfactant production using PCO as substrate.

Source	Sum of Squares	DF	*F* Value	Prob > *F*
Model	377.91	9	14.81	0.0001 ***
A	1.91	1	0.67	0.4306
B	11.72	1	4.13	0.0694
C	0.06	1	0.02	0.8895
AB	3.30	1	1.17	0.3058
AC	4.26	1	1.50	0.2482
BC	14.84	1	5.24	0.0452 *
A^2	152.00	1	53.61	<0.0001 ***
B^2	99.82	1	35.20	0.0001 ***
C^2	156.55	1	55.21	<0.0001 ***
Residual	28.35	10		
Lack of Fit	22.16	5	3.57	0.0942
Pure Error	6.20	5		
Cor Total	406.26	19		
R-squared	0.930		Pred *R*-squared	0.558
Adj *R*-squared	0.867		Adeq Precision	10.071

A: salinity (% *w*/*v*), B: pH, C: PCO concentration (% *v*/*v*). * $p < 0.05$, *** $p < 0.001$.

As illustrated in the 3D response surface plots (Figure 8), high concentrations of WCO (1.40 to 1.70%) with low salinity (0.06 to 0.13%) were predicted to lead to high biosurfactant production (13.44 mg/mL). Low WCO and high salt concentration decreased the ability of BS14 bacterial consortium to produce biosurfactants. The optimum conditions for biosurfactant production in PCO media were predicted to be a medium pH, 7.20–7.30, and an initial PCO substrate concentration of 1.10–1.14% (*w*/*v*), yielding a highest biosurfactant production of 14.07 mg/mL.

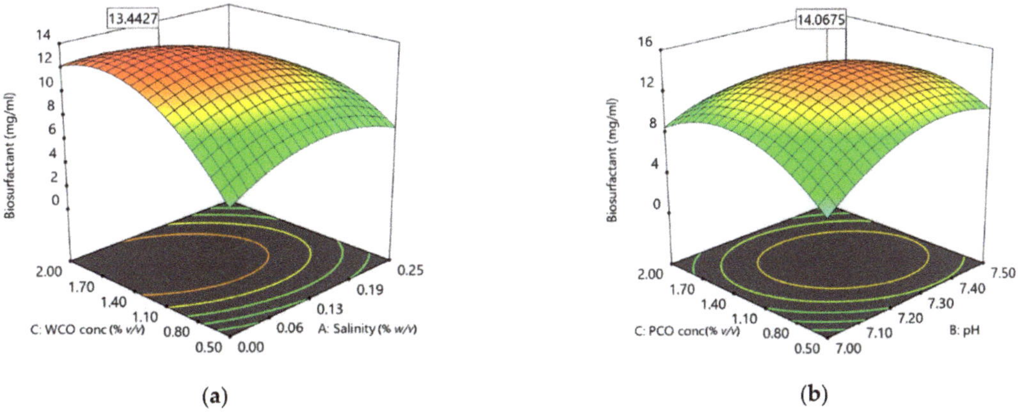

Figure 8. Three-dimensional (3D) response surface plots for the significant factors identified in CCD for biosurfactant production using WCO, (**a**): salinity and WCO concentration; and PCO (**b**): pH and PCO concentration.

The perturbation plots (Figure 9) illustrate the comparative effects of all factors on biosurfactant production. Although ANOVA only identified one significant interaction for both conditions, all factors still have a major contribution to biosurfactant production: salinity (A), pH (B) and substrate concentration (C). The curvature of the perturbation plot factors shows that the response was sensitive and influenced by all the factors.

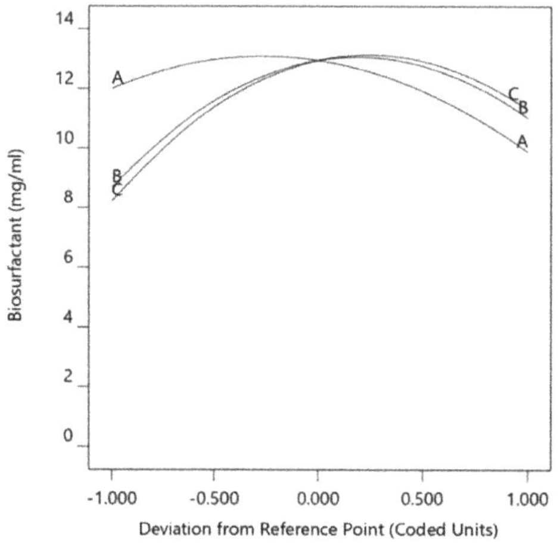

(a)

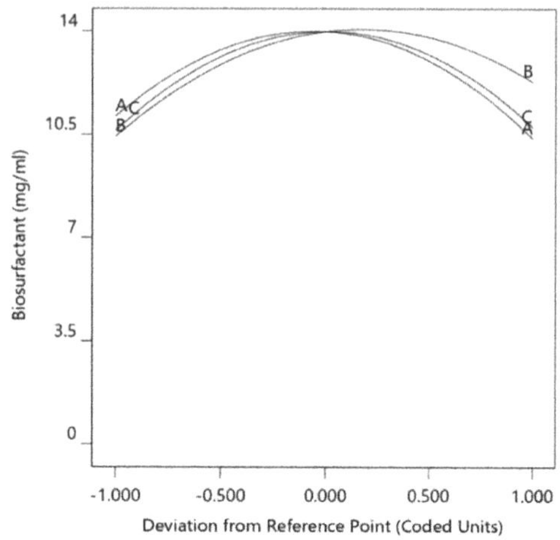

(b)

Figure 9. Perturbation plots for biosurfactant production using (**a**) WCO and (**b**) PCO as substrate.

The results of the responses in CCDs were validated experimentally using the specific conditions generated by the software. Table 10 shows the results for WCO and PCO media, with no significant difference between the actual and predicted values in either case, validating the model.

Table 10. Validation of predicted response surface model.

Substrate	Biosurfactant (mg/mL)		p Value
	Expected Value	Actual Value	
WCO	12.65	11.97	0.787
PCO	10.94	10.78	0.954

4. Discussion

4.1. Mathematical Modelling

The BS14 Antarctic bacterial consortium was confirmed to produce biosurfactants in the presence of WCO or PCO as a substrate, showing good ability to degrade both WCO and PCO. To our knowledge, there have been no previous studies modelling the biodegradation kinetics of WCO and PCO using bacterial consortia isolated from Antarctic soils, and only one study addressing the kinetics of bacterial growth by Antarctic bacteria has been reported [14].

The relationship between degradation rate and initial concentration of substrate is important for large-scale bioremediation applications. Our data indicated that the BS14 bacterial consortium achieved better degradation of PCO than WCO. Bacteria generally cannot degrade highly concentrated oil due to its properties and toxicity. Cooking oil changes in composition and properties when heated above a certain temperature [32]. For example, the oxidative stability of canola oil can lead to peroxide formation in three days under accelerated conditions at 60–65 °C [33]. Lipid oxidation during cooking can lead to the production of aldehydes, epoxides, hydroxyketones and dicarboxylic compounds. These compounds can react with amino acids to produce the carcinogenic compound acrylamide [34].

The complex structure of the PCO molecule could also inhibit the bacterial community from degrading the PCO at high concentrations. PCO has a homogenous fatty acid composition, with 95% unsaturated and saturated fatty acids, including oleic acid, linoleic acid, linolenic acid, stearic acids and palmitic acids [35,36]. Generally, canola oil has 18 carbon fatty acids in its total unsaturated fatty acids and antioxidants. The biodegradability of long-chain fatty acids decreases with an increasing carbon chain length and a decreasing degree of unsaturation [37]. Several studies have confirmed low degradation at high concentrations of cooking oil [38–40].

Of the different kinetic models examined here, the Haldane, Aiba, Yano and Teissier–Edwards models predicted inhibition at high substrate concentrations. The Monod model assumed a critical inhibitor concentration, above which bacterial cells cannot grow and function. The Haldane model is widely used in kinetic studies due to its ability to integrate substrate and growth inhibition constants as well as its simplicity [40]. Generally, the Haldane model is considered an extension of the Monod model, introducing a third constant parameter known as the inhibition constant (K_i). A previous study on phenol degradation and the growth kinetics of *P. aeruginosa* demonstrated that the best fit was provided by the Haldane model [41], as was also found using *P. putida* in the biodegradation of phenol [42]. Free cyanide removal by *Acinetobacter courvalinii* was also best-fitted using the Haldane model [43]. Wang et al. [44] similarly reported a good fit using the Haldane model in the removal of ammonia/nitrate using *Acinetobacter* sp.

The BS14 bacterial consortium could function and degrade 0.5 and 1% initial concentrations of WCO and PCO, respectively. However, more than 1.5% initial concentration of either media reduced the percentage degradation of the initial concentration achieved. The mathematical best fit was provided by the Haldane model, with high substrate concentrations reducing the rate of activity [45]. The practical application of such models lies in the assessment of product formation kinetics, and in their utility in monitoring biological processes in the real world [10].

4.2. *Preliminary Test on Biosurfactant Production*

The preliminary test on biosurfactant production in this study was used to confirm the BS14 consortium's ability to produce biosurfactants. The haemolytic test has long been used in the preliminary screening of biosurfactant-producing microorganisms [26,28, 46,47]. Surfactants are generally able to lyse the erythrocytes [48] by altering the structure of membrane lipids and proteins. cell lysis is caused by osmotic and solubilisation haemolysis [49,50]. of 89 strains collected from seawater samples, 24% showed a positive result in the haemolysis test, with 10% of these showing α-haemolytic activity and 24% showing β-haemolytic activity [51]. Nonetheless, the haemolytic test has a limitation in that lytic enzymes may also be present in the consortium, which can also cause clearing zones [52]. As a result, this method is only recommended as a preliminary screening step [53], which should be supplemented by other techniques, such as those described below.

The hydrophobicity of the microbial cell surface can be determined using a MATH assay, which is a simple and rapid method [54]. Microorganisms can adsorb on the surface of an organic pollutant by increasing the hydrophobicity of their cell surface, which takes place through the remodelling of the cell's outer layers [55]. The positive results obtained here confirmed the affinity of cells towards the hydrocarbons. The process of cell adherence to hydrophobic compounds is considered an indirect method to screen bacteria for the production of biosurfactants. This is because the cells are able to attach themselves to the hydrocarbons by producing surface-active compounds called biosurfactants [56]. The effectiveness of this method has been proven by studies in which W-28 and ISL-01 bacterial isolates from hydrocarbon-rich soil and effluent showed a high hydrophobicity value of 58% [57], *Pseudomonas aeruginosa* MS3 isolated from oil-contaminated marine sediments showed 50% hydrophobicity [58] and *Bacillus subtilis* isolated from a pharmaceutical effluent sample was able to adhere to kerosene up to 58% [59].

The size of the clearing zone in the oil spreading test indicates the concentration of the biosurfactant solution [60] and that the interfacial tension between the hydrophobic part and biosurfactant is reduced [61]. Previously, *Streptomyces* sp., *Microbacterium* sp., *Rhodococcus* sp., *Arthrobacter* sp. and *Bacillus subtilis* have been shown to produce clearing zone diameters of 33, 23, 16, 29 and 35 mm, respectively, using this test [24]. These bacterial strains were also reported to give positive results in the drop-collapse test. A study on lactic acid bacteria including *Lactobacillus fermentum* strain SHU6343, *Pediococcus dextrinicus* strain SHU1593, *L. plantarum* strain SHU3455, *L. rhamnosus* strain SHU1904 and *L. dextrinicus* strain SHU68 confirmed positive results in both oil-spreading and drop-collapse tests in the screening for the production of biosurfactants [62]. Both assays are widely used for screening purposes because they are rapid, simple to perform, do not require specialised equipment and require a small number of samples [63].

The E_{24} test is a well-known method for the screening of microorganisms that can produce biosurfactants, because one of the key characteristics of biosurfactants is to emulsify hydrocarbons. Amodu et al. [64] reported that the biosurfactants produced by *Bacillus licheniformis* strain STK and *B. subtilis* strain STK 02 can emulsify better in oils such as diesel and lubricant oil (60 to 90%) compared to the organic solvent hydrocarbon, kerosene (< 20%). In contrast, the biosurfactant produced by *Rhodococcus erythropolis* strain AQ5-07 showed higher E_{24} in organic solvents rather than in diesel oil, with hexane, hexadecane and tetrahexadecane being emulsified more than 80% while only 60% of diesel was emulsified [13]. *Arthrobacter* spp. strains AQ5-05 and AQ5-06 from Antarctica also exhibited high E_{24} values in hexadecane (60% and 67%, respectively), though there was only a thin layer for both strains when using diesel [65].

4.3. Biosurfactant Production

In our study, the production of biosurfactants was measured over 10 days in WCO and PCO media, with the highest production being observed on day seven for both substrates, at 7 and 8 mg/mL, respectively. Different microorganisms might require different incubation times to produce the maximum amount of biosurfactant. *Virgibacillus salaries* was reported to require between 3 to 6 days to produce 2.8 to 2.9 g/L of biosurfactant [66], while the marine bacterium *Nocardiopsis* sp. strain B4 required 7 to 9 days to yield its maximum amount of biosurfactant [67].

The highest biosurfactant production phase can also be estimated by observing the pattern of bacterial growth in the presence of the substrate. In general, the biosurfactant starts to be produced during the stationary phase and continues until the cell biomass starts to decrease or the death phase is reached. However, in some cases, the biosurfactant itself may be utilised by the bacteria during the survival–starvation state [68].

The important factors in biosurfactant production were optimised using response surface methodology. The factors were screened first using PBD to eliminate non-significant factors. In CCDs, the use of a polynomial model in this study allowed the prediction reliability of the model to be improved, eliminating non-significant predictor factors and reducing model error [69].

The optimum conditions for the bacteria to produce high biosurfactant concentrations in WCO and PCO media can be analysed using 3D contour plots. Bacteria such as *Acinetobacter* sp., *P. aeruginosa*, *B. subtilis* and *B. licheniformis* show the greatest production in low-salinity conditions [32,70,71], similar to the current study. *P. aeruginosa* strain UKMP14T and *Klebsiella* sp. strain KOD36 also required 1% initial substrate concentration to yield a high percentage of surface tension reduction and E_{24} activities [72,73]. In some cases, salinity can reduce soil microorganism activity due to osmotic stress and the presence of toxic ion concentrations. Soluble salts may increase the osmotic potential of water in the soil, drawing water out of the cells and even eventually killing microorganisms through plasmolysis [74]. Numerous studies have shown that salinity can lead to reduced microbial biomass as well as changing the microbial community structure and activity [75–78]. A similar study, Amodu et al. [62], reported that biosurfactant activity was determined through surface tension reduction. The central point of the CCD based on the contour plot identifies the optimum point of the response. In the current study this was between 6.5 to 8.0 for pH and 4 to 8% for initial substrate concentration. *B. subtilis* strain UKMP-4M5 can similarly produce a biosurfactant and the optimum pH at pH 6 to 8 yielded 32 to 34 mN/m surface tension [60]. In another study, biosurfactant production in *Lactococcus lactis* strain CECT-4434 was tested in two conditions, with or without pH control. With pH control, a high production of biosurfactant was measured through surface tension analysis [72]. This shows the importance of studying pH in environmental microbiology [79]. Variation in environmental pH can also affect metabolic activity in natural communities, as different microorganisms prefer different pH conditions. Variation in pH can interfere with microbial metabolism and modulate thermodynamic and redox reactions [80]. The slightly alkaline conditions in the current study contributed to the optimisation of biosurfactant production.

The yield of the biosurfactant generally increases with the concentration of substrate [81]. Here, low biosurfactant yield was obtained at low concentration of WCO during the optimisation process. Other studies have reported similar results, such as the ability of *B. subtilis* strains ATCC 21332 and MSH1 to produce up to 450 and 200 mg/L of biosurfactant, respectively, at a high concentration of glucose (around 30 g/L) [82]. However, in certain cases high substrate concentration can reduce biosurfactant production [83]. In such cases, the high substrate concentration could inhibit the bacterial growth and metabolism.

Although biosurfactants have higher surface activity with high tolerance to various factors and can withstand up to extreme conditions (acidity or basicity of an aqueous solution, temperature, salt concentration, ionic strength and demulsifying as well as emulsifying ability) [84], several important factors need to be optimised in order for bacteria to produce biosurfactants during the biodegradation of oil, as carried out in this study. Microorganisms increase the bioavailability of potentially biodegradable nutrients, such as oil, by producing biosurfactants [85]. Biosurfactants influence microbial cell surface properties, causing significant alterations of surface hydrophobicity and surface functional groups. The cell surface hydrophobicity influences the tendency of cells to adhere to the hydrophobic part, which is commonly analysed through a MATH assay. Interactions of cells with their environment are also influenced by carboxyl, phosphate and amino functional groups present on the cell surface. Changes in cell surface characteristics appear to be correlated with cell adhesion to pollutants [86].

5. Conclusions

The mathematically best-fitting kinetic model for the biodegradation of WCO and PCO using the BS14 bacterial consortium was the Haldane model, although several models gave closely similar fits to the experimental data. The consortium also proved to be a strong biosurfactant producer when using WCO and PCO as substrates. The biosurfactant produced gave positive results in the haemolytic test, oil-spreading test, drop-collapse test and had a high percentage of hydrophobicity and emulsification index. The critical factors in biosurfactant production included salinity, pH, temperature and initial substrate concentration, which were optimised to produce the highest yield. This study provides important information for the development of bioremediation protocols for the degradation of oils, such as canola oil, in the Antarctic. The presence of biosurfactants in the bioremediation of oil could enhance the rate of biodegradation activity efficiently, as biosurfactants could increase the hydrocarbons' mobility and bioavailability. Thus, the biosurfactant produced during the degradation process has potential for use in other applications such as the in situ or ex situ bioremediation of other hydrocarbons, including diesel and petroleum-derived oils.

Supplementary Materials: The following are available online at https://www.mdpi.com/article/10.3390/foods10112801/s1, Table S1: Kinetic modelling data on WCO and PCO degradation.

Author Contributions: Conceptualization, S.A.A., K.A.K., C.G.-F. and A.Z.; methodology, S.A.A., K.N.M.Z., K.A.K. and S.S.; software, K.N.M.Z. and K.A.K.; validation, S.A.A., S.S., C.G.-F., A.Z., K.A.K., P.C. and S.L.; formal analysis, K.N.M.Z., S.A.A. and K.A.K.; investigation, K.N.M.Z.; resources, S.A.A. and C.G.-F.; Writing—original draft preparation, K.N.M.Z.; Writing—review and editing S.A.A., K.A.K., C.G.-F., A.Z., S.S., P.C. and S.L.; supervision, S.A.A., K.A.K., C.G.-F., A.Z., S.S. and P.C.; project administration, S.A.A. and C.G.-F.; funding acquisition, S.A.A., C.G.-F. and S.L. All authors have read and agreed to the published version of the manuscript.

Funding: This project was financially supported by Putra–IPM fund under the research grant attached to S.A. Ahmad (GPM–2018/9660000, GPM–2019/9678900) disbursed by Universiti Putra Malaysia (UPM) and YPASM Smart Partnership Initiative (6300247) by Sultan Mizan Antarctic Research Foundation (YPASM) C. Gomez-Fuentes is supported by Centro de Investigacion y Monitoreo Ambiental Antàrctico (CIMAA). S. Lim is supported by the National Research Foundation of Korea (NRF) grant (No. 2021R1G1A1010154) funded by the Korea government (MSIT). P. Convey is supported by NERC core funding to the BAS 'Biodiversity, Evolution and Adaptation' Team.

Institutional Review Board Statement: Not applicable.

Informed Consent Statement: Not applicable.

Data Availability Statement: The datasets generated for this study are available on request to the corresponding author.

Acknowledgments: The authors would like to thank Universiti Putra Malaysia, Centro de Investigacion y Monitoreo Ambiental Antàrctico (CIMAA), Centro de Asuntos Antárticos del Ejército, the Antarctic General Bernardo O'Higgins Station staff 2018 especially the Comandante de la Base O'Higgins; Teniente Coronel Jose Ignacio Alvarado Camps, the Comandante de la sección de exploracion y rescate O'higgins; Capitan René Salgado Rebolledo and the staff; especially; Suboficial Juan David Sandoval Navarrete and Sargento Juan Eduardo Cortínez Padovani and Sultan Mizan Antarctic Research Foundation (YPASM).

Conflicts of Interest: The authors declare no conflict of interest.

References

1. Australian Antarctic Division. Pollution and Waste, Department of the Environmental and Energy. 2012. Available online: https://www.antarctica.gov.au (accessed on 30 December 2018).
2. Bharti, P.K.; Sharma, B.; Singh, R.K.; Tyagi, A.K. Waste generation and management in Antarctica. *Procedia Environ. Sci.* **2016**, *35*, 40–50. [CrossRef]
3. Kariminia, S.; Ahmad, S.S.; Hashim, R.; Ismail, Z. Environmental consequences of Antarctic tourism from a global prospective. *Procedia Soc. Behav. Sci.* **2013**, *105*, 781–791. [CrossRef]
4. Cripps, G.C.; Shears, J. The fate in the marine environment of a minor diesel fuel spill from an Antarctic research station. *Environ. Monit. Assess.* **1996**, *46*, 221–232. [CrossRef]
5. Aislabie, J.M.; Balks, M.R.; Fought, J.M.; Waterhouse, E.J. Hydrocarbon spills on Antarctic soils: Effects and management. *Environ. Sci. Technol.* **2004**, *38*, 1265–1274. [CrossRef] [PubMed]
6. Giudice, A.L.; Bruni, V.; De Domenico, M.; Michaud, L. *Psychrophiles-Cold-Adapted Hydrocarbons-Degrading Microorganisms*; Springer: Berlin/Heidelberg, Germany, 2010; pp. 1898–1921. [CrossRef]
7. Koshlaf, E.; Ball, A. Soil bioremediation approaches for petroleum hydrocarbon polluted environments. *AIMS Microbiol.* **2017**, *3*, 25–49. [CrossRef]
8. Miri, S.; Naghdi, M.; Rouissi, T.; Brar, S.K.; Martel, R. Recent biotechnological advances in petroleum hydrocarbons degradation under cold climate conditions: A review. *Crit. Rev. Environ. Sci. Technol.* **2018**, *49*, 553–586. [CrossRef]
9. Zahri, K.N.M.; Zulkharnain, A.; Fuentes, C.; Sabri, S.; Khalil, K.A.; Convey, P.; Ahmad, S.A. The use of response surface methodology as a statistical tool for the optimisation of the waste and pure canola oil biodegradation by Antarctic soil bacteria. *Life* **2021**, *11*, 456. [CrossRef]
10. Sakthiselvan, P.; Meenambiga, S.S.; Madhumathi, R. Kinetic Studies on cell growth, cell growth. *IntechOpen* **2019**, *13*. [CrossRef]
11. Pacwa-Plociniczak, M.; Plaza, G.A.; Piotrowska-Seget, Z.; Cameotra, S.S. Environmental applications of biosurfactants: Recent advances. *Int. J. Mol. Sci.* **2011**, *12*, 633–654. [CrossRef]

12. Kubicki, S.; Bollinger, A.; Katzke, N.; Jaeger, K.; Loeschcke, A.; Thies, S. Marine biosurfactants: Biosynthesis, structural diversity and biotechnology applications. *Mar. Drugs* **2019**, *17*, 408. [CrossRef]
13. Ibrahim, S.; Zulkharnain, A.; Zahri, K.N.M.; Lee, G.L.Y.; Convey, P.; Gomez-Fuentes, C.; Sabri, S.; Khalil, K.; Alias, S.; Gonzalez-Rocha, G.; et al. Effect of heavy metals and other xenobiotics on biodegradation of waste canola oil by cold-adapted *Rhodococcus* sp. strain AQ5-07. *Rev. Mex. Ing. Quím.* **2020**, *19*, 1041–1052. [CrossRef]
14. Zahri, K.N.M.; Zulkharnain, A.; Gomez-Fuentes, C.; Sabri, S.; Ahmad, S.A. Study of growth kinetics of Antarctic bacterial community for biodegradation of waste canola oil. *Desalin. Water Treat.* **2021**, *213*, 128–138. [CrossRef]
15. Monod, J. The growth of bacterial cultures. *Annu. Rev. Microbiol.* **1949**, *3*, 371–394. [CrossRef]
16. Haldane, J.S.B. *Enzymes*; Longmans, Green & Co, MIT Press: Cambridge, UK, 1930.
17. Edward, V.H. The influence of high substrate concentration on microbial kinetics. *Biotechnol. Bioeng.* **1970**, *12*, 679–712. [CrossRef]
18. Aiba, S.; Shoda, M.; Nagatani, M. Kinetics of product inhibition in alcohol fermentation. *Biotechnol. Bioeng.* **1968**, *10*, 845–864. [CrossRef]
19. Yano, T.; Koga, S. Dynamic behavior the chemostat subject to substrate inhibition. *Biotechnol. Bioeng.* **1969**, *11*, 139–153. [CrossRef]
20. Shahaby, A.F.; Alharthi, A.A.; El Tarras, A.E. Bioremediation of petroleum oil by potential biosurfactant-producing bacteria using gravimetric assay. *Int. J. Microbiol. Appl. Sci.* **2015**, *4*, 390–403.
21. Patowary, K.; Patowary, R.; Kalita, M.C.; Deka, S. Development of an efficient bacterial consortium for the potential remediation of hydrocarbons from contaminated sites. *Front. Microbiol.* **2016**, *14*, 1092. [CrossRef]
22. Priyadharshini, A.D.; Latha, D. Biosurfactant production from *Bacillus* sp. and its application in the medical field. *Int. Res. J. Pharm. Biosci.* **2016**, *3*, 1–8.
23. Nayarisseri, A.; Singh, P.; Singh, S.K. Screening, isolation and characterisation of biosurfactant producing *Bacillus subtilis* strain ANSKLAB03. *J. Biomed. Inform.* **2018**, *14*, 304–318. [CrossRef]
24. Rani, M.; Weadge, J.T.; Jabaji, S. Isolation and characterisation of biosurfactant-producing bacteria from oil well batteries with antimicrobial activities against food-borne and plant pathogens. *Front. Microbiol.* **2020**, *11*, 64. [CrossRef] [PubMed]
25. Chittepu, O.R. Isolation and characterisation of biosurfactant producing bacteria from groundnut oil cake dumping site for the control of foodborne pathogens. *Grain Oil Sci. Technol.* **2019**, *2*, 15–20. [CrossRef]
26. Sohail, R.; Jamil, N. Isolation of biosurfactant producing bacteria from Potwar oil fields: Effect of non-fossil fuel based carbon sources. *Green Process Synth.* **2020**, *9*, 77–86. [CrossRef]
27. Almeida, D.G.; Soares da Silva, R.C.F.; Luna, J.M.; Rufino, R.D.; Santos, V.A.; Sarubbo, L.A. Response surface methodology for optimising the production of biosurfactant by *Candida tropicalis* on industrial waste substrate. *Front. Microbiol.* **2017**, *8*, 157. [CrossRef]
28. Anandaraj, B.; Thivakaran, P. Isolation and production of biosurfactant producing organism from oil spilled oil. *J. Biosci. Technol.* **2010**, *1*, 120–126.
29. Maan, P.; Bharti, A.K.; Gautam, S.; Dutt, D. Screening of important factors for xylanase and cellulose production from the fungus *C. cinerea* RM-1 NFCCI-3086 through placket-Burman experimental design. *Bioresources* **2016**, *11*, 8268–8276. [CrossRef]
30. Sadhukhan, B.; Mondal, N.K.; Chattoraj, S. Optimisation using central composite design (CCD) and the desirability function for sorption of methylene blue from aqueous solution onto *Lemna major*. *Karbala Int. J. Modern Sci.* **2016**, *2*, 145–155. [CrossRef]
31. Chen, J.; Huang, P.T.; Zhang, K.Y.; Ding, F.R. Isolation of biosurfactant producers, optimisation and properties of biosurfactant produced by *Acinetobacter* sp. from petroleum-contaminated soil. *J. Appl. Microbiol.* **2012**, *112*, 660–671. [CrossRef]
32. Kumar, S.; Mathur, A.; Signh, V.; Nandy, S.; Khare, S.K.; Negi, S. Bioremediation of waste cooking oil using a novel lipase produced by *Penicillium chrysogenum* SNP5 grown in solid medium containing waste grease. *Biosour. Technol.* **2012**, *120*, 300–304. [CrossRef]
33. Malcolmson, L.; Vaisey-Genser, M. *Canola Oil: Performance Properties of Canola Oil*; Canola Council of Canada: Winnipeg, MB, Canada, 2001.
34. Choe, E.; Min, D.B. Chemistry of deep-fat frying oils. *J. Food Sci.* **2007**, *72*, 77–86. [CrossRef]
35. Przybylski, R.; Mag, T.; Eskin, N.A.M.; McDonald, B.E. *Canola Oil: Physical and Chemical Properties*; John Wiley and Sons Inc.: New York, NY, USA, 2005.
36. Batista, A.C.F.; Viera, A.T.; Rodrigues, H.S.; Silva, T.A.; Assumcao, R.M.N.; Beluomini, M.A.; Rezende, H.P.; Hernandez-Terrones, M.G. Production and physiochemical characterisation of methylic and ethylic biodiesel from canola oil. *Brazillian J. Biosyst, Eng.* **2014**, *8*, 289–298. [CrossRef]
37. Chipasa, K.B.; Medrzycka, K. Behavior of lipids in biological wastewaster treatment processes. *J. Ind. Microbiol. Biotechnol.* **2006**, *33*, 635–645. [CrossRef]
38. Sugimori, D.; Utsue, T. A study of the efficiency of edible oils degraded in alkaline conditions by *Pseudomonas aeruginosa* SS-219 and *Acinetobacter* sp. SS-192 bacteria isolated from Japanese soil. *World J. Microbiol. Biotechnol.* **2011**, *28*, 841–848. [CrossRef]
39. Affandi, I.Z.; Suratman, N.H.; Abdullah, S.; Ahamd, W.A.; Zakaria, Z.A. Degradation of oil and grease from high-strength industrial effluents using locally isolated aerobic biosurfactant-producing bacteria. *Int. Biodeter. Biodegr.* **2014**, *95*, 33–40. [CrossRef]
40. Ibrahim, S.; Zahri, K.N.M.; Convey, P.; Khalil, K.A.; Gomez-Fuentes, C.; Zulkharnain, A.; Alias, S.A.; Gonzalez-Rocha, G.; Ahmad, S.A. Optimisation of biodegradation conditions for waste canola oil by cold-adapted *Rhodococcus* sp. AQ5-07 from Antarctica. *Electron. J. Biotechnol.* **2020**, *48*, 1–12. [CrossRef]

41. Aggary, S.E.; Solomon, B.O.; Audu, T.O.K. Substrate utilization and inhibition kinetics: Batch degradation of phenol by indigenous monoculture of Pseudomonas aeruginosa. *Int. J. Biotechnol. Mol. Biol. Res.* **2010**, *1*, 22–30.
42. Aggary, S.E.; Solomon, B.O.; Layokun, S.K. Kinetics of batch microbial degradation of phenols by indigenous binary mixed culture of *Pseudomonas aeruginosa* and *Pseudomonas fluorescence*. *African J. Biotechnol.* **2008**, *7*, 2417–2423. [CrossRef]
43. Mpongwana, N.; Ntwampe, S.K.O.; Omodanisi, E.I.; Chisi, B.S.; Razanamahandry, L.C.; Dlangamandla, C.; Mukansi, M.R. Bio-kinetics of simultaneous nitrification and aerobic denitrification (SNaD) by a cyanide-degrading bacterium under cyanide-laden conditions. *Appl. Sci.* **2020**, *10*, 4823. [CrossRef]
44. Wang, X.; Wang, W.; Zhang, Y.; Zhang, J.; Li, J.; Wang, S.; Chen, G. Isolation and characterisation of *Acinetobacter* sp. JQ1004 and evaluation of its inhibitory kinetics by free ammonia. *Desalin. Water Treat.* **2019**, *147*, 316–325. [CrossRef]
45. Tazdait, D.; Abdi, N.; Grib, H.; Lounici, H.; Pauss, A.; Mameri, N. Comparison of different models of substrate inhibition in aerobic batch biodegradation of malathion. *Turkish J. Eng. Environ. Sci.* **2013**, *37*, 221–230. [CrossRef]
46. Morikawa, M.; Daido, H.; Takao, T.; Murata, S.; Shimonishi, Y.; Imanaka, T. A new lipopeptide biosurfactant produced by *Arthrobacter* sp. strain MIS38. *J. Bacteriol.* **1993**, *175*, 6459–6466. [CrossRef] [PubMed]
47. Adebajo, S.O.; Akintokun, A.K.; Balogun, S.A. Screening and characterisation of biosurfactant producing bacteria from soil samples in Ogun-State, Nigeria. *Microbiol. Res. J. Int.* **2017**, *18*, 1–12.
48. Shalel, S.; Streichman, S.; Marmur, A. The use of hemolysis kinetics to evaluate erythrocyte-bound surfactant. *Colloids Surf. B.* **2003**, *27*, 215–222. [CrossRef]
49. Shalel, S.; Streichman, S.; Marmur, A. The mechanism of hemolysis by surfactants: Effect of solution composition. *J. Colloid Interface Sci.* **2002**, *252*, 66–76. [CrossRef] [PubMed]
50. Manaargadoo-Catin, M.; Ali-Cherif, A.; Pougnas, J.; Perrin, C. Hemolysis by surfactants—A review. *Adv. Colloid Interface Sci.* **2016**, *228*, 1–16. [CrossRef] [PubMed]
51. Shoeb, E.; Ahmed, A.; Akhter, J.; Badar, U.; Siddqui, K.; Ansari, F.A.; Waqar, M.; Imtiza, S.; Akhtar, N.; Shaikh, Q.A.; et al. Screening and characterisation of biosurfactant-producing bacteria isolated from the Arabian Sea coast of Karachi. *Turkish J. Biol.* **2015**, *39*, 210–216. [CrossRef]
52. Jain, D.; Collins-Thompson, D.; Lee, H.; Trevors, J.T. A drop-collapsing test for screening surfactant-producing microorganisms. *J. Microbiol. Methods.* **1991**, *13*, 271–279. [CrossRef]
53. Mulligan, C.; Cooper, D.; Neufeld, R. Selection of microbes producing biosurfactants in media without hydrocarbons. *J. Fermentation Technol.* **1984**, *62*, 311–314.
54. Rosenberg, E.; DeLong, E.F.; Lory, S.; Stackebrandt, E.; Thompson, F. *The Prokaryotes: Applied Bacteriology and Biotechnology*; Springer: Berlin, Germany, 2013.
55. Kaczorek, E.; Pacholak, A.; Zdarta, A.; Smulek, W. The impact of biosurfactants on microbial cell properties leading to hydrocarbon bioavailability increase. *Colloids Interfaces* **2018**, *2*, 35. [CrossRef]
56. Thavasi, R.; Nambaru, V.R.M.S.; Jayalakshmi, S.; Balasubramanian, T.; Banat, I.M. Biosurfactant production by *Pseudomonas aeruginosa* from renewable resources. *Indian J. Microbiol.* **2011**, *51*, 30–36. [CrossRef]
57. Pendse, A.; Aruna, K. Use of various screening methods for isolation of potential biosurfactant producing microorganism from oil-contaminated soil samples. *J. Pharm. Res.* **2018**, *12*, 599–605.
58. Sumathi, R.; Yogananth, N. Isolation and identification of biosurfactant producing *Pseudomonas aeruginosa* from marine sediment samples and its antimicrobial properties. *Int. J. Adv. Res. Biol. Sci.* **2016**, *3*, 200–212.
59. Akintokun, A.K.; Adebajo, S.O.; Akinremi, C.A. Potential biosurfactant-producing bacteria from pharmaceutical wastewater using simple screening methods in south-west Nigeria. *App. Environ. Res.* **2017**, *39*, 41–54. [CrossRef]
60. Syahriansyah, U.K.M.; Hamzah, A. Determination of optimum conditions and stability study of biosurfactant produced by *Bacillus subtilis* UKMP-4M5. *Malaysian J. Anal. Sci.* **2016**, *20*, 986–1000.
61. Mouafo, T.H.; Mbawala, A.; Ndjouenkeu, R. Effect of different carbon sources on biosurfactants production by three strains of *Lactobacillus* spp. *Biomed. Res. Int.* **2018**, *2018*, 5034783. [CrossRef]
62. Ghasemi, A.; Moosavi-Nasab, M.; Setoodeh, P.; Mesbahi, G.; Yousefi, G. Biosurfactant production by lactic acid bacterium *Pediococcus dextrinicus* SHU1593 grown on different carbon source: Strain screening followed by product characterisation. *Sci. Rep.* **2019**, *9*, 5287. [CrossRef]
63. Walter, V.; Syldatk, C.; Hausmann, R. Screening concepts for the isolation of biosurfactant producing microorganisms. Screening concepts for the isolation of biosurfactant producing microorganisms. In *Madame Curie Bioscience Database*; Landes Bioscience: Austin, TX, USA, 2010. Available online: https://www.ncbi.nlm.nih.gov/books/NBK6189/ (accessed on 20 February 2021).
64. Amodu, O.S.; Ntwampe, S.K.; Ojumu, T.V. Emulsification of hydrocarbons by biosurfactant: Exclusive use of agrowaste. *Bioresources* **2014**, *9*, 3508–3525. [CrossRef]
65. Abdulrasheed, M.; Zakaria, N.N.; Roslee, A.F.A.; Shukor, M.Y.; Zulkharnain, A.; Napis, S.; Convey, P.; Alias, S.A.; Gonzalez-Rocha, G.; Ahmad, S.A. Biodegradation of diesel oil by cold-adapted bacterial strains of *Arthrobacter* spp. from Antarctica. *Antarct. Sci.* **2020**, *32*, 341–353. [CrossRef]
66. Elazzazy, A.M.; Abdelmoneim, T.S.; Almaghrabi, O.A. Isolation and characterisation of biosurfactant production under extreme environmental conditions by alkali-halo-thermophilic bacteria from Saudi Arabia. *Saudi J. Biol. Sci.* **2015**, *22*, 466–475. [CrossRef]
67. Khopade, A.; Biao, R.; Liu, X.; Mahadik, K.; Zhang, L.; Kokare, C. Production and stability studies of the biosurfactant isolated from marine *Nocardiopsis* sp. B4. *Desalination* **2012**, *285*, 198–204. [CrossRef]

68. Rodrigues, L.; Banat, I.M.; Teixeira, J.; Oliveira, R. Biosurfactants: Potential applications in medicine. *J. Antimicrob. Chemother.* **2006**, *57*, 609–618. [CrossRef] [PubMed]
69. Arora, J.S. *Chapter 20—Additional Topics on OPTIMUM Design*; Academic Press; Elsevier Inc.: Iowa City, IA, USA, 2012.
70. Souza, E.C.; de Azevedo, P.O.S.; Dominguez, J.M.; Converti, A.; Oliveira, R.P.S. Influence of temperature and pH on the production of biosurfactant, bacteriocin and lactic acid by *Lactococcus lactis* CECT-4434. *J. Food* **2017**, *15*, 525–530. [CrossRef]
71. Ikhwani, A.Z.N.; Nurlaila, H.S.; Ferdinand, F.D.K.; Fachria, R.; Hasan, A.E.Z.; Yani, M.; Setyawati, I. Preliminary study: Optimisation of pH and salinity for biosurfactant production from *Pseudomonas aeruginosa* in diesel fuel and crude oil medium. In Proceedings of the IOP Conference Series: Earth and Environmental Science, Bogor, Indonesia, 4 November 2016; IOP Publishing: Bristol, UK, 2017; Volume 58, p. 012056. [CrossRef]
72. Hamzah, A.; Sabturani, N.; Radiman, S. Screening and optimisation of biosurfactant production by the hydrocarbon-degrading bacteria. *Sains Malays.* **2013**, *42*, 615–623.
73. Ahmad, Z.; Zhang, X.; Imran, M.; Zhong, H.; Andleeb, S.; Zulekha, R.; Liu, G.; Ahmad, I.; Coulon, F. Production, functional stability, and effect of rhamnolipid biosurfactant from *Klebsilla* sp. on phenanthrene degradation in various medium system. *Ecotoxicol. Environ. Safety* **2021**, *207*, 111514. [CrossRef] [PubMed]
74. Yan, N.; Marschner, P.; Cao, W.; Zuo, C.; Qin, W. Influence of salinity and water content on soil microorganisms. *Int. Soil Water Conserv. Res.* **2015**, *3*, 316–323. [CrossRef]
75. Dhasayan, A.; Selvin, J.; Kiran, S. Biosurfactant production from marine bacteria associated with sponge *Callyspongia diffusa*. *3 Biotech* **2015**, *5*, 443–454. [CrossRef]
76. Hisham, N.H.M.B.; Ibrahim, M.F.; Ramli, N.; Abd-Aziz, S. Production of biosurfactant produced from used cooking oil by *Bacillus* sp., HIP3 for heavy metals removal. *Molecules* **2019**, *24*, 2617. [CrossRef]
77. Roslee, A.F.A.; Zakaria, N.N.; Convey, P.; Zulkharnain, A.; Lee, G.L.Y.; Gomez-Fuentes, C.; Ahmad, S.A. Statistical optimisation of growth conditions and diesel degradation by the Antarctic bacterium, *Rhodococcus* sp. strain AQ5-07. *Extremophiles* **2020**, *24*, 277–291. [CrossRef]
78. Phulpoto, I.A.; Yu, Z.; Hu, B.; Wang, Y.; Ndayisenga, F.; Li, J.; Liang, H.; Qazi, M.A. Production and characterisation of surfactin-like biosurfactant produced by novel strain *Bacillus nealsonii* S2MT and its potential for oil contaminated soil remediation. *Microbial Cell Fact.* **2020**, *19*, 145. [CrossRef]
79. Kemmitt, S.J.; Wright, D.; Goulding, K.W.T.; Jones, D.L. pH regulation of carbon and nitrogen dynamics in two agricultural soils. *Soil Biol. Biochem.* **2006**, *38*, 898–911. [CrossRef]
80. Jin, Q.; Kirk, M.F. pH as a primary control in environmental microbiology: 1. Thermodynamic perspective. *Front. Environ. Sci.* **2018**, *6*, 21. [CrossRef]
81. Banat, I.M.; Satpute, S.K.; Cameotra, S.S.; Patil, R.; Nyayanit, N.V. Cost effective technologies and renewable substrates for biosurfactants' production. *Front. Microbiol.* **2014**, *5*, 697. [CrossRef]
82. Mubarak, M.Q.E.; Jufri, S.H.M.; Zahar, S.M.N.S.; Kalil, M.S.; Hamid, A.A.; Isa, M.H.M. Kinetics of surfactin production by *Bacillus subtilis* in a 5 L stirred-tan bioreactor. *Sains Malays.* **2017**, *46*, 1541–1548. [CrossRef]
83. Heryani, H.; Putra, M.D. Kinetic study and modelling of biosurfactant production using *Bacillus* sp. *Electron. J. Biotechnol.* **2017**, *27*, 49–54. [CrossRef]
84. Karlapudi, A.P.; Venkateswarulu, T.C.; Jahnavi, T.; Lohit, K.; Bharath, K.R.; Vijaya, R.D.; Vidya, P. Role of biosurfactants in bioremediation of oil pollution-a review. *Petroleum* **2018**, *4*, 241–249. [CrossRef]
85. Bognolo, G. Biosurfactants as emulsifying agents for hydrocarbons. *Colloids Surf. A Physicochem. Eng. Asp.* **1999**, *152*, 41–52. [CrossRef]
86. Bai, N.; Wang, S.; Abuduaini, R.; Zhang, M.; Zhu, X.; Zhao, Y. Rhamnolipid-aided biodegradation of carbendazim by *Rhodococcus* sp. D-1: Characteristics, products, and phytotoxicity. *Sci. Total Environ.* **2017**, *590–591*, 343–351. [CrossRef]

foods

MDPI

Article

Potential Activity of Abrantes Pollen Extract: Biochemical and Cellular Model Studies

Ana C. Gonçalves [1], Radhia Aitfella Lahlou [2,3,4], Gilberto Alves [1], Cristina Garcia-Viguera [5], Diego A. Moreno [5,*] and Luís R. Silva [1,6,*]

1 CICS–UBI–Health Sciences Research Centre, Department of Medical Science, University of Beira Interior, 6201-506 Covilhã, Portugal; anacarolinagoncalves@sapo.pt (A.C.G.); gilberto@fcsaude.ubi.pt (G.A.)
2 Chemistry Department, University of Beira Interior, 6201-001 Covilhã, Portugal; ra.aitfella@ubi.pt
3 Fiber Materials and Environmental Technologies (FibEnTech), University of Beira Interior, 6201-001 Covilhã, Portugal
4 Laboratory of Valorisation and Conservation of Biological Resources, Biology Department, Faculty of Sciences, University M'Hamed Bougara, Boumerdes 35000, Algeria
5 Phytochemistry and Healthy Foods Laboratory, Food Science and Technology Department, CEBAS-CSIC, Campus Universitario de Espinardo, 30100 Espinardo, Murcia, Spain; cgviguera@cebas.csic.es
6 CPIRN-UDI-IPG–Research Unit for Inland Development, Center for Potential and Innovation of Natural Resources, Polythecnic of Guarda, 6300-559 Guarda, Portugal
* Correspondence: dmoreno@cebas.csic.es (D.A.M.); luisfarmacognosia@gmail.com (L.R.S.)

Abstract: The aim of this study was to determine the grain composition and (poly)phenolic profile of pollen from Abrantes (Portugal), as well as its antioxidative and antidiabetic properties, and abilities to protect human erythrocytes against induced hemoglobin oxidation, lipid peroxidation, and hemolysis. The phytochemical profile of the Abrantes' bee pollen revealed twenty phenolic compounds, identified by high-performance liquid chromatography with electrospray ionization mass spectrometry coupled with photodiode array detection. Among them, quercetin derivatives were the most abundant. Concerning the biological potential, the pollen extract showed notable capacity for 2,2-diphenyl-1-picrylhydrazyl, nitric oxide, and superoxide radicals, as well as for inhibition of α-glucosidase action, and protection of human erythrocytes against oxidative damage. Non-cytotoxic effects regarding the NHDF normal cell line, human adenocarcinoma Caco-2, and human liver HepG2 cells were observed. The results obtained contributed to further research on modes of action related to oxidative damage and metabolic health problems, to generate deeper knowledge of potential health-promoting effects to develop novel pharmaceutical drugs, nutraceuticals, and dietary supplements.

Keywords: bee pollen; phenolic characterization; biological potential; functional food; antioxidant effects

1. Introduction

Nowadays, there is an interest in more balanced, functional, and healthier food. In this context, natural plants and their derivatives have received a lot of attention owing to their health-promoting properties with little or no adverse effects [1]. Bee pollen, a hive-derived product, is one of these products that is gaining more attention [2]. In fact, it has been used since early times in complementary and folk medicine to attenuate several pathological conditions, such as anemia, allergies, colds, colitis, enteritis, flu, and ulcers [3]. It consists of flower pollen collected from various selected botanical sources by the honeybee *Apis mellifera*, with the purpose of feeding larvae in their early stages of development [4,5].

Bee pollen is constituted of about 200 biologically active substances, collected from plants and transported to the hive in the form of pollen loads [2]. The general composition of pollens roughly consists of polysaccharides (50%), simple sugars (mainly fructose and glucose at percentages varying from 4 to 10%), fats and lipids (1–20%), proteins (6–28%), and amino acids (6%), along with several secondary plant metabolites such as phenolic compounds (flavonoids and phenolic acids), which are the main contributors of their color and bitter taste, carotenoids, and terpenes [2]. Additionally, pollen constituents can be useful chemotaxonomic markers, as well as quality indicators [5].

Regarding phenolics, pollen has higher concentrations of flavonols, namely quercetin derivatives, with values ranging from 24.0 (quercetin) to 956 μg/g (quercetin 3-*O*-rutinoside) [6]. Furthermore, it is also high in several essential elements, including calcium, copper, magnesium, iron, and zinc [7]. In fact, bee pollen has been widely acclaimed owing to its nutritional and therapeutic properties, and is already commercialized as a functional food and recognized as a medicinal product by the German Federal Board of Health [2,8]. In general, it can be consumed fresh, dried, or incorporated in beverages, tonics, bars, granules, tablets, and nutraceuticals [2]. Its notable antioxidant, anti-inflammatory, antimicrobial, antifungal, antiviral, anti-mutagenic, and anti-allergic potential are mainly attributed to the presence of phenolics, which are considered the main active compounds in this foodstuff [3].

Therefore, given the current interest in pollen and the existence of few studies on this natural product, the aim of this work was to determine pollen grain composition and the phenolic profile of multifloral bee pollen from Abrantes (Portugal). This city is located in the province of Ribatejo and belongs to Santarém district. Considering its favorable edaphoclimatic conditions and flora variety (e.g., *Castanea sativa, Cytisus, Echium, Quercus* and *Rubus* plants) which confer to this regional honey unique characteristics, including a dark color and intense flavor, while not being too sweet, it is not surprising that honey production is increasing in this region [9,10]. Furthermore, its biological potential was also evaluated, namely the capacity to scavenge 2,2-diphenyl-1-picrylhydrazyl (DPPH$^\bullet$), nitric oxide ($^\bullet$NO), and superoxide ($O_2^{\bullet-}$) radicals and to inhibit the α-glucosidase enzyme. Additionally, its protective effects on different cellular models were also studied: (1) human erythrocytes, studying hemoglobin oxidation, lipid peroxidation, and hemolysis induced by peroxyl species (ROO$^\bullet$); (2) human colon carcinoma cells (Caco-2); and (3) liver carcinoma cells (HepG2). Erythrocytes were selected not only for being the most abundant cells in vertebrates, but also for their crucial role in immune reactions and respiratory gas exchange, which in turn, makes them highly susceptible targets to oxidation and inflammatory conditions. On the other hand, Caco-2 and HepG2 cancer cells were chosen as they present many morphological characteristics of intestinal epithelium and liver parenchymal cells, respectively. For comparative purposes, the effects on primary normal human dermal fibroblasts (NHDF) were studied.

2. Materials and Methods

2.1. Chemical Reagents

All chemicals used were of analytical grade.

4-Nitrophenyl-alpha-D-glucopyranoside (pNPG), N-(1-naphthyl)ethylenediamine dihydrochloride, sodium nitroprusside dihydrate (SNP), and sulfanilamide were acquired from Alfa Aesar (Karlsruhe, Germany). Acetonitrile and methanol were purchased from Fisher Chemical (Leicestershire, UK). The remaining ones were from Sigma-Aldrich (St. Louis, MO, USA). Water was deionized using a Milli-Q water purification system (Millipore Ibérica, S.A.U., Madrid). Caco-2, HepG2 and NHDF cell lines were from American Type Culture Collection (Manassas, VA, USA).

2.2. Pollen Samples

The bee pollen sample (REF: M08AG006), composed of at least 500 pollen grains, was directly provided by a local company (Colmeicentro) in Alferrarede, Abrantes region, in 2017. The sample was stored in the dark under desiccating conditions to avoid alterations until use.

2.3. Pollen Analysis

The botanical origin of the sample was determined according to the melissopalynological method described by Louveaux, Maurizio, & Vorwohl [11]. Briefly, 10 g of the sample was diluted with bidistilled water (50 mL) and centrifuged at 3900× *g* for 10 min in order to separate the pollen grains. Then, the obtained sediment was again dissolved in water and centrifuged for 5 min. Finally, two aliquots of the sediment were examined microscopically at 45×, using a bright-field microscope (Olympus, Tokyo). Each aliquot was composed of a minimum of 800 pollen grains. The results were expressed as percentages.

2.4. Pollen Extract Preparation

The extract was prepared according to Moita et al. [3]. Briefly, 0.2 g of bee pollen were thoroughly mixed in 1 mL of ethanol:water (70:30, *v/v*), ultrasonicated for 1 h and centrifuged at 2900× *g* during 10 min at room temperature. Then, the supernatant was evaporated under reduced pressure to complete dryness at 40 °C. The resulting concentrate residue was stored at −20 °C, and protected from light until use. The obtained extraction yield from the starting dry material was 64.5 ± 0.16%. The extractions were performed in triplicate.

2.5. Identification of Phenolic Compounds via HPLC-DAD-ESI/MS^n

The identification of phenolic compounds from bee pollen was performed according to Gonçalves et al. [12]. They were tentatively identified based on their ultraviolet-visible and mass spectra features, elution order, and retention times as compared to authentic standards analyzed the under same conditions (Table 1), and also with data available in the literature [12–16]. Injections were performed in triplicate.

Table 1. Retention time (Rt), wavelengths of maximum absorption in the visible region (λ_{max}) used for quantification, mass fragmentation data, and tentative identification of quantified peaks of compounds (μg/g of dry weight) in pollen.

Peak	Compounds Identification	HPLC-DAD-ESI-MS^n Data				Quantification
		R_t (min)	λ_{max} (nm)	Molecular Ion [M-H] (*m/z*)	Fragments MS/MS (*m/z*)	
1	Caffeoyl di-hexoside	10.0	320	635	341, 179	nq
2	Coumaroyl hexose	13.8	320	325	145, 163, 205, 235	nq
3	Caffeoyl hexose	14.0	320	341	179, 135	0.056 ± 0.0057
4	Quercetin 7-glucoside-3-*O*-rutinoside	24.0	350	771	609, 301	0.35 ± 0.020
5	Kaempferol di-hexoside	24.3	350	609	447, 285	nq
6	Myricetin rhamno-hexoside	25.0	350	625	316, 271, 287	nq
7	Quercetin 3-*O*-rutinoside	25.9	350	609	271, 301	0.76 ± 0.037
8	Kaempferol 3-*O*-rutinoside-*O*-hexoside	26.2	350	755	593, 447, 285	nq
9	Quercetin derivative 1	26.3	350	639	314, 301, 150	0.49 ± 0.031
10	Myricetin derivative	26.4	350	521	316; 271	nq
11	Isorhamnetin 3-*O*-rutinoside 1	26.8	350	623	315	nq
12	Kaempferol 3-*O*-rutinoside	27.3	350	447	285, 256	nq
13	Quercetin 3-*O*-glucoside	28.4	350	463	300/301, 271	nq
14	Isorhamnetin 3-*O*-rutinoside 2	28.5	350	623	315	nq
15	Quercetin derivative 2	28.6	350	609	315, 300, 271	nq
16	Quercetin acetyl rhamnoside	29.3	350	505	463, 301	1.33 ± 0.022

Table 1. *Cont.*

Peak	Compounds Identification	HPLC-DAD-ESI-MS^n Data				Quantification
		R_t (min)	λ_{max} (nm)	Molecular Ion [M-H] (m/z)	Fragments MS/MS (m/z)	
17	Isorhamnetin acetyl hexoside	31.7	350	519	315	0.82 ± 0.013
18	Kaempferol acetyl hexoside	31.9	350	447	235	nq
19	Kaempferol hexoside	32.6	350	489	447, 285	nq
20	Quercetin hexoside	35.0	350	463	301	0.36 ± 0.032
		Σ				4.17

Values are expressed as mean ± standard deviation of three assays. Σ, sum of the determined phenolic compounds; nq, not quantified.

2.6. *Quantification of Phenolic Compounds via HPLC-DAD*

Concerning phenolics quantification, 20 μL extract was injected on an LC model Agilent 1260 system (Agilent, Santa Clara, CA, USA), according to Gonçalves et al. [5]. Spectral data from all peaks were accumulated in a range varying from 200 to 600 nm and chromatograms were recorded at 320 and 350 nm for hydroxycinnamic acids and flavonols, respectively. Compounds were identified by comparing their retention times and ultraviolet-visible spectra with those of authentic standards. Peak purity was checked by software contrast facilities. Caffeoyl di-hexose and caffeoyl hexose were quantified as caffeic acid, and coumaroyl hexose as *ρ*-coumaric acid. Quercetin 7-glucoside-3-*O*-rutinoside, quercetin derivatives 1 and 2, quercetin hexoside and quercetin acethylrhamnoside were quantified as quercetin. Kaempferol di-hexoside, kaempferol 3-*O*-rutinoside-*O*-heoxide, kaempferol acethylhexoside, myrcetin rhamno-hexoside, and myrcetin derivative were quantified as kaempferol 3-*O*-rutinoside. On the other hand, isorhamnetin 3-*O*-rutinoside and isorhamnetin acetyl hexoside were quantified as quercetin. Analyses were carried out in triplicate.

2.7. *Antioxidant Capacity Experiments*

The capacity of bee pollen hydroethanolic extracts to scavenge free radicals, namely DPPH•, •NO and $O_2^{\bullet -}$, was performed spectrophotometrically through in vitro microassays using 96-well plates. Microplate reader Bio-Rad Xmark spectrophotometer (Bio-Rad Laboratories, Hercules, CA, USA) was used to measure the absorbances. The results were expressed as 25% and 50% maximal inhibitory concentration (IC_{25} and IC_{50}, respectively) values (μg/mL). Each experiment was performed in triplicate, and seven different concentrations were tested.

2.7.1. 1,1-Diphenyl-2-Picrylhydrazyl Radical (DPPH•) Assay

The capability of bee pollen hydroethanolic extract to quench DPPH• was performed according to Gonçalves et al. [1]. Each well contained 25 μL extract dissolved in methanol and 200 μL methanolic DPPH (150 mM). Control was composed of replacing the sample with water, while blank was composed of water, or diluted extract, and methanol. The absorbances were taken at 515 nm. Ascorbic acid was used as positive control.

2.7.2. Nitric Oxide Radical (•NO) Assay

The effects of bee pollen extracts in capturing •NO were based on the work of Silva & Teixeira [17]. Briefly, each well was composed of 100 μL each extract dissolved in potassium phosphate buffer (100 mM, pH 7.4) and 100 μL of sodium nitroprusside dihydrate (20 mM). Blanks and controls contained phosphate buffer and sodium nitroprusside dihydrate. The plates were incubated at room temperature for one hour, under light. Then, 100 μL Griess reagent (1% sulfanilamide and 0.1% naphthylethylenediamine in 2% H_3PO_4) was added to each well and incubated for 10 min in a dark (blanks received 100 μL H_3PO_4). After this time, the absorbance was read at 560 nm. Ascorbic acid was used as positive control.

2.7.3. Superoxide Radical ($O_2^{\bullet-}$) Assay

The possibility of pollen extracts being able to reduce $O_2^{\bullet-}$ generated by the NADH/ phenazine methosulfate system was done according to a previous method [1]. Each well contained 50 μL each diluted extract dissolved in potassium phosphate buffer (19 mM, pH 7.4), 50 μL β-nicotinamide adenine dinucleotide (NADH), 150 μL nitrotetrazolium blue chloride (NBT), and 50 μL phenazine methosulfate (PMS). The control was composed of phosphate buffer instead of diluted extract. After the addition of PMS, each plate was immediately collocated in the spectrophotometer and absorbance readings were collected every 10 seg over 2 min at 562 nm. Ascorbic acid was used as a positive control.

2.8. α-Glucosidase Inhibitory Activity

The inhibition of α-glucosidase was determined based on a previous method [17]. Seven different concentrations were done. Each well contained 150 μL potassium phosphate buffer (19 mM, pH 7.4), 50 μL each concentration dissolved in buffer, and 100 μL 4-nitrophenyl-α-D-glucopyranoside (PNP-G). Control was only composed of phosphate buffer and PNP-G. The reaction was initiated by the addition of 25 μL α-glucosidase in each well, followed by a period of incubation of 10 min of incubation at 37 °C. Finally, and after that time, the absorbance of 4-nitrophenol released from PNP-G was read at 405 nm. Acarbose was used as positive control. Three experiments were performed in triplicate.

2.9. Biological Activity on Cellular Models

2.9.1. Isolation of Human Erythrocytes

Venous human blood was collected from randomized patients of Cova da Beira Hospital Centre (Covilhã, Portugal) by antecubital venipuncture into K_3EDTA vacuum tubes, in accordance with the protocol approved by the Health Ethical Commission of Cova da Beira Hospital Centre (no. 25/2018) without the need for written informed consent of the patients. Therefore, erythrocytes from blood collected in the early morning already processed by the hospital and without any need for more studies. Then, they were isolated according to Gonçalves et al. [1]. Briefly, 4 mL of a collected sample was transferred to sterile conic tubes, mixed with 6 mL PBS at pH 7.4 and centrifuged at 1500× *g* for 5 min at 4 °C. This procedure was repeated three more times, and the resulting supernatant from each centrifugation was always discarded. The number of cells (cells/mL) and viability (always above 98%) were obtained by the Trypan blue exclusion method. The suspensions of isolated erythrocytes were kept on ice until use.

2.9.2. Peroxyl Radical (ROO$^{\bullet}$)-Induced Oxidative Damage in Human Erythrocytes

For evaluating the protective effects of pollen hydroethanolic extract against ROO$^{\bullet}$ species in human erythrocytes, six different concentrations were prepared and dissolved in PBS. Five experiments were performed in duplicate in 48-well microplates.

2.9.3. Inhibition of Hemoglobin Oxidation

The capacity of pollen extracts to prevent methemoglobin oxidation induced by 2,2′-azobis(2-amidinopropane) dihydrochloride (AAPH) was based on a previous work [1]. Six dilutions of the extract were prepared in PBS. The reaction was composed of 100 μL sample solution and 200 μL erythrocytes solution (1250 × 10^6 cells/mL, final density). Controls and blanks were performed by replacing the sample with 100 μL PBS. The mixture was then incubated in a water bath for 30 min at 37 °C, under slow agitation. After the incubation, 200 μL AAPH (50 mM, final concentration) prepared in PBS were added to the media (except in the blank), followed by another incubation using the same conditions described above for 4 h. After that time, the entire volume of the reaction mixture was transferred to conic Eppendorf tubes and centrifuged at 1500× g for 5 min at 4 °C. Finally, 300 µL supernatant was placed in a 96-well plate and the absorbance was measured at 630 nm. Quercetin was used as a positive control.

2.9.4. Inhibition of Lipid Peroxidation

The ability of the extract to avoid lipid peroxidation was indirectly monitored through the formation of thiobarbituric-acid-reactive substances (TBARS) [18]. Six different concentrations were prepared in PBS. First, 100 µL of each concentration was mixed with 200 µL of erythrocyte suspension (500 × 10^6 cells/mL, final density) and incubated at 37 °C for 30 min, under slow agitation. After this period, 200 µL *tert*-BHP (0.2 mM, final concentration) was added to the media, and each well was further incubated under the same conditions for 30 min. Finally, the entire content was transferred to conic Eppendorfs, mixed with 250 µL trichloroacetic acid 28% (*w/v*) to incite protein precipitation, and centrifuged at 16,000× g for 10 min at 18 °C. Then, the supernatant was collected in conical test tubes (with screw caps) and 125 µL thiobarbituric acid 1% (*w/v*) was added to generate TBARS. The resulting mixture was heated for 15 min at 100 °C in a water bath. Finally, test tubes were cooled in ice and the absorbance was taken at 532 nm. Quercetin was used as a positive control.

2.9.5. Inhibition of Hemolysis

The capacity of pollen extracts to avoid erythrocyte lysis induced by AAPH was evaluated by monitoring the release of hemoglobin to the media after membrane disruption, according to a previous work [19]. Briefly, six different dilutions were prepared in PBS. The mixture consisted of a 200 µL suspension of human erythrocytes (1775 × 10^6 cells/mL, final density) and 100 µL of each concentration. Controls and blanks were composed of PBS. In a first step, the reaction mixture was incubated at 37 °C in a water bath for 30 min, under slow agitation. After this period, 200 µL AAPH (17 mM, final concentration) was added to the mixture (except in the blank), followed by another period of incubation for 3 h, using the aforementioned conditions. Lastly, the entire volume was then transferred to a conic Eppendorf and centrifuged at 1500× g, for 5 min at 4 °C. After that, 300 µL supernatant was transferred to 96-well plates and the absorbance was read at 540 nm. Quercetin was used as a positive control.

2.10. Cancer Cell Models

2.10.1. Cell Culture Conditions and Treatments

To understand the possible cytotoxicity effects, Caco-2, HepG2, and NHDF cell lines were cultured in 75 cm^2 culture flasks and incubated at 37 °C in a humidified atmosphere of 5% CO_2. Caco-2 and HepG2 cells were cultured as a monolayer in DMEM containing 10% FBS, 1% penicillin/streptomycin, while NHDF cells were cultured in RPMI 1640 medium supplemented with FBS (10%), l-glutamine (2 mM), 4-(2-hydroxyethyl)-1-piperazineethanesulfonic acid (10 mM), sodium pyruvate (1 mM), and penicillin/ streptomycin (1%). The medium was changed every 2 days. After a few passages, cells were washed twice with PBS and detached by gentle trypsinization. Then, viable cells were counted and suitably diluted in the adequate complete culture medium (25,000 cells/mL for Caco-2 cells, and 10,000 cells/mL for HepG2 and NHDF cells) [1,20]. After counting the cells, 200 µL of prepared cellular suspension was seeded in 96-well plates and incubated for one day before carrying out the viability assays. After 24 h of incubation, 200 µL of different concentrations of pollen extracts were dissolved in medium containing 0.5% (*v/v*) DMSO (the final concentration of DMSO did not affect cellular viability (data not shown)) and plates were again incubated for a further 24 h. At the end of this period, the metabolic activity of cells was evaluated concerning their ability to reduce yellow tetrazolium 3-(4,5-dimethylthiazol-2-yl)-2,5-diphenyltetrazolium bromide (MTT) to a blue formazan product and to ensure membrane integrity. Cell culture conditions and procedures were consistent through all assays. All studies were conducted when cells were in the logarithmic growth phase.

2.10.2. MTT Assay

After 24 h of cell treatment with different concentrations of pollen extracts, the medium was rejected, and each well was washed with PBS. Then, cultures were incubated with 200 μL MTT solution (dissolved in appropriate serum-free medium) for 4 h at 37 °C. After this period, the medium containing MTT was removed, and the purple crystals of formazan were dissolved in DMSO. The absorbance of the different solutions was read at 570 nm. Cell proliferation values were expressed as percentages from the relative absorbance measured in the treated wells versus control wells [21]. A total of six independent experiments per extract were performed. Untreated cells were used as control.

2.10.3. Membrane Integrity Assay

In order to evaluate the release of the stable cytosolic enzyme lactate dehydrogenase (LDH) into the medium, after 24 h of cells incubation with the extracts, 50 μL culture medium was placed to a 96-well plate. Then, 200 μL NADH (252.84 mM) and 25 μL pyruvate (14.99 mM) were added to each well, according to Gonçalves et al. [1]. Both pyruvate and NADH solutions were prepared in PBS (pH 7.4). The LDH released was determined at 340 nm due to NADH oxidation during the conversion of pyruvate to lactate, in a kinetic mode. A decrease in absorbance is directly related to the quantity of LDH released by the cells in the culture environment. A total of six independent experiments were performed. Untreated cells were used as control.

2.11. Statistical Analysis

Statistical comparison was determined using one-way ANOVA and the means were classified by Tukey's test at a 95% level of significance. Differences were considered significant for $p < 0.05$. To determine the contribution of the total phenols, on their antioxidant activity, Pearson's correlation coefficients were calculated. All analyses were carried out using Graph Pad Prism Version 6.01 (GraphPad Software, Inc., San Diego, CA, USA).

3. Results and Discussion

3.1. Pollen Composition

As expected, grains from different plant species were identified, which allows the pollen sample to be classified as heterofloral [8] (Figure 1). Particularly, thirteen different grain types were distinguished in the studied sample. Among them, *Echium plantagineum*, *Cistus ladanifer* and *Quercus* species were the predominant ones, with relative abundance percentages of 26, 23, and 19%, respectively. Other pollen types detected include *Rubus* spp., *Eriobotrya* spp., *Raphanus raphanistrum*, *Crepis capillaris*, *Anthyllis* spp., *Trifolium* spp., *Cytisus* spp., *Lavandula stoechas*, *Calendula arvensis*, *Salix* spp. And *Lotus criticus*. The obtained results are in line with the wide plant diversity observed in the Abrantes region. In fact, this area has a mixed forest of *Echium*, *Castanea sativa* and *Quercus* spp., Leguminosae scrubs (*Cytisus* spp.), and *Rubus* plants, which often colonize abandoned agricultural fields [9]. Additionally, the presence of these grain types would contribute to the dark blue color exhibited by honeys from this region [10]. These pollen grains have also been identified in other Portuguese samples, but at different proportions, depending on the region [2,5,9]. Particularly, honeys from Trás-os-Montes and Serra de Aire e Candeeiros are also richer in *Echium plantagineum* pollen grains [9,22], while honeys from Castelo Branco are mainly composed of *Lavandula* and *Echium* types (93.75 and 81.25%, respectively) [10]. On the other hand, Caramulo and Luso regions showed higher amounts of *Eucalyptus* spp. [5,23], while *C. sativa* is the most abundant grain type in Bragança samples [24]. Regarding other countries, *Rubus*, *C. sativa*, and *Cytisus* species are the predominant pollen grains in Spanish honeys [25], while Brazilian ones mainly consist of *Arecaceae, Asteraceae* spp., and *Brassicaceae* grains [8]. In addition to those already mentioned, it is important to note pollen composition is not only strongly influenced by floral species, but also by growth conditions and cultivation techniques [8]. On the other hand, its quality mainly depends on the harvesting, cleanliness, dryness, and storage procedures applied by beekeepers to increase shelf-life [2].

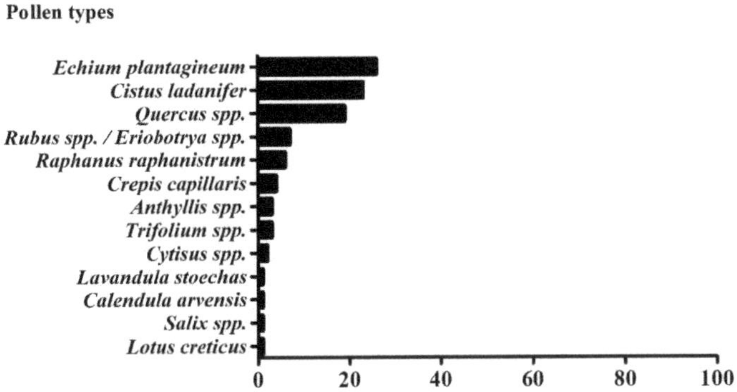

Figure 1. Frequency of the different pollen types in the investigated sample.

3.2. *Phenolic Characterization*

In this study, twenty phenolic compounds were tentatively identified by the interpretation of their fragmentation patterns obtained from mass spectra and by comparison with literature. Particularly, three hydroxycinnamic acids and seventeen flavonols were detected, as listed in Table 1. The total amount was 4.17 μg/g of dry weight (fw). As far as we know, this is the first study concerning the phenolic profile of bee-collected pollen from the Abrantes region (Portugal). Phenolic presence contributes to bee products appearance and functional properties, boosting antioxidant, anti-mutagenic, antimicrobial, and anti-inflammatory effects [5,26]. Their levels are strongly influenced by floral sources, geographical origin, bee species, and environmental factors, such as temperature, humidity, soil type, and storage time [7].

3.2.1. Hydroxycinnamic Acids

Three hydroxycinnamic acids were found (Table 1, peaks 1–3). Regarding caffeoyl-derivatives, two signal peaks were detected (peaks 1 and 3), accounting for less than 1.34% of total phenolic compounds identified. Both showed characteristic fragments at *m/z* 179, 342 and 135, which indicates the presence of a caffeic acid and the consequent decarboxylation of this moiety [12,15]. As far as we know, this is the first time that these compounds have been reported in bee-pollen samples. Even so, the presence of caffeic acid was already reported in pollen (1.05 μg/mL of dw) [14] and honey (ca. 2.42 μg/mL of dw) [5,9,25].

Additionally, one coumaroyl hexose was found (peak 2), exhibiting a molecular ion at *m/z* 325 and a fragment pattern at *m/z* 163, which is characteristic of coumaroyl compounds [12,16]. Other researchers already reported the presence of coumaroyl acids in pollen and honey samples at residual amounts [9,14,16].

In accordance with the literature, it is possible to state that hydroxycinnamic acids, even in vestigial amounts, increase the antioxidant potential of bee-pollen products, essentially due to their CH = CH-COOH group [27].

3.2.2. Flavonol Glycosides

Seventeen different flavonol glycosides were analyzed, including seven quercetin glycosides (peaks 4, 7, 9, 13, 15, 16 and 20), five kaempferols (peaks 5, 8, 12, 18 and 19), two myricetins (peaks 6 and 10), and three glycosides of isorhamnetin (peaks 11, 14 and 17).

Seven isomers of quercetin were identified. The majority displayed molecular ions at m/z 463 or 609 and fragment ion at m/z 300/301, which corresponds to quercetin aglycon losses [12–16]. These compounds were the most abundant phenolic compound found in the studied sample, which is in line with previous reports [8,13,14,16]. Trace amounts of quercetin derivatives were also found in honey samples [13].

Additionally, five kaempferol derivatives were found. Most of them showed molecular ions and fragment ions at m/z 447 and 285, respectively, which are characteristics of this compound, as previously reported [12,14–16]. The presence of kaempferol derivatives in bee pollens is in agreement with existing literature [8,14,16]. Vestigial amounts of kaempferol were also previously detected in honey samples [5,25].

Two myricetin compounds were also identified. Both showed fragment ions at m/z 316 and 271, which suggest a myricetin aglycon losses [15]. Myricetin was also found in other bee-pollen products at vestigial amounts (0.04 μg/g of dw) [8,14] and also in honey of *Castanea sativa* [25].

Finally, three forms of isorhamnetin were detected, exhibiting a molecular peak at m/z 623 (519 for isorhamnetin acetyl hexoside) and a fragment pattern at m/z 315, which is coherent with isorhamnetin aglycon losses [12–14]. Their identification in bee pollen was already described in other works [14], as well as in honey [13].

The abundance of flavonols in pollen, namely quercetin derivatives, is in accordance with previous reports, being considered the main reason for the biological potential exhibited by this matrix, including abilities to capture free radicals and reactive species, chelate metals, inhibit cytochrome P450 action, and increase antioxidant defenses [28,29].

3.3. Antioxidant Capacity

We decided to perform a general screening of the antioxidative effects of the pollen extract against DPPH•. This revealed an ability to reduce the radical in a dose-dependent manner (Table 2 and Figure 2A), exhibiting an IC_{50} value of 695.99 ± 1.69 μg/mL. Even so, the extract was 46 times less efficient than ascorbic acid (IC_{50} = 15.18 ± 0.47 μg/mL).

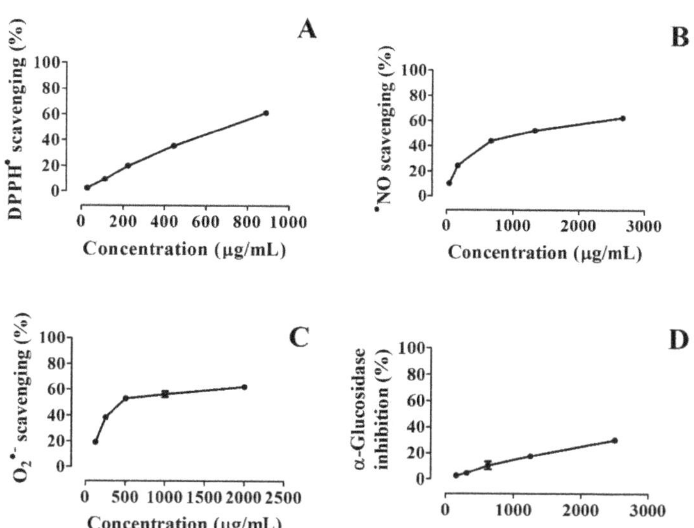

Figure 2. Effects of pollen hydroethanolic extracts against (**A**) 1,1-diphenyl-2-picrylhydrazyl (DPPH•), (**B**) nitric oxide (•NO), (**C**) superoxide (O2•−), and (**D**) α-glucosidase inhibition activity.

Table 2. The 25% inhibitory concentration (IC_{25}) and half maximal inhibitory concentration (IC_{50}) (µg/mL) values found in the antioxidant activity, α-glucosidase, hemoglobin oxidation, and hemolysis assays of pollen hydroethanolic extracts.

	DPPH•	•NO	$O_2^{•-}$	α-Glucosidase	Haemoglobin Oxidation	Haemolysis	Lipid Peroxidation
Extract	695.99 ± 1.69	1115.28 ± 5.32	449.04 ± 2.01	**1192.71 ± 8.14**	311.50 ± 1.37	**103.48 ± 2.23**	**277.03 ± 2.52**

Values are expressed as mean ± standard deviation of three assays concerning the antioxidant capacity against 1,1-diphenyl-2-picrylhydrazyl, nitric oxide and superoxide radicals (DPPH•, •NO and $O_2^{•-}$, respectively), α-glucosidase inhibitory activity, hemoglobin oxidation, hemolysis and lipid peroxidation. **IC_{25}: 25% inhibitory concentration**; IC_{50}: 50% inhibitory concentration.

Additionally, we also tested its capacity against •NO and $O_2^{•-}$. Although these radicals are products of normal cellular metabolism, their combination promotes the generation of more radicals, including peroxynitrite, which in turn oxidizes low-density lipoprotein and causes irreversible cell damage [1]. The extract also showed potential to scavenge •NO and $O_2^{•-}$ species in a dose-dependent manner (IC_{50} scores of 1115.28 ± 5.32 and 449.04 ± 2.01 µg/mL, respectively) (Table 2 and Figure 2B,C). Even so, the capturing abilities presented by the extract were lower than those found for the ascorbic acid control (IC_{50} values of 62.77 ± 0.42 and 28.13 ± 0.47 µg/mL for •NO and $O_2^{•-}$, respectively).

As far as we know, this is the first report concerning the pollen potential to scavenge •NO and $O_2^{•-}$ species. Regarding DPPH• potential, pollen collected on Parque Nacional da Peneda of Gerês (Portugal) revealed more effectiveness in scavenging DPPH• (IC_{50} = 5.87 µg/mL) [30], as well as Thai and Indian pollens (IC_{50} values of 428.6 and 6.09 µg/mL, respectively) [4,31]. On the other hand, Brazilian pollen showed weaker activity (IC_{50} = 860 µg/mL, respectively) [32]. Besides, Abrantes pollen showed better capturing potential than honey bees (IC_{50} values ranging from 7.5 and 109.0 mg/mL, depending on the country) [33]. The obtained differences are mainly due to different local origins, which influences phenolic levels and consequent biological potential [2].

In fact, the antioxidant activities exhibited by pollen are mainly due to its richness in phenolics, which have well-known abilities due to their ease of transferring hydrogen atoms or electrons to free radicals and reactive species, neutralizing them [1]. This health property is essentially due to their structure, predominantly due to the catechol group on the B ring, the 2,3-double bond, and the conjugated group on the C ring [27]. Besides, this activity is enhanced by a number of hydroxyl groups [1]. These facts are supported by the high correlations found using Pearson's test. Particularly strong correlations were verified in the DPPH• experiment, caffeoyl hexose ($r = 0.8217$, $p < 0.05$, $n = 1$), quercetin 7-glucoside-3-*O*-rutinoside ($r = 0.9994$, $p < 0.05$, $n = 1$), quercetin derivative 1 ($r = 0.8869$, $p < 0.05$, $n = 1$), quercetin acetyl rhamnoside ($r = 0.8092$, $p < 0.05$, $n = 1$), and also with total phenolic compounds and total flavonoids ($r > 0.7560$, $p < 0.05$, $n = 1$). On the other hand, regarding •NO scavenging potential, high correlations were found between this and caffeoyl hexose ($r = 0.8769$, $p < 0.05$, $n = 1$), quercetin 7-glucoside-3-*O*-rutinoside ($r = 0.9620$, $p < 0.05$, $n = 1$), quercetin derivative 1 ($r = 0.7960$, $p < 0.05$, $n = 1$), quercetin acetyl rhamnoside ($r = 0.8334$, $p < 0.05$, $n = 1$), and total phenolic compounds ($r = 0.7581$, $p < 0.05$, $n = 1$). These results are not surprising, since it was already reported that the CH = CH-COOH group presented by caffeoyl hexose, together with the catechol group on the B ring, the double bound and 4-oxo-group on the C ring, and the several hydroxyl groups found in quercetin derivatives make these compounds powerful natural antioxidants [1,27]. Furthermore, a high correlation was also reported between DPPH• and •NO antioxidant assays ($r = 0.9694$, $p < 0.05$, $n = 1$). Even so, negative correlations were found between the both experiments; isorhamnetin acetyl hexoside and quercetin hexoside ($r > -0.8054$, $p < 0.05$, $n = 1$). Concerning $O_2^{•-}$ scavenging potential, mild correlations were found between this assay, isorhamnetin acetyl hexoside ($r = 0.5385$, $p < 0.05$, $n = 1$), quercetin 3-*O*-rutinoside ($r = 0.4862$, $p < 0.05$, $n = 1$), and quercetin hexoside ($r = 0.5385$, $p < 0.05$, $n = 1$), while negative correlations were verified between this experiment, total phenolics ($r = -0.8197$, $p < 0.05$, $n = 1$), flavonols ($r = -0.7693$, p

< 0.05, $n = 1$), quercetin 7-glucoside-3-O-rutinoside ($r > -0.9354$, $p < 0.05$, $n = 1$), and quercetin derivative 1 ($r = -0.9961$, $p < 0.05$, $n = 1$), and also with DPPH$^{\bullet}$ ($r = -0.9233$, $p < 0.05$, $n = 1$) and $^{\bullet}$NO ($r = -0.8330$, respectively, $p < 0.05$, $n = 1$) assays. The obtained correlations are in accordance with others works [4,29,32].

Nevertheless, it is also important to consider the presence of other natural antioxidant compounds, such as vitamins, organic acids, and carotenoids not determined in this work, whose presence also increase pollen's biological potential [8].

3.4. Antidiabetic Capacity

The pollen hydroethanolic extract also exhibits the capacity to inhibit α-glucosidase activity, an important enzyme involved in carbohydrate digestion. In this study, this activity was dependent on the concentration, revealing an IC_{25} score of 1192.71 $\pm$ 8.14 µg/mL (Table 2 and Figure 2D). However, it was ten times less effective than the acarbose control (IC_{25} = 113.81 $\pm$ 1.00 µg/mL).

The antidiabetic effects of pollen have already been described by Daudu et al. [34], who conducted a study involving aqueous extracts of pollen and reported IC_{50} values of 4510 and 600 µg/mL regarding α-amylase and α-glucosidase inhibitory activities, respectively. Comparatively to other bee products, propolis hydroethanolic extracts (75:25, *v/v*) already showed ability to reduce baker's yeast α-glucosidase action, and rat intestinal sucrase and maltase enzymes (IC_{50} values of 7.24, 32.34 and 71.82 µg/mL) [35]. Besides, pure honey also showed antidiabetic properties. Indeed, diabetic rats which were fed with 1 mg/kg of honey over 6 weeks showed lower levels of serum glucose, creatinine, cholesterol, malondialdehyde, triglycerides, aspartate transaminase, and aspartate aminotransferase, and higher levels of insulin as compared to the untreated group [36]. Furthermore, thirty-two type 2 diabetes mellitus patients who ingested 25 g of honey over 4 months had reduced cholesterol and glycated hemoglobin levels [37].

Indeed, phenolics have already been proven to protect pancreatic β-cells from oxidative damage, incentive insulin production [36], and interact with carbohydrate enzyme substrates in both competitive and non-competitive ways, slowing down the breakdown of sugars, and therefore reducing glucose levels in the bloodstream [35]. The mentioned effects are enhanced by phenolic levels and also by their structures; namely the presence of hydroxyl and carbonyl groups [12,38]. In pollen's case, this capacity is strictly related to the presence of quercetin derivatives. Indeed, the notable ability of quercetin in stopping α-glucosidase activity was already described, revealing inhibitory effects of 91% at 200 µM [38]. This fact is corroborated by the mild correlation found between quercetin 3-O-rutinoside and α-glucosidase inhibitory action ($r = 0.5362$, $p < 0.05$, $n = 1$).

3.5. Protective Effects on Human Erythrocytes

In this study, pollen extracts inhibited hemoglobin oxidation induced by AAPH dose-dependently, with an IC_{50} score of 311.50 $\pm$ 1.37 µg/mL (Table 2 and Figure 3A). Still, it was not more efficient than the quercetin control (IC_{50} = 2.61 $\pm$ 0.15 µg/mL). Furthermore, it also prevented lipid peroxidation and hemolysis in a concentration-dependent manner, exhibiting 25% inhibitory concentration values of 277.03 $\pm$ 2.52 and 103.48 $\pm$ 2.23 µg/mL, respectively (Table 2 and Figure 3B,C). Once again, this was less than the quercetin control, whose IC_{25} values obtained were 1.00 $\pm$ 0.15 and 0.60 $\pm$ 0.15 µg/mL for lipid peroxidation and hemolysis, respectively.

Although this is the first study regarding the capacity of pollen to protect human erythrocytes against hemolysis and lipid peroxidation, Barbieri et al. [39] already reported that 50 µg/mL of bee pollen extract can protect these cells against AAPH-induced oxidation. Regarding other bee products, monofloral honeys were able to diminish lipid peroxidation in human erythrocytes by around 70% at concentrations of 80 µg/mL, and also avoided hemolysis, displaying IC_{50} values varying from 24.68 to 62.56 µg/mL [13]. As far as we know, honey did not show any preventive effects in protecting human erythrocytes against hemoglobin oxidation [40].

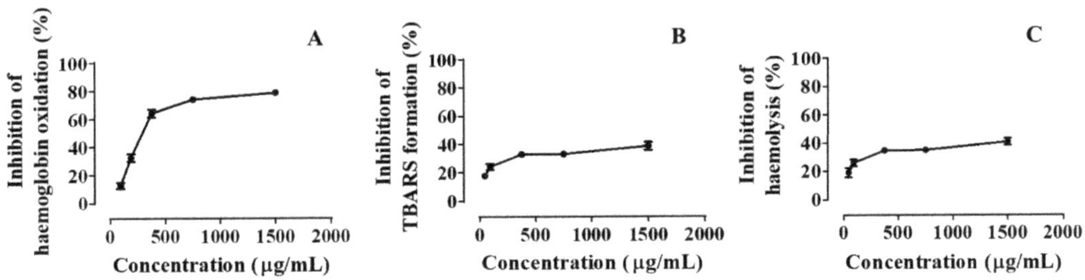

Figure 3. Inhibition of (**A**) hemoglobin oxidation, (**B**) lipid peroxidation, and (**C**) hemolysis of human erythrocytes by extracts of pollen hydroethanolic extracts.

Indeed, phenolic compounds, mainly due to their hydroxyl groups and liposolubility, can penetrate the cytoplasm of erythrocytes, and hence scavenge free radicals before they can attack erythrocyte membranes, enhancing their integrity and preventing oxidative injury [1,13,39]. Between phenolics, Peyrat-Maillard, Cuvelier, & Berset [41] already reported the protective effects displayed by caffeic acid, quercetin 3-*O*-rutinoside and quercetin against AAPH induced oxidation with values of 40, 41 and 55 μM/min. Moreover, quercetin and quercetin 3-*O*-rutinoside also showed capacity to prevent hemolysis in bovine erythrocytes, revealing IC_{50} scores of 31 and 37 μM [42], and avoided lipid peroxidation at concentrations below 10 μmol/L [43]. Moreover, the ability of phenolic acids, including caffeic acid, to block lipid peroxidation initiated by metmyoglobin has also been reported [44]. These facts are in line with the positive correlations observed involving hemoglobin oxidation, hemolysis, and quercetin 3-*O*-rutinoside ($r > 0.7355$, $p < 0.05$, $n = 1$), and between lipid peroxidation, caffeoyl hexose ($r = 0.7352$, $p < 0.05$, $n = 1$), quercetin rutinoside ($r = 0.9000$, $p < 0.05$, $n = 1$), and quercetin acetyl rhamnoside ($r = 0.7557$, $p < 0.05$, $n = 1$). Furthermore, notable correlations were also described regarding the $O_2^{\bullet-}$ scavenging potential of pollen, anti-hemolytic effects ($r = 0.9493$, $p < 0.05$, $n = 1$), and between hemoglobin oxidation, lipid peroxidation ($r = 0.8096$, $p < 0.05$, $n = 1$), and hemolysis ($r = 0.7091$, $p < 0.05$, $n = 1$). Mild correlations were verified among hemoglobin oxidation and $O_2^{\bullet-}$ assay ($r = 0.5387$, $p < 0.05$, $n = 1$). Even so, negative correlations were obtained between hemoglobin oxidation and quercetin derivative 1 ($r = -0.6037$, $p < 0.05$, $n = 1$), and between hemolysis, total phenolics ($r = -0.6770$, $p < 0.05$, $n = 1$), total flavonols ($r = -0.6928$, $p < 0.05$, $n = 1$), quercetin 7-glucoside-3-*O*-rutinoside ($r = -0.7769$, $p < 0.05$, $n = 1$), quercetin derivative 1 ($r = -0.9724$, $p < 0.05$, $n = 1$), DPPH$^{\bullet}$ ($r = -0.7557$, $p < 0.05$, $n = 1$), and $^{\bullet}$NO experiments ($r = -0.6354$, $p < 0.05$, $n = 1$). Concerning lipid peroxidation, negative correlations were obtained with isorhamnetin acetyl hexoside ($r = -0.7596$, $p < 0.05$, $n = 1$) and quercetin hexoside ($r = -0.7454$, $p < 0.05$, $n = 1$). The obtained correlations reinforce previous results which report that the antioxidant ability of phenolics is strongly influenced by the catechol residue, and number and position of hydroxyl groups, and less by the glycosylation pattern [27]. In fact, the increase of hydroxyl groups, together with the presence of the catechol group, improves this activity, facilitating neutralization of free radicals and reactive species [1,38].

3.6. *Effect of Pollen Extracts in Mitochondrial Activity and Membrane Integrity*

As already mentioned before, it was already reported that pollen has antidiabetic properties and is able to prevent lipid peroxidation [13,34,35]. Therefore, the capacity of the pollen extract to interfere with Caco-2 and HepG2 cancer cells growth was also tested. These cell lines were selected since they are considered models for intestinal epithelium, human toxicology, and metabolism studies, respectively [1,29]. Although pollen extract did not show any selective cytotoxicity for the tested cancer cells (i.e., Caco-2 and HepG2 cells), nor with the NHDF normal cell line (cells viability > 90%), its use as an antidiabetic agent is encouraged by these results combined with its ability to interact with α-glucosidase activity [17].

Considering other studies, Sousa et al. [28] verified that the pollen fraction enriched in flavonols protects human adenocarcinoma Caco-2 cells against oxidative damage triggered by *tert*-Butyl hydroperoxide. Furthermore, Oyarzún et al. [29] reported that pollen extracts also showed protective effects against oxidative injury induced by AAPH on Hepa1-6 hepatic cells, finding a mild correlation between this potential and quercetin content (r = 0.64).

4. Conclusions

Given the current interest regarding natural products, the present study provides new insights about Abrantes multifloral bee pollen. This regional pollen was portrayed here for the first time and showed notable antioxidant capacity, inhibition of α-glucosidase enzyme action, and protection of human erythrocytes against hemoglobin oxidation, lipid peroxidation, and hemolysis. These health-promoting activities are strongly correlated with phenolic content, highlighting the presence of quercetin derivatives. Furthermore, and although non-anticancer effects were observed, the results obtained herein, namely regarding its antidiabetic action, prove that pollen may be used safely as an antidiabetic agent and incorporated in new pharmaceutical drugs, nutraceuticals, and dietary supplements. The obtained data, together with the verified correlations, can be extrapolated to other pollens with similar compositions. Notwithstanding, further investigation, namely in vivo studies, should be conducted to assess the full biological potential of this bee derivative and to reveal a safe dosage.

Author Contributions: Conceptualization, G.A. and L.R.S.; methodology, all authors; software, A.C.G., R.A.L., L.R.S., C.G.-V. and D.A.M.; validation, L.R.S., G.A., C.G.-V. and D.A.M.; formal analysis, A.C.G., R.A.L., L.R.S. and D.A.M.; investigation, all authors; resources, all authors; data curation, R.A.L., L.R.S., C.G.-V. and D.A.M.; writing—original draft preparation, A.C.G.; writing—review and editing, A.C.G., R.A.L., G.A., L.R.S., C.G.-V. and D.A.M.; visualization, all authors; supervision, G.A., L.R.S., C.G.-V. and D.A.M.; project administration, G.A.; funding acquisition, L.R.S., G.A. All authors have read and agreed to the published version of the manuscript.

Funding: This work is supported by FEDER funds through the POCI-COMPETE 2020-Operational Programme Competitiveness and Internationalisation in Axis I—Strengthening research, technological development and innovation (Project POCI-01-0145-FEDER-007491), Operational Program of the Center (Project CENTRO-01-0247-FEDER-017547) and National Funds by FCT—Foundation for Science and Technology (Project UID/Multi/00709/2013). Ana Carolina was initially supported by the R&D Business Project (Project CENTRO-01-0247-FEDER-017547), co-financed by the European Regional Development Fund (ERDF) through the Regional Operational Program of the Center (Portugal 2020). The authors are grateful to the Foundation for Science and Technology (FCT), Ministry of Science, Technology and Higher Education (MCTES), European Social Fund (EFS) and Europe Union (EU) for the PhD fellowship of Ana C. Gonçalves (2020.04947.BD). Part of this work was also supported by the Grant for Research Group of Excellence—Fundación Seneca, Murcia Regional Agency for Science and Technology, Project 19900/GERM/15 at CEBAS-CSIC.

Institutional Review Board Statement: Not applicable.

Informed Consent Statement: Not applicable.

Data Availability Statement: Data are contained within this article.

Conflicts of Interest: The authors declare no conflict of interest.

References

1. Gonçalves, A.C.; Bento, C.; Nunes, A.R.; Simões, M.; Alves, G.; Silva, L.R. Multitarget protection of *Pterospartum tridentatum* phenolic-rich extracts against a wide range of free radical species, antidiabetic activity and effects on human colon carcinoma (Caco-2) cells. *J. Food Sci.* **2020**, *85*, 4377–4388. [CrossRef] [PubMed]
2. Thakur, M.; Nanda, V. Composition and functionality of bee pollen: A review. *Trends Food Sci. Technol.* **2020**, *98*, 82–106. [CrossRef]

3. Moita, E.; Sousa, C.; Andrade, P.B.; Fernandes, F.; Pinho, B.R.; Silva, L.R.; Valentão, P. Effects of *Echium plantagineum* L. bee pollen on basophil degranulation: Relationship with metabolic profile. *Molecules* **2014**, *19*, 10635–10649. [CrossRef] [PubMed]
4. Chantarudee, A.; Phuwapraisirisan, P.; Kimura, K.; Okuyama, M.; Mori, H.; Kimura, A.; Chanchao, C. Chemical constituents and free radical scavenging activity of corn pollen collected from *Apis mellifera* hives compared to floral corn pollen at Nan, Thailand. *BMC Complementary Altern. Med.* **2012**, *12*, 45. [CrossRef] [PubMed]
5. Silva, L.R.; Gonçalves, A.C.; Nunes, A.R.; Alves, G. Authentication of honeys from Caramulo region (Portugal): Pollen spectrum, physicochemical characteristics, mineral content, and phenolic profile. *J. Food Sci.* **2020**, *85*, 374–385. [CrossRef]
6. Kaškonienė, V.; Ruočkuvienė, G.; Kaškonas, P.; Akuneca, I.; Maruška, A. Chemometric analysis of bee pollen based on volatile and phenolic compound compositions and antioxidant properties. *Food Anal. Methods* **2015**, *8*, 1150–1163. [CrossRef]
7. Adaškevičiūtė, V.; Kaškonienė, V.; Kaškonas, P.; Barčauskaitė, K.; Maruška, A. Comparison of physicochemical properties of bee pollen with other bee products. *Biomolecules* **2019**, *9*, 819. [CrossRef]
8. de Florio Almeida, J.; dos Reis, A.S.; Heldt, L.F.; Pereira, D.; Bianchin, M.; de Moura, C.; Plata-Oviedo, M.V.; Haminiuk, C.W.; Ribeiro, I.S.; da Luz, C.F.; et al. Lyophilized bee pollen extract: A natural antioxidant source to prevent lipid oxidation in refrigerated sausages. *LWT-Food Sci. Technol.* **2017**, *76*, 299–305. [CrossRef]
9. Moreira, R.; Fernandes, F.; Valentão, P.; Pereira, D.M.; Andrade, P.B. *Echium plantagineum* L. honey: Search of pyrrolizidine alkaloids and polyphenols, anti-inflammatory potential and cytotoxicity. *Food Chem.* **2020**, *328*, 127169. [CrossRef]
10. Silva, L.R.; Sousa, A.; Taveira, M. Characterization of Portuguese honey from Castelo Branco region according to their pollen spectrum, physicochemical characteristics and mineral contents. *J. Food Sci. Technol.* **2017**, *54*, 2551–2561. [CrossRef]
11. Louveaux, J.; Maurizio, A.; Vorwohl, G. Methods of Melissopalynology. *Bee World* **1978**, *59*, 139–157. [CrossRef]
12. Gonçalves, A.C.; Campos, G.; Alves, G.; Garcia-Viguera, C.; Moreno, D.A.; Silva, L.R. Physical and phytochemical composition of 23 Portuguese sweet cherries as conditioned by variety (or genotype). *Food Chem.* **2021**, *335*, 127637. [CrossRef] [PubMed]
13. Alvarez-Suarez, J.M.; Giampieri, F.; González-Paramás, A.M.; Damiani, E.; Astolfi, P.; Martinez-Sanchez, G.; Bompadre, S.; Quiles, J.L.; Santos-Buelga, C.; Battino, M. Phenolics from monofloral honeys protect human erythrocyte membranes against oxidative damage. *Food Chem. Toxicol.* **2012**, *50*, 1508–1516. [CrossRef] [PubMed]
14. De-Melo, A.A.; Estevinho, L.M.; Moreira, M.M.; Delerue-Matos, C.; Freitas, A.D.; Barth, O.M.; Almeida-Muradian, L.B. Phenolic profile by HPLC-MS, biological potential, and nutritional value of a promising food: Monofloral bee pollen. *J. Food Biochem.* **2018**, *42*, e12536. [CrossRef]
15. Tang, J.; Dunshea, F.R.; Suleria, H.A.R. LC-ESI-QTOF/MS characterization of phenolic compounds from medicinal plants (Hops and Juniper berries) and their antioxidant activity. *Foods* **2020**, *9*, 7. [CrossRef]
16. Zhang, H.; Liu, R.; Lu, Q. Separation and characterization of phenolamines and flavonoids from Rape bee pollen, and comparison of their antioxidant activities and protective effects. *Molecules* **2020**, *25*, 1264. [CrossRef] [PubMed]
17. Silva, L.R.; Teixeira, R. Phenolic profile and biological potential of *Endopleura uchi* extracts. *Asian Pac. J. Trop. Med.* **2015**, *8*, 889–897. [CrossRef]
18. Chisté, R.C.; Freitas, M.; Mercadante, A.Z.; Fernandes, E. Carotenoids inhibit lipid peroxidation and hemoglobin oxidation, but not the depletion of glutathione induced by ROS in human erythrocytes. *Life Sci.* **2014**, *99*, 52–60. [CrossRef]
19. Jesus, F.; Gonçalves, A.C.; Alves, G.; Silva, L.R. Exploring the phenolic profile, antioxidant, antidiabetic and anti-hemolytic potential of *Prunus avium* vegetal parts. *Food Res. Int.* **2018**, *116*, 600–610. [CrossRef]
20. Forbes-Hernández, T.Y.; Gasparrini, M.; Afrin, S.; Cianciosi, D.; González-Paramás, A.M.; Santos-Buelga, C.; Mezzetti, B.; Quiles, J.L.; Battino, M.; Giampieri, F.; et al. Strawberry (cv. *Romina*) methanolic extract and anthocyanin-enriched fraction improve lipid profile and antioxidant status in HepG2 cells. *Int. J. Mol. Sci.* **2017**, *18*, 1149. [CrossRef]
21. Gonçalves, A.C.; Rodrigues, M.; Santos, A.O.; Alves, G.; Silva, L.R. Antioxidant status, antidiabetic properties and effects on Caco-2 cells of colored and non-colored enriched extracts of sweet cherry fruits. *Nutrients* **2018**, *10*, 1688. [CrossRef] [PubMed]
22. Estevinho, L.M.; Feás, X.; Seijas, J.A.; Pilar Vázquez-Tato, M. Organic honey from Trás-Os-Montes region (Portugal): Chemical, palynological, microbiological and bioactive compounds characterization. *Food Chem. Toxicol.* **2012**, *50*, 258–264. [CrossRef] [PubMed]
23. Silva, L.R.; Videira, R.; Monteiro, A.P.; Valentão, P.; Andrade, P.B. Honey from Luso region (Portugal): Physicochemical characteristics and mineral contents. *Microchem. J.* **2009**, *93*, 73–77. [CrossRef]
24. Karabagias, I.K.; Maia, M.; Karabagias, V.K.; Gatzias, I.; Badeka, A.V. Quality and origin characterisation of Portuguese, Greek, Oceanian, and Asian honey, based on poly-parametric analysis hand in hand with dimension reduction and classification techniques. *Eur. Food Res. Technol.* **2020**, *246*, 987–1006. [CrossRef]
25. Escuredo, O.; Silva, L.R.; Valentão, P.; Seijo, M.C.; Andrade, P.B. Assessing Rubus honey value: Pollen and phenolic compounds content and antibacterial capacity. *Food Chem.* **2012**, *130*, 671–678. [CrossRef]
26. Bento, C.; Gonçalves, A.C.; Silva, B.; Silva, L.R. Assessing the phenolic profile, antioxidant, antidiabetic and protective effects against oxidative damage in human erythrocytes of peaches from Fundão. *J. Funct. Foods* **2018**, *43*, 224–233. [CrossRef]
27. Bento, C.; Gonçalves, A.C.; Jesus, F.; Simões, M.; Silva, L.R. Phenolic compounds: Sources, properties and applications. In *Bioactive Compounds: Sources, Properties and Applications*; Porter, R., Parker, N., Eds.; Nova Science Publishers: New York, NY, USA, 2017; pp. 271–299.

28. Sousa, C.; Moita, E.; Valentão, P.; Fernandes, F.; Monteiro, P.; Andrade, P.B. Effects of colored and noncolored phenolics of *Echium plantagineum* L. bee pollen in Caco-2 cells under oxidative stress induced by tert -butyl hydroperoxide. *J. Agric. Food Chem.* **2015**, *63*, 2083–2091. [CrossRef]
29. Oyarzún, J.E.; Andia, M.E.; Uribe, S.; Núñez Pizarro, P.; Núñez, G.; Montenegro, G.; Bridi, R. Honeybee pollen extracts reduce oxidative stress and steatosis in hepatic cells. *Molecules* **2020**, *26*, 6. [CrossRef] [PubMed]
30. Morais, M.; Moreira, L.; Feás, X.; Estevinho, L.M. Honeybee-collected pollen from five Portuguese Natural Parks: Palynological origin, phenolic content, antioxidant properties and antimicrobial activity. *Food Chem. Toxicol.* **2011**, *49*, 1096–1101. [CrossRef]
31. Thakur, M.; Nanda, V. Screening of Indian bee pollen based on antioxidant properties and polyphenolic composition using UHPLC-DAD-MS/MS: A multivariate analysis and ANN based approach. *Food Res. Int.* **2021**, *140*, 110041. [CrossRef]
32. Duarte, A.W.F.; Vasconcelos, M.R.D.S.; Oda-Souza, M.; Oliveira, F.F.; López, A.M.Q. Honey and bee pollen produced by Meliponini (Apidae) in Alagoas, brazil: Multivariate analysis of physicochemical and antioxidant profiles. *Food Sci. Technol.* **2018**, *38*, 493–503. [CrossRef]
33. Stagos, D.; Soulitsiotis, N.; Tsadila, C.; Papaeconomou, S.; Arvanitis, C.; Ntontos, A.; Karkanta, F.; Adamou-Androulaki, S.; Petrotos, K.; Spandidos, D.A.; et al. Antibacterial and antioxidant activity of different types of honey derived from Mount Olympus in Greece. *Int. J. Mol. Med.* **2018**, *42*, 726–734. [CrossRef]
34. Daudu, O.M. Bee pollen extracts as potential antioxidants and inhibitors of α-amylase and α-glucosidase enzymes-In vitro assessment. *J. Apic. Sci.* **2019**, *63*, 315–325. [CrossRef]
35. Zhang, H.; Wang, G.; Beta, T.; Dong, J. Inhibitory properties of aqueous ethanol extracts of propolis on alpha-glucosidase. *Evid. -Based Complementary Altern. Med.* **2015**, *2015*, 1–7. [CrossRef] [PubMed]
36. Hassan, A.K.; El-Kotby, D.A.; Tawfik, M.M.; Badr, R.E.; Bahgat, I.M. Antidiabetic effect of the Egyptian honey bee (*Apis mellifera*) venom in alloxan-induced diabetic rats. *J. Basic Appl. Zool.* **2019**, *80*, 58. [CrossRef]
37. Enginyurt, O.; Cakir, L.; Karatas, A.; Cankaya, S.; Kaya, Y.; Handan Tugcu, H.; Dogan Iscanli, M.; Cankaya, N.; Yarilgac, S. The role of pure honey in the treatment of diabetes mellitus. *Biomed. Res.* **2017**, *28*, 3305–3312.
38. Tadera, K.; Minami, Y.; Takamatsu, K.; Matsuoka, T. Inhibition of α-glucosidase and α-amylase by flavonoids. *J. Nutr. Sci. Vitaminol.* **2006**, *52*, 149–153. [CrossRef]
39. Blasa, M.; Candiracci, M.; Accorsi, A.; Piacentini, M.P.; Piatti, E. Honey flavonoids as protection agents against oxidative damage to human red blood cells. *Food Chem.* **2007**, *104*, 1635–1640. [CrossRef]
40. Barbieri, D.; Gabriele, M.; Summa, M.; Colosimo, R.; Leonardi, D.; Domenici, V.; Pucci, L. Antioxidant, nutraceutical properties, and fluorescence spectral profiles of bee pollen samples from different botanical origins. *Antioxidants* **2020**, *9*, 1001. [CrossRef] [PubMed]
41. Peyrat-Maillard, M.N.; Cuvelier, M.E.; Berset, C. Antioxidant activity of phenolic compounds in 2,2′-Azobis (2-amidinopropane) Dihydrochloride (AAPH)-induced oxidation: Synergistic and antagonistic effects. *J. Am. Oil Chem. Soc.* **2003**, *80*, 1007–1012. [CrossRef]
42. Kitagawa, S.; Sakamoto, H.; Tano, H. Inhibitory effects of flavonoids on free radical-induced hemolysis and their oxidative effects on hemoglobin. *Chem. Pharm. Bull.* **2004**, *52*, 999–1001. [CrossRef] [PubMed]
43. Peng, I.W.; Kuo, S.M. Flavonoid structure affects the inhibition of lipid peroxidation in Caco-2 intestinal cells at physiological concentrations. *J. Nutr.* **2003**, *133*, 2184–2187. [CrossRef] [PubMed]
44. Laranjinha, J.; Vieira, O.; Almeida, L.; Madeira, V. Inhibition of metmyoglobin/H2O2-dependent low density lipoprotein lipid peroxidation by naturally occurring phenolic acids. *Biochem. Pharmacol.* **1996**, *51*, 395–402. [CrossRef]

foods

MDPI

Communication

Tubular Cellulose from Orange Juice By-Products as Carrier of Chemical Preservatives; Delivery Kinetics and Microbial Stability of Orange Juice

Athanasia Panitsa [1], Theano Petsi [1], Panagiotis Kandylis [2,*], Maria Kanellaki [1] and Athanasios A. Koutinas [1]

[1] Food Biotechnology Group, Department of Chemistry, University of Patras, Rio, 26504 Patras, Greece; sissy_panitsa@hotmail.com (A.P.); thpetsi@upatras.gr (T.P.); M.Kanellaki@upatras.gr (M.K.); A.A.Koutinas@upatras.gr (A.A.K.)

[2] Laboratory of Oenology and Alcoholic Beverages, Department of Food Science and Technology, School of Agriculture, Aristotle University of Thessaloniki, P.O. Box 235, 54124 Thessaloniki, Greece

* Correspondence: pkandylis@agro.auth.gr; Tel.: +0030-2310-991-678

Abstract: The quality and safety of juices are assured mainly through heat treatments and chemical preservatives. However, there is a growing trend in the food industry for lowering energy and water demands, and the chemicals and additives that may have negative effects on human health. Following that trend, in the present study, the reduced use of chemical preservatives in orange juice is proposed by using encapsulated sodium benzoate (SB) in tubular cellulose (TC), derived from orange pulp. The effects of SB concentration and contact time on SB encapsulation were evaluated. The use of the wet impregnation method, 12% *w/v* SB solution and 2 h of contact proved to be ideal for application in the juice industry. The use of starch gel resulted in a more stable composite (TC/SB-SG) with a slower SB delivery, showing its potential for future controlled delivery applications. Furthermore, similar delivery rates of SB in juice were noted at 25 and 2 °C. The TC/SB-SG proved capable of inhibiting the growth and reducing the numbers of spoilage microorganisms (yeasts and lactic acid bacteria). The results of the present study are promising for potential applications; however, more research is needed in order to evaluate the controlled delivery of SB in juice.

Keywords: juice; sodium benzoate; *Saccharomyces cerevisiae*; *L. plantarum*; *L. casei*; delivery; orange pulp

1. Introduction

The purpose of adding preservatives to food is the avoidance of microbial spoilage or alterations in their chemical composition and, therefore, the extension of their lifespan [1, 2]. There are various preservatives that are widely used in the industry, and among them are benzoic acid and its salts. Due to their antimicrobial activity, they can be added to carbonated drinks, fruit juice, vinegar and, in general to foods and drinks with low pH values [1,3,4]. Sodium benzoate (SB) is among the most commonly used preservatives in food, and especially in fruit juices [1,2,4–7]. SB was the first preservative that the FDA allowed to be used in food products [6].

Fruits and fruit juices due to their high content of organic acids have, in most cases, low pH values (2.0–4.5). It is therefore more difficult for them to be spoiled by microorganisms. However, microbial degradation can occur when low-pH-resistant microorganisms grow, such as acid-tolerant bacteria (for example, lactic, acetic and propionic bacteria), yeasts (such as sugar-fermenting yeasts) and molds [8,9]. Such alterations may lead to creation of off-flavor compounds, discoloration, changes in texture and appearance; thus, the products are distorted and become unsuitable for commercial use [8]. In addition, microorganism spoilage can cause the production of carbon dioxide that may lead to distortion or leakage of cans or bottles. Furthermore, several cases of microbial contamination by pathogens in foods such as fruit juices have been reported. Microorganisms such as *Escherichia coli* O157:H7, *Salmonella* and the parasite *Cryptosporidium* can be transmitted from raw fruit juice, causing several illnesses [8,9].

However, despite its effective antimicrobial activity, SB also has some side effects on the human body. For example, it can burden the liver, cause sensitization and affect children's behavior. The mutagenic and cytotoxic effects of SB are proven in vitro on human lymphocytes, and they may cause cancer [2]. SB is metabolized within the mitochondria, producing hippurate, which is then cleared by the kidneys. Ingestion of SB at doses generally regarded as safe (GRAS) can lead to a high increased hippurate level in the plasma. Side effects of benzoate and hippurate on glucose homeostasis in cells and animal models have been reported [5]. Beverages with benzoate salts contain low levels of the carcinogen benzene, which may be formed during storage after decarboxylation of the benzoate through hydroxyl radicals [10].

In order to avoid excessive quantities of SB in food, an intermediate carrier, in which SB can be encapsulated and released gradually, can be used. This work aims at the preparation of a suitable material for the encapsulation of SB and its gradual release in drinks such as orange juice. Although the addition of SB in fresh juices is continuously reducing, in carbonated and non-carbonated soft drinks based on juices, SB is necessary. Tubular cellulose (TC) is a natural material with great potential as an encapsulation/immobilization carrier. In several studies, TC from various origins has already been used successfully for the encapsulation of enzymes (TC from wood sawdust) [11] and microbial cells (TC from Indian mango, sal wood and rice husk) [12] and for the preservative delivery in meat products (TC from leaf celery and spinach) [13]. Following this trend, the aim of the present study is the evaluation of TC from edible orange pulp as a suitable material for the delivery of SB in orange juice. Encapsulation of SB could lead to (i) delay of the delivery of increased amounts of SB in juices, (ii) maintenance of its antimicrobial activity and (iii) reduction in SB side effects on human health. Furthermore, the similar nature of orange pulp to that of orange juice is expected to have no effect on the characteristics of the final product.

2. Materials and Methods

2.1. Raw Materials, Microorganisms and Chemicals

Navelina oranges were purchased from a local market. The oranges were squeezed and the pulp and orange peel pith were kept for treatment and the encapsulation of SB.

For the delivery of SB, concentrated orange juice was obtained from the local industry (Loux S.A., Patra, Greece) and after appropriate dilution was used in the delivery of the preservative experiments (23 g/100 mL sugars, pH 4.27, total acidity 12.8 g citric acid/L, volatile acidity 0.08 g acetic acid/L).

NaOH pellets were obtained from Lachner, food grade SB was purchased from Sigma Aldrich, and food grade corn starch was purchased from Wintersun Chemical.

For the microbial stability of juices, commercial orange juice purchased from a local market was used (*Μαράτα*, natural orange juice, Sklavenitis Group, Athens, Greece). Its composition per 100 mL was as follows: 9.0 g sugars, 0.6 g edible fibers, 0.6 g proteins, 0 g fats, 0 g salt and 40 mg vitamin C. The following microorganisms were used: *Lacticaseibacillus* (former *Lactobacillus*) *casei* ATCC 393 (DSMZ, Braunschweig, Germany), *Lactiplantibacillus plantarum* 2035 (from the collection of the Laboratory of Food Microbiology and Hygiene, Department of Food Science and Technology, Aristotle University of Thessaloniki, Greece), *Saccharomyces cerevisiae* CR51 (Mangrove Jacks) and *Saccharomyces cerevisiae* EC1118 (Lalvin, Lallemand Brewing).

2.2. Tubular Cellulose from Orange Pulp

Delignification was carried out by the treatment of orange pulp with 1% *w/v* NaOH and heating at 70 °C for 3 h (Table 1) [13]. The delignified orange pulp (TC) was dried for approximately 48 h, at −45 °C (temperature of condenser) under vacuum pressure of 15×10^{-3} mbar using the freeze-drying process (Labtech Freeze Dryer, LFD 5508, Daihan Labtech Co., Ltd, Namyangju, South Korea).

Table 1. Experimental design.

1. Tubular Cellulose (TC) from Orange Pulp
Orange pulp (pulp and peel pith) → delignification (1% *w/v* NaOH + 70 °C for 3 h) → TC (freeze drying; FD)
2. Sodium Benzoate Encapsulation
I. TC/SB-WI (Wet Impregnation Method)
TC (1.5 g) + SB solution (30 mL; 1.5–49% *w*/*v*) + stirring (0.5–5 h)
II. TC/SB-SG (Starch Gel Method)
7 mL of deionized water was mixed with corn starch and 5% *w/v* SB solution + TC
3. Sodium Benzoate Delivery in Orange Juice
2 g FD TC/SB-WI and FD TC/SB-SG (containing 0.13 g of encapsulated SB) + 100 mL orange juice
Delivery after 0.5–5 h and 0.5–60 days; 2 °C and 25 °C
4. Microbiological Stability
Contaminated juice with *L. casei*, *L. planatrum* and *S. cerevisiae* (storage 0, 2, 5 and 15 days; 5 °C)

The surface characterization of orange pulp was performed using a Micromeritics Tristar surface area and porosity analyzer and scanning electron microscope as described previously [13].

2.3. Sodium Benzoate Encapsulation

Encapsulation of SB using the wet impregnation method (WI) (TC/SB-WI) was conducted by adding TC in SB solution with continuous stirring. In order to study the effect of contact time on the final SB amount encapsulated, 30 mL of 15 and 30% *w/v* SB solution was added to 1.5 g TC, and samples were taken at 0.5, 2, 3.5 and 5 h. In order to study the effect of initial concentration of SB solution, different concentrations of SB (1.5–49% *w*/*v*) were prepared in 30 mL of deionized water, and 1.5 g TC was added. The mixtures were left under stirring for 2 h.

TC/SB-SG (tubular cellulose/sodium benzoate substrates using the starch gel method (SG)) was prepared following the method described by Panitsa et al. [13] with slight modifications (7 mL of deionized water was mixed with corn starch and 5% *w/v* SB solution).

2.4. Sodium Benzoate Delivery in Orange Juice

Two grams of freeze dried TC/SB-WI and TC/SB-SG (containing 0.13 g of encapsulated SB) was placed in 100 mL of juice. Several samples (100 mL each) were placed at 2 °C and at room temperature (25 °C) without stirring for up to 60 days. In several time intervals (up to 60 days), a sample was taken for SB analysis.

2.5. Microbiological Stability

L. casei, *L. plantarum* and *S. cerevisiae* were activated prior to their usage in the contamination experiments based on the method described by Katsaros et al. [14] with some modifications. The two strains of lactic acid bacteria from glycerol stock cultures were inoculated into MRS broth tubes and incubated at 30 °C for *L. plantarum* and 37 °C for *L. casei* for 24 h [15,16]. After activation, 100 µL of each was transferred into the MRS broth and was incubated at the same temperatures for 24 h. Activation of the two strains of yeasts from freeze-dried cultures was performed by inoculating cells into a yeast extract peptone glucose broth (YPGB) composed of (*w*/*v*) 0.5% yeast extract, 0.1% peptone and 2.0% glucose, at 25 °C for 48 h. After activation, 100 µL was also transferred into the YPGB and incubated at 25 °C for 24 h. Additional enrichment of the cultures (lactic acid bacteria and yeasts) was performed, following the same procedure and incubation for 18 h–20 h.

Contamination of the juices was performed by transferring 1 mL of each microorganism to 100 mL of juice, separately, with and without encapsulated SB (0.13 g of SB). They were stored at 5 °C for up to 15 days.

Juice samples were taken at 2, 5 and 15 days after contamination and serial diluted in Ringers' solution (1/4 strength); then, solid cultures were prepared on MRS agar for *L. casei* (incubated at 37 °C for at least 48 h) and *L. plantarum* (incubated at 30 °C for at least 48 h) and on potato dextrose agar for *S. cerevisiae* (incubated at 30 °C for at least 72 h).

2.6. Chemical Analyses

The pH was measured directly in the juice sample with an automatic electronic pH meter (SensoDirect pH 110, Aqualytic (Lovibond), Dortmund, Germany).

Total acidity (expressed as g of citric acid/L juice) was measured by the titration of 10 mL of orange juice with 0.1 M NaOH solution and phenolphthalein as the indicator.

Volatile acidity (expressed as g of acetic acid/L juice) was determined after the distillation of 50 mL of juice. More specifically, 200 mL of distillate was collected and titrated with 0.1 M NaOH solution and phenolphthalein as the indicator.

The determination of sugars (sucrose, glucose and fructose) was performed using high-performance liquid chromatography (HPLC). A Shimadzu chromatograph was used, consisting of an LC-9A high-pressure pump, a Nucleogel Ion 300 OA column (300 × 7.8 mm id, 10 μm particle size), a CTO-10A oven, an RID-6A refractive index detector connected to a C-R6 integrator and a DGU-2A degassing unit. The analysis was performed using isocratic separation at 33 °C with a mobile phase of 0.017 N H_2SO_4 and a flow rate of 0.55 mL/min. Ten milliliters of juice was added to a 100 mL volumetric flask and filled with deionized water. From this solution, 500 μL was placed in a 25 mL volumetric flask, 500 μL of propanol-2 1% *v*/*v*, was added as an internal standard, and ultra-pure water was added. The samples after the above preparation were filtered with a 0.22 μm microfilter and a volume of 40 μL was injected into the chromatograph. The sugar concentrations were determined using standard curves.

The determination of SB was performed using high-performance liquid chromatography (HPLC). A Shimadzu Corporation chromatograph was used, consisting of a high pressure pump LC-20AT, a Supelco Discovery C8 column (250 × 4.6 mm id, 5 μm particle size), a CTO-10AC oven and an array detector of an M20A-DGA-single-diode SGD-A unit. The analysis was performed using isocratic separation at 30 °C with mobile phase 80% acetate buffer/20% acetonitrile, a flow rate of 0.8 mL/min and an SIL-20AC sampler. Five milliliters of each sample was placed in a 50 mL volumetric flask, and it was filled with 50% aqueous acetonitrile (solution 1). Ten milliliters of solution 1 was placed in a 50 mL volumetric flask and filled with 50% aqueous acetonitrile solution (solution 2). Then, 1 mL of solution 2 was mixed with 5 mL of mobile phase (solution 3). Then, 1 mL of solution 3 was mixed with 10 mL of mobile phase (solution 4). Solution 4 was centrifuged for 15 min at 4000 rpm, the supernatant was filtered through a 0.22 μm microfilter, and 5 μL from each sample was loaded on the chromatograph. SB was determined using standard curves.

2.7. Statistical Analysis

Duplicate experiments were conducted, and duplicate or triplicate samples were used for the analyses. Experimental data were evaluated for their significance ($p <$ 0.05, 0.01 and 0.001) with analysis of variance (ANOVA) and Tukey's honest significant difference (HSD) test using Statistica version 12.0 (StatSoft Inc., Tulsa, OK, USA).

3. Results and Discussion

3.1. Delignification and Characterization of Orange Pulp

The first step of the present study was the preparation of an appropriate support for the encapsulation of chemical preservatives for applications in the food industry. In order to achieve an increased and sufficient encapsulation, the support should have the appropriate properties to guarantee this encapsulation. Delignification has been proven an ideal method for the pretreatment of lignocellulosic materials prior to the enzymatic process and hydrolysis, since the removal of lignin increases the available surface area of material and also increases the pore volume [17,18]. However, delignified materials have also been successfully used for the immobilization of microorganisms for several food applications [12,19,20]. Recently, our team reported the preparation of an edible tubular cellulose-based support for the delivery of chemical preservative in meat products [13]. Following that trend, in the present study, orange juice by-products such as orange pulp (orange pulp and orange peel pith) were delignified and evaluated as support for SB for applications in the juice and beverage industry. Delignification resulted in more than a two-times increase in specific surface area, a five-times increase in pore volume and a two-times increase in pore diameter (Table 2). These results are lower than those reported in the previous study with TC from leaf celery and spinach [13], which is attributed to the different composition of each material.

Table 2. Porosimetry analyses of untreated and delignified orange pulp.

Analyses	Untreated Orange Pulp	Delignified Orange Pulp
Specific surface area (BET [1]) (m^2/g)	0.30 ± 0.03	0.67 ± 0.05
Pore volume (BJH) ($\times 10^{-3}$ cm^3/g)	1.3 ± 0.4	6.7 ± 0.5
Average pore diameter (BET) (Å)	47.0 ± 5.6	115.3 ± 13.2

[1] BET: Brunauer–Emmett–Teller; BJH: Barrett–Joyner–Halenda.

3.2. *Preparation of Tubular Cellulose/Sodium Benzoate (TC/SB) Substrates*

The encapsulation of SB in TC was achieved by adding dried TC to a SB solution. In order to increase the rate of encapsulation, stirring was applied. The effect of contact time on encapsulation efficacy was not significant ($p > 0.05$) in the case of 30% *w/v* SB solution up to 3.5 h, while a reduction was observed in 5 h (Figure 1). Similar results were noted in the case of 15% *w/v* SB solution. For this reason, 2 h was selected as the ideal time of stirring/contact in order to ensure a high encapsulation rate.

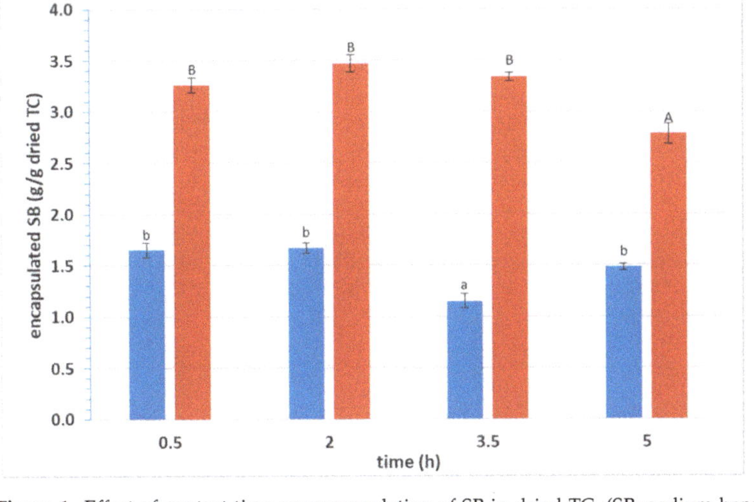

Figure 1. Effect of contact time on encapsulation of SB in dried TC. (SB: sodium benzoate; TC: tubular cellulose; (■) 15% *w/v* SB; (■) 30% *w/v* SB; means within the same SB concentration with different superscripts differ significantly ($p < 0.05$)).

Another important factor for the encapsulation of SB in TC was the concentration of the SB solution used. The results of Figure 2 clearly report that the increase in concentration of the SB solution resulted in a proportional increase in the encapsulated SB on the TC. More specifically, SB concentrations up to 9% *w/v* resulted in similar values ($p > 0.05$), while in higher than 9% *w/v*, the encapsulated SB was significantly higher ($p < 0.05$). Indeed, an encapsulation of higher than 6 g/g dried TC can be achieved using a 49% *w/v* SB solution. This proportional increase in the encapsulation ratio with the SB solution used is important since it is possible to prepare TC/SB composites with different concentrations of SB for different applications in the food industry.

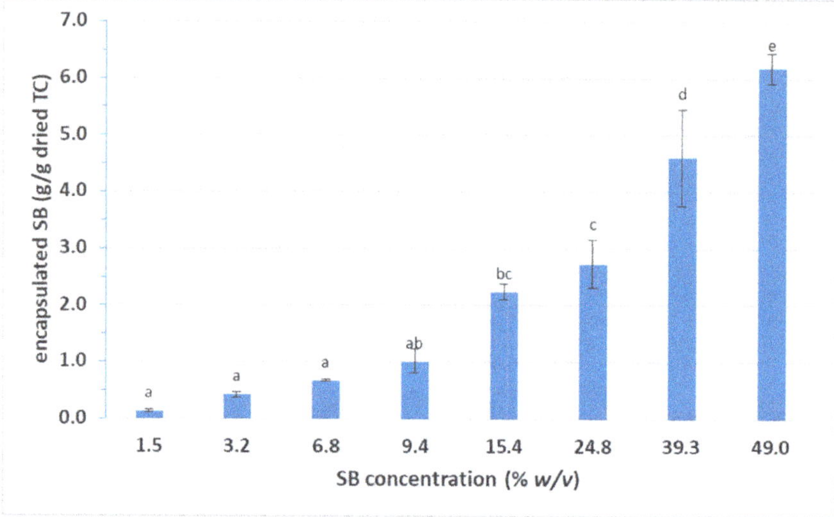

Figure 2. Effect of SB concentration on encapsulation of SB in dried TC. (SB: sodium benzoate; TC: tubular cellulose; means with different superscripts differ significantly ($p < 0.05$)).

3.3. Sodium Benzoate Delivery

The effect of the encapsulation method, temperature and time on SB delivery in orange juice is presented in Table 3. The effect of temperature was significant mainly after 360 h. This observation may be important since the delivery of SB is the same even at low refrigerator temperatures, which will also help the increased shelf life of the juice. Regarding the effect of encapsulation method, the use of SG seems more appropriate since slower SB delivery is achieved. The presence of SG strongly affected the SB release, thus suggesting the potential of SG encapsulation to control the release mechanism of active compounds, as was also reported in the case of alginate beads [21]. This may be attributed to the better encapsulation of SB and to the starch gel matrix that permits low SB mobility. However, both encapsulation methods could have practical use depending on the rate of SB delivery required.

Table 3. Effect of time on sodium benzoate delivery (%) in orange juice at 2 °C and 25 °C.

Time (h)	2 °C		25 °C		Significance	
	SG	WI	SG	WI	WI/SG	Temperature
0.5	10.6 ± 4.2 ^{a}	9.9 ± 2.1 ^{a}	16.9 ± 0.2 ^{a}	9.0 ± 0.6 ^{a}	ns	ns
2	12.0 ± 0.5 ^{ab,A}	16.6 ± 1.2 ^{ab,B}	16.2 ± 1.2 ^{a,AB}	16.3 ± 1.4 ^{ab,AB}	*	ns
5	17.9 ± 2.7 ^{abc}	17.9 ± 2.1 ^{abc}	18.8 ± 2.1 ^{ab}	18.7 ± 3.4 ^{bc}	ns	ns
12	17.3 ± 5.7 ^{abc}	18.3 ± 1.4 ^{abc}	17.6 ± 0.3 ^{a}	21.0 ± 1.1 ^{bcd}	ns	ns
24	17.8 ± 0.7 ^{abc,A}	19.3 ± 0.5 ^{bc,AB}	20.9 ± 0.7 ^{ab,AB}	24.5 ± 1.2 ^{bcde,B}	*	*
48	18.4 ± 1.5 ^{abc,A}	25.4 ± 1.5 ^{cde,B}	21.8 ± 0.7 ^{ab,AB}	25.8 ± 0.7 ^{cde,B}	**	ns
72	19.3 ± 0.9 ^{abc}	24.4 ± 2.8 ^{bcd}	21.2 ± 3.7 ^{ab}	25.9 ± 0.7 ^{cde}	*	ns
96	21.9 ± 0.7 ^{bcd}	25.8 ± 0.7 ^{cde}	24.9 ± 3.5 ^{abc}	26.0 ± 0.9 ^{cde}	ns	ns
120	20.9 ± 2.0 ^{abcd,A}	28.4 ± 1.4 ^{de,B}	26.0 ± 0.1 ^{abc,AB}	28.4 ± 1.4 ^{de,B}	**	ns
240	21.9 ± 0.7 ^{bcd}	28.6 ± 3.8 ^{de}	28.1 ± 1.8 ^{bcd}	31.0 ± 2.3 ^{ef}	*	ns
360	23.5 ± 2.2 ^{cd,A}	29.5 ± 3.1 ^{de,AB}	32.8 ± 4.5 ^{cde,AB}	39.2 ± 3.5 ^{fg,B}	ns	*
720	27.6 ± 5.2 ^{cd}	34.0 ± 2.8 ^{ef}	37.1 ± 3.5 ^{de}	41.8 ± 3.5 ^{g}	ns	*
1440	31.5 ± 0.7 ^{d,A}	38.2 ± 2.1 ^{f,AB}	40.7 ± 3.3 ^{e,AB}	48.2 ± 4.5 ^{g,B}	*	*

$^{a–g}$ Means within a column with different lowercase superscripts differ significantly ($p < 0.05$); $^{A–B}$ means within a row with different uppercase superscripts differ significantly ($p < 0.05$); SG: tubular cellulose/sodium benzoate substrates by starch gel method; WI: tubular cellulose/sodium benzoate substrates by wet impregnation method; ns: not significant; * $p < 0.05$; ** $p < 0.01$.

3.4. *Antimicrobial Effectiveness of TC/SB Susbstrates in Orange Juice*

The quality and safety of juices is assured mainly through heat treatments due to their effectiveness and traditional nature. However, these treatments usually have negative impacts on the final nutritional quality and fresh-like characteristics of the juice. In order to overcome such problems, several non-thermal approaches have been evaluated in recent years [22]. Furthermore, there is a growing trend in the food industry for lowering energy and water demands, and the chemicals and additives that may have negative effects on human health, so-called "green consumerism" [23]. Following this trend, in the present study, the reduced use of chemical preservatives in orange juice is proposed by using TC/SB-SG. Its effectiveness was evaluated after contamination of orange juice with yeasts and lactic acid bacteria, and the results are presented in Table 4.

Table 4. Effect of encapsulated sodium benzoate (TC/SB-SG) on inactivation of spoilage microorganisms (log CFU/mL) of orange juices during 15 days of storage at 5 °C.

Time (d)	*L. casei*		*L. plantarum*		*S. cerevisiae*			
					EC1118		CR51	
	C	SB	C	SB	C	SB	C	SB
0	8.31 ± 0.01 ^{ab}	8.31 ± 0.01 ^{a}	8.32 ± 0.01 ^{a}	8.32 ± 0.01 ^{a}	7.41 ± 0.04 ^{ab}	7.41 ± 0.04 ^{a}	7.38 ± 0.03 ^{ab}	7.38 ± 0.03 ^{a}
2	7.69 ± 0.01 ^{b}	7.68 ± 0.68 ^{a}	8.54 ± 0.06 ^{b}	7.69 ± 0.34 ^{a}	7.49 ± 0.02 ^{bc}	4.75 ± 0.38 ^{b}	7.54 ± 0.01 ^{a}	5.57 ± 0.03 ^{b}
5	8.35 ± 0.07 ^{ab}	7.35 ± 0.01 ^{a}	8.63 ± 0.06 ^{b}	7.48 ± 0.02 ^{a}	7.66 ± 0.05 ^{c}	2.78 ± 0.67 ^{c}	7.59 ± 0.12 ^{a}	3.93 ± 0.11 ^{c}
15	9.97 ± 1.07 ^{a}	1.05 ± 0.07 ^{b}	8.34 ± 0.03 ^{a}	5.70 ± 0.28 ^{b}	7.30 ± 0.06 ^{a}	1.70 ± 0.28 ^{c}	7.19 ± 0.02 ^{b}	4.00 ± 0.07 ^{c}
Significance	ns	***	**	**	**	***	*	***

$^{a–c}$ Means within the same microorganism with different lowercase superscripts differ significantly ($p < 0.05$); C: control orange juice; SB: orange juice with encapsulated sodium benzoate; ns: not significant; * $p < 0.05$; ** $p < 0.01$; *** $p < 0.001$.

Saccharomyces cerevisiae is a common spoilage yeast in juices; therefore, it is usually used in studies with contaminated juices [24–29]. In the case of lactic acid bacteria, *L. plantarum* and *L. casei* are also reference spoilage strains used in orange and other juices [26,27,30,31]. Although relatively high initial contamination levels were selected (8.3 log CFU/mL in lactic acid bacteria and 7.4 log CFU/mL in yeasts), the encapsulated SB resulted in significantly lower contents compared to the control juice. SB was more effective in the case of yeasts during the first days of storage (2–2.5 log CFU reduction). After the 5th day and up to 15th day, the reduction was the highest in all cases due to the gradual release of SB in the orange juice. These results are promising considering the fact that the usual levels of contamination are significantly lower than those tested in the present study. However, more research is needed in order to evaluate more microorganisms and the possibility to control the release of SB.

4. Conclusions

A new method for the potential controlled delivery of SB in juices was reported by using encapsulation. The encapsulated methods that were evaluated resulted in different rates of SB delivery, while the use of different SB solutions gave different encapsulation rates. All of these provide a variety of TC/SB composites with different potential application in the food industry. The use of the WI method, 12% *w/v* SB solution and 2 h of stirring/contact seems ideal for application in the juice industry. However, the use of starch gel resulted in a more stable composite (TC/SB-SG) with slower SB delivery. The delivery proved capable of inhibiting the growth of potential spoilage microorganisms and even significantly reducing their numbers. Although the results are promising for potential applications, more research is needed in order to evaluate the controlled delivery of SB in juice or similar matrices and the possible effects on the organoleptic (flavor, aroma and color) properties of the product.

Author Contributions: Conceptualization, A.A.K.; methodology, A.A.K. and T.P.; validation, A.P., T.P. and P.K.; formal analysis, P.K.; investigation, A.P.; resources, A.A.K. and M.K.; writing—original draft preparation, A.P. and P.K.; writing—review and editing, T.P., P.K., M.K. and A.A.K.; visualization, A.P., T.P. and P.K.; supervision, M.K. and A.A.K.; project administration, M.K. and A.A.K.; funding acquisition, M.K. and A.A.K. All authors have read and agreed to the published version of the manuscript.

Funding: We acknowledge support of this work by the project "Research Infrastructure on Food Bioprocessing Development and Innovation Exploitation—Food Innovation RI" (MIS 5027222), which is implemented under the Action "Reinforcement of the Research and Innovation Infrastructure", funded by the Operational Program "Competitiveness, Entrepreneurship and Innovation" (NSRF 2014–2020) and co-financed by Greece and the European Union (European Regional Development Fund).

Institutional Review Board Statement: Not applicable.

Informed Consent Statement: Not applicable.

Data Availability Statement: All data are contained within the article.

Conflicts of Interest: The authors declare no conflict of interest. The funders had no role in the design of the study; in the collection, analyses, or interpretation of data; in the writing of the manuscript; or in the decision to publish the results.

References

1. Pongsavee, M. Effect of Sodium Benzoate Preservative on Micronucleus Induction, Chromosome Break, and Ala40Thr Superoxide Dismutase Gene Mutation in Lymphocytes. *BioMed Res. Int.* **2015**, *2015*, 1–5. [CrossRef]
2. Raposa, B.; Pónusz, R.; Gerencsér, G.; Budán, F.; Gyöngyi, Z.; Tibold, A.; Hegyi, D.; Kiss, I.; Koller, Á.; Varjas, T. Food Additives: Sodium Benzoate, Potassium Sorbate, Azorubine, and Tartrazine Modify the Expression of NFκB, GADD45α, and MAPK8 Genes. *Physiol. Int.* **2016**, *103*, 334–343. [CrossRef] [PubMed]
3. European Commission. Opinion on Benzoic Acid and Sodium Benzoate, Scientific Committee on Consumer Products (SCCP). 2005. Available online: https://ec.europa.eu/health/ph_risk/committees/04_sccp/docs/sccp_o_015.Pdf (accessed on 2 June 2021).
4. Chen, H.; Zhong, Q. Antibacterial Activity of Acidified Sodium Benzoate Against *Escherichia coli* O157:H7, *Salmonella enterica*, and *Listeria monocytogenes* in Tryptic Soy Broth and on Cherry Tomatoes. *Int. J. Food Microbiol.* **2018**, *274*, 38–44. [CrossRef] [PubMed]
5. Nan, A. Miscellaneous Drugs, Materials, Medical Devices and Techniques. In *Side Effects of Drugs Annual*; Elsevier BV: Amsterdam, The Netherlands, 2016; Volume 38, pp. 523–532. [CrossRef]
6. Shahmihammadi, M.; Javadi, M.; Nassiri-Asl, M. An Overview on the Effects of Sodium Benzoate as a Preservative in Food Products. *Biotechnol. Health Sci.* **2016**, *3*, 7–11. [CrossRef]
7. Kumar, M.; Bala, R.; Gondil, V.S.; Pandey, S.; Chhibber, S.; Jain, D.V.S.; Sharma, R.K.; Wangoo, N. Combating Food Pathogens Using Sodium Benzoate Functionalized Silver Nanoparticles: Synthesis, Characterization and Antimicrobial Evaluation. *J. Mater. Sci.* **2017**, *52*, 8568–8575. [CrossRef]

8. Aneja, K.R.; Dhiman, R.; Aggarwal, N.K.; Kumar, V.; Kaur, M. Microbes Associated with Freshly Prepared Juices of Citrus and Carrots. *Int. J. Food Sci.* **2014**, *2014*, 1–7. [CrossRef]
9. Salomão, B.D.C.M. Pathogens and Spoilage Microorganisms in Fruit Juice. In *Fruit Juices*; Elsevier BV: Amsterdam, The Netherlands, 2018; pp. 291–308. [CrossRef]
10. Piper, J.; Piper, P. Benzoate and Sorbate Salts: A Systematic Review of the Potential Hazards of These Invaluable Preservatives and the Expanding Spectrum of Clinical Uses for Sodium Benzoate. *Compr. Rev. Food Sci. Food Saf.* **2017**, *16*, 868–880. [CrossRef] [PubMed]
11. Barouni, E.; Petsi, T.; Kanellaki, M.; Bekatorou, A.; Koutinas, A.A. Low Volume Bioreactor Development in Dairy Industry Based on Encapsulated Rennin in Tubular Cellulose/Starch Gel Composite. *Food Bioprocess. Technol.* **2018**, *11*, 194–200. [CrossRef]
12. Kumar, M.; Gialleli, A.-I.; Masson, J.B.; Kandylis, P.; Bekatorou, A.; Koutinas, A.A.; Kanellaki, M. Lactic Acid Fermentation by Cells Immobilised on Various Porous Cellulosic Materials and their Alginate/Poly-Lactic Acid Composites. *Bioresour. Technol.* **2014**, *165*, 332–335. [CrossRef] [PubMed]
13. Panitsa, A.; Petsi, T.; Kandylis, P.; Nigam, P.S.; Kanellaki, M.; Koutinas, A.A. Chemical Preservative Delivery in Meat Using Edible Vegetable Tubular Cellulose. *LWT* **2021**, *141*, 111049. [CrossRef]
14. Katsaros, G.; Tsevdou, M.; Panagiotou, T.; Taoukis, P.S. Kinetic Study of High Pressure Microbial and Enzyme Inactivation and Selection of Pasteurisation Conditions for Valencia Orange Juice. *Int. J. Food Sci. Technol.* **2010**, *45*, 1119–1129. [CrossRef]
15. Dimitrellou, D.; Kandylis, P.; Kokkinomagoulos, E.; Hatzikamari, M.; Bekatorou, A. Emmer-Based Beverage Fortified with Fruit Juices. *Appl. Sci.* **2021**, *11*, 3116. [CrossRef]
16. Kandylis, P.; Dimitrellou, D.; Gousi, M.; Kordouli, E.; Kanellaki, M. Effect of Immobilization Support and Fermentation Temperature on Beer and Fermented Milk Aroma Profiles. *Beverages* **2021**, *7*, 47. [CrossRef]
17. Kandylis, P.; Bekatorou, A.; Pissaridi, K.; Lappa, K.; Dima, A.; Kanellaki, M.; Koutinas, A.A. Acidogenesis of Cellulosic Hydrolysates for New Generation Biofuels. *Biomass Bioenergy* **2016**, *91*, 210–216. [CrossRef]
18. Kandylis, P.; Kanellaki, M. Effect of Alkaline Pretreatment on Corn Leaves. *Ecol. Saf.* **2017**, *11*, 246–254.
19. Koutinas, A.A.; Sypsas, V.; Kandylis, P.; Michelis, A.; Bekatorou, A.; Kourkoutas, Y.; Kordulis, C.; Lycourghiotis, A.; Banat, I.; Nigam, P.; et al. Nano-Tubular Cellulose for Bioprocess Technology Development. *PLoS ONE* **2012**, *7*, e34350. [CrossRef] [PubMed]
20. Gialleli, A.-I.; Ganatsios, V.; Terpou, A.; Kanellaki, M.; Bekatorou, A.; Koutinas, A.A.; Dimitrellou, D. Technological Development of Brewing in Domestic Refrigerator Using Freeze-Dried Raw Materials. *Food Technol. Biotechnol.* **2017**, *55*, 325–332. [CrossRef]
21. Conte, A.; Lecce, L.; Iannetti, M.; Del Nobile, M.A. Study on the Influence of Bio-Based Packaging System on Sodium Benzoate Release Kinetics. *Foods* **2020**, *9*, 1010. [CrossRef]
22. Bevilacqua, A.; Petruzzi, L.; Perricone, M.; Speranza, B.; Campaniello, D.; Sinigaglia, M.; Corbo, M.R. Nonthermal Technologies for Fruit and Vegetable Juices and Beverages: Overview and Advances. *Compr. Rev. Food Sci. Food Saf.* **2018**, *17*, 2–62. [CrossRef]
23. Bevilacqua, A.; Corbo, M.R.; Sinigaglia, M. Green Consumerism and Alternative Approaches for Food Preservation: An Intro-Duction. In *Application of Alternative Food-Preservation Technologies to Enhance Food Safety and Stability*; Bevilacqua, A., Corbo, M.R., Sinigaglia, M., Eds.; Bentham Publisher: Sharjah, United Arab Emirates, 2010; pp. 1–3.
24. Samani, B.H.; Khoshtaghaza, M.H.; Lorigooini, Z.; Minaei, S.; Zareiforoush, H. Analysis of the Combinative Effect of Ultrasound and Microwave Power on *Saccharomyces cerevisiae* in Orange Juice Processing. *Innov. Food Sci. Emerg. Technol.* **2015**, *32*, 110–115. [CrossRef]
25. Shi, J.; Zhu, X.; Lu, Y.; Zhao, H.; Lu, F.; Lu, Z. Improving Iturin A Production of *Bacillus amyloliquefaciens* by Genome Shuffling and Its Inhibition Against *Saccharomyces cerevisiae* in Orange Juice. *Front. Microbiol.* **2018**, *9*, 2683. [CrossRef]
26. Hashemi, S.M.B.; Mahmoudi, M.R.; Roohi, R.; Torres, I.; Saraiva, J.A. Statistical Modeling of the Inactivation of Spoilage Microorganisms During Ohmic Heating of Sour Orange Juice. *LWT* **2019**, *111*, 821–828. [CrossRef]
27. Fenoglio, D.; Ferrario, M.; Schenk, M.; Guerrero, S. Effect of Pilot-Scale UV-C Light Treatment Assisted by Mild Heat on *E. coli*, *L. plantarum* and *S. cerevisiae* Inactivation in Clear and Turbid Fruit Juices. Storage Study of Surviving Populations. *Int. J. Food Microbiol.* **2020**, *332*, 108767. [CrossRef] [PubMed]
28. Niu, L.; Wu, Z.; Yang, L.; Wang, Y.; Xiang, Q.; Bai, Y. Antimicrobial Effect of UVC Light-Emitting Diodes Against *Saccharomyces cerevisiae* and Their Application in Orange Juice Decontamination. *J. Food Prot.* **2021**, *84*, 139–146. [CrossRef] [PubMed]
29. Anapi, G.R.; Aba, R.P.M.; Gabriel, A.A. Screening for Heat-Resistant Reference Yeast Isolate in Orange Juice. *Food Microbiol.* **2021**, *94*, 103639. [CrossRef] [PubMed]
30. Gialleli, A.-I.; Bekatorou, A.; Kanellaki, M.; Nigam, P.; Koutinas, A.A. Apple Juice Preservation through Microbial Adsorption by nano/Micro-Tubular Cellulose. *Innov. Food Sci. Emerg. Technol.* **2016**, *33*, 416–421. [CrossRef]
31. Peña, N.; Panagopoulos, V.; Kanellaki, M.; Koutinas, A.A.; Ruiz-Rico, M.; Fernández-Segovia, I.; Barat, J.M. Non-Thermal Treatment for the Stabilisation of Liquid Food Using a Tubular Cellulose Filter from Corn Stalks. *Food Control.* **2020**, *112*, 107164. [CrossRef]

foods

MDPI

Article

Effects of Hay, Baleage, and Soybean Hulls Waste Used as Supplemental Feeds on the Nutritional Profile of Grass-Finished Beef

Lucas Krusinski [1], Isabella C. F. Maciel [2], Selin Sergin [1], Vijayashree Jambunathan [1], Esha Garg [1], Andrea J. Garmyn [1,2], Sukhdeep Singh [3], Chad A. Bitler [4], Jason E. Rowntree [2] and Jenifer I. Fenton [1,*]

[1] Department of Food Science and Human Nutrition, Michigan State University, 469 Wilson Rd, East Lansing, MI 48824, USA
[2] Department of Animal Science, Michigan State University, 474 S Shaw Ln, East Lansing, MI 48824, USA
[3] Department of Plant, Soil, and Microbial Sciences, Michigan State University, 1066 Bogue St, East Lansing, MI 48824, USA
[4] Greenacres Foundation Inc., Cincinnati, OH 45242, USA
* Correspondence: imigjeni@msu.edu; Tel.: +1-(517)-353-3342

Abstract: Grass-finished beef (GFB) has demonstrated wide nutritional variations with some GFB having a considerably higher *n*-6:*n*-3 ratio compared to grain-finished beef. To better understand these variations, the current study investigated the effects of commonly used supplemental feeds on the nutritional profile of GFB. This two-year study involved 117 steers randomly allocated to one of four diets: (1) grass+hay (G-HAY), (2) grass+baleage (G-BLG), (3) grass+soybean hulls (G-SH), and (4) baleage+soybean hulls in feedlot (BLG-SH). Feed samples were analyzed for their nutritional value, and beef samples underwent analysis for fatty acids (FAs), vitamin E, minerals, lipid oxidation, and shear force. FAs were measured by GC-MS, vitamin E was analyzed chromatographically, minerals were analyzed by ICP-MS, and lipid oxidation was measured via a thiobarbituric acid reactive substances (TBARS) assay. G-SH beef had the highest *n*-6:*n*-3 ratio ($p < 0.001$), while BLG-SH beef contained less vitamin E ($p < 0.001$) and higher TBARS values ($p < 0.001$) compared to the other groups. G-HAY beef contained more long-chain *n*-3 polyunsaturated FAs compared to the other groups ($p < 0.001$). In conclusion, G-HAY beef had the most beneficial nutritional profile, while soybean hulls increased the *n*-6:*n*-3 ratio of beef.

Keywords: cattle feed; conserved forages; fatty acids; grassland; meat quality; minerals; pasture; sustainability

1. Introduction

The market for grass-finished beef (GFB) is growing with retail sales of pasture-raised beef increasing from $17 million in 2012 to $272 million in 2016 [1,2]. The consumer reasoning for these growing sales is complex. However, sustainable food production is driving food choices for savvy consumers [3,4]. GFB appeals to consumers who are interested in a healthier product for human consumption and the environmental considerations [5]. Leading organizations and platforms such as the EAT Lancet commission recommend to drastically reduce red meat consumption for health and environmental reasons [6]; however, the type of production system is generally ignored in this recommendation [2]. GFB remains an underexplored alternative to attain sustainability goals [7]. For one, GFB is more consistent with health recommendations. Compared to conventional grain-finished beef, GFB contains more omega-3 (*n*-3) fatty acids (FAs) including eicosapentaenoic acid (EPA) and docosahexaenoic acid (DHA), twice as much conjugated linoleic acid (CLA), and 25% more polyunsaturated FAs (PUFAs) [3,8–11]. GFB also contains less omega-6 (*n*-6) PUFAs, less total fat, and less cholesterol-raising saturated FAs (SFAs) [12,13]. GFB has an

n-6:n-3 ratio of 1.5:1 compared to 7.7:1 for conventional grain-finished beef [14,15]. Foods containing an n-6:n-3 ratio closer to 1:1 are recommended by human nutrition health professionals [16,17]. Additionally, consumption of foods containing higher concentrations of phytochemicals, typically fruits and vegetables, is also a key recommendation. In fact, GFB has important antioxidant properties that also contribute to its healthfulness [18,19]. GFB contains three times more vitamin E and 1.5–10 times more β-carotene than grain-finished beef [15,20–22]. Some studies suggest that grass-finishing enhances the phenolic content of beef, especially when cattle are grazing on phytochemically biodiverse pastures [2,19,23]. Properly managed complex biodiverse pastures also have the merit to sequester more carbon and to enhance fresh-water systems [24].

The nutritional composition of beef is highly dependent on the feeding system, yet these differences in nutritional quality are generally not reflected on food labels [2,19]. GFB typically means that cattle were fattened solely on grass and forages before slaughter [25]. According to the American Grassfed Association standards, all cattle must be pasture based meaning that grass and forage must be consumed throughout the lifetime of the animal except for milk consumed before weaning. Hay, baleage, and silages may be consumed by the animal when fresh grass and forages are not available due to inclement weather for instance [26]. Supplemental feeds might be needed in some regions where fresh forages are not available year-round. Season, soil composition, weather, and light exposure all play crucial roles in feed quality and availability [19,27–29].

Interestingly, a nutritional survey of commercially available GFB published by our group highlighted important differences among beef from grass-finishing systems [21]. The n-6:n-3 ratio varied from 1.8:1 to as high as 28.3:1 and some GFB was devoid of β-carotene. These differences were hypothesized to be due to a wide variety of feeding practices that were reflected in the nutritional profile of beef [21]. Feeding fresh forages to cattle usually results in the most beneficial nutritional profile of beef [19,21]. However, producers may rely on conserved forages and other supplemental feeds when fresh grass is not available [30]. Unfortunately, conserved forages made by drying (hay) or fermentation (baleage) often display lower nutritional quality compared to fresh forages with lower concentrations of antioxidants and phenolic compounds [31,32]. The processing of fresh forages into conserved feeds results in oxidation of PUFAs and an increase in palmitic acid [33,34]. These changes in the nutritional profile of feeds modify FA metabolism in the rumen, resulting in variations in the FA content of beef [35,36]. Although not allowed by the American Grassfed Association [26], soybean hulls are also used as supplemental feed by some producers in the U.S. [19,21]. The effects of feeding soybean hulls to cattle remain controversial in the literature. Some studies found no differences in CLA, *trans* vaccenic acid (TVA), n-3 PUFAs, and the n-6:n-3 ratio among cattle fed soybean hulls or fresh forage in the finishing phase [20,21]. On the other hand, another study reported that cattle supplemented with soybean hulls had more total fat, less n-3 PUFAs, and a higher n-6:n-3 ratio compared to cattle fed only fescue [37].

Krusinski et al. [19] highlighted that the nutritional profile of GFB is highly variable and depends on a multitude of factors ranging from supplemental feeds to seasonal variations. The present study builds on the work of Bronkema et al. [21] in an attempt to explain the large variations reported among GFB, especially regarding the n-6:n-3 ratio. With growing interest in assessing the nutritional impact and sustainability of food systems, determining the accurate nutritional value of foods is crucial. The objective of this study was to compare the FA and micronutrient content of GFB fed a diverse pasture mixture and commonly used supplemental feeds to better understand the effects of different feeds on the nutritional profile of GFB.

2. Materials and Methods

The animal protocol was reviewed and approved by the Michigan State University Institutional Animal Care and Use Committee (IACUC #202000054).

2.1. Experimental Design, Animals, and Diets

This two-year study (2020 and 2021) took place at the Michigan State University Kellogg Biological Station (KBS) located in Hickory Corners, MI (latitude: 42°24′38″ N, longitude: 85°22′45″ W, elevation: 282 m). Sixty steers for each year were randomly allocated to one of four diets: grass supplemented with hay (G-HAY), grass supplemented with baleage (G-BLG), grass supplemented with soybean hulls (G-SH), or baleage and soybean hulls in feedlot (BLG-SH). Three groups for each diet were formed (n = 5 animals/replicate; 3 replicates/diet; 15 animals/diet) for each year. Animals were randomly stratified and allocated to one of the three groups in each diet.

In April of each year, 60 Simmental-Angus influenced feeder cattle weighing on average 387 kg (±47 kg) were purchased from the same producer and shipped from Oklahoma to KBS. Upon their arrival at KBS, initial weights were collected, and steers were randomly stratified and assigned to the diets. Steers allocated to the three diets containing grass were kept on pasture and had ad libitum access to a diverse pasture mixture (GRASS) and 4.5 kg of supplemental feed (dry matter; DM) per day per head. Steers kept in the feedlot group had ad libitum access to baleage (BLG) and 4.5 kg of soybean hulls (SH) per day per head. GRASS was a five-species mix of alfalfa (*Medicago sativa*), red clover (*Trifolium pratense* L.), white clover (*Trifolium repens* L.), orchard grass (*Dactylis glomerata* L.), and endophyte-free tall fescue (*Festuca arundinacea*). Dry hay (HAY) was composed of alfalfa (*Medicago sativa*), orchard grass (*Dactylis glomerata* L.), and tall fescue (*Festuca arundinacea*). BLG was a mixture of alfalfa (*Medicago sativa*) and orchard grass (*Dactylis glomerata* L.). Each subgroup for each grass-containing diet was allocated a fenced paddock. In total, each diet was allocated three paddocks, each containing five animals. Each paddock was further divided into sub-paddocks to give time to the pasture to rest and regrow. Animals were rotated three times per week within their paddocks to fresh parcels of grass. The 15 steers in the BLG-SH diet were treated as feedlot cattle and were separated into three pens each containing five animals. One animal died during the first year of the experiment, and two carcasses were misplaced by the slaughterhouse during the second year of the study, bringing the total number of animals for the entire study to 117 (n = 117).

2.2. Sample Collection and Preparation

2.2.1. Feed Samples

Samples from grazing areas and supplemental feeds were collected every two weeks. The sample collection started in July and ended in late October of each year. No sample collection occurred between April and July 2020 because of COVID-19 restrictions. To stay consistent, the sample collection period was kept the same for 2021. GRASS samples were gathered every two weeks in each sub-paddock immediately before animals had access to the pasture (n = 63 for each year, n = 126 in total). GRASS samples were collected by randomly cutting three 0.25 m^2 quadrats to a 5 cm stubble using hand grass clipper scissors. HAY, BLG, and SH were sampled monthly before being distributed to the steers (n = 4 of each supplemental feed for each year, n = 8 of each in total). Supplemental feeds were sampled monthly instead of bi-weekly because less variations over time in the nutritional profile of these feeds were expected. For proximate analysis, wet weights were recorded, and samples were dried in a forced-air oven (72 h, 55 °C) and ground through a 1 mm screen in a Wiley mill (Arthur H. Thomas, Philadelphia, PA, USA). For FA and phytochemical analysis, a 30 g sub-sample was packed in a Whirl-Pak bag and frozen at −20 °C immediately after collection. In order to obtain representative samples, bag contents were mixed, and 10 g of each replicate was taken before being combined. Feed samples were stored at −20 °C for the length of the trial before being stored at −80 °C once they were brought back to the laboratory. Before further analysis, samples were freeze-dried in a freeze dryer (Harvest Right, North Salt Lake, UT, USA) for 18.5 h, and ground in a Wiley mill (1 mm screen) (Arthur H. Thomas, Philadelphia, PA, USA) with dry ice.

Weather conditions were reported according to Krusinski et al. [38] using the Michigan State University Enviroweather platform at KBS. 30-year normal temperature and precipitation (1991–2020) were reported according to the National Centers for Environmental Information: National Oceanic and Atmospheric Administration website (Gull Lake, MI meteorological station).

2.2.2. Meat Samples

In November of each year, before going to slaughter, steers were weighed again to obtain total weight gain and average daily gain (ADG). Steers were slaughtered in a USDA facility at 18–20 months of age. Body performance and carcass traits (ribeye area, 12th rib back fat, USDA yield grade, and marbling score) were collected by trained personnel 48 h after slaughter. Simultaneously, meat samples (approximately 7.5 to 10 cm in length) were collected from the left-side longissimus lumborum (between 13th rib and first two lumbar vertebra). For FA analysis and thiobarbituric acid reactive substances (TBARS), one steak per carcass was cut into 1 × 1 cm cubes before being flash frozen with liquid nitrogen, put into Whirl-Pak bags, and stored at −80 °C until analysis. For Warner-Bratzler shear force (WBSF), another 2.54 cm-thick steak was cut, vacuum packed, and stored at 4 °C until 14 days postmortem. At 14 days postmortem, the steaks were frozen at −20 °C until WBSF analysis was performed.

2.3. Feed Chemical Analysis

2.3.1. Proximate Analysis

The protocol for the feed proximate analysis was previously described by Maciel et al. [39]. Samples were dried at 105 °C in a forced-air oven for 8 h. To determine the ash content, feed samples were oxidized at 500 °C for 6 h in a muffle furnace. Neutral detergent fiber (NDF) was determined according to Mertens [40] with the addition of amylase and sodium sulfite. The protocol described in AOAC [41] was used to determine acid detergent fiber (ADF). Crude protein (CP) was measured according to Hach et al. [42] and gross energy was measured by bomb calorimeter.

2.3.2. Phytochemical Analysis

Chlorophyll A and B were determined as described previously [43]. Briefly, 2 g of lyophilized and ground feed was added to 70% aqueous acetone. The mixture was shaken for 30 min and centrifuged for 20 min (4 °C, 2500 RPM). The upper layer was transferred to a new tube, and the extraction was repeated twice. Compounds were measured using a UV-Vis Double Beam Spectrophotometer (VWR, Radnor, PA, USA) in cuvettes. Readings were recorded at 663 and 646 nm and were used in the following equations:

$$\text{Chlorophyll A } (C_a) = 12.21A_{663} - 2.81A_{646}$$
$$\text{Chlorophyll B } (C_b) = 20.13A_{646} - 5.03A_{663}$$

To extract phenolic compounds, a modified protocol from Nimalaratne et al. [44] was performed. First, 2 g of freeze-dried and ground feed was added to 20 mL of methanol:distilled water:acetic acid (70:28:2, *v*/*v*/*v*). The mixture was shaken for 30 min and centrifuged for 20 min (4 °C, 2500 RPM). The upper layer was recovered and transferred to a new tube. An additional 20 mL of acetone:distilled water:acetic acid (70:28:2, *v*/*v*/*v*) was added to the original tube before being shaken for 10 min and centrifuged for 15 min (4 °C, 2500 RPM). Both supernatants were combined and stored at 4 °C. The Folin-Ciocalteu assay adapted from Singleton and Rossi [45] was used to measure total phenolic content. A standard curve was made using a 1 mg/mL gallic acid stock solution in methanol. A serial dilution was performed by a factor of two to obtain concentrations ranging from 1 mg/mL to 0.002 mg/mL. Next, 100 µL of Folin-Ciocalteu reagent and 800 µL of 5% sodium bicarbonate were added to the standard curve and to 100 µL of supernatant. The standard curve and the samples were heated at 40 °C for 30 min. Samples were allowed to cool down to room temperature before being plated in triplicates in a 96-well plate. Samples were scanned at 765 nm and compared against the gallic acid standard curve. Values were reported as mg of gallic acid equivalents (GAE)/g of feed.

2.4. Fatty Acid Analysis of Feed and Meat

All chemicals were purchased from Sigma-Aldrich (St. Louis, MO, USA) unless otherwise noted.

The FA analysis for feed and meat samples was conducted according to Sergin et al. [46]. A microwave-assisted extraction protocol was performed to extract FAs as reported by Bronkema et al. [21] using a CEM Mars 6 microwave (CEM Corp., Matthews, NC, USA). For this step, 400 mg of ground feed sample or minced meat was added to a microwave vessel containing 8 mL of 4:1 (*v*/*v*) ethyl acetate:methanol solution with 0.1% BHT. The samples underwent extraction with the following settings: 55 °C for 15 min with initial ramp of 2 min at 400 W. Samples were then filtered in another tube containing 3.5 mL of HPLC water before being centrifuged (6 min, 2500 RPM). The upper layer was removed and dried under nitrogen. To resuspend the oil, a 4:1 (*v*/*v*) dichloromethane:methanol solution with 0.1% BHT was used to bring the concentration of each sample to 20 mg of oil/mL.

For the creation of FA methyl esters (FAMEs), a modified version of the protocol by Jenkins [47] was applied. Briefly, 2 mg of oil (100 µL) was resuspended in toluene with 20 µg of internal standard (methyl 12-tridecenoate, U-35M, Nu-Chek Prep, Elysian, MN, USA). Then, 2 mL of 0.5 N anhydrous potassium methoxide was added and the samples were heated at 50 °C for 10 min. Next, 3 mL of methanolic HCl (5%) was added to the samples before being heated at 80 °C for 10 min. Once cool, 2 mL of HPLC water and 2 mL of hexane were added, and samples were centrifuged for 5 min at 2500 RPM. The top layer was moved to another tube and dried under nitrogen to obtain FAMEs. FAMEs were then resuspended in 1 mL of isooctane to get a concentration of 2mg/mL. Samples were transferred to gas chromatography-mass spectrometry (GC-MS) vials with glass inserts.

The PerkinElmer (Waltham, MA, USA) 680/600S GC-MS in electron impact mode (70 eV) equipped with an Agilent Technologies (Santa Clara, CA, USA) HP-88 column (100 m, 0.25 mm ID, 0.2 µM film thickness) was used for the quantification of FAMEs. For feed samples, one µL of sample was injected with the GC temperature set at 250 °C. For meat samples, one µL was injected twice (20:1 split) at two different GC temperatures (175 °C and 150 °C). The temperature settings for both feed and meat samples were as follows: initial temperature at 80 °C for 4 min; ramp 13 °C/min to 175 °C; hold 27 min; ramp 4 °C/min to 215 °C; hold 35 min, and then an initial temperature at 80 °C for 4 min; ramp 13 °C/min to 150 °C; hold 47 min; ramp 4 °C/min to 215 °C; hold 35 min. For meat samples, a third injection followed in splitless mode (0.75 min splitless hold time, 40 mL/min flow exiting the vent). This GC-MS method was adapted from Kramer et al. [48]. Helium was the carrier gas (flow rate of 1 mL/min). MS data were recorded in full scan mode (mass range of m/z 70-400 amu) and the MS transfer line and ion source temperature were set at 180 °C.

MassLynx V4.1 SCN 714 (Water Corporation, Milford, MA, USA) was used for the identification of FAMEs. FAs were identified by retention time and EI mass fragmentation compared to the reference standard containing the Supelco 37 Component FAME Mix with mead acid, docosatetraenoic acid, *n*-3 DPA, *n*-6 DPA, and palmitelaidic acid purchased from Cayman Chemical (Ann Arbor, MI, USA). CLA isomers were identified using the CLA reference standard UC-59M (Nu-Chek Prep, Elysian, MN, USA). FAs not included in the reference standard were identified by elution order and confirmed by the EI mass fragmentation [48]. FAs were quantified using a standard curve including the reference and internal standards. Each FAME concentration was calculated by using the internal standard peak area and analyte peak area compared to the standard curve. C18:1 4*t* and C18:1 5*t* were below the limit of detection, and C18:2 9*c*,12*t* and C18:2 9*t*,12*c* were not separated from the C18:2 11*t*,15*c* peak. Eicosatetraenoic acid (C20:4 *n*-3) was not included in our reference standard and was therefore not reported. FAs were reported in mg/100 g of beef in this manuscript and in percent of total FAs in the Supplementary Tables S1 and S2.

2.5. Vitamin E and Mineral Analysis

Protocols by Rettenmaier and Schüep [49] were followed for vitamin E analysis. In brief, 1 g of beef was homogenized in 5 mL of water before being frozen. For extraction, samples were thawed, and a measured aliquot was pipetted out. To precipitate the protein, ethanol was added, and vitamins were extracted with hexane. After being centrifuged, part of the hexane layer was removed and dried under reduced pressure in a vortexing chamber (10 min, 35 °C, 300 mBar vacuum). What remained after evaporation was solubilized in the chromatographic mobile phase and placed in vials. A calibration curve (six points) was made as follows: a vitamin E solution (Sigma-Aldrich, St. Louis, MO, USA) diluted with ethanol (containing BHT) underwent serial dilutions (from 50 µg/mL to 0.2 µg/mL). For the chromatography analysis, a Waters Acquity system and Water Empower Pro Chromatography Manager software (Water Corporation, Milford, MA, USA) were used. An isocratic elution was performed using a mobile phase of acetonitrile:methylene chloride:methanol (70:20:10, *v/v/v*) and a Symmetry C18, 1.7 µm, 2.1 × 50 mm analytical column (Waters Corporation, Milford, MA, USA). The flow rate was 0.5mL/min and the detection was performed by UV absorption at 292 nm.

Mineral analysis was performed as previously described [50,51]. Briefly, beef samples underwent drying and digestion in an oven (95 °C, overnight) using 10 times the dry tissue mass of nitric acid. A dilution with water to 100 times the dried tissue mass followed. An Agilent 7900 Inductivity Coupled Plasma–Mass Spectrometer (ICP-MS) (Agilent Technologies Inc., Santa Clara, CA, USA) was used for the analysis. A six-point calibration curve was used. Standards of bovine liver and mussels (National Institute of Standards and Technology, Gaithersburg, MD, USA) were used as controls.

2.6. Thiobarbituric Acid Reactive Substances (TBARS)

The TBARS assay for food and beverages (Oxford Biomedical Research, Oxford, MI, USA) adapted for a 96-well plate reader was used. First, an eight-point standard curve was created by serial dilution ranging from 0 (only HPLC water) to 3 mg/L malondialdehyde (MDA) (MDA stock solution provided in the kit). Then, 500 mg of minced beef sample was added to 5 mL of HPLC water. Samples were homogenized to obtain a smooth solution. In a microcentrifuge tube, 250 µL of sample solution and 250 µL of the indicator solution (thiobarbituric acid (TBA) and acid solution) were mixed. The indicator solution was also added to the standard curve, and the samples and the curve were set aside for 60 min for the reaction to occur. Samples were then centrifuged at 11,000 RPM for 5 min at room temperature. The aqueous layer was removed and plated in duplicates on a 96-well plate, next to the standard curve. Absorbance was read at 532 nm on a Bio-Tek Synergy HT spectrophotometer (Bio-Tek Instruments, Inc., Winooski, VT, USA). The standard curve was plotted and the MDA concentration for the samples (mg MDA/L) was calculated according to the manufacturer's instructions.

2.7. Warner-Bratzler Shear Force (WBSF)

The protocol for WBSF was previously reported [39]. Briefly, the steaks were cooked to an internal temperature of 71 °C using a preheated clamshell electric grill (George Foreman, Beachwood, OH, USA). The steaks were then cooled down overnight at 4 °C. Six to eight 1.27 cm diameter cores were cut from each steak by paying close attention to cut parallel to the muscle and fibers using a drill mounted corer. Shear force was measured using the TA-XT Texture Analyzer (Stable Micro System Ltd., Godalming, UK) with a V-shaped Warner-Bratzler blade. The blade was moving down at a speed of 20cm/min and cut the sample across the muscle fiber. The purpose of the shear force testing was to measure how much force is required to cut through cooked meat. This should be a representative measure of the ease or difficulty a consumer would have chewing a cooked steak. Most consumer prefer steaks cooked between medium rare and medium well, which is why an internal temperature of 71 °C was chosen for the analysis. The mean of the cores for each sample were used for the statistical analysis.

2.8. Statistical Analysis

SAS version 9.4 (SAS Institute Inc., Cary, NC, USA) was used to perform the statistical analysis. Mixed model analysis was performed to test the effect of diet on response variables. In the model, the fixed effect was diet, and the random effects were year and pen nested within year x diet. Each pen was the experimental unit. Post hoc comparison was performed using Tukey's adjustment, and results were considered significant at $p < 0.05$. Outliers were removed for chlorophyll A and chlorophyll B after running an outlier test. The data satisfied model's normality and equal variance assumptions. Data are shown as mean ± standard error across mean (SEM).

3. Results

3.1. Weather Conditions

Weather conditions at the experimental site for the length of the study are shown in Figure 1. The hottest month in 2020 was July with an average of 23.96 °C. The average temperature in August, September, and October 2020 were all below the 30-year normal. Every month in 2020 was below the 30-year normal for rainfall. In 2021, August was the hottest month with an average temperature of 23.10 °C. July, September, and October 2021 were above the 30-year normal temperature. September 2021 showed unusually high rainfall with 338.87 mm compared to the 30-year normal precipitation of 88.39 mm.

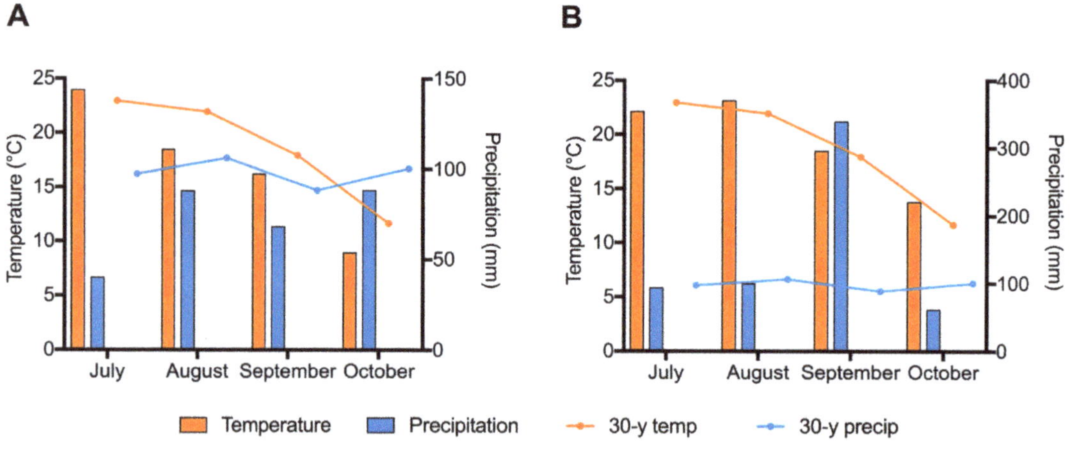

Figure 1. Monthly weather conditions at the experimental site. (**A**) 2020 and (**B**) 2021. 30-y temp: 30-year normal temperature at the experimental site; 30-y precip: 30-year normal precipitation at the experimental site.

3.2. Feeds

3.2.1. Proximate Composition of Feeds

The proximate composition of the feeds is displayed in Table 1. No significant differences were observed between feed types regarding dry matter (DM) ($p = 0.159$). GRASS and BLG had the highest values for ash ($p < 0.001$) and CP ($p < 0.001$), while SH had the lowest values for ash ($p < 0.001$) and HAY had the lowest values for CP ($p < 0.001$). Regarding NDF, SH was higher than GRASS, HAY, and BLG ($p = 0.004$). For ADF, SH was highest while GRASS was lowest ($p < 0.001$). Finally, SH had the lowest amount of energy compared to the other three feed types ($p < 0.001$).

Table 1. Mean proximate composition of the feeds.

	GRASS	HAY	BLG [1]	SH [2]	p-Value
DM [3]	57.52 ± 22.47	85.82 ± 25.01	82.64 ± 25.02	89.98 ± 25.01	0.159
Ash *	9.20 ± 0.46 [a]	7.14 ± 0.61 [b]	8.38 ± 0.63 [a,b]	4.74 ± 0.61 [c]	<0.001
CP [4] *	15.65 ± 0.39 [a]	7.15 ± 1.14 [c]	13.48 ± 1.19 [a,b]	9.47 ± 1.19 [b,c]	<0.001
NDF [5] *	54.91 ± 3.30 [b]	66.19 ± 4.01 [a,b]	54.23 ± 4.04 [b]	68.29 ± 4.01 [a]	0.004
ADF [6] *	30.84 ± 0.72 [c]	37.98 ± 1.35 [b]	33.51 ± 1.39 [b,c]	51.72 ± 1.35 [a]	<0.001
Energy [7]	4566.41 ± 50.99 [a]	4405.00 ± 90.76 [a]	4465.41 ± 50.99 [a]	3709.96 ± 90.76 [b]	<0.001

Values reported as means ± standard error. Different letters denote statistical significance at $p < 0.05$ (mixed model analysis, post hoc comparison performed using Tukey's adjustment). $n = 126$ for GRASS, and $n = 8$ for the other three feeds. [1] BLG: baleage; [2] SH: soybean hulls; [3] DM: dry matter (%), * reported in %DM; [4] CP: crude protein; [5] NDF: neutral detergent fiber; [6] ADF: acid detergent fiber; [7] Energy (cal/g).

3.2.2. Fatty Acid Composition of the Feeds

The FA profile of the feeds is reported in Table 2. Palmitic acid (C16:0) made up most of the SFA content of the feeds. HAY contained the highest concentration of C16:0, while GRASS and SH contained the lowest ($p < 0.001$). Regarding stearic acid (C18:0), SH and HAY contained the most and GRASS contained the least ($p < 0.001$). The total SFA content was significantly higher in HAY and was lower in GRASS and SH ($p < 0.001$). Total MUFA content was significantly higher in SH and lower in GRASS. BLG and HAY values were in between and were significantly different than SH and GRASS ($p < 0.001$). Regarding PUFAs, the linoleic acid (LA) content was the highest in SH and was lower in GRASS and HAY ($p < 0.001$), while the α-linolenic acid (ALA) content followed the opposite trend with GRASS containing the most and SH containing the least ($p < 0.001$). GRASS contained the highest concentration of total n-3 PUFAs and the lowest concentration of n-6 PUFAs. SH contained the most n-6 PUFAs and the least n-3 PUFAs ($p < 0.001$). This resulted in SH having the highest n-6:n-3 ratio and GRASS having the lowest ($p < 0.001$).

Table 2. Mean fatty acid composition of the diets (% total fatty acids).

	GRASS	HAY	BLG [1]	SH [2]	p-Value
C10:0	0.16 ± 0.26	1.16 ± 0.38	0.14 ± 0.38	0.03 ± 0.38	0.116
C12:0	0.50 ± 0.22 [b]	1.16 ± 0.23 [a]	0.49 ± 0.23 [b]	0.10 ± 0.23 [c]	<0.001
C13:0	0.01 ± 0.02	0.07 ± 0.03	0.02 ± 0.03	0.00 ± 0.03	0.193
C14:0	0.53 ± 0.29 [b]	2.14 ± 0.45 [a]	0.62 ± 0.45 [a,b]	0.21 ± 0.45 [b]	0.041
C15:0	0.11 ± 0.06 [b]	0.49 ± 0.09 [a]	0.33 ± 0.09 [a,b]	0.14 ± 0.09 [b]	0.010
C16:0	14.29 ± 1.98 [c]	32.16 ± 2.37 [a]	23.77 ± 2.38 [b]	14.95 ± 2.37 [c]	<0.001
C16:1 9*c*	0.23 ± 0.02 [b]	0.51 ± 0.04 [a]	0.31 ± 0.04 [b]	0.23 ± 0.04 [b]	0.001
C16:1 7*c*	1.19 ± 0.27 [a]	1.21 ± 0.29 [a]	1.52 ± 0.29 [a]	0.11 ± 0.29 [b]	<0.001
C17:0	0.21 ± 0.02 [c]	0.57 ± 0.03 [a]	0.36 ± 0.03 [b]	0.28 ± 0.03 [b,c]	<0.001
C18:0	1.57 ± 0.13 [c]	3.80 ± 0.20 [a]	2.78 ± 0.20 [b]	4.33 ± 0.20 [a]	<0.001
C18:1 9*c*	1.84 ± 0.07 [c]	3.45 ± 0.23 [b]	2.43 ± 0.24 [b,c]	12.61 ± 0.23 [a]	<0.001
C18:1 11*c*	0.61 ± 0.03 [c]	1.04 ± 0.09 [b]	0.72 ± 0.10 [b,c]	2.43 ± 0.09 [a]	<0.001
C18:2 *n*-6 (LA) [3]	12.22 ± 0.35 [c]	14.86 ± 0.88 [b,c]	16.42 ± 0.91 [b]	48.45 ± 0.88 [a]	<0.001
C18:3 *n*-3 (ALA) [4]	64.66 ± 1.74 [a]	32.48 ± 2.57 [c]	46.47 ± 2.64 [b]	15.00 ± 2.57 [d]	<0.001
C20:0	0.58 ± 0.08 [b]	2.72 ± 0.16 [a]	1.05 ± 0.16 [b]	0.41 ± 0.16 [b]	<0.001
C20:3 *n*-3	0.08 ± 0.03	0.03 ± 0.03	0.05 ± 0.03	0.01 ± 0.03	0.052
C22:0	0.59 ± 0.16 [b,c]	1.90 ± 0.20 [a]	1.04 ± 0.21 [b]	0.29 ± 0.20 [c]	<0.001
C24:0	0.61 ± 0.32 [b]	1.78 ± 0.36 [a]	1.40 ± 0.36 [a]	0.24 ± 0.36 [b]	0.001
∑ SFA [5]	19.17 ± 2.41 [c]	48.03 ± 3.28 [a]	31.86 ± 3.29 [b]	21.01 ± 3.28 [b,c]	<0.001
∑ OCFA [6]	0.33 ± 0.06 [c]	1.11 ± 0.10 [a]	0.71 ± 0.10 [a,b]	0.44 ± 0.10 [b,c]	<0.001
∑ MUFA [7]	3.87 ± 0.30 [c]	6.18 ± 0.39 [b]	4.96 ± 0.40 [b]	15.40 ± 0.39 [a]	<0.001
∑ PUFA [8]	76.96 ± 2.72 [a]	45.79 ± 3.56 [c]	63.11 ± 3.58 [b]	63.51 ± 3.56 [b]	<0.001
∑ *n*-6 [9]	12.22 ± 0.35 [c]	14.86 ± 0.88 [b,c]	16.42 ± 0.91 [b]	48.45 ± 0.88 [a]	<0.001

Table 2. *Cont.*

	GRASS	HAY	BLG [1]	SH [2]	*p*-Value
∑ *n*-3 [10]	64.73 ± 1.71 [a]	32.51 ± 2.55 [c]	46.52 ± 2.62 [b]	15.02 ± 2.55 [d]	<0.001
n-6:*n*-3 ratio [11]	0.19 ± 0.04 [c]	0.53 ± 0.07 [b]	0.38 ± 0.07 [b,c]	3.20 ± 0.07 [a]	<0.001

Values reported as means ± standard error. Different letters denote statistical significance at $p < 0.05$ (mixed model analysis, post hoc comparison performed using Tukey's adjustment). $n = 126$ for GRASS, and $n = 8$ for the other three feeds. [1] BLG: baleage; [2] SH: soybean hulls; [3] LA: linoleic acid; [4] ALA: α-linolenic acid; [5] ∑ SFA: total saturated FAs; [6] ∑ OCFA: total odd chain FAs; [7] ∑ MUFA: total monounsaturated FAs; [8] ∑ PUFA: total polyunsaturated FAs; [9] ∑ *n*-6: LA; [10] ∑ *n*-3: ALA + C20:3 *n*-3; [11] *n*-6:*n*-3 ratio: ∑ *n*-6/∑ *n*-3.

3.2.3. Phytochemical Content of Feeds

The chlorophyll A, chlorophyll B, and total phenols content of the feeds are shown in Figure 2. BLG and GRASS contained the highest levels of chlorophyll A. SH contained the least ($p < 0.001$). GRASS contained more chlorophyll B than HAY and SH, and BLG contained more chlorophyll B than SH ($p < 0.001$). Finally, BLG and GRASS contained more phenols than SH ($p = 0.010$).

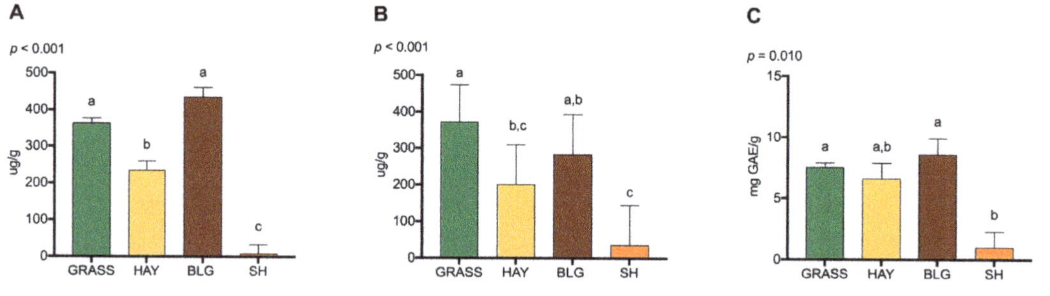

Figure 2. Phytochemical content of feeds. Chlorophyll A (**A**), chlorophyll B (**B**), and total phenols (**C**) found in feed samples. Values reported at means ± standard error. Different letters denote statistical significance at $p < 0.05$ (mixed model analysis, post hoc comparison performed using Tukey's adjustment, $n = 126$ for GRASS, $n = 8$ for each of the other feeds). GRASS: fresh pasture; HAY: dry hay; BLG: baleage; SH; soybean hulls; GAE: gallic acid equivalent.

3.3. *Animal Performance and Carcass Traits*

Performance and carcass traits are shown in Table 3. Initial weight did not differ between diet groups, which was the goal when assigning animals to each group. Final weight, total gain, and average daily gain (ADG) were all higher in the BLG-SH and G-SH groups, while they were lower in the G-BLG and G-HAY groups ($p < 0.001$). A similar trend was seen regarding hot carcass weight (HCW) with higher weights observed in the G-SH and BLG-SH groups and lower weights observed in the G-HAY and G-BLG groups ($p = 0.003$). Backfat significantly differed by diet ($p = 0.011$). BLG-SH and G-SH had more backfat compared to beef from G-BLG but did not differ from G-HAY. G-HAY had the smallest ribeye area compared to the other three groups ($p = 0.012$). Regarding USDA yield grade, G-BLG had a lower numerical yield grade than G-SH but did not differ from the remaining groups ($p = 0.032$). G-BLG had a lower marbling score than G-SH and BLG-SH but was similar to G-HAY ($p = 0.004$).

Table 3. Mean animal performance and carcass traits by diet.

	G-HAY [1]	G-BLG [2]	G-SH [3]	BLG-SH [4]	p-Value
Growth (kg)					
Initial BW [5]	388.70 ± 30.30	390.77 ± 30.33	388.77 ± 30.36	378.27 ± 30.30	0.560
Final BW	483.27 ± 8.63 [c]	493.74 ± 8.77 [b,c]	524.33 ± 8.92 [a,b]	536.31 ± 8.63 [a]	<0.001
Total gain	94.57 ± 25.64 [b]	103.02 ± 25.65 [b]	135.77 ± 25.66 [a]	158.04 ± 25.64 [a]	<0.001
ADG [6]	0.61 ± 0.10 [b]	0.66 ± 0.10 [b]	0.88 ± 0.10 [a]	1.03 ± 0.10 [a]	<0.001
Carcass					
HCW [7] (kg)	281.85 ± 5.53 [c]	287.05 ± 5.62 [b,c]	311.11 ± 5.72 [a]	306.53 ± 5.53 [a,b]	0.003
Backfat (mm)	7.15 ± 0.75 [a,b]	5.91 ± 0.76 [b]	9.18 ± 0.77 [a]	9.38 ± 0.75 [a]	0.011
Ribeye area (cm^2)	68.28 ± 2.19 [b]	75.99 ± 2.20 [a]	75.57 ± 2.22 [a]	76.41 ± 2.19 [a]	0.012
USDA yield grade	2.80 ± 0.30 [a,b]	2.10 ± 0.31 [b]	3.29 ± 0.31 [a]	2.80 ± 0.30 [a,b]	0.032
Marbling score [8]	348.00 ± 11.66 [a,b]	332.55 ± 11.81 [b]	387.39 ± 11.97 [a]	392.00 ± 11.66 [a]	0.004

Values reported as means ± standard error. Different letters denote statistical significance at $p < 0.05$ (mixed model analysis, post hoc comparison performed using Tukey's adjustment, $n = 117$). [1] G-HAY: grass and hay diet; [2] G-BLG: grass and baleage diet; [3] G-SH: grass and soybean hulls diet; [4] BLG-SH: baleage and soybean hulls diet; [5] BW: body weight; [6] ADG: average daily gain; [7] HCW: hot carcass weight; [8] Marbling score: 300-Slight-00 and 400-Small-00.

3.4. *Beef Fatty Acids*

3.4.1. Saturated and Monounsaturated Fatty Acids

The saturated and monounsaturated FA content of beef is presented in Table 4. No significant differences were observed by diet for total SFAs ($p = 0.400$). Individual SFAs ranging from C10:0 to C20:0 did not differ between groups ($p > 0.05$), but C22:0 was higher in G-HAY compared to the other three groups ($p < 0.001$). No significant differences between groups were observed for total branched-chain FA (BCFA) content or for individual BCFAs ($p > 0.05$). The same trend was observed for total MUFA content and individual *cis*-MUFAs ($p > 0.05$). Regarding *trans*-MUFAs, the only significant difference seen was for C16:1 9*t* which was lower in the BLG-SH group compared to the other three groups ($p = 0.003$). No differences were observed for the total FA content between groups ($p > 0.05$).

Table 4. Mean concentrations of saturated and monounsaturated fatty acids by diet (mg per 100 g beef).

	G-HAY [1]	G-BLG [2]	G-SH [3]	BLG-SH [4]	p-Value
∑ SFA [5]	275.08 ± 40.89	272.87 ± 41.46	332.71 ± 42.05	356.22 ± 40.89	0.400
C10:0	1.75 ± 1.92	2.54 ± 1.92	3.17 ± 1.92	3.28 ± 1.92	0.163
C12:0	0.57 ± 0.31	0.60 ± 0.31	0.73 ± 0.31	0.70 ± 0.31	0.442
C13:0	0.09 ± 0.07	0.11 ± 0.07	0.12 ± 0.07	0.11 ± 0.07	0.247
C14:0	12.40 ± 2.29	12.04 ± 2.33	14.95 ± 2.36	15.98 ± 2.29	0.566
C15:0	2.00 ± 0.31	2.11 ± 0.31	1.96 ± 0.32	2.02 ± 0.31	0.988
C16:0	162.66 ± 24.15	162.29 ± 24.49	202.57 ± 24.83	220.59 ± 24.15	0.259
C17:0	4.65 ± 0.79	4.86 ± 0.81	5.52 ± 0.82	6.01 ± 0.79	0.617
C18:0	87.03 ± 14.33	84.13 ± 14.49	99.28 ± 14.66	103.06 ± 14.33	0.694
C19:0	1.91 ± 1.54	2.81 ± 1.55	2.84 ± 1.55	2.99 ± 1.54	0.270
C20:0	0.78 ± 0.35	0.67 ± 0.35	0.81 ± 0.35	0.81 ± 0.35	0.305
C22:0	1.23 ± 0.39 [a]	0.73 ± 0.39 [b]	0.79 ± 0.39 [b]	0.84 ± 0.39 [b]	<0.001
∑ BCFA [6]	11.11 ± 1.63	11.41 ± 1.65	13.08 ± 1.68	12.29 ± 1.63	0.831
C14:0 *iso*	0.15 ± 0.06	0.19 ± 0.06	0.17 ± 0.06	0.16 ± 0.06	0.863
C15:0 *iso*	0.73 ± 0.16	0.77 ± 0.16	0.94 ± 0.16	0.84 ± 0.16	0.771
C15:0 *anteiso*	0.69 ± 0.15	0.72 ± 0.15	0.73 ± 0.15	0.67 ± 0.15	0.979
C16:0 *iso*	0.77 ± 0.25	0.77 ± 0.25	0.82 ± 0.25	0.80 ± 0.25	0.989
C17:0 *iso*	4.11 ± 0.45	4.16 ± 0.46	4.81 ± 0.47	4.21 ± 0.45	0.689
C17:0 *anteiso*	4.06 ± 0.68	4.27 ± 0.69	4.96 ± 0.70	4.96 ± 0.68	0.717
C18:0 *iso*	0.60 ± 0.19	0.54 ± 0.20	0.65 ± 0.20	0.65 ± 0.19	0.819

Table 4. *Cont.*

	G-HAY [1]	G-BLG [2]	G-SH [3]	BLG-SH [4]	*p*-Value
∑ MUFA [7]	313.55 ± 39.66	313.17 ± 40.22	381.51 ± 40.81	371.90 ± 39.66	0.489
∑ *c*MUFA [8]	276.91 ± 36.21	272.67 ± 36.72	342.84 ± 37.26	342.10 ± 36.21	0.358
C14:1 9*c*	2.67 ± 0.52	2.78 ± 0.53	3.36 ± 0.53	3.13 ± 0.52	0.724
C16:1 9*c*	36.73 ± 4.92	35.81 ± 5.00	45.02 ± 5.09	44.15 ± 4.92	0.444
C16:1 10*c*	4.17 ± 1.55	5.19 ± 1.55	4.51 ± 1.55	4.16 ± 1.55	0.554
C16:1 11*c*	2.00 ± 1.39	2.52 ± 1.39	2.66 ± 1.39	2.76 ± 1.39	0.328
C17:1 9*c*	3.76 ± 0.43	3.98 ± 0.43	4.35 ± 0.44	4.50 ± 0.43	0.490
C18:1 9*c*	199.97 ± 34.69	196.83 ± 35.01	252.79 ± 35.35	255.23 ± 34.69	0.309
C18:1 11*c*	11.87 ± 1.50	10.09 ± 1.51	12.75 ± 1.52	11.57 ± 1.50	0.321
C18:1 12*c*	2.14 ± 0.61	2.12 ± 0.61	2.44 ± 0.61	2.46 ± 0.61	0.560
C18:1 13*c*	2.33 ± 1.60	3.11 ± 1.61	3.64 ± 1.61	3.72 ± 1.60	0.065
C18:1 14*c*	0.91 ± 0.26	0.83 ± 0.26	0.90 ± 0.26	0.90 ± 0.26	0.878
C18:1 15*c*	1.29 ± 0.83	1.74 ± 0.83	1.81 ± 0.83	1.95 ± 0.83	0.153
C20:1 9*c*	2.82 ± 1.70	2.67 ± 1.71	3.08 ± 1.71	2.92 ± 1.70	0.718
C20:1 11*c*	6.26 ± 1.44	4.88 ± 1.44	5.51 ± 1.44	4.63 ± 1.44	0.179
∑ *t*MUFA [9]	36.64 ± 9.96	40.71 ± 9.99	38.88 ± 10.02	29.80 ± 9.96	0.479
C16:1 9*t*	6.06 ± 1.66 ^{a}	6.68 ± 1.66 ^{a}	6.11 ± 1.67 ^{a}	3.82 ± 1.66 ^{b}	0.003
C16:1 10,11,12*t*	5.49 ± 2.51	6.99 ± 2.52	6.83 ± 2.52	6.63 ± 2.51	0.373
C18:1 6-8*t*	1.56 ± 0.99	2.07 ± 0.99	1.83 ± 0.99	2.08 ± 0.99	0.181
C18:1 9*t*	1.62 ± 1.19	2.44 ± 1.19	2.51 ± 1.19	2.52 ± 1.19	0.221
C18:1 10*t*	1.35 ± 1.23	2.49 ± 1.23	2.17 ± 1.23	1.83 ± 1.23	0.149
C18:1 11*t*	13.77 ± 2.73	12.37 ± 2.77	11.73 ± 2.80	5.64 ± 2.73	0.196
C18:1 12*t*	1.36 ± 0.51	1.43 ± 0.51	1.53 ± 0.51	1.26 ± 0.51	0.755
C18:1 13,14*t*	2.92 ± 0.31	2.72 ± 0.31	2.73 ± 0.32	2.51 ± 0.31	0.832
C18:1 15*t*	1.44 ± 1.65	2.24 ± 1.65	2.60 ± 1.65	2.34 ± 1.65	0.079
C18:1 16*t*	1.36 ± 0.35	1.29 ± 0.35	1.27 ± 0.35	1.17 ± 0.35	0.918
∑ FA [10]	729.92 ± 83.69	698.46 ± 84.92	833.36 ± 86.21	840.46 ± 83.69	0.550

Values reported as means ± standard error. Different letters denote statistical significance at $p < 0.05$ (mixed model analysis, post hoc comparison performed using Tukey's adjustment, n = 117). [1] G-HAY: grass and hay diet; [2] G-BLG: grass and baleage diet; [3] G-SH: grass and soybean hulls diet; [4] BLG-SH: baleage and soybean hulls diet; [5] ∑ SFA: total saturated FAs; [6] ∑ BCFA: total branched chain FAs; [7] ∑ MUFA: total monounsaturated FAs; [8] ∑ *c*MUFA: total *cis*-monounsaturated FAs; [9] ∑ *t*MUFA: total *trans*-monounsaturated FAs; [10] ∑ FA: all FAs.

3.4.2. Polyunsaturated Fatty Acids and Biohydrogenation Intermediates

The PUFA, CLA, and atypical dienes (AD) content of beef are displayed in Table 5. The total PUFA content of beef was higher in the G-HAY group compared to the other three groups ($p < 0.001$). The same was true for the total *n*-3 PUFA content ($p < 0.001$). More specifically, the ALA content of beef was highest in G-HAY and lowest in BLG-SH ($p = 0.014$). All long-chain *n*-3 PUFAs including EPA, DPA, and DHA were higher in beef from the G-HAY group compared to the other three groups ($p < 0.001$). The sum of EPA+DHA was also higher in beef from G-HAY (11.59 mg/100 g of beef) compared to the other three groups, but all four groups had lower amounts of EPA+DHA compared to European Union standards to consider a food "a source of *n*-3 FAs" [52] (Figure 3). No significant differences between groups were observed for *n*-6 PUFAs ($p > 0.05$) except for C22:4 *n*-6 which was higher in beef from G-HAY compared to the other groups ($p < 0.001$). Significant differences in the *n*-6:*n*-3 ratio were also seen; the lowest ratio was seen in the G-HAY group while the highest ratio was seen in beef from G-SH ($p < 0.001$). No differences were observed between the groups for ADs and conjugated linolenic acid (CLnA) ($p > 0.05$). Significant differences were reported for the individual CLA C18:2 9*c*,11*t*/9*c*,7*t* where beef from G-HAY contained the most and beef from BLG-SH contained the least ($p = 0.015$).

Table 5. Mean concentration of polyunsaturated fatty acids by diet (mg per 100 g beef).

	G-HAY [1]	G-BLG [2]	G-SH [3]	BLG-SH [4]	*p*-Value
∑ PUFA [5]	98.31 ± 3.50 [a]	72.54 ± 3.56 [b]	76.57 ± 3.63 [b]	73.69 ± 3.50 [b]	<0.001
∑ *n*-3 [6]	47.29 ± 2.97 [a]	29.04 ± 2.99 [b]	26.57 ± 3.01 [b]	27.20 ± 2.97 [b]	<0.001
C18:3 *n*-3 (ALA) [7]	10.63 ± 1.26 [a]	10.39 ± 1.26 [a,b]	8.62 ± 1.27 [a,b]	8.31 ± 1.26 [b]	0.014
C20:3 *n*-3	0.89 ± 0.41	0.68 ± 0.41	0.74 ± 0.41	0.70 ± 0.41	0.235
C20:5 *n*-3 (EPA) [8]	9.26 ± 0.38 [a]	5.68 ± 0.39 [b]	5.04 ± 0.39 [b]	5.00 ± 0.38 [b]	<0.001
C22:5 *n*-3 (DPA) [9]	24.18 ± 2.82 [a]	11.09 ± 2.82 [b]	10.88 ± 2.83 [b]	11.59 ± 2.82 [b]	<0.001
C22:6 *n*-3 (DHA) [10]	2.33 ± 0.54 [a]	1.43 ± 0.54 [b]	1.51 ± 0.54 [b]	1.60 ± 0.54 [b]	<0.001
∑ *n*-6 [11]	47.64 ± 2.84	41.67 ± 2.86	48.26 ± 2.88	44.67 ± 2.84	0.108
C18:2 *n*-6 (LA) [12]	28.23 ± 3.73	25.42 ± 3.73	30.14 ± 3.74	27.37 ± 3.73	0.121
C18:3 *n*-6	0.72 ± 0.46	0.69 ± 0.46	0.76 ± 0.46	0.76 ± 0.46	0.733
C20:2 *n*-6	1.01 ± 0.32	0.76 ± 0.32	0.84 ± 0.32	0.90 ± 0.32	0.087
C20:3 *n*-6	2.20 ± 0.31	2.21 ± 0.31	2.65 ± 0.31	2.53 ± 0.31	0.123
C20:4 *n*-6	10.43 ± 1.19	10.02 ± 1.20	10.63 ± 1.20	10.02 ± 1.19	0.903
C22:4 *n*-6	5.05 ± 1.36 [a]	2.68 ± 1.36 [b]	3.35 ± 1.36 [b]	3.29 ± 1.36 [b]	<0.001
n-6:*n*-3 ratio [13]	1.03 ± 0.23 [c]	1.49 ± 0.23 [b]	1.89 ± 0.23 [a]	1.70 ± 0.23 [a,b]	<0.001
C20:3 *n*-9	3.38 ± 0.93 [a]	1.87 ± 0.93 [b]	1.79 ± 0.93 [b]	1.83 ± 0.93 [b]	<0.001
∑ CLnA [14]	1.59 ± 1.15	1.56 ± 1.15	1.76 ± 1.15	1.63 ± 1.15	0.690
C18:3 9*c*,11*t*,15*t*	0.80 ± 0.57	0.82 ± 0.57	0.91 ± 0.57	0.84 ± 0.57	0.676
C18:3 9*c*,11*t*,15*c*	0.79 ± 0.58	0.74 ± 0.58	0.85 ± 0.58	0.79 ± 0.58	0.648
∑ AD [15]	18.30 ± 6.84	17.25 ± 6.84	17.56 ± 6.85	15.84 ± 6.84	0.833
C18:2 11*t*,15*t*	4.26 ± 0.84	3.27 ± 0.84	3.22 ± 0.85	2.65 ± 0.84	0.207
C18:2 9*t*,12*t*	1.82 ± 1.09	2.27 ± 1.09	2.37 ± 1.09	2.21 ± 1.09	0.377
C18:2 9*c*,14*t*/9*c*,13*t*	2.60 ± 1.27	2.69 ± 1.27	2.78 ± 1.27	2.56 ± 1.27	0.944
C18:2 11*t*,15*c*	4.20 ± 0.82	3.48 ± 0.82	3.04 ± 0.83	2.52 ± 0.82	0.189
C18:2 9*c*,16*t*	1.90 ± 0.88	1.87 ± 0.88	2.07 ± 0.88	2.00 ± 0.88	0.715
C18:2 9*c*,15*c*	2.10 ± 1.29	2.35 ± 1.29	2.59 ± 1.29	2.39 ± 1.29	0.587
C18:2 12*c*,15*c*	1.42 ± 0.80	1.31 ± 0.80	1.48 ± 0.80	1.51 ± 0.80	0.589
∑ CLA [16]	10.45 ± 3.43	8.35 ± 3.44	9.03 ± 3.44	7.11 ± 3.43	0.107
C18:2 9*c*,11*t*/9*c*,7*t*	6.26 ± 1.00 [a]	4.41 ± 1.00 [a,b]	4.66 ± 1.00 [a,b]	3.05 ± 1.00 [b]	0.015
C18:2 11*t*,13*c*	1.72 ± 0.89	1.51 ± 0.89	1.64 ± 0.89	1.49 ± 0.89	0.570
C18:2 11*t*,13*t*	1.27 ± 0.83	1.28 ± 0.83	1.44 ± 0.83	1.36 ± 0.83	0.580
C18:2 *t*,*t*	1.20 ± 0.80	1.15 ± 0.80	1.27 ± 0.80	1.21 ± 0.80	0.799

Values reported as means ± standard error. Different letters denote statistical significance at $p < 0.05$ (mixed model analysis, post hoc comparison performed using Tukey's adjustment, $n = 117$). [1] G-HAY: grass and hay diet; [2] G-BLG: grass and baleage diet; [3] G-SH: grass and soybean hulls diet; [4] BLG-SH: baleage and soybean hulls diet; [5] ∑ PUFA: total polyunsaturated FAs; [6] ∑ *n*-3: total omega-3 FAs; [7] ALA: α-linolenic acid; [8] EPA: eicosapentaenoic acid; [9] DPA: docosapentaenoic acid; [10] DHA: docosahexaenoic acid; [11] ∑ *n*-6: total omega-6 FAs; [12] LA: linoleic acid; [13] *n*-6:*n*-3 ratio: ∑ *n*-6/∑ *n*-3; [14] ∑ CLnA: total conjugated linolenic acid isomers; [15] ∑ Atypical Dienes: total non-conjugated linoleic acid isomers; [16] ∑ CLA: total conjugated linoleic acid isomers.

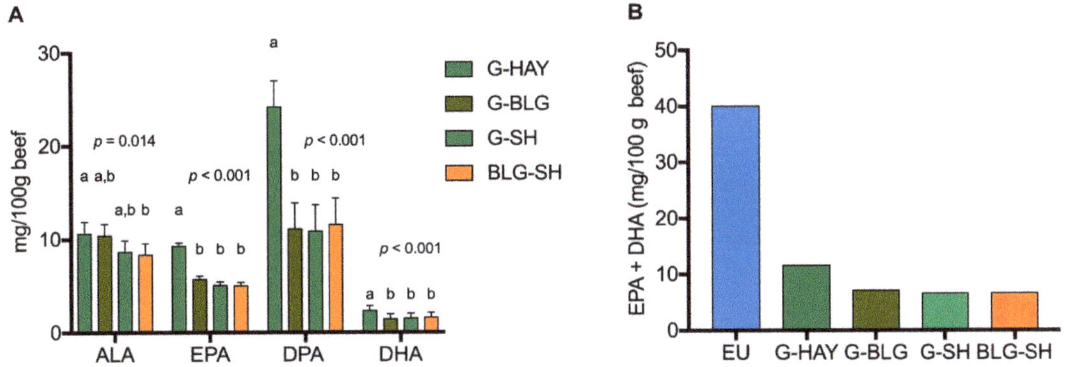

Figure 3. Long-chain *n*-3 PUFAs in beef by diet. (**A**) α-linolenic acid (ALA), eicosapentaenoic acid (EPA), docosapentaenoic acid (DPA), and docosahexaenoic acid (DHA) content of beef by diet. (**B**) Sum

of EPA+DHA in beef by diet compared to the European Union (EU) standard to consider a food "a source of n-3 PUFAs" [52]. Data shown as means ± standard error. Different letters denote statistical significance at $p < 0.05$ (mixed model analysis, post hoc comparison performed using Tukey's adjustment, $n = 117$). G-HAY: grass and hay diet; G-BLG: grass and baleage diet; G-SH: grass and soybean hulls diet; BLG-SH: baleage and soybean hulls diet; EU: European Union standard for a food to be considered "a source of n-3 fatty acids".

3.5. Vitamin E and Minerals in Beef

The vitamin E and minerals in beef are presented in Table 6. Vitamin E was significantly lower in beef from the BLG-SH group compared to the other three diets ($p < 0.001$). Selenium, iron, copper, and zinc did not differ between diets ($p > 0.05$). Manganese was higher in beef from G-HAY and G-BLG, and lower in beef from BLG-SH ($p = 0.002$).

Table 6. Mean concentrations of vitamin E and minerals by diet (μg per g of beef).

	G-HAY [1]	G-BLG [2]	G-SH [3]	BLG-SH [4]	p-Value
Vitamin E	29.93 ± 1.44 ^{a}	28.86 ± 1.46 ^{a}	25.62 ± 1.47 ^{a}	13.83 ± 1.44 ^{b}	<0.001
Selenium	0.44 ± 0.04	0.42 ± 0.04	0.44 ± 0.04	0.45 ± 0.04	0.561
Iron	59.87 ± 7.61	59.65 ± 7.61	60.41 ± 7.62	56.94 ± 7.61	0.422
Copper	1.98 ± 0.07	2.09 ± 0.07	2.07 ± 0.07	1.93 ± 0.07	0.117
Zinc	126.31 ± 3.35	123.35 ± 3.40	123.80 ± 3.45	119.36 ± 3.35	0.545
Manganese	0.92 ± 0.02 ^{a}	0.90 ± 0.02 ^{a,b}	0.85 ± 0.02 ^{b,c}	0.84 ± 0.02 ^{c}	0.002

Values reported as means ± standard error. Different letters denote statistical significance at $p < 0.05$ (mixed model analysis, post hoc comparison performed using Tukey's adjustment, $n = 117$). [1] G-HAY: grass and hay diet; [2] G-BLG: grass and baleage diet; [3] G-SH: grass and soybean hulls diet; [4] BLG-SH: baleage and soybean hulls diet.

3.6. Thiobarbituric Acid Reactive Substances and Warner-Bratzler Shear Force Values for Beef

The TBARS and WBSF values for beef by diet are displayed in Figure 4. Beef from the BLG-SH group showed higher TBARS values compared to the other three groups ($p < 0.001$). Regarding WBSF values, beef from the BLG-SH group displayed lower values compared to beef from G-HAY and G-BLG ($p = 0.017$).

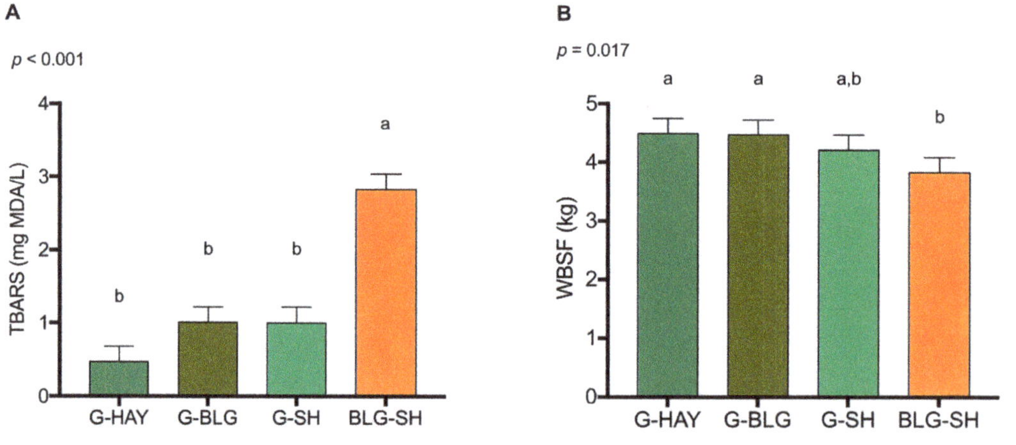

Figure 4. (**A**) Thiobarbituric acid reactive substances (TBARS) and (**B**) Warner-Bratzler Shear Force (WBSF) values of beef by diet. Data shown as means ± standard error. Different letters denote statistical significance at $p < 0.05$ (mixed model analysis, post hoc comparison performed using Tukey's adjustment, $n = 117$). G-HAY: grass and hay diet; G-BLG: grass and baleage diet; G-SH: grass and soybean hulls diet; BLG-SH: baleage and soybean hulls diet.

4. Discussion

4.1. Feeds

4.1.1. Proximate Composition of the Feeds

The proximate analysis of the feeds indicates that SH are higher in fiber than GRASS, HAY, and BLG. This finding is supported by the literature [53,54]. SH are low in lignin and have a high digestibility potential for ruminants [54]. Therefore, SH provide energy without the management problems associated with high grain diets [53]. In the present study, SH provided less energy than the other three feed types. However, SH were not consumed by the animals in isolation but as a combination with either GRASS or BLG. Poore et al. [54] noted that energy levels of SH were variable in various studies, but because of an associative effect on forage digestion, SH appear to have an effective energy value. GRASS contained the most CP and gross energy compared to the other feeds. These findings confirm previously published results by Krusinski et al. [38] showing that pastures contain more CP and gross energy than conserved forages.

4.1.2. Fatty Acid Profile of the Feeds

Numerous differences in the FA content of the feeds were observed in the current study. GRASS contained more *n*-3 PUFAs in the form of ALA than all the other feeds. This was expected since grasses contain higher concentrations of ALA (50-75% of total FAs) [19,21,38,55]. High levels of PUFAs are found in chloroplasts of green plants (i.e., grasses), which may explain the higher concentrations of ALA [56]. Levels of *n*-3 PUFAs drastically decreased in conserved forages (HAY and BLG). While they still contained more PUFAs than SH, conserved forages usually have reduced nutritional quality compared to fresh grasses. This is due to the drying process of HAY and the fermenting process of BLG which result in the oxidation of PUFAs, especially ALA. More specifically, PUFAs are released from the plant membranes and are then oxidized with exposure to air by lipoxygenases [33]. This process is generally followed by an increase in levels of palmitic acid (C16:0) since SFAs are less prone to oxidation [33,34]. This was confirmed in the present study with HAY and BLG containing more palmitic acid than GRASS. We also reported a higher *n*-6:*n*-3 ratio in SH compared to the other feeds. Bronkema et al. [21] indicated that SH have a higher LA content, thus increasing the *n*-6:*n*-3 ratio. The results presented in the current study confirm these statements since we found higher levels of *n*-6 PUFAs and lower levels of *n*-3 PUFAs in SH compared to the other feeds, thus increasing the *n*-6:*n*-3 ratio. Interestingly, SH contained more MUFA than the other types of feeds (mainly as oleic acid). O'Callaghan et al. [57] showed that adding SH to a concentrate diet decreased levels of oleic acid. Ensiled forages such as BLG have advantages compared to HAY. Ensiling does not greatly impact the FA profile [33,36]. Ensiling forages protects FAs from oxidation, explaining why oxidation in HAY is generally more prevalent [58]. Our results confirm the more beneficial FA profile of BLG compared to HAY; BLG contained more *n*-3 PUFAs and had a lower *n*-6:*n*-3 ratio compared to HAY. However, it is important to note that the feed composition plays a major role in the nutritional profile of feeds. Different plant species have different effects on the FA profile of feeds [32]. A limitation of the current study is the lack of information about the proportion of plant species present in the feeds. Krusinski et al. [38] showed that individual plant species affect the FA and antioxidant profiles of pastures.

4.1.3. Phytochemical Content of the Feeds

GRASS and BLG contained the most chlorophyll A, chlorophyll B, and total phenols while SH contained the least of these compounds. There is a strong positive correlation between chlorophyll A and B and ALA in grasses [59]. Green forages are also known to contain vitamins with antioxidant properties such as vitamin E [56]. The high total phenols levels found in GRASS were expected. It was previously reported that the total phenolic content is higher in grasses than in seeds [60]. In a study comparing a complex pasture mixture to a grain diet, the authors reported higher levels of chlorophyll A, chlorophyll B, and total phenols in pasture [38]. Surprisingly, levels of these antioxidant compounds were not lower in BLG. Drying and fermenting usually decrease concentrations of antioxidants and phenolics [19,31,32]. Tripathi et al. [61] noted that haymaking may cause more leaf dropping and shattering compared to BLG making (leaves are the most nutritious parts of the plant), which may explain why concentrations of these compounds were reduced in HAY but not in BLG. While SH have been investigated for their antioxidant potential [62], our results indicate that SH have low levels of total phenols when compared to GRASS or BLG. However, it appears that the growth stage of the soybean plant affects its phenolic concentration [63].

4.2. Animal Performance and Carcass Traits

Results in the present study demonstrate that the diet has an impact on animal growth, carcass traits, and meat quality. Initial body weight did not differ between groups, which may be attributed to pre-trial management. The addition of SH to either GRASS or BLG led to higher final body weight, higher total gain, and higher ADG. While the BLG-SH group was expected to be higher than the other groups for these variables, it was interesting to see a similar trend in the G-SH group which was out on pasture. The BLG-SH group was treated as feedlot cattle. Previous studies showed that cattle finished in feedlots have higher final body weight, total weight gain, and ADG compared to pasture-finished cattle [39,64]. Besides diet, another explanation may be that cattle in feedlots also have less exercise than cattle out on pasture, which reduces their maintenance requirements. The G-SH group was also out on pasture and consumed mostly GRASS, but differences with the BLG-SH group were not significant. Neel et al. [64] showed that increasing the amount of soybean meal and SH in the cattle's diet led to higher final body weight and ADG. Dennis et al. [65] reported that animals consuming a diet of only HAY showed higher final body weight and ADG than animals consuming a BLG diet. In the present study, no significant differences were seen between the G-HAY and the G-BLG groups. This might be due to animals consuming HAY and BLG as supplemental feeds while eating mostly GRASS. The plant species used might also differ compared to other studies.

Regarding carcass traits, a similar trend was observed for hot carcass weight with the G-SH and the BLG-SH groups weighing more than the other two groups. This finding aligns with previous results about weight gain. Maciel et al. [39] reported that animals in feedlots finished on grain have greater backfat, ribeye size, USDA yield grade, and marbling scores than animals finished on pasture. In the present study, the BLG-SH group showed the same trends, and the same was observed for the G-SH group (which can be attributed to the inclusion of SH). It was previously shown that increasing the amount of soybean meal and SH in the diet increased fat thickness and the USDA yield grade [64]. Supplementing grass-finished cattle with BLG seems to reduce backfat, USDA yield grade, and marbling score. There is limited evidence in the literature demonstrating how feeding conserved forages affects carcass traits and the nutritional profile of beef [19]. The present study indicates that including SH in the diet increases weight gain, while HAY and BLG may reduce weight gain, yield grade, and marbling scores.

4.3. Beef Fatty Acids

4.3.1. Saturated Fatty Acid Content of Beef

No differences in SFAs between groups were seen. Red meat and especially beef are criticized for their high SFA content [66]. SFAs increase low-density lipoprotein (LDL) cholesterol, which may increase risks of coronary heart diseases [67]. Based on this, dietary guidelines in the U.S. recommend limiting the intake of SFAs to 10% of daily caloric intake [68]. However, not all SFAs have the same health effects. For instance, palmitic acid (C16:0) has a strong LDL cholesterol-raising effect, while stearic acid (C18:0) has a neutral effect on LDL-cholesterol [69,70]. While no differences were observed in this study, Baublits et al. [37] found that supplementing with SH increases C16:0 levels in beef. Based on the feed FA profile, it was expected that HAY and BLG supplementation would increase the SFA content of beef, especially C16:0. The processing of forages into HAY and BLG generally result in a loss of PUFAs accompanied with an increase in palmitic acid [33,34]. Nevertheless, the lack of significant differences found in the present study might be due to cattle consuming mostly fresh pasture. One limitation of this study is that we did not record the intake of supplemental feeds. Even if the FA profile of the feeds can give us an idea of what to expect in the meat, the gross transfer of dietary SFAs into ruminant products is variable. For example, the transfer of dietary C16:0 into milk fat ranges from 12% to 50% [71].

4.3.2. Monounsaturated Fatty Acid Content of Beef

The MUFA content of beef did not differ between groups. MUFAs make up almost half of beef fat (mostly as oleic acid) [72]. Oleic acid consumption has the potential to lower LDL-cholesterol and blood pressure in humans [69]. Grain-finished beef usually contains up to 70% more MUFAs than GFB [8,73]. In the present study, oleic acid (C18:1 9*c*) was the most abundant FA. The only difference observed was in concentrations of C16:1 9*t* which were lower in beef from the BLG-SH group compared to the other three groups. O'Callaghan et al. [57] observed that adding SH to a concentrate diet resulted in lower MUFA content in milk. Further, Baublits et al. [37] found that supplementing cattle diet with SH led to lower levels of C16:1 9*t*. Ruminant *trans*-FAs are produced by the isomerization of MUFAs in the rumen [74], and grass feeding generally leads to a more favorable rumen pH which allows for more efficient biohydrogenation and isomerization [19,75,76]. Thus, the higher amount of C16:1 9*t* found in the groups with fresh GRASS was expected. The health effects of ruminant *trans*-FAs remain unclear. Some studies reported the antiatherogenic and anticarcinogenic effects of ruminant *trans*-FAs [33], while others reported potential negative health effects [77,78].

4.3.3. Polyunsaturated Fatty Acid Content of Beef

More differences between groups were observed for PUFAs. Overall, beef from the G-HAY group contained more PUFAs than the other three groups. The concentration of *n*-6 PUFAs did not differ between groups, so the variations in PUFA content were due to differences in *n*-3 PUFAs. Both *n*-6 and *n*-3 PUFAs are of interest for human health. Consumption of long-chain *n*-3 PUFAs have anti-inflammatory potential, while *n*-6 PUFAs are generally considered pro-inflammatory [17]. This makes the *n*-6:*n*-3 ratio a crucial metric to determine the health effects of a food [16,17]. In the present study, beef from BLG-SH contained less ALA than beef from G-HAY. This was expected since fresh forages contain 50–75% *n*-3 PUFAs, mostly as ALA [79]. It appears that the addition of SH to the cattle diet reduced the amount of ALA in beef. This finding is supported by results published by Baublits et al. [37]. They reported that the addition of SH to forages resulted in a decrease in *n*-3 PUFAs, especially ALA. Regarding long-chain *n*-3 PUFAs (EPA, DPA, DHA), beef from the G-HAY group contained higher levels of these beneficial FAs than the three other groups. The European Commission considers a food "a source of *n*-3 PUFAs" if 100 g of the food contains at least 40 mg of EPA+DHA or 0.3 g of ALA [52]. Even if the EPA+DHA content in beef of all four groups were below the limit to qualify as a "source of *n*-3 PUFAs", beef from G-HAY was the closest to meet these standards and can contribute to the intake of these long-chain *n*-3 PUFAs, especially for individuals who have limited access to marine foods [80]. EPA and DHA are linked to healthier cardiovascular, immune, and cognitive functions [81,82]. DPA has been shown to improve cognitive functions, lower cholesterol, and reduce inflammation [83]. Our results indicate that consuming GFB supplemented with HAY provides higher levels of these long-chain *n*-3 PUFAs compared to the other groups. While lower levels of *n*-3 PUFAs in the groups fed SH were expected, it was surprising to see lower levels of these FAs in the groups fed BLG. Haymaking generally results in the loss of PUFAs because of oxidation and the dropping of leaves compared to BLG [19,61]. One explanation might be that animals in the G-BLG group consumed more of their supplemental feed than animals in the G-HAY group. Cattle seem to prefer BLG over HAY [84]. If animals consumed more GRASS in the G-HAY group than the G-BLG group, it might explain the differences in long-chain *n*-3 PUFAs.

The Western diet is generally high in *n*-6 PUFAs and low in *n*-3 PUFAs, leading to increased risks of diseases [16]. The *n*-6:*n*-3 ratio in the Western diet is estimated to be between 15:1 and 20:1. The optimal *n*-6:*n*-3 ratio to benefit human health is between 1:1 and 4:1 [16,17]. In the present study, the *n*-6:*n*-3 ratio was higher in beef from G-SH and BLG-SH, and lower in beef from G-HAY. Even though we noted significant differences between groups, the *n*-6:*n*-3 ratio was still below 2:1 for all of them. The higher *n*-6:*n*-3 ratio in the groups containing SH was expected due to lower amounts of *n*-3 PUFAs and higher levels of *n*-6 PUFAs in SH compared to the other feeds. Duckett et al. [20] found no differences in the *n*-6:*n*-3 ratio when feeding SH to cattle before forage finishing. Baublits et al. [37], on the other hand, reported a greater *n*-6:*n*-3 ratio when cattle where supplemented with SH. However, the addition of any of the supplemental feeds tested cannot explain the wide variations in GFB found by Bronkema et al. [21]. Increasing the *n*-3 PUFA and CLA content while decreasing the SFA and *n*-6 content are priorities to improve the nutritional quality of beef [85].

4.3.4. Biohydrogenation Intermediates in Beef

Biohydrogenation intermediates including CLA, CLnA, and ADs are formed when LA and ALA undergo biohydrogenation in the rumen (70-95% and 85-100%, respectively) [86]. No differences in CLnA and AD were observed in this study. The only difference was seen in levels of C18:2 *9c,11t/9c,7t*, with beef from G-HAY containing the most and beef from BLG-SH containing the least. This was expected since GRASS contains more PUFAs than SH. Feeding mostly GRASS to cattle results in a more favorable rumen pH, leading to more efficient biohydrogenation [75,76]. However, SH has the potential to increase the CLA content of beef compared to grain-diets, mainly because of their high fiber content resulting in a more optimal rumen pH for biohydrogenation to occur [87].

4.4. Vitamin E, TBARS, and WBSF

Beef from the three groups fed fresh GRASS contained more vitamin E than beef from BLG-SH. Vitamin E is of interest for human health due to its antioxidant properties [14]. Duckett et al. [20] found no difference in vitamin E levels when pasture was supplemented with SH or not. However, GFB generally contains up to three times more vitamin E than grain-finished beef [15,20,22,51]. The amount of vitamin E found in GFB is enough to protect the beef from oxidation and extend the shelf-life of meat [14,88]. TBARS is an effective assay to measure lipid oxidation. In the present study, TBARS values were higher in beef from BLG-SH compared to the other three groups. Untrained panelists usually do not detect oxidation flavors until oxidation values reach 2.0 mg MDA/kg of tissue [89]. In the current study, only beef from the BLG-SH group exceeded this threshold. Feeding GRASS to cattle usually leads to reduced TBARS values in beef compared to concentrate diets [90]. The higher amounts of antioxidants (including vitamin E) present in GFB might explain the better oxidative stability and lower TBARS values [91].

Shear force values were the lowest for beef from BLG-SH compared to beef from G-HAY and G-BLG. Maciel et al. [39] reported that GFB has higher WBSF values compared to grain-finished beef. These findings indicate that grass-finishing affects the tenderness of beef. Marbling may be a contributing factor to increased meat tenderness (lower WBSF values). Baublits et al. [37] found no difference in shear force values when including SH to a grass diet. Based on the results of the present study, supplementing the diet of cattle with SH might help with meat tenderness, especially in GFB.

5. Conclusions

Based on our findings, we can conclude that SH caused more weight gain in cattle, increased the marbling score of beef, and improved the tenderness of GFB. SH did increase the *n*-6:*n*-3 ratio in beef, but it remained under 2:1. The use of SH as a supplemental feed increased TBARS values as well. Feeding GFB fresh GRASS and HAY resulted in a higher PUFA content, especially higher levels of long-chain *n*-3 PUFAs including EPA, DPA, and DHA. Vitamin E concentrations were also increased in beef on fresh pasture, likely contributing to lower TBARS values. In conclusion, none of the supplemental feeds tested in the current study increased the *n*-6:*n*-3 ratio to the values observed previously by Bronkema et al. [21] in their nutritional survey of commercially available GFB. Future research should investigate other feeds (with different plant species) and determine what ingredients cause large increases in the *n*-6:*n*-3 ratio of GFB. As observed here, the *n*-6:*n*-3 ratio of GFB should remain under 4:1 to benefit human health. GFB has the potential to provide beneficial bioactive compounds for human health including long-chain *n*-3 PUFAs, phenols, and vitamin E.

Supplementary Materials: The following supporting information can be downloaded at: https://www.mdpi.com/article/10.3390/foods11233856/s1, Table S1: Mean concentrations of saturated and monounsaturated fatty acids by diet (% total fatty acids); Table S2: Mean concentrations of polyunsaturated fatty acids by diet (% total fatty acids).

Author Contributions: Conceptualization, L.K., I.C.F.M., C.A.B., J.E.R. and J.I.F.; methodology, L.K., S.S. (Selin Sergin), V.J., E.G., A.J.G., S.S. (Sukhdeep Singh), J.E.R. and J.I.F.; software, L.K., S.S. (Selin Sergin), S.S. (Sukhdeep Singh) and J.I.F.; validation, C.A.B., J.E.R. and J.I.F.; formal analysis, L.K., S.S. (Sukhdeep Singh) and J.I.F.; investigation, L.K., I.C.F.M., C.A.B., J.E.R. and J.I.F.; resources, C.A.B., J.E.R. and J.I.F.; data curation, L.K. and S.S. (Sukhdeep Singh); writing—original draft preparation, L.K. and J.I.F.; writing—review and editing, L.K., I.C.F.M., S.S. (Selin Sergin), V.J., E.G., A.J.G., S.S. (Sukhdeep Singh), C.A.B., J.E.R. and J.I.F.; visualization, L.K., S.S. (Selin Sergin) and S.S. (Sukhdeep Singh); supervision, A.J.G., J.E.R. and J.I.F.; project administration, C.A.B., J.E.R. and J.I.F.; funding acquisition, C.A.B., J.E.R. and J.I.F. All authors have read and agreed to the published version of the manuscript.

Funding: This research was funded by the Greenacres Foundation, grant number G0488F.

Institutional Review Board Statement: The animal study protocol was approved by the Michigan State University Institutional Animal Care and Use Committee (IACUC #202000054, from 24 March 2020 to 31 August 2022).

Data Availability Statement: The original contributions presented in the study are included in the article/Supplementary Material. Further inquiries can be directed to the corresponding author.

Acknowledgments: The authors would like to thank the Michigan State University Kellogg Biological Station in Hickory Corners, MI for allowing the field trial on their land. The authors would also like to thank the Michigan State University College of Agriculture and Natural Resources Statistical Consulting Center (CANR SCC) for their help with analyzing the data for this study. The authors also express their gratitude for the financial support provided by the Greenacres Foundation (G0488F) that enabled them to complete this work. The graphical abstract was created with BioRender.com.

Conflicts of Interest: C.A.B. is part of the Greenacres Foundation. J.I.F. and J.E.R. have not accepted personal honoraria from any organization to prevent undue influence in the eye of the public. The remaining authors declare that the research was conducted in the absence of any commercial or financial relationships that could be construed as a potential conflict of interest. The funders had no role in the design of the study; in the collection, analyses, or interpretation of data; in the writing of the manuscript; or in the decision to publish the results.

References

1. Cheung, R.; McMahon, P.; Norell, E.; Kissel, R.; Benz, D. *Back to Grass: The Market Potential for U.S. Grassfed Beef*; Stone Barns Center for Food and Agriculture: New York, NY, USA, 2017.
2. Provenza, F.D.; Kronberg, S.L.; Gregorini, P. Is grassfed meat and dairy better for human and environmental health? *Front. Nutr.* **2019**, *6*, 1–13. [CrossRef] [PubMed]
3. Butler, G.; Ali, A.M.; Oladokun, S.; Wang, J.; Davis, H. Forage-fed cattle point the way forward for beef? *Future Foods* **2021**, *3*, 100012. [CrossRef]
4. Stampa, E.; Schipmann-Schwarze, C.; Hamm, U. Consumer perceptions, preferences, and behavior regarding pasture-raised livestock products: A review. *Food Qual. Prefer.* **2020**, *82*, 103872. [CrossRef]

5. Xue, H.; Mainville, D.; You, W.; Nayga, R.M. Consumer preferences and willingness to pay for grass-fed beef: Empirical evidence from in-store experiments. *Food Qual. Prefer.* **2010**, *21*, 857–866. [CrossRef]
6. Willett, W.; Rockström, J.; Loken, B.; Springmann, M.; Lang, T.; Vermeulen, S.; Garnett, T.; Tilman, D.; DeClerck, F.; Wood, A.; et al. Food in the Anthropocene: The EAT–Lancet Commission on healthy diets from sustainable food systems. *Lancet* **2019**, *393*, 447–492. [CrossRef]
7. Davis, H.; Magistrali, A.; Butler, G.; Stergiadis, S. Nutritional benefits from fatty acids in organic and grass-fed beef. *Foods* **2022**, *11*, 646. [CrossRef]
8. Van Elswyk, M.E.; McNeill, S.H. Impact of grass/forage feeding versus grain finishing on beef nutrients and sensory quality: The U.S. experience. *Meat Sci.* **2014**, *96*, 535–540. [CrossRef]
9. Berthelot, V.; Gruffat, D. Fatty acid composition of muscles. In *INRA Feeding System for Ruminants*; Wageningen Academic Publishers: Wageningen, NL, USA, 2018; p. 640. ISBN 978-690-8686-8292-8684.
10. Lenighan, Y.M.; Nugent, A.P.; Moloney, A.P.; Monahan, F.J.; Walton, J.; Flynn, A.; Roche, H.M.; Mcnulty, B.A. A modelling approach to investigate the impact of consumption of three different beef compositions on human dietary fat intakes. *Public Health Nutr.* **2019**, *23*, 2373–2383. [CrossRef]
11. Prache, S.; Martin, B.; Coppa, M. Review: Authentication of grass-fed meat and dairy products from cattle and sheep. *Animal* **2020**, *14*, 854–863. [CrossRef]
12. Ponnampalam, E.N.; Mann, N.J.; Sinclair, A.J. Effect of feeding systems on omega-3 fatty acids, conjugated linoleic acid and trans fatty acids in Australian beef cuts: Potential impact on human health. *Asia Pac. J. Clin. Nutr.* **2006**, *15*, 21–29.
13. Alfaia, C.P.M.; Alves, S.P.; Martins, S.I.V.; Costa, A.S.H.; Fontes, C.M.G.A.; Lemos, J.P.C.; Bessa, R.J.B.; Prates, J.A.M. Effect of the feeding system on intramuscular fatty acids and conjugated linoleic acid isomers of beef cattle, with emphasis on their nutritional value and discriminatory ability. *Food Chem.* **2009**, *114*, 939–946. [CrossRef]
14. Daley, C.A.; Abbott, A.; Doyle, P.S.; Nader, G.A.; Larson, S. A review of fatty acid profiles and antioxidant content in grass-fed and grain-fed beef. *Nutr. J.* **2010**, *9*, 1–12. [CrossRef] [PubMed]
15. Pighin, D.; Pazos, A.; Chamorro, V.; Paschetta, F.; Cunzolo, S.; Godoy, F.; Messina, V.; Pordomingo, A.; Grigioni, G. A contribution of beef to human health: A review of the role of the animal production systems. *Sci. World J.* **2016**, *2016*, 8681491. [CrossRef] [PubMed]
16. Simopoulos, A.P. The importance of the ratio of omega-6/omega-3 essential fatty acids. *Biomed. Pharmacother.* **2002**, *56*, 365–379. [CrossRef]
17. Simopoulos, A.P. Evolutionary aspects of diet, the omega-6/omega-3 ratio and genetic variation: Nutritional implications for chronic diseases. *Biomed. Pharmacother.* **2006**, *60*, 502–507. [CrossRef]
18. van Vliet, S.; Provenza, F.D.; Kronberg, S.L. Health-promoting phytonutrients are higher in grass-fed meat and milk. *Front. Sustain. Food Syst.* **2021**, *4*, 1–20. [CrossRef]
19. Krusinski, L.; Sergin, S.; Jambunathan, V.; Rowntree, J.E.; Fenton, J.I. Attention to the details: How variations in U.S. grass-fed cattle-feed supplementation and finishing date influence human health. *Front. Sustain. Food Syst.* **2022**, *6*, 851494. [CrossRef]
20. Duckett, S.K.; Neel, J.P.S.; Fontenot, J.P.; Clapham, W.M. Effects of winter stocker growth rate and finishing system on: III. Tissue proximate, fatty acid, vitamin, and cholesterol content. *J. Anim. Sci.* **2009**, *87*, 2961–2970. [CrossRef]
21. Bronkema, S.M.; Rowntree, J.E.; Jain, R.; Schweihofer, J.P.; Bitler, C.A.; Fenton, J.I. A nutritional survey of commercially available grass-finished beef. *Meat Muscle Biol.* **2019**, *3*, 116–126. [CrossRef]
22. Logan, B.G.; Hopkins, D.L.; Schmidtke, L.; Morris, S.; Fowler, S.M. Preliminary investigation into the use of Raman spectroscopy for the verification of Australian grass and grain fed beef. *Meat Sci.* **2020**, *160*, 107970. [CrossRef]
23. van Vliet, S.; Bain, J.R.; Muehlbauer, M.J.; Provenza, F.D.; Kronberg, S.L.; Pieper, C.F.; Huffman, K.M. A metabolomics comparison of plant-based meat and grass-fed meat indicates large nutritional differences despite comparable Nutrition Facts panels. *Sci. Rep.* **2021**, *11*, 13828. [CrossRef] [PubMed]
24. Godfray, H.C.J.; Aveyard, P.; Garnett, T.; Hall, J.W.; Key, T.J.; Lorimer, J.; Pierrehumbert, R.T.; Scarborough, P.; Springmann, M.; Jebb, S.A. Meat consumption, health, and the environment. *Science* **2018**, *361*, eaam5324. [CrossRef] [PubMed]
25. Mathews, K.H.; Johnson, R.J. Alternative beef production systems: Issues and implications. *Econ. Res. Serv.* **2013**, 1–41. Available online: https://www.ers.usda.gov/webdocs/outlooks/37473/36491_ldpm-218-01.pdf?v=7815.8 (accessed on 23 September 2022).
26. American Grassfed Association. American Grassfed Association Grassfed Ruminant Standards–March 2022. 2022, pp. 1–19. Available online: https://www.americangrassfed.org/aga-grassfed-ruminant-standards/ (accessed on 15 October 2022).
27. Dewhurst, R.J.; Scollan, N.D.; Youell, S.J.; Tweed, J.K.S.; Humphreys, M.O. Influence of species, cutting date and cutting interval on the fatty acid composition of grasses. *Grass Forage Sci.* **2001**, *56*, 68–74. [CrossRef]
28. Khan, N.A.; Cone, J.W.; Hendriks, W.H. Stability of fatty acids in grass and maize silages after exposure to air during the feed out period. *Anim. Feed Sci. Technol.* **2009**, *154*, 183–192. [CrossRef]
29. Garcia, P.T.; Pordomingo, A.; Perez, C.D.; Rios, M.D.; Sancho, A.M.; Volpi Lagreca, G.; Casal, J.J. Influence of cultivar and cutting date on the fatty acid composition of forage crops for grazing beef production in Argentina. *Grass Forage Sci.* **2016**, *71*, 235–244. [CrossRef]
30. Gwin, L. Scaling-up sustainable livestock production: Innovation and challenges for grass-fed beef in the U.S. *J. Sustain. Agric.* **2009**, *33*, 189–209. [CrossRef]

31. Owens, F.N.; Secrist, D.S.; Hill, W.J.; Gill, D.R. The effect of grain source and grain processing on performance of feedlot cattle: A review. *J. Anim. Sci.* **1997**, *75*, 868–879. [CrossRef]
32. Butler, G. Manipulating dietary PUFA in animal feed: Implications for human health. *Proc. Nutr. Soc.* **2014**, *73*, 87–95. [CrossRef]
33. Kalač, P.; Samková, E. The effects of feeding various forages on fatty acid composition of bovine milk fat: A review. *Czech J. Anim. Sci.* **2010**, *55*, 521–537. [CrossRef]
34. Van Ranst, G.; Fievez, V.; Vandewalle, M.; De Riek, J.; Van Bockstaele, E. Influence of herbage species, cultivar and cutting date on fatty acid composition of herbage and lipid metabolism during ensiling. *Grass Forage Sci.* **2009**, *64*, 196–207. [CrossRef]
35. Buccioni, A.; Decandia, M.; Minieri, S.; Molle, G.; Cabiddu, A. Lipid metabolism in the rumen: New insights on lipolysis and biohydrogenation with an emphasis on the role of endogenous plant factors. *Anim. Feed Sci. Technol.* **2012**, *174*, 1–25. [CrossRef]
36. Glasser, F.; Doreau, M.; Maxin, G.; Baumont, R. Fat and fatty acid content and composition of forages: A meta-analysis. *Anim. Feed Sci. Technol.* **2013**, *185*, 19–34. [CrossRef]
37. Baublits, R.T.; Brown, A.H., Jr.; Pohlman, F.W.; Rule, D.C.; Johnson, Z.B.; Onks, D.O.; Murrieta, C.M.; Richards, C.J.; Loveday, H.D.; Sandelin, B.A.; et al. Fatty acid and sensory characteristics of beef from three biological types of cattle grazing cool-season forages supplemented with soyhulls. *Meat Sci.* **2006**, *72*, 100–107. [CrossRef] [PubMed]
38. Krusinski, L.; Maciel, I.C.F.; Sergin, S.; Goeden, T.; Ali, H.; Kesamneni, S.; Jambunathan, V.; Cassida, K.A.; Singh, S.; Medina-Meza, I.G.; et al. Evaluation of fatty acid and antioxidant variation in a complex pasture system as compared to standard cattle feed in the Great Lakes Region. *Front. Sustain. Food Syst.* **2022**, *6*, 945080. [CrossRef]
39. Maciel, I.C.F.; Schweihofer, J.P.; Fenton, J.I.; Hodbod, J.; McKendree, M.G.S.; Cassida, K.; Rowntree, J.E. Influence of beef genotypes on animal performance, carcass traits, meat quality, and sensory characteristics in grazing or feedlot-finished steers. *Transl. Anim. Sci.* **2021**, *5*, txab214. [CrossRef]
40. Mertens, D.R. Gravimetric determination of amylase-treated neutral detergent fiber in feeds with refluxing in beakers or crucibles: Collaborative study. *J. AOAC Int.* **2002**, *85*, 1217–1240. [CrossRef]
41. AOAC. *Official Methods of Analysis*, 17th ed.; The Association of Official Analytical Chemists: Gaithersburg, MD, USA, 2000.
42. Hach, C.C.; Bowden, B.K.; Kopelove, A.B.; Brayton, S.V. More powerful peroxide Kjeldahl digestion method. *J. Assoc. Off. Anal. Chem.* **1987**, *70*, 783–787. [CrossRef]
43. Lichtenthaler, H.K.; Wellburn, A.R. Determinations of total carotenoids and chlorophylls a and b of leaf extracts in different solvents. *Biochem. Soc. Trans.* **1983**, *11*, 591–592. [CrossRef]
44. Nimalaratne, C.; Lopes-Lutz, D.; Schieber, A.; Wu, J. Free aromatic amino acids in egg yolk show antioxidant properties. *Food Chem.* **2011**, *129*, 155–161. [CrossRef]
45. Singleton, V.L.; Rossi, J.A. Colorimetry of total phenolics with phosphomolybdic-phosphotungstic acid reagents. *Am. J. Enol. Vitic.* **1965**, *16*, 144–158.
46. Sergin, S.; Goeden, T.; Krusinski, L.; Kesamneni, S.; Ali, H.; Bitler, C.A.; Medina-Meza, I.G.; Fenton, J.I. Fatty acid and antioxidant composition of conventional compared to pastured eggs: Characterization of conjugated linoleic acid and branched chain fatty acid isomers in eggs. *Food Sci. Technol.* **2021**, *1*, 260–267. [CrossRef]
47. Jenkins, T.C. Technical note: Common analytical errors yielding inaccurate results during analysis of fatty acids in feed and digesta samples. *J. Dairy Sci.* **2010**, *93*, 1170–1174. [CrossRef] [PubMed]
48. Kramer, J.K.; Hernandez, M.; Cruz-Hernandez, C.; Kraft, J.; Dugan, M.E. Combining results of two GC separations partly achieves determination of all cis and trans 16:1, 18:1, 18:2 and 18:3 except CLA isomers of milk fat as demonstrated using Ag-ion SPE fractionation. *Lipids* **2008**, *43*, 259–273. [CrossRef] [PubMed]
49. Rettenmaier, R.; Schüep, W. Determination of vitamins A and E in liver tissue. *Int. J. Vitam. Nutr. Res.* **1992**, *62*, 312–317.
50. Wahlen, R.; Evans, L.; Turner, J.; Hearn, R. The use of collision/reaction cell ICP-MS for the determination of elements in blood and serum samples. *Spectroscopy* **2005**, *20*, 84–89.
51. Krusinski, L.; Maciel, I.C.F.; Sergin, S.; Goeden, T.; Schweihofer, J.P.; Singh, S.; Rowntree, J.E.; Fenton, J.I. Fatty acid and micronutrient profile of longissimus lumborum from Red Angus and Red Angus x Akaushi cattle finished on grass or grain. *Foods* **2022**, *11*, 3451. [CrossRef]
52. Commission Regulation of European Union. Amending Regulation (EC) No 1924/2006 of the European Parliament and of the Council with regard to the list of nutrition claims. *Off. J. Eur. Union* **2010**, *116*. Available online: https://eur-lex.europa.eu/LexUriServ/LexUriServ.do?uri=OJ:L:2010:037:0016:0018:EN:PDF (accessed on 29 September 2022).
53. Booth, J.; Howard, T.; Miller-Auwerda, P.; Moore, J. *Soybean Hulls as an Alternative Feed for Horses*; Animal Industry Report; Iowa State University: Ames, IA, USA, 2004; p. ASL R1931.
54. Poore, M.H.; Johns, J.T.; Burris, W.R. Soybean hulls, wheat middlings, and corn gluten feed as supplements for cattle on forage-based diets. *Vet. Clin. N. Am. Food Anim. Pract.* **2002**, *18*, 213–231. [CrossRef]
55. Scollan, N.D.; Dannenberger, D.; Nuernberg, K.; Richardson, I.; MacKintosh, S.; Hocquette, J.F.; Moloney, A.P. Enhancing the nutritional and health value of beef lipids and their relationship with meat quality. *Meat Sci.* **2014**, *97*, 384–394. [CrossRef]
56. Elgersma, A.; Søegaard, K.; Jensen, S.K. Fatty acids, α-tocopherol, β-carotene, and lutein contents in forage legumes, forbs, and a grass-clover mixture. *J. Agric. Food Chem.* **2013**, *61*, 11913–11920. [CrossRef]
57. O'Callaghan, T.F.; Mannion, D.; Apopei, D.; McCarthy, N.A.; Hogan, S.A.; Kilcawley, K.N.; Egan, M. Influence of supplemental feed choice for pasture-based cows on the fatty acid and volatile profile of milk. *Foods* **2019**, *8*, 137. [CrossRef] [PubMed]

58. Kalač, P. The effects of feeding fresh forage and silage on some nutritional attributes of beef: An overview. *J. Agrobiol.* **2011**, *28*, 1–13. [CrossRef]
59. Khan, N.A.; Cone, J.W.; Fievez, V.; Hendriks, W.H. Causes of variation in fatty acid content and composition in grass and maize silages. *Anim. Feed Sci. Technol.* **2012**, *174*, 36–45. [CrossRef]
60. Niroula, A.; Khatri, S.; Khadka, D.; Timilsina, R. Total phenolic contents and antioxidant activity profile of selected cereal sprouts and grasses. *Int. J. Food Prop.* **2019**, *22*, 427–437. [CrossRef]
61. Tripathi, H.P.; Singh, A.P.; Upadhyay, V.S.; Kessels, H.P.P.; Harika, A.S.; Singh, S.; Ibrahim, M.N.M. Forage conservation, storage and feeding. In *Handbook for Straw Feeding System*; Singh, K., Schiere, J.B., Eds.; ICAR: New Delhi, India, 1995; pp. 303–323.
62. Liu, S.X.; Chen, D.; Singh, M.; Xu, J. Extraction of proteins and pasting and antioxidant properties of soybean hulls. *J. Food Res.* **2019**, *8*, 66–77. [CrossRef]
63. Peiretti, P.; Karamać, M.; Janiak, M.; Longato, E.; Meineri, G.; Amarowicz, R.; Gai, F. Phenolic composition and antioxidant activities of soybean (Glycine max (L.) Merr.) plant during growth cycle. *Agronomy* **2019**, *9*, 153. [CrossRef]
64. Neel, J.P.; Fontenot, J.P.; Clapham, W.M.; Duckett, S.K.; Felton, E.E.; Scaglia, G.; Bryan, W.B. Effects of winter stocker growth rate and finishing system on: I. Animal performance and carcass characteristics. *J. Anim. Sci.* **2007**, *85*, 2012–2018. [CrossRef]
65. Dennis, T.S.; Tower, J.E.; Nennich, T.D. Effects of feeding hay and baleage to prepubertal dairy heifers during the grower period. *Prof. Anim. Sci.* **2012**, *28*, 648–656. [CrossRef]
66. Casperson, S.L.; Conrad, Z.; Raatz, S.K.; Derner, J.; Roemmich, J.N.; Jahns, L.; Picklo, M.J. Impact of beef consumption on saturated fat intake in the United States adult population: Insights from modeling the influences of bovine genetics and nutrition. *Meat Sci.* **2020**, *169*, 108225. [CrossRef]
67. Billingsley, H.E.; Carbone, S.; Lavie, C.J. Dietary fats and chronic noncommunicable diseases. *Nutrients* **2018**, *10*, 1385. [CrossRef]
68. U.S. Department of Agriculture; U.S. Department of Health and Human Services. *Dietary Guidelines for Americans, 2020–2025*, 9th ed.; U.S. Department of Health and Human Services: Washington, DC, USA, 2020.
69. Calder, P.C. Functional roles of fatty acids and their effects on human health. *J. Parenter. Enteral Nutr.* **2015**, *39*, 18S–32S. [CrossRef] [PubMed]
70. FAO. Fats and fatty acids in human nutrition: Report of an expert consultation. *FAO Food Nutr. Pap.* **2010**, *91*, 1–166.
71. Loften, J.R.; Linn, J.G.; Drackley, J.K.; Jenkins, T.C.; Soderholm, C.G.; Kertz, A.F. Invited review: Palmitic and stearic acid metabolism in lactating dairy cows. *J. Dairy Sci.* **2014**, *97*, 4661–4674. [CrossRef] [PubMed]
72. Leheska, J.M.; Thompson, L.D.; Howe, J.C.; Hentges, E.; Boyce, J.; Brooks, J.C.; Shriver, B.; Hoover, L.; Miller, M.F. Effects of conventional and grass-feeding systems on the nutrient composition of beef. *J. Anim. Sci.* **2008**, *86*, 3575–3585. [CrossRef]
73. Duckett, S.K.; Neel, J.P.S.; Lewis, R.M.; Fontenot, J.P.; Clapham, W.M. Effects of forage species or concentrate finishing on animal performance, carcass and meat quality. *J. Anim. Sci.* **2013**, *91*, 1454–1467. [CrossRef] [PubMed]
74. Vargas-bello-pérez, E.; Garnsworthy, P.C. Trans fatty acids and their role in the milk of dairy cows. *Cien. Inv. Agr.* **2013**, *40*, 449–473. [CrossRef]
75. French, P.; Stanton, C.; Lawless, F.; O'Riordan, E.G.; Monahan, F.J.; Caffrey, P.J.; Moloney, A.P. Fatty acid composition, including conjugated linoleic acid, of intramuscular fat from steers offered grazed grass, grass silage, or concentrate-based diets. *J. Anim. Sci.* **2000**, *78*, 2849–2855. [CrossRef]
76. Kraft, J.; Kramer, J.K.G.; Schoene, F.; Chambers, J.R.; Jahreis, G. Extensive analysis of long-chain polyunsaturated fatty acids, CLA, trans-18-1 isomers, and plasmalogenic lipids in different retail beef types. *J. Agric. Food Chem.* **2008**, *56*, 4775–4782. [CrossRef]
77. Gebauer, S.K.; Destaillats, F.; Dionisi, F.; Krauss, R.M.; Baer, D.J. Vaccenic acid and trans fatty acid isomers from partially hydrogenated oil both adversely affect LDL cholesterol: A double-blind, randomized controlled trial. *Am. J. Clin. Nutr.* **2015**, *102*, 1339–1346. [CrossRef]
78. Verneque, B.J.F.; Machado, A.M.; de Abreu Silva, L.; Lopes, A.C.S.; Duarte, C.K. Ruminant and industrial trans-fatty acids consumption and cardiometabolic risk markers: A systematic review. *Crit. Rev. Food Sci. Nutr.* **2020**, *62*, 2050–2060. [CrossRef]
79. Dewhurst, R.J.; Shingfield, K.J.; Lee, M.R.F.; Scollan, N.D. Increasing the concentrations of beneficial polyunsaturated fatty acids in milk produced by dairy cows in high-forage systems. *Anim. Feed Sci. Technol.* **2006**, *131*, 168–206. [CrossRef]
80. Howe, P.; Meyer, B.; Record, S.; Baghurst, K. Dietary intake of long-chain ω-3 polyunsaturated fatty acids: Contribution of meat sources. *Nutrition* **2006**, *22*, 47–53. [CrossRef]
81. Parolini, C. Effects of fish n-3 PUFAs on intestinal microbiota and immune system. *Mar. Drugs* **2019**, *17*, 374. [CrossRef]
82. Mendivil, C.O. Dietary fish, fish nutrients, and immune function: A review. *Front. Nutr.* **2021**, *7*, 1–9. [CrossRef]
83. Byelashov, O.A.; Sinclair, A.J.; Kaur, G. Dietary sources, current intakes, and nutritional role of omega-3 docosapentaenoic acid. *Lipid Technol.* **2015**, *27*, 79–82. [CrossRef]
84. Hunde, Y.; Patton, J.; Padgitt, D.; Vlieger, D. Particle size influences preferential consumption of hay and baleage in total mixed dairy rations. *Trans. Mo. Acad. Sci.* **2008**, *42*, 45–49. [CrossRef]
85. Scollan, N.D.; Hocquette, J.F.; Nuernberg, K.; Dannenberger, D.; Richardson, I.; Moloney, A. Innovations in beef production systems that enhance the nutritional and health value of beef lipids and their relationship with meat quality. *Meat Sci.* **2006**, *74*, 17–33. [CrossRef]
86. Lock, A.L.; Harvatine, K.J.; Drackley, J.K.; Bauman, D.E. Concepts in fat and fatty acid digestion in ruminants. In Proceedings of the Intermountain Nutrition Conference, Utah State University, Logan, UT, USA, January 2006; pp. 85–100.

87. Kiesling, D.D. *Performance, Carcass Traits and Fatty Acid Profiles of Yearling Beef Cattle Supplemented with Self-Fed Byproducts on Pasture*; Iowa State University: Ames, IA, USA, 2013.
88. De la Fuente, J.; Diaz, M.T.; Alvarez, I.; Oliver, M.A.; Font, I.F.M.; Sanudo, C.; Campo, M.M.; Montossi, F.; Nute, G.R.; Caneque, V. Fatty acid and vitamin E composition of intramuscular fat in cattle reared in different production systems. *Meat Sci.* **2009**, *82*, 331–337. [CrossRef]
89. Greene, B.E.; Cumuze, T.H. Relationship between TBA numbers and inexperienced panelists' assessments of oxidized flavor in cooked beef. *J. Food Sci.* **1981**, *47*, 52–54. [CrossRef]
90. Nuernberg, K.; Dannenberger, D.; Nuernberg, G.; Ender, K.; Voigt, J.; Scollan, N.D.; Wood, J.D.; Nute, G.R.; Richardson, R.I. Effect of a grass-based and a concentrate feeding system on meat quality characteristics and fatty acid composition of longissimus muscle in different cattle breeds. *Livest. Prod. Sci.* **2005**, *94*, 137–147. [CrossRef]
91. Alothman, M.; Hogan, S.A.; Hennessy, D.; Dillon, P.; Kilcawley, K.N.; O'Donovan, M.; Tobin, J.; Fenelon, M.A.; O'Callaghan, T.F. The "grass-fed" milk story: Understanding the impact of pasture feeding on the composition and quality of bovine milk. *Foods* **2019**, *8*, 350. [CrossRef]

foods

MDPI

Article

Fatty Acid and Antioxidant Profile of Eggs from Pasture-Raised Hens Fed a Corn- and Soy-Free Diet and Supplemented with Grass-Fed Beef Suet and Liver

Selin Sergin [1], Vijayashree Jambunathan [1], Esha Garg [1], Jason E. Rowntree [2] and Jenifer I. Fenton [1,*]

[1] Department of Food Science and Human Nutrition, Michigan State University, 469 Wilson Rd, East Lansing, MI 48824, USA
[2] Department of Animal Science, Michigan State University, 474 S Shaw Ln, East Lansing, MI 48824, USA
* Correspondence: imigjeni@msu.edu; Tel.: +1-517-353-3342

Abstract: There is increasing interest in using grass-fed beef (GFB) by-products to augment the nutrient profile of eggs among local pasture-raising systems in the US. The objective of this study was to characterize egg yolk fatty acid and antioxidant profiles using eggs from pasture-raised hens fed a corn- and soy-free diet and supplemented with GFB suet and liver compared to eggs from pasture-raised hens fed a corn and soy layer hen feed and commercially obtained cage-free eggs. The egg yolk vitamin and mineral profile was also assessed by a commercial laboratory. Both pasture-raised groups had twice as much carotenoid content, three times as much omega-3 fatty acid content, and a 5–10 times lower omega-6:omega-3 fatty acid ratio compared to the cage-free eggs ($p < 0.001$). Eggs from hens fed a corn- and soy-free feed and GFB by-products had half as much omega-6 fatty acid content and five times more conjugated linoleic acid, three times more odd-chain fatty acid, and 6–70 times more branched-chain fatty acid content ($p < 0.001$). Feeding pasture-raised hens GFB suet and liver reduces agricultural waste while producing improved egg products for consumers, but further research is needed to quantify optimal supplementation levels and the efficacy of corn- and soy-free diets.

Keywords: poultry; yolk; omega-3; conjugated linoleic acid; GC-MS; branched-chain

1. Introduction

Poultry farmers in the United States (US) are increasingly utilizing pasture systems to raise and feed hens for egg production [1]. Pasture-raising systems are distinct from "free-range" production systems in that producers must provide pasture for consumption, though these designations are sometimes used interchangeably [2]. Pasture-raised hens are able to forage on phytochemically diverse grasslands and scavenge for invertebrates, allowing them access to nutrients such as fats, vitamins and minerals, and secondary compounds including antioxidants that benefit both animal and human health [3]. Pasture-raised egg systems contribute to regenerative agriculture systems which aim to improve human and ecosystem health, farm production, and food system resiliency by linking soil and livestock management [4]. Pasture and hens play mutually beneficial roles: planting diverse forages supports good soil quality, while hens can help control insects and weeds without pesticides and spread manure that fertilizes the soil. In contrast, within both caged and cage-free production systems, hens are not provided outdoor access and are fed a typical layer hen diet consisting mainly of corn and soy, high in omega-6 (n-6) fatty acids and low in omega-3 (n-3) fatty acids [2,5–7].

Pasture-raised egg systems not only benefit the environment but also typically produce eggs that have a more favorable nutrient profile for human health [1,8–10]. Eggs are a widely available source of proteins, fat, vitamins, and minerals [11]. Notably, the bioavailability of carotenoids is greater in eggs compared to other sources such as green leafy vegetables, likely due to the lipid content of egg yolks [11,12]. Further, egg yolks from pasture-raised hens typically have a healthier fatty acid and nutrient profile compared to egg yolks from caged hens [10]. A study comparing egg yolk *n*-3 fatty acids from pasture-raised to caged hens found that yolks from pasture-raised hens had 4.5 times more alpha-linolenic acid (ALA) and 2 times more eicosapentaenoic acid (EPA) and docosahexaenoic acid (DHA) than yolks from caged hens [1]. Providing pasture to hens yields egg yolks with a lower omega-6:omega-3 (*n*-6:*n*-3) ratio and significantly greater vitamin A, vitamin E, and carotenoid content compared to egg yolks from caged hens [1,9,10]. Further, egg yolks from pasture-raised hens contain significantly greater branched-chain fatty acids (BCFAs) compared to yolks from commercial free-range and caged eggs [10].

Livestock serves an important role in reducing food waste by converting inedible foods into high-value products, yet by-products of the livestock industry including fat and internal organs contribute to the accumulation of waste, economic losses, and environmental pollution [13,14]. About 27.5% of animal weight is wasted in beef slaughter [14]. While previous studies have examined the impact of both vegetable and animal by-product supplementation for layer hens on the egg nutrient profile, there is increasing interest in using by-products of grass-fed beef (GFB) production to augment the nutrient profile of eggs among local pasture-raising systems in the US [15–20]. By feeding hens otherwise under-utilized GFB suet and liver, pasture-raised egg and GFB systems can work in tandem to reduce the environmental impact of beef production and produce eggs that may be more beneficial for human health given that modifying the composition of hen diets can change the fatty acid, vitamin, and mineral content of eggs [11]. Beef fat is about 19% stearic acid, a saturated fatty acid (SFA) known to have a neutral effect on human LDL cholesterol [21,22]. Further, grass-fed cattle have consumed only grass or other forages prior to slaughter [23]. Accordingly, GFB generally has a lower *n*-6:*n*-3 ratio and higher conjugated linoleic acid (CLA) content compared to grain-fed beef [24]. GFB liver contains 92 mg ALA, 151 mg EPA, and 83 mg DHA per 100 g of liver; these fatty acid levels are significantly higher than those found in grain-fed cattle liver [25]. Additionally, GFB liver is a rich source of vitamin A, iron, zinc, and vitamin B12 [26]. While it should be noted that this supplementation practice may not reflect what is considered acceptable to feed layer hens by regulatory bodies in other countries, supplementing hen diets with these by-products among small-scale farms in the US may increase the deposition of these nutrients in eggs and thus confer health benefits to consumers.

The objective of this study was to characterize egg yolk fatty acid and antioxidant profiles using eggs from pasture-raised hens fed a corn- and soy-free diet and supplemented with GFB suet and liver in comparison to eggs from pasture-raised hens fed a standard corn and soy layer hen feed and commercially obtained cage-free eggs. In addition, egg yolk vitamin and mineral profiles were assessed.

2. Materials and Methods

2.1. Chemicals

The gas chromatography–mass spectrometry (GC-MS) reference standard curve contained a mixture of Supelco 37 Component FAME Mix (Sigma-Aldrich, St. Louis, MO, USA) with mead acid, docosatetraenoic acid, *n*-3 docosapentaenoic acid (DPA), *n*-6 DPA, and palmitelaidic acid (Cayman Chemical, Ann Arbor, MI, USA). BCFAs were compared to Mixture BR 3 (Larodan AB, Solna, Sweden), while CLA isomers were compared to CLA reference standard UC-59M (Nu-Chek Prep, Elysian, MN, USA). Dichloromethane was purchased from VWR Chemicals (Radnor, PA, USA). All other chemicals were purchased from Sigma-Aldrich (St. Louis, MO, USA) unless otherwise noted.

2.2. Diet Characteristics and Collection

Pasture-raised, corn- and soy-free feed with beef by-products (PBB): Eggs were collected from hens on a farm in Southwest MI where hens were mobile pasture-raised in a 180-ft enclosed circle rotated every 3–4 days to move the hens to fresh pasture. The flock consisted of approximately 50 hens that were a mix of 1-year-old ISA Brown and Barred Rock hens. Hens were fed a custom feed of peas, barley, alfalfa, and calcium. Exact details regarding the composition of the custom layer hen feed cannot be shared to protect the farm's custom recipe. However, feed samples were provided for fatty acid and antioxidant analysis. Hens were supplemented with GFB suet, GFB liver, and grit (crushed stone to support hen digestion) twice a week. The GFB suet and liver were obtained from a local farm that rotationally grazed cattle on a complex pasture mixture, moving cattle to fresh forage each day. The suet was obtained from fat around the internal organs of the cattle and was provided in raw, chopped pieces. Hens consumed about 30–60 g of liver per week. The suet was provided ad libitum in bins. Samples of the custom layer hen feed ($n = 3$) and the suet ($n = 3$) and liver ($n = 3$) were obtained from the farm. A dozen eggs were randomly collected on one day from the nest box, and from this, $n = 6$ eggs were randomly selected for analysis upon arrival to the laboratory. Layer hen feed samples were ground to pass a 1 mm screen in a Wiley mill (Arthur H. Thomas, Philadelphia, PA, USA) with dry ice and stored at −80 °C under nitrogen. The liver was prepared by mincing and finely grinding it in a crucible over dry ice, while the suet was finely minced before sampling.

Pasture-raised, corn, and soy feed comparison (PCS): Eggs were collected for comparison from hens from a farm in Central-West MI where hens were managed similarly to the PBB group. Hens were mobile pasture-raised in a 180-ft enclosed circle rotated every 3–4 days to move the hens to fresh pasture. The flock consisted of approximately 50 hens that were a mix of 1-year-old ISA Brown and Barred Rock hens. Distinct from the PBB group, these hens were fed only a standard layer hen feed consisting of corn, soy, calcium, and grit obtained from a local feed mill in Michigan in addition to what they consumed from the pasture. Samples of the layer hen feed ($n = 3$) were obtained from the farm. A dozen eggs were randomly collected on one day from the nest box, and from this, $n = 6$ eggs were randomly selected for analysis upon arrival to the laboratory. Layer hen feed samples were ground to pass a 1 mm screen in a Wiley mill (Arthur H. Thomas, Philadelphia, PA, USA) with dry ice and stored at −80 °C under nitrogen.

Cage-free comparison (CFC): Cage-free eggs were collected for comparison from a local grocery store ($n = 6$ eggs). These eggs were labeled as "cage-free" and were from hens that would not have been provided pasture, according to USDA requirements for "cage-free" eggs [5]. It was not possible to obtain samples of the commercial layer hen diet.

2.3. Chemical Composition of Feeds

Chemical composition of the PBB and PCS layer hen feeds was assessed by Dairy One Forage Laboratory (Ithaca, NY, USA) using respective AOAC methods.

2.4. Physical Characteristics of Eggs

Physical characteristics were measured upon collection of the eggs as previously described [10]. Briefly, albumen, yolk, and shell weights were determined, and albumen height was measured using a micrometer. Haugh units were calculated using albumen height and egg weight (Haugh unit = $100 \times \log(\text{albumen height} + 7.57 - 1.7 \times \text{egg weight}^{0.37})$ [27]. Egg yolk color was determined using the DSM Yolk Color Fan (DSM Nutritional Products, Basel, Switzerland) with values ranging from 1 for pale yellow to 14 for dark orange. After being analyzed for physical characteristics, egg yolks were immediately freeze-dried, ground, and stored at −80 °C under nitrogen.

2.5. Phenolic Analysis of Eggs and Feeds

Total phenolic content was assessed using spectrophotometric screening methods as previously described by Sergin et al. [10] using a modified extraction method based on Nimalaratne et al. [28] and the Folin–Ciocalteu assay modified from Chen et al. [29]. Briefly, two extractions, first with 20 mL of a methanol/distilled water/acetic acid solvent [80:18:2 ($v/v/v$)], and second with 20 mL of an acetone/distilled water/acetic acid solvent [80:18:2 ($v/v/v$)], were used to extract phenolic compounds from 2 g of lyophilized egg yolk sample or ground layer hen feed. Tubes were shaken and centrifuged (840× g, 4 °C) and supernatants were combined following the addition of each solvent as previously described [10]. Next, 100 µL Folin–Ciocalteu reagent and 800 µL 5% sodium bicarbonate were added to a gallic acid standard curve (1 mg/mL to 0.002 mg/mL) and to 100 µL of the supernatant. These samples were heated at 40 °C for 30 min, cooled at room temperature for 10 min, and plated in triplicate in a 96-well plate. Samples were then scanned in a microplate reader (Bio-Tek, Winooski, VT, USA) at 765 nm, compared against the standard curve, and reported as milligrams of gallic acid equivalents (GAE) per gram of fresh egg yolk or feed.

2.6. Egg Carotenoid Analysis

Egg yolk total carotenoid content was determined using spectrophotometric screening methods as described by Sergin et al. [10] using methods adapted from Chen et al. [29] and Rodriguez-Amaya and Kimura [30]. Briefly, 0.5 g of lyophilized egg yolk sample and 5 mL of cold acetone (0.05% butylated hydroxytoluene (BHT)) were combined and homogenized. Samples were vortexed for 2 min, placed in an ultrasound water bath for 5 min, and centrifuged for 15 min (1200× g, 4 °C). The supernatant was evaluated in a UV–Vis Double Beam Spectrophotometer (VWR, Radnor, PA, USA) at 450 nm against an acetone blank. Total carotenoid content was calculated according to Biehler et al. [31] using an ε of 140,663 L/mol for beta-carotene in acetone and was expressed as micrograms of beta-carotene per gram of fresh egg yolk.

2.7. Layer Hen Feed Carotenoid Analysis

Total carotenoids of the layer hen feeds were determined using spectrophotometric screening methods adapted from Lichtenthaler and Wellburn [32]. In a conical tube, 2 g of ground layer hen feed was combined with 20 mL of 70% aqueous acetone. The tubes were shaken for 30 min and centrifuged for 20 min at 840× g and 4 °C. The supernatant was recovered in a new tube. The extraction was repeated with an additional 20 mL of 70% aqueous acetone and the supernatants were pooled. Using the spectrophotometer, carotenoid and chlorophyll content of the supernatants were assessed in glass cuvettes at three wavelengths (663, 646, and 470 nm). Chlorophyll A, chlorophyll B, and total carotenoids were calculated using the following equations:

$$\text{Chlorophyll a } (C_a) = 12.21A_{663} - 2.81A_{646}$$

$$\text{Chlorophyll b } (C_b) = 20.13A_{646} - 5.03A_{663}$$

$$\text{Total carotenoids} = \frac{1000A_{470} - 3.27C_a - 104C_b}{229}$$

Total carotenoids were reported as micrograms per gram of feed.

2.8. Fatty Acid Analysis

The fatty acid profile of the lyophilized egg yolks, ground layer hen feed, and grass-fed beef suet and liver tissue was assessed as previously described by Sergin et al. [10]. Briefly, a modified version of the microwave-assisted extraction method by Bronkema et al. [33] was used to extract fatty acids using 400 mg of each sample and 8 mL of a 4:1 (v/v) ethyl acetate/methanol solution with 0.1% BHT as an antioxidant. Fatty acids were extracted in a CEM Mars 6 microwave (CEM Corp., Matthews, NC, USA) using the following microwave parameters: 55 °C for 15 min with an initial ramp of 2 min at 400 W maximum power. Samples were then filtered and prepared as previously described [10].

Methylation described by Sergin et al. [10] modified from Jenkins [34] was conducted to create fatty acid methyl esters (FAMEs). Two milligrams of extracted oil were combined with 500 µL toluene and 20 µg of methyl-12-tridecenoate (U-35M, Nu-Chek Prep, Elysian, MN, USA) as an internal standard. Base-catalyzed methylation was conducted using 2 mL of anhydrous potassium methoxide (0.5 N) at 50 °C for 10 min. Then, acid-catalyzed methylation was conducted using 3 mL of methanolic HCl (5%) at 80 °C for 10 min. Two milliliters of HPLC water were added, then FAMEs were extracted twice using 2 mL of hexane. Extracted FAMEs were resuspended in 1 mL of isooctane and stored at −20 °C until GC-MS analysis.

FAMEs in the samples were separated using the HP-88 column with a 100 m length, 0.25 mm inner diameter, and 0.2 µm film thickness (Agilent Technologies, Santa Clara, CA, USA) on a Perkin Elmer 680/600 GC-MS (Waltham, MA, USA) in the electron impact mode with helium as the carrier gas (1 mL/min). For improved separation of fatty acid isomers, column temperature parameters described by Kramer et al. [35] were used as follows: initial temperature of 80 °C for 4 min, ramp at a rate of 13.0 °C/min to 175 °C, held for 27 min, ramp at a rate of 4.0 °C/min to 215 °C, and held for 35 min. Two different injections with a 1 µL injection volume and 250 °C injection temperature were conducted to measure both lower- and higher-concentration analytes. These were a 30:1 split injection and a splitless injection (0.75 min splitless hold time, 40 mL/min flow exiting the vent). Regarding the MS, electron energy was 70 eV, and the transfer line and ion source temperature were set to 180 °C. MS data were recorded in full scan mode (*m/z* 70–400).

Data were analyzed using MassLynx (4.1 SCN 714; Waters Corp., Milford, MA, USA). Retention time and EI mass fragmentation of each analyte were compared to those in the reference standard (described above). Fatty acids not included in the reference standard were identified by elution order as reported by Kramer et al. [35] and confirmed with EI mass fragmentation. Fatty acids were quantified using extracted ion chromatograms of the corresponding quantitative ions utilizing a standard curve constructed from our reference and internal standard. To calculate each FAME concentration, the internal standard peak area and analyte peak area in each sample were compared to those of the standard curve. Fatty acids were reported as percent of total fatty acids quantified and in gram amounts based upon the total oil yield from the microwave-assisted extraction.

2.9. Egg Yolk Vitamin and Mineral Analysis

For the egg vitamin and mineral analysis, remaining lyophilized egg yolk powder was pooled to form three replicates per group consisting of two egg yolks each. Egg yolk vitamin and mineral analyses were conducted by a commercial laboratory (Creative Proteomics, Shirley, NY, USA). Vitamins were analyzed using targeted metabolite quantitation using LC-MRM/MS and minerals were quantified using ICP-MS.

2.10. Statistical Analysis

Based on a power analysis conducted using changes in the *n*-6:*n*-3 ratio of egg yolks from Sergin et al. [10], $n = 6$ egg yolks was determined to be sufficient to detect significant differences. The egg characteristics, feed chemical composition, phenolic, and carotenoid data were analyzed using Prism Mac 7.0d (GraphPad Software, La Jolla, CA, USA). The fatty acid and vitamin and mineral data were analyzed using R version 3.6.1. Independent samples *t*-tests were performed for all the feed analyses. Group comparisons for all the egg analyses were performed using one-way analysis of variance (ANOVA) and Tukey's HSD with correction for multiple comparisons. Values under the lower limit of detection were treated as zeroes. Statistical significance was set at $p < 0.05$ for all analyses.

3. Results and Discussion

3.1. Nutrient Composition of the Layer Hen Diets

The chemical composition and antioxidant profile of the PBB and PCS layer hen feeds are shown in Table 1.

Table 1. Chemical composition and antioxidant profile of the layer hen feeds [1].

Parameters	PBB Feed	PCS Feed	CFC Feed	*p*-Value [2]
Dry matter (%)	86.8	88.5	N/A	-
Crude protein (% DM)	17.2 ± 0.1	19.7 ± 0.2	N/A	<0.001
Metabolizable energy (Mcal/kg)	3.4 ± 0.0	3.5 ± 0.0	N/A	0.002
Total carotenoids (μg/g feed)	5.38 ± 0.48	7.20 ± 0.37	N/A	0.007
Total phenols (mg of GAE/g feed)	0.84 ± 0.05	1.06 ± 0.13	N/A	0.057

[1] Means ± standard deviation (n = 3 per group). [2] Results of independent samples t-test between PBB feed and PCS feed. PBB feed, custom feed of peas, barley, alfalfa, and calcium; PCS feed, standard corn and soy feed; CFC feed, cage-free comparison; DM, dry matter, GAE, gallic acid equivalents; N/A, no data.

The main fatty acid profile of the GFB liver, GFB suet, and layer hen feeds are reported in Tables 2 and S1.

Table 2. Fatty acid profile of grass-fed beef by-products and layer hen feeds (% of total fatty acids) [1].

Fatty Acid	Beef Liver	Beef Suet	PBB Feed	PCS Feed	*p*-Value [2]
10:0	<LLOD	0.05 ± 0.01	0.03 ± 0.00	0.02 ± 0.00	0.006
12:0	0.01 ± 0.00	0.07 ± 0.01	0.01 ± 0.00	0.01 ± 0.00	0.016
14:0	0.32 ± 0.05	3.89 ± 0.07	0.21 ± 0.01	0.06 ± 0.00	0.002
14:1	0.03 ± 0.01	0.29 ± 0.02	<LLOD	<LLOD	-
16:0	14.36 ± 0.28	26.33 ± 0.21	17.70 ± 0.47	12.01 ± 0.42	<0.001
16:1 *n*-7	0.50 ± 0.01	1.64 ± 0.09	0.12 ± 0.03	0.07 ± 0.01	0.120
16:1 *n*-7 *t*	0.21 ± 0.04	0.12 ± 0.02	<LLOD	<LLOD	-
16:1 *n*-9	0.25 ± 0.06	0.37 ± 0.06	0.06 ± 0.01	0.04 ± 0.01	0.135
18:0	35.61 ± 0.62	23.74 ± 0.58	2.14 ± 0.08	2.57 ± 0.06	0.003
18:1 *n*-7	0.84 ± 0.08	0.98 ± 0.07	0.65 ± 0.04	0.97 ± 0.11	0.023
18:1 *n*-9	8.08 ± 0.26	30.21 ± 0.68	19.43 ± 0.40	20.93 ± 0.08	0.020
18:1 *n*-9 *t*	1.53 ± 0.13	5.42 ± 0.35	<LLOD	<LLOD	-
18:2 *n*-6	4.11 ± 0.14	0.80 ± 0.04	51.10 ± 0.91	56.14 ± 0.40	0.004
18:3 *n*-3	2.18 ± 0.20	0.79 ± 0.10	7.06 ± 0.17	6.31 ± 0.61	0.162
18:3 *n*-6	0.26 ± 0.05	<LLOD	<LLOD	<LLOD	-
20:0	0.11 ± 0.03	0.14 ± 0.02	0.32 ± 0.02	0.21 ± 0.02	0.003
20:1 *n*-9	<LLOD	0.25 ± 0.01	0.42 ± 0.04	0.19 ± 0.02	0.003
20:2 *n*-6	0.24 ± 0.11	<LLOD	<LLOD	<LLOD	-
20:3 *n*-6	3.39 ± 0.16	<LLOD	<LLOD	<LLOD	-
20:4 *n*-6	7.12 ± 0.43	<LLOD	<LLOD	<LLOD	-
20:5 *n*-3	3.28 ± 0.12	<LLOD	<LLOD	<LLOD	-
22:0	<LLOD	0.07 ± 0.01	0.16 ± 0.02	0.18 ± 0.02	0.406
22:4 *n*-6	2.32 ± 0.40	<LLOD	<LLOD	<LLOD	-
22:5 *n*-3	9.83 ± 0.55	<LLOD	<LLOD	<LLOD	-
22:6 *n*-3	2.55 ± 0.05	<LLOD	<LLOD	<LLOD	-
24:0	<LLOD	<LLOD	0.31 ± 0.02	0.18 ± 0.03	0.003
Total SFA	51.90 ± 0.49	56.28 ± 0.35	21.16 ± 0.63	15.34 ± 0.39	<0.001
Total MUFA	11.68 ± 0.53	39.69 ± 0.21	20.68 ± 0.45	22.21 ± 0.09	0.023
Total PUFA	35.50 ± 0.40	1.85 ± 0.16	58.16 ± 0.74	62.45 ± 0.32	0.004
Total *n*-6	17.44 ± 0.28	0.80 ± 0.04	51.10 ± 0.91	56.14 ± 0.40	0.004
Total *n*-3	17.85 ± 0.44	0.79 ± 0.10	7.06 ± 0.17	6.31 ± 0.61	0.162
n-6:*n*-3 ratio	1.00 ± 0.03	1.01 ± 0.11	7.25 ± 0.31	8.96 ± 0.88	0.065

[1] Means ± standard deviation (n = 3 per group). [2] Results of independent samples t-test between PBB feed and PCS feed. PBB feed, custom feed of peas, barley, alfalfa, and calcium; PCS feed, standard corn, and soy feed; <LLOD, below lower limit of detection; SFA, saturated fatty acids; MUFA, monounsaturated fatty acids, PUFA, polyunsaturated fatty acids.

3.2. *Egg Physical Characteristics and Antioxidant Profiles*

Egg physical characteristics are presented in Table 3 and egg yolk antioxidant profiles are presented in Table 4.

Table 3. Physical characteristics of the eggs [1].

Parameters	PBB Eggs	PCS Eggs	CFC Eggs	*p*-Value [2]
Egg weight (g)	56.98 ± 5.11	63.04 ± 8.49	64.27 ± 1.88	0.100
Yolk weight (g)	16.36 ± 1.78	17.94 ± 0.95	17.30 ± 1.45	0.191
Shell weight (g)	5.48 ± 0.77	5.70 ± 0.93	5.98 ± 0.40	0.515
Albumen weight (g)	35.15 ± 4.33	39.40 ± 8.50	40.99 ± 1.10	0.202
Haugh units	73.05 ± 6.33	81.51 ± 4.29	73.81 ± 7.39	0.057
Yolk color fan	14.00 ± 0.00	>14 [3]	8.17 ± 0.98	<0.001 [4]

[1] Means ± standard deviation (n = 6 per group). [2] Results of one-way ANOVA. [3] All PCS egg yolks were darker orange than the color scale used. [4] Comparison between PBB and CFC using an independent samples t-test. PBB eggs, pasture-raised eggs with beef by-products and corn/soy free feed; PCS eggs, pasture-raised eggs with corn/soy feed; CFC eggs, cage-free comparison eggs.

Table 4. Antioxidant profile of the egg yolks [1].

Parameters	PBB Eggs	PCS Eggs	CFC Eggs	*p*-Value [2]
Total carotenoids (µg/g fresh yolk)	41.54 ± 7.07 a	48.99 ± 12.35 a	18.23 ± 3.23 b	<0.001
Total phenols (mg of GAE/g fresh yolk)	0.60 ± 0.51 b	1.67 ± 0.82 a	1.80 ± 0.49 a	0.008

[1] Means ± standard deviation (n = 6 per group). [2] Results of one-way ANOVA. a, b Means within a row with different letters significantly differ ($p < 0.05$). PBB eggs, pasture-raised eggs with beef by-products and corn/soy free feed; PCS eggs, pasture-raised eggs with corn/soy feed; CFC eggs, cage-free comparison eggs, GAE, gallic acid equivalents.

Egg weight, yolk weight, shell weight, albumen weight, and Haugh units were not significantly different among the egg groups. Egg yolk color was significantly more orange in PBB and PCS eggs compared to CFC eggs. Carotenoids, which have important antioxidant properties, give egg yolks their orange–yellow color [11]. Accordingly, the screening analysis revealed that PBB and PCS egg yolks had significantly greater carotenoid content than CFC egg yolks ($p < 0.001$). Egg yolks from pasture-raised hens contained higher total carotenoids due to access to green vegetation, a source of a variety of carotenoids [1,9,10,36]. Current recommendations gathered from previous studies advise the consumption of at least 7 mg of beta-carotene per day to ensure adequate vitamin A intake [37]. According to this recommendation, consuming two PBB or PCS egg yolks provides approximately 20% of recommended intake, while two CFC egg yolks only provide about 8%.

In general, all three egg groups had a low total phenolic content, ranging from 0.60–1.80 mg GAE/g of fresh yolk as shown by the screening analysis (Table 4). Similar studies have observed a total phenolic content in eggs between 1.1–1.7 mg GAE/g of fresh yolk [10,38]. The higher phenolic content of the PCS and CFC egg yolks is likely due to phenolic compounds that are commonly found in cereal grains, including corn, which is a component of standard layer hen feeds [28]. Given that the deposition of phenolic compounds into egg yolks is low and that average polyphenol intakes of diets across varying dietary patterns are estimated to range from 300–1700 mg/day, egg consumption is unlikely to make a large contribution to polyphenol intake [28,39].

3.3. *Egg Yolk Fatty Acid Profiles*

The differences in the egg yolk fatty acid profile among the egg groups, shown in Table 5, can be partially attributed to differences in the hen diets. There was no difference in percent SFA among the egg groups, but there were significant differences in percent MUFA and PUFA. In general, the proportions of SFA, MUFA, and PUFA were consistent with other studies comparing egg yolks from hens fed conventional and pasture-containing diets, except for the low percent PUFA in PBB egg yolks [7,40]. An additional table displaying absolute quantification of the fatty acid profile in the egg yolks in grams of fatty acids per 100 g of fresh yolk is available in Table S2.

Table 5. Fatty acid profile of the egg yolks (% of total fatty acids) [1].

Fatty Acid	PBB Eggs	PCS Eggs	CFC Eggs	*p*-Value [2]
10:0	0.02 ± 0.00	0.02 ± 0.01	0.02 ± 0.00	0.482
12:0	0.01 ± 0.00	0.01 ± 0.00	0.01 ± 0.00	0.442
14:0	0.56 ± 0.13 a	0.29 ± 0.10 b	0.27 ± 0.08 b	<0.001
14:1	0.13 ± 0.05 a	0.03 ± 0.01 b	0.02 ± 0.01 b	<0.001
16:0	26.60 ± 1.90	28.12 ± 1.80	28.53 ± 1.64	0.177
16:1 *n*-7	3.24 ± 0.88 a	2.54 ± 0.63 ab	1.71 ± 0.44 b	0.005
16:1 *n*-7 *t*	0.13 ± 0.04 a	0.09 ± 0.03 b	0.08 ± 0.01 b	0.004
16:1 *n*-9	0.70 ± 0.16 a	0.51 ± 0.09 b	0.37 ± 0.07 b	0.001
18:0	7.43 ± 0.37	7.00 ± 1.33	7.83 ± 1.45	0.479
18:1 *n*-7	1.97 ± 0.29 a	1.74 ± 0.21 a	1.14 ± 0.15 b	<0.001
18:1 *n*-9	46.62 ± 2.16 a	39.02 ± 2.74 b	35.18 ± 3.34 b	<0.001
18:1 *n*-9 *t*	0.32 ± 0.04 a	0.19 ± 0.04 b	0.17 ± 0.01 b	<0.001
18:2 *n*-6	7.08 ± 1.46 c	16.19 ± 3.91 b	21.76 ± 4.63 a	<0.001
18:3 *n*-3	0.40 ± 0.10 ab	0.62 ± 0.27 a	0.16 ± 0.04 b	0.001
18:3 *n*-6	0.11 ± 0.01 b	0.15 ± 0.05 ab	0.17 ± 0.01 a	0.015
20:0	0.04 ± 0.00	0.04 ± 0.02	0.04 ± 0.00	0.882
20:1 *n*-9	0.34 ± 0.06 a	0.19 ± 0.07 b	0.18 ± 0.02 b	<0.001
20:2 *n*-6	0.07 ± 0.04 b	0.19 ± 0.05 a	0.25 ± 0.08 a	<0.001
20:3 *n*-6	0.13 ± 0.01	0.16 ± 0.06	0.17 ± 0.01	0.171
20:4 *n*-6	0.61 ± 0.14	0.65 ± 0.13	0.79 ± 0.07	0.050
20:5 *n*-3	<LLOD	<LLOD	<LLOD	-
22:0	<LLOD	<LLOD	<LLOD	-
22:4 *n*-6	0.30 ± 0.10	0.33 ± 0.21	0.16 ± 0.14	0.176
22:5 *n*-3	0.33 ± 0.04 a	0.19 ± 0.12 b	<LLOD c	<0.001
22:5 *n*-6	0.26 ± 0.05 b	0.27 ± 0.08 b	0.38 ± 0.03 a	0.006
22:6 *n*-3	0.77 ± 0.10 a	0.93 ± 0.23 a	0.30 ± 0.05 b	<0.001
24:0	<LLOD	<LLOD	<LLOD	-
Total SFA	35.21 ± 2.10	35.74 ± 2.27	36.88 ± 2.27	0.431
Total MUFA	53.84 ± 1.66 a	44.42 ± 2.82 b	38.92 ± 3.68 c	<0.001
Total PUFA	10.36 ± 1.65 b	19.74 ± 3.72 a	24.19 ± 4.66 a	<0.001
Total *n*-6	8.55 ± 1.62 c	17.94 ± 3.68 b	23.67 ± 4.59 a	<0.001
Total *n*-3	1.50 ± 0.12 a	1.74 ± 0.39 a	0.47 ± 0.08 b	<0.001
n-6:*n*-3 ratio	5.72 ± 1.12 c	10.79 ± 3.30 b	50.63 ± 4.21 a	<0.001

[1] Means ± standard deviation (n = 6 per group). [2] Results of one-way ANOVA. a–c Means within a row with different letters significantly differ ($p < 0.05$). PBB eggs, pasture-raised eggs with beef by-products and corn/soy free feed; PCS eggs, pasture-raised eggs with corn/soy feed; CFC eggs, cage-free comparison eggs; <LLOD, below lower limit of detection; SFA, saturated fatty acids; MUFA, monounsaturated fatty acids, PUFA, polyunsaturated fatty acids.

3.3.1. Saturated Fatty Acids

Despite the inclusion of beef suet, high in SFAs, in the PBB diet, there was no difference in total SFAs by percentage or absolute quantification (Tables 5 and S2). Only myristic acid was significantly higher in PBB egg yolks. These results are consistent with previous studies showing little to no differences in SFA concentration in hen egg yolks regardless of hen diet composition. One study compared the effects of the inclusion of seal blubber oil, high in *n*-3 PUFAs and MUFAs, to a control diet containing tallow, high in SFAs, on egg yolk fatty acid profile [15]. They observed SFA concentration ranging from 35–37% in all treatment groups, similar to the present study. The concentration of myristic acid was greater and stearic acid was lower in the egg yolks from hens fed 5% seal blubber oil compared to the egg yolks from hens fed 5% tallow. Despite these individual fatty acid differences, there were no observed differences in total SFA concentration among the egg yolks from hens fed different diets [15]. Another study compared the effects of fish oil, high in PUFAs, and tallow supplementation on the egg yolk fatty acid profile [16]. Egg yolk percent SFA ranged from 39–42%. The yolks from hens fed 100% of fat from tallow had significantly lower total SFA content than the yolks from hens fed 100% of fat from fish oil. Similarly, Celebi and Macit [17] and Grobas et al. [18] found that increasing the amount of tallow supplementation decreased the concentration of egg yolk SFAs. However, these differences, as stated by Baucells et al. [16], are negligible, demonstrating how the degree of saturation in egg yolks is maintained within very narrow margins [16].

3.3.2. Mono- and Poly-Unsaturated Fatty Acids

Percent MUFAs were significantly higher in PBB eggs, while percent PUFAs were significantly higher in PCS and CFC eggs compared to PBB eggs (Table 5). This difference in percent total MUFAs was due to the lower gram amount of PUFAs in PBB eggs and not due to an actual increase in grams of MUFAs. Though the percent total MUFA and percent oleic acid were significantly greater in the PBB eggs compared to PCS and CFC eggs, there were no significant differences for grams of total MUFA or grams of oleic acid per 100 g of egg yolk (Table S2). This finding is consistent with previous studies demonstrating the limited ability of hens to increase egg yolk oleic acid content [16,41]. PBB eggs contained about half as many grams of total PUFA per 100 g of egg yolk compared to PCS and CFC eggs. This large difference was due to the significantly lower grams of linoleic acid (LA) in PBB egg yolks. Normally LA contributes > 80% of total PUFA in egg yolks, seen in PCS and CFC eggs with about 81% and 90%, respectively [42]. However, LA constituted only 68% of total PUFA in the PBB eggs (Table S2). The corn- and soy-free PBB feed contained significantly fewer grams of total PUFA and LA compared to the corn- and soy-based PCS feed, which helps to explain the differences between the PBB and PCS egg yolks (Table S1).

The differences in n-6 PUFA and n-3 PUFA concentration among the egg groups further demonstrates the impact of the production system on egg yolk fatty acid profile. While the egg yolk ALA concentration was significantly higher in PCS eggs compared to CFC eggs, the egg yolk DPA and DHA concentrations were significantly higher in both PBB and PCS eggs compared to CFC eggs. Hens convert ALA obtained from their diet into these long-chain n-3 PUFAs via liver elongases and desaturases. The same enzymes can convert LA into arachidonic acid (C20:4 n-6) [43]. Though LA concentration was significantly lower in PBB eggs, there was no difference in arachidonic acid concentration compared to the other groups. PUFA metabolism, including whether the conversion of ALA or LA is favored, is affected by diet, genotype, age, and rearing system [43]. The hen's access to forages in the PBB and PCS groups provided greater access to n-3 precursors than would have been available in the CFC-rearing system [44]. This difference likely contributed to greater conversion of ALA into DPA and DHA in the PBB and PCS egg groups. It should be noted that EPA was below the lower limit of detection in all the egg groups, indicating that EPA was converted into the longer chain n-3 PUFAs detected in the PBB and PCS groups. Numerous studies have found egg yolk EPA concentration to be below the limit of detection when feeding standard layer hen diets, and that supplementation with additional exogenous sources of ALA or EPA from supplements such as flaxseed or fish oil is necessary to increase egg yolk EPA concentration [45]. Though the GFB liver provided about 60 mg of EPA per 100 g to the PBB group, this was not sufficient to increase egg yolk EPA.

Further, percent n-6 PUFAs were lowest in PBB eggs, while percent n-3 PUFAs were significantly greater in the PBB and PCS eggs compared to CFC eggs. Thus, the n-6:n-3 ratios were significantly different among the egg groups, with a 51:1 ratio observed in CFC eggs compared to 6:1 and 11:1 in PBB and PCS eggs, respectively, from hens with unlimited access to fresh pasture ($p < 0.001$) (Table 5). This is congruent with previous studies showing that pasture access increases n-3 PUFA concentration and decreases the n-6:n-3 ratio of hen egg yolks [1,10,40,46]. Consuming diets with a lower n-6:n-3 ratio, around 4:1, is associated with a decreased risk of many chronic diseases including obesity and cardiovascular disease [47]. Thus, PBB and PCS egg yolks, compared to CFC egg yolks, are more aligned with nutrition recommendations to reduce n-6 PUFA and increase n-3 PUFA consumption [48,49].

3.3.3. Conjugated Linoleic Acid

Conjugated linoleic acid is reported to possess potential anti-atherogenic and anti-obesogenic properties, have cancer-protective effects and modify immunity by reducing pro-inflammatory cytokines, though the health effects of CLA are still under investigation [50,51]. Percent conjugated linoleic acid in the beef liver and suet, layer hen feeds, and egg yolks are displayed in Table 6. Beef liver and suet were sources of CLA in the PBB hen diet. Thus, PBB eggs had a significantly higher CLA concentration compared to PCS and CFC eggs from hens who consumed corn and soy feed. A previous study comparing pasture-raised, free-range, and conventional egg yolks found no differences in CLA concentration by production system [10]. Similarly, there was no difference in CLA concentration between the pasture-raised PCS eggs and the conventionally raised CFC eggs. Thus, it is likely that the increased CLA in the PBB eggs was due to the hens' animal-product consumption. Similarly, Shinn, et al. [52] demonstrated that feeding CLA-rich soy oil to hens increased the CLA concentration in egg yolks. CLA-rich oil appears to be more effective at increasing egg yolk CLA compared to adding animal products to hen diets, achieving 2% CLA compared to 0.3% CLA in PBB egg yolks in the present study [52]. However, adding CLA-rich oil to hen diets also undesirably increased egg yolk SFA content and decreased MUFA content [52,53]. This effect was not seen with the addition of CLA from beef by-products in this study. Despite the beneficial effect of beef by-product supplementation on egg yolk CLA concentration, the effects on human health may be negligible. To yield health benefits, the CLA content of PBB eggs is relatively low with about 0.05 g of CLA per 100 g of fresh yolk (Table S2). Previous studies have estimated that consumption of 620 and 441 mg of CLA/day for men and women, respectively, was necessary to achieve a cancer-protective effect, and at least 3 g of CLA/day was needed to achieve body fat reduction [54,55].

Table 6. Conjugated linoleic acid, odd-chain fatty acid, and branched-chain fatty acid content of grass-fed beef by-products, layer hen feeds, and egg yolks (% of total fatty acids) [1].

Fatty Acid	Beef Liver	Beef Suet	PBB Feed	PCS Feed	*p*-Value [2]	PBB Eggs	PCS Eggs	CFC Eggs	*p*-Value [3]
Conjugated linoleic acid 9*c*, 11*t* 18:2	0.21 ± 0.03	0.26 ± 0.04	<LLOD	<LLOD	-	0.30 ± 0.13 a	0.06 ± 0.02 b	0.05 ± 0.00 b	<0.001
Odd-chain									
13:0	0.010 ± 0.003	0.021 ± 0.002	0.010 ± 0.002	0.004 ± 0.001	0.012	0.004 ± 0.001	0.004 ± 0.002	0.004 ± 0.000	0.541
15:0	0.29 ± 0.04	0.87 ± 0.11	0.11 ± 0.00	0.02 ± 0.00	<0.001	0.15 ± 0.04 a	0.06 ± 0.02 b	0.05 ± 0.01 b	<0.001
17:0	1.19 ± 0.05	1.10 ± 0.09	0.15 ± 0.01	0.09 ± 0.00	0.009	0.40 ± 0.11 a	0.18 ± 0.05 b	0.13 ± 0.02 b	<0.001
17:1	0.25 ± 0.05	0.40 ± 0.04	<LLOD	<LLOD	-	0.39 ± 0.13 a	0.13 ± 0.04 b	0.08 ± 0.01 b	<0.001
Total OCFA	1.74 ± 0.12	2.40 ± 0.24	0.26 ± 0.02	0.12 ± 0.01	0.001	0.95 ± 0.27 a	0.36 ± 0.12 b	0.26 ± 0.03 b	<0.001
Branched-chain									
15:0-*iso*	0.10 ± 0.01	0.42 ± 0.04	<LLOD	<LLOD	-	0.06 ± 0.03 a	0.02 ± 0.01 b	0.002 ± 0.005 b	<0.001
15:0-*anteiso*	0.08 ± 0.01	0.42 ± 0.04	<LLOD	<LLOD	-	0.02 ± 0.01 a	0.01 ± 0.01 b	0.004 ± 0.006 b	0.003
16:0-*iso*	0.08 ± 0.01	0.24 ± 0.03	<LLOD	<LLOD	-	0.05 ± 0.02 a	0.01 ± 0.00 b	<LLOD b	<0.001
17:0-*iso*	0.23 ± 0.00	0.44 ± 0.05	<LLOD	<LLOD	-	0.12 ± 0.04 a	0.02 ± 0.02 b	<LLOD b	<0.001
17:0-*anteiso*	0.37 ± 0.01	0.58 ± 0.05	<LLOD	<LLOD	-	0.29 ± 0.11 a	0.03 ± 0.01 b	0.002 ± 0.006 b	<0.001
18:0-*iso*	0.07 ± 0.01	0.09 ± 0.01	<LLOD	<LLOD	-	0.04 ± 0.02 a	<LLOD b	<LLOD b	<0.001
Total *iso*-BCFA	0.47 ± 0.02	1.20 ± 0.13	<LLOD	<LLOD	-	0.27 ± 0.10 a	0.06 ± 0.03 b	0.002 ± 0.005 b	<0.001
Total *anteiso*-BCFA	0.44 ± 0.01	1.00 ± 0.09	<LLOD	<LLOD	-	0.32 ± 0.12 a	0.04 ± 0.02 b	0.01 ± 0.01 b	<0.001
Total BCFA	0.91 ± 0.04	2.19 ± 0.22	<LLOD	<LLOD	-	0.59 ± 0.22 a	0.10 ± 0.05 b	0.008 ± 0.016 b	<0.001

[1] Means ± standard deviation (n = 3 per group for feed, n = 6 per group for eggs). [2] Results of independent samples *t*-test between PBB feed and PCS feed. [3] Results of one-way ANOVA. a, b Means within a row with different letters significantly differ ($p < 0.05$). PBB feed, custom feed of peas, barley, alfalfa, and calcium; PCS feed, standard corn and soy feed; PBB eggs, pasture-raised eggs with beef by-products and corn/soy free feed; PCS eggs, pasture-raised eggs with corn/soy feed; CFC eggs, cage-free comparison eggs; <LLOD, below lower limit of detection; OCFA, odd-chain fatty acids; BCFA, branched-chain fatty acids.

3.3.4. Odd-Chain Fatty Acids

Generally, OCFAs constitute a small amount of total fatty acids in foods; milk, which is one of the main dietary sources of OCFAs, contains 1% C15:0 and 0.5% C17:0 [56]. In contrast to even-chain saturated fatty acids, dietary intake of odd-chain saturated fatty acids is associated with a lower risk of inflammation, cardiovascular disease, type 2 diabetes, and other chronic diseases according to several human epidemiological studies [56–60]. Intake of OCFAs was also associated with lower total mortality in both males and females in a 14-year prospective cohort study with 14,000 participants [61]. Venn-Watson, et al. [56] demonstrated that OCFAs have biological effects that align with these health benefits in vitro and in vivo in animal studies. The substitution of 0.1% of energy or about 220 mg of PUFAs or MUFAs in a 2000 kilocalorie diet with 220 mg of OCFAs was associated with 15–19% lower total mortality in men and women [61].

Odd-chain fatty acid concentration is not frequently reported in eggs. Limited evidence indicates egg yolk OCFA concentration ranges from 0.24–0.43%, with no significant differences observed among the production systems studied, similar to the OCFA concentration in PCS and CFC eggs in this study (Table 6) [10,62]. However, PBB eggs contained 0.95% OCFAs (Table 6). Further, per yolk, PBB eggs contained significantly more OCFAs than PCS or CFC eggs (0.15 g vs. 0.06 g or 0.04 g per 100 g of fresh yolk: Table S2). The beef liver, beef suet, and PBB feed were all sources of OCFAs in the PBB hen diet, and the PBB layer hen feed had a significantly higher concentration of OCFAs than the PCS feed (Table 6). This suggests that the inclusion of beef by-products and the PBB feed contributed to increasing the OCFA concentration in PBB egg yolks rather than pasture access. Interestingly, Evans and LaMont [63] demonstrated that the inclusion of microalgae to create an OCFA-enriched layer hen feed can increase egg yolk OCFAs to at least 3% of total fatty acids. Effective methods to increase the OCFA content of hen eggs need to be further studied to help consumers increase their OCFA intake and achieve the observed health benefits of OFCA consumption.

3.3.5. Branched-Chain Fatty Acids

To our knowledge, this is one of the few studies to characterize BCFA isomers in egg yolks. BCFAs are found primarily in animal products such as beef and dairy, at around 1.89% and 2.05% of total fatty acids, respectively [64]. Health information about BCFAs is limited, but they are thought to function similarly to *cis*-unsaturated fatty acids due to their branched structure [64]. BCFAs reduced the incidence of necrotizing enterocolitis and increased the expression of anti-inflammatory cytokines in a rat model, demonstrated anticancer properties in vitro, and modulated gut microbiota composition in vivo [65–68].

PBB eggs contained greater amounts of all BCFA isomers compared to PCS and CFC eggs (Tables 6 and S2). BCFAs were not detected in either the PBB or PCS feed, but both the GFB suet and liver were sources of BCFA isomers (Table 6). A previous study characterizing egg yolk BCFA isomers found that pasture-raised eggs had a higher BCFA concentration than free-range and commercially raised eggs, hypothesizing that the higher BCFA concentration may be related to differences in pasture access and microbe deposition from the hens' microbiota, containing BCFAs, during egg formation [10]. However, egg yolks from pasture-raised hens in that study only contained about 0.05% BCFAs [10]. PBB egg yolks from hens with both pasture access and animal product consumption contained 0.59% BCFAs, around ten times that of the previous study. This suggests that feeding PBB hens animal products increased the BCFA content of the egg yolks in this study. PBB eggs contained about 0.09 g of BCFAs per 100 g of fresh yolk, a small portion of the typical daily BCFA intake in Western diets of around 0.4–0.5 g (Table S2) [64,67].

3.4. *Egg Yolk Vitamin and Mineral Profiles*

The vitamin profile of the egg yolks is shown in Table 7. Due to the high lipid content of egg yolks, lipid-soluble vitamins are easily incorporated into egg yolks [11]. There were no significant differences in egg yolk vitamin A, D3, or E content in this study. Previously, pasture access significantly increased the vitamin A and E content of hen egg yolks [1,9,10]. Similar to the present study, Anderson [69] found that free-range and caged hens had similar vitamin A content, while carotenoids were higher in free-range eggs. However, it is surprising that PBB eggs did not contain more vitamin A given that beef liver is a significant source of vitamin A, providing roughly 2000 retinol activity equivalents per ounce [70]. Since vitamin A is a recommended component of layer hen feeds, it is possible that the layer hen diets provided to the PCS and CFC groups provided a similar vitamin A content to that in the PBB diet [71]. The lack of differences in vitamin E content in this study may be due to the assessment of free alpha-tocopherol rather than total alpha-tocopherol which is typically reported in eggs [72,73]. Eggs are one of the few food sources rich in vitamin D due to the efficient transfer of vitamin D from hen diets into egg yolks and increasing hen dietary concentrations of vitamin D increases egg yolk vitamin D content accordingly [74]. The lack of differences in vitamin D3 content in the present study indicates there were similar vitamin D3 contents among PBB, PCS, and CFC diets. There were no differences in vitamin K2 content, while vitamin K1 was greater in PCS eggs compared to CFC eggs. Further, vitamin K1 was only detected in the pasture-raised egg groups. Access to green vegetation, a source of vitamin K1, may contribute to this difference [75].

Table 7. Vitamin profile of the egg yolks (μg/100 g fresh yolk) [1].

Vitamin	PBB Eggs	PCS Eggs	CFC Eggs	*p*-Value [2]
Vitamin B1	296.32 ± 19.61 b	401.68 ± 38.89 a	234.88 ± 27.31 b	0.001
Vitamin B2	692.33 ± 109.72	705.05 ± 65.99	685.24 ± 93.17	0.965
Vitamin B3	16.64 ± 0.66 a	11.21 ± 1.76 b	10.06 ± 1.20 b	0.002
Vitamin B5	8077.25 ± 241.75 a	7260.07 ± 654.03 a	4190.95 ± 544.52 b	<0.001
Vitamin B9	0.15 ± 0.05	0.10 ± 0.01	0.18 ± 0.03	0.071
Vitamin B12	0.45 ± 0.06 b	0.49 ± 0.01 b	0.71 ± 0.03 a	<0.001
Vitamin A	246.24 ± 56.29	340.36 ± 24.07	221.94 ± 91.05	0.130
Vitamin D3	20.61 ± 2.94	21.75 ± 1.22	16.91 ± 5.58	0.318
Vitamin E [3]	107.13 ± 60.77	93.16 ± 11.15	44.89 ± 19.30	0.182
Vitamin K1	0.53 ± 0.21 ab	0.97 ± 0.63 a	<LLOD b	0.055
Vitamin K2	0.57 ± 0.14	0.71 ± 0.15	0.87 ± 0.10	0.081

[1] Means ± standard deviation ($n = 3$ per group). Each replicate was composed of two pooled egg yolks. [2] Results of one-way ANOVA. a, b Means within a row with different letters significantly differ ($p < 0.05$). [3] Vitamin E is reported as free alpha-tocopherol. PBB eggs, pasture-raised eggs with beef by-products and corn/soy free feed; PCS eggs, pasture-raised eggs with corn/soy feed; CFC eggs, cage-free comparison eggs; <LLOD, below lower limit of detection.

The B-vitamin profile is not frequently reported in egg yolks. There were no differences in vitamin B2 or B9 content of the eggs (Table 7). Vitamin B1, found in higher amounts in corn and soy compared to beef liver, was greater in PCS eggs compared to PBB and CFC eggs [76]. The difference in B1 content between PCS and CFC eggs could be due to differences in the quality or preparation of the corn and soy feed provided in either diet [76]. Further, beef liver is a richer source of vitamin B3 compared to corn and soy, potentially explaining the higher B3 content of PBB eggs [76]. Regarding B5, corn and soy contain significantly less B5 than beef liver, yet PBB and PCS eggs had similar B5 content, nearly twice that of CFC eggs [76,77]. Given that vitamin B5 can be synthesized microbially, pasture access may have played a role in altering hen microbial communities and thus increasing vitamin B5 content [76,78]. Lastly, the B12 content was higher in CFC eggs compared to the other groups.

There were no significant differences in the mineral profile of the egg groups except for manganese content which was greater in the PCS eggs compared to the other groups ($p < 0.001$) (Table 8). The lack of differences in mineral content was surprising since beef liver is a rich source of iron, phosphorus, and zinc [79]. These minerals were expected to be greater in PBB eggs due to the inclusion of beef liver in the hen diets. Assessing a greater sample size, particularly for iron content, may be necessary to better elucidate any significant differences.

Table 8. Mineral profile of the egg yolks (mg/100 g fresh yolk) [1].

Mineral	PBB Eggs	PCS Eggs	CFC Eggs	*p*-Value [2]
Sodium	36.70 ± 4.91	35.26 ± 6.02	45.86 ± 0.65	0.055
Magnesium	16.65 ± 1.42	17.02 ± 1.27	16.43 ± 0.74	0.830
Phosphorus	505.61 ± 45.77	502.22 ± 64.55	458.06 ± 13.74	0.428

Table 8. *Cont.*

Mineral	PBB Eggs	PCS Eggs	CFC Eggs	p-Value [2]
Potassium	103.25 ± 12.71	90.64 ± 5.09	84.31 ± 2.70	0.070
Manganese	0.06 ± 0.01 b	0.11 ± 0.01 a	0.07 ± 0.00 b	<0.001
Iron	3.22 ± 0.45	2.53 ± 0.18	2.38 ± 0.39	0.060
Selenium	0.27 ± 0.02	0.33 ± 0.06	0.33 ± 0.12	0.635
Calcium	171.73 ± 18.43	161.87 ± 20.38	147.32 ± 5.29	0.254
Copper	0.13 ± 0.03	0.17 ± 0.07	0.13 ± 0.02	0.514
Zinc	3.00 ± 0.29	3.28 ± 0.47	3.94 ± 1.41	0.452

[1] Means ± standard deviation (n = 3 per group). Each replicate was composed of two pooled egg yolks. [2] Results of one-way ANOVA. a, b Means within a row with different letters significantly differ ($p < 0.05$). PBB eggs, pasture-raised eggs with beef by-products and corn/soy free feed; PCS eggs, pasture-raised eggs with corn/soy feed; CFC eggs, cage-free comparison eggs.

4. Conclusions

This study is limited by the lack of specific layer hen diet composition information for PBB and PCS eggs and the inability to obtain diet information for CFC eggs. We were also unable to measure total phenolics and carotenoids using a more sensitive HPLC methodology. Further, this study was unable to connect egg yolk vitamin and mineral profiles to hen dietary factors. Since most previous egg studies mainly investigate tallow supplementation rather than suet and do not typically specify the source of the fat provided, it was difficult to make comparisons to the GFB suet provided in the present study. It is not clear whether understanding these distinctions would make a significant difference in this discussion. Lastly, since this study characterizes the nutrient profile of eggs from hens fed animal by-products in a small-scale farm in the US, it may not necessarily reflect what is allowed by regulatory bodies in other countries or in a commercial setting. This should be taken into account in future studies on this supplementation practice.

However, to our knowledge, this is the first study to report the nutrient profile of egg yolks from hens fed GFB suet and liver in a local, pasture-raised setting. In this study, both groups with unlimited access to fresh pasture had a higher egg yolk n-3 PUFA and carotenoid content and a lower n-6:n-3 ratio than commercially obtained eggs. Likely due to the corn- and soy-free diet, PBB eggs contained significantly less n-6 PUFAs than eggs from the other groups in this study. Moreover, feeding GFB suet and liver to the PBB group increased CLA, OCFA, and BCFA content without an increase in total SFA content but had a limited impact on the vitamin and mineral profile. These results demonstrate that providing GFB by-products and exploring alternative layer hen feeds have beneficial impacts on the egg nutrient profile that favor human health and reduce agricultural waste. Future studies should investigate differences in sources of beef tallow or suet and liver and the optimal quantity to achieve desired effects on the egg nutrient profile. Further investigation is needed to determine the efficacy of using alternatives to corn and soy in layer hen diets, particularly in pasture-raised systems.

Supplementary Materials: The following supporting information can be downloaded at: https://www.mdpi.com/article/10.3390/foods11213404/s1, Table S1: Fatty acid profile of grass-fed beef by-products and layer hen feeds (g fatty acid/100 g tissue or feed); Table S2: Fatty acid profile of the egg yolks (g fatty acid/100 g fresh yolk); Data S1: Data presented in this study.

Author Contributions: Conceptualization, J.I.F., J.E.R. and S.S.; methodology, J.I.F. and S.S.; software, S.S.; validation, J.I.F. and S.S.; formal analysis, S.S., V.J. and E.G.; investigation, S.S., V.J. and E.G.; resources, J.I.F. and J.E.R.; data curation, J.I.F. and S.S.; writing—original draft preparation, J.I.F. and S.S.; writing—review and editing, J.I.F., S.S., V.J. and E.G.; visualization, S.S., V.J. and E.G.; supervision, J.I.F.; project administration, J.I.F.; funding acquisition, J.I.F. and J.E.R. All authors have read and agreed to the published version of the manuscript.

Funding: Financial support for the vitamin and mineral analysis was provided by the Ancestral Health Foundation. The Ancestral Health Foundation had no role in the planning of this study.

Institutional Review Board Statement: Not applicable.

Informed Consent Statement: Not applicable.

Data Availability Statement: The data presented in this study are available in Data S1.

Conflicts of Interest: The authors declare no conflict of interest. The funders had no role in the design of the study; in the collection, analyses, or interpretation of data; in the writing of the manuscript; or in the decision to publish the results.

References

1. Karsten, H.; Patterson, P.H.; Stout, R.; Crews, G. Vitamins A, E and fatty acid composition of the eggs of caged hens and pastured hens. *Renew. Agric. Food Syst.* **2010**, *25*, 45–54. [CrossRef]
2. Jacob, J.P.; Pescatore, A.J.; Anderson, K.E.; McCrea, B.; Shaw, D.P. *Impact of Free-Range Poultry Production Systems on Animal Health, Human Health, Productivity, Environment, Food Safety, and Animal Welfare Issues*; Council for Agricultural Science and Technology (CAST): Ames, IA, USA, 2018; p. 20.
3. van Vliet, S.; Provenza, F.D.; Kronberg, S.L. Health-promoting phytonutrients are higher in grass-fed meat and milk. *Front. Sustain. Food Syst.* **2021**, *4*, 555426. [CrossRef]
4. Spratt, E.; Jordan, J.; Winsten, J.; Huff, P.; van Schaik, C.; Jewett, J.G.; Filbert, M.; Luhman, J.; Meier, E.; Paine, L. Accelerating regenerative grazing to tackle farm, environmental, and societal challenges in the upper Midwest. *J. Soil Water Conserv.* **2021**, *76*, 15A–23A. [CrossRef]
5. United States Department of Agriculture. Questions and Answers—USDA Shell Egg Grading Service. Available online: https://www.ams.usda.gov/publications/qa-shell-eggs (accessed on 31 March 2022).
6. McReynolds, J.L.; Moore, R.W.; Kubena, L.F.; Byrd, J.A.; Woodward, C.L.; Nisbet, D.J.; Ricke, S.C. Effect of various combinations of alfalfa and standard layer diet on susceptibility of laying hens to Salmonella enteritidis during forced molt. *Poult. Sci.* **2006**, *85*, 1123–1128. [CrossRef]
7. Kop-Bozbay, C.; Akdag, A.; Bozkurt-Kiraz, A.; Gore, M.; Kurt, O.; Ocak, N. Laying performance, egg quality characteristics, and egg yolk fatty acids profile in layer hens housed with free access to chicory- and/or white clover-vegetated or non-vegetated areas. *Animals* **2021**, *11*, 1708. [CrossRef] [PubMed]
8. Provenza, F.D.; Meuret, M.; Gregorini, P. Our landscapes, our livestock, ourselves: Restoring broken linkages among plants, herbivores, and humans with diets that nourish and satiate. *Appetite* **2015**, *95*, 500–519. [CrossRef]
9. Hammershøj, M.; Johansen, N.F. Review: The effect of grass and herbs in organic egg production on egg fatty acid composition, egg yolk colour and sensory properties. *Livest. Sci.* **2016**, *194*, 37–43. [CrossRef]
10. Sergin, S.; Goeden, T.; Krusinski, L.; Kesamneni, S.; Ali, H.; Bitler, C.A.; Medina-Meza, I.G.; Fenton, J.I. Fatty acid and antioxidant composition of conventional compared to pastured eggs: Characterization of conjugated linoleic acid and branched chain fatty acid isomers in eggs. *ACS Food Sci. Technol.* **2021**, *1*, 260–267. [CrossRef]
11. Nimalaratne, C.; Wu, J. Hen egg as an antioxidant food commodity: A review. *Nutrients* **2015**, *7*, 8274–8293. [CrossRef]
12. Chung, H.Y.; Rasmussen, H.M.; Johnson, E.J. Lutein bioavailability is higher from lutein-enriched eggs than from supplements and spinach in men. *J. Nutr.* **2004**, *134*, 1887–1893. [CrossRef]
13. Ominski, K.; McAllister, T.; Stanford, K.; Mengistu, G.; Kebebe, E.G.; Omonijo, F.; Cordeiro, M.; Legesse, G.; Wittenberg, K. Utilization of by-products and food waste in livestock production systems: A Canadian perspective. *Anim. Front.* **2021**, *11*, 55–63. [CrossRef]
14. Jayathilakan, K.; Sultana, K.; Radhakrishna, K.; Bawa, A.S. Utilization of byproducts and waste materials from meat, poultry and fish processing industries: A review. *J. Food Sci. Technol.* **2012**, *49*, 278–293. [CrossRef]
15. Schreiner, M.; Hulan, H.W.; Razzazi-Fazeli, E.; Böhm, J.; Iben, C. Feeding laying hens seal blubber oil: Effects on egg yolk incorporation, stereospecific distribution of omega-3 fatty acids, and sensory aspects. *Poult. Sci.* **2004**, *83*, 462–473. [CrossRef]
16. Baucells, M.D.; Crespo, N.; Barroeta, A.C.; López-Ferrer, S.; Grashorn, M.A. Incorporation of different polyunsaturated fatty acids into eggs. *Poult. Sci.* **2000**, *79*, 51–59. [CrossRef]
17. Celebi, S.; Macit, M. The effects of sources of supplemental fat on performance, egg quality, and fatty acid composition of egg yolk in laying hens. *J. Sci. Food Agric.* **2008**, *88*, 2382–2387. [CrossRef]
18. Grobas, S.; Méndez, J.; Lázaro, R.; de Blas, C.; Mateo, G.G. Influence of source and percentage of fat added to diet on performance and fatty acid composition of egg yolks of two strains of laying hens. *Poult. Sci.* **2001**, *80*, 1171–1179. [CrossRef]
19. Vlaicu, P.A.; Panaite, T.D.; Turcu, R.P. Enriching laying hens eggs by feeding diets with different fatty acid composition and antioxidants. *Sci. Rep.* **2021**, *11*, 20707. [CrossRef]
20. Batkowska, J.; Drabik, K.; Brodacki, A.; Czech, A.; Adamczuk, A. Fatty acids profile, cholesterol level and quality of table eggs from hens fed with the addition of linseed and soybean oil. *Food Chem.* **2021**, *334*, 127612. [CrossRef]
21. Denke, M.A. Role of beef and beef tallow, an enriched source of stearic acid, in a cholesterol-lowering diet. *Am. J. Clin. Nutr.* **1994**, *60*, 1044s–1049s. [CrossRef]
22. Elswyk, M.E.V.; McNeill, S.H. Impact of grass/forage feeding versus grain finishing on beef nutrients and sensory quality: The U.S. experience. *Meat Sci.* **2014**, *96*, 535–540. [CrossRef]

23. Mathews, K.H.; Johnson, R.J. Alternative Beef Production Systems: Issues and Implications LDPM-218-01. Available online: https://www.ers.usda.gov/webdocs/outlooks/37473/36491_ldpm-218-01.pdf?v=7815.8 (accessed on 31 March 2022).
24. Daley, C.A.; Abbott, A.; Doyle, P.S.; Nader, G.A.; Larson, S. A review of fatty acid profiles and antioxidant content in grass-fed and grain-fed beef. *Nutr. J.* **2010**, *9*, 10. [CrossRef]
25. Enser, M.; Hallett, K.G.; Hewett, B.; Fursey, G.A.J.; Wood, J.D.; Harrington, G. The polyunsaturated fatty acid composition of beef and lamb liver. *Meat Sci.* **1998**, *49*, 321–327. [CrossRef]
26. Duizer, L.M.; Diana, A.; Rathomi, H.S.; Luftimas, D.E.; Rahmannia, S.; Santi, W.; Nugraha, G.I.; Haszard, J.J.; Gibson, R.S.; Houghton, L.A. An acceptability trial of desiccated beef liver and meat powder as potential fortifiers of complementary diets of young children in Indonesia. *J. Food Sci.* **2017**, *82*, 2206–2212. [CrossRef]
27. Eisen, E.J.; Bohren, B.B.; McKean, H.E. The Haugh unit as a measure of egg albumen quality. *Poult. Sci.* **1962**, *41*, 1461–1468. [CrossRef]
28. Nimalaratne, C.; Lopes-Lutz, D.; Schieber, A.; Wu, J. Free aromatic amino acids in egg yolk show antioxidant properties. *Food Chem.* **2011**, *129*, 155–161. [CrossRef]
29. Chen, Y.-S.; Aluwi, N.A.; Saunders, S.R.; Ganjyal, G.M.; Medina-Meza, I.G. Metabolic fingerprinting unveils quinoa oil as a source of bioactive phytochemicals. *Food Chem.* **2019**, *286*, 592–599. [CrossRef]
30. Rodriguez-Amaya, D.B.; Kimura, M. *HarvestPlus Handbook for Carotenoid Analysis*; International Food Policy Research Institute (IFPRI) and International Center for Tropical Agriculture (CIAT): Washington, DC, WA, USA, 2004.
31. Biehler, E.; Mayer, F.; Hoffmann, L.; Krause, E.; Bohn, T. Comparison of 3 spectrophotometric methods for carotenoid determination in frequently consumed fruits and vegetables. *J. Food Sci.* **2010**, *75*, C55–C61. [CrossRef]
32. Lichtenthaler, H.K.; Wellburn, A.R. Determinations of total carotenoids and chlorophylls a and b of leaf extracts in different solvents. *Biochem. Soc. Trans.* **1983**, *11*, 591–592. [CrossRef]
33. Bronkema, S.M.; Rowntree, J.E.; Jain, R.; Schweihofer, J.P.; Bitler, C.A.; Fenton, J.I. A nutritional survey of commercially available grass-finished beef. *Meat Muscle Biol.* **2019**, *3*, 116–126. [CrossRef]
34. Jenkins, T.C. Technical note: Common analytical errors yielding inaccurate results during analysis of fatty acids in feed and digesta samples. *J. Dairy Sci.* **2010**, *93*, 1170–1174. [CrossRef]
35. Kramer, J.K.G.; Hernandez, M.; Cruz-Hernandez, C.; Kraft, J.; Dugan, M.E.R. Combining results of two GC separations partly achieves determination of all cis and trans 16:1, 18:1, 18:2 and 18:3 except CLA isomers of milk fat as demonstrated using Ag-ion SPE fractionation. *Lipids* **2008**, *43*, 259–273. [CrossRef]
36. Popova, T.; Petkov, E.; Ayasan, T.; Ignatova, M. Quality of eggs from layers reared under alternative and conventional system. *Rev. Bras. Cienc. Avic.* **2020**, *22*, 001–008. [CrossRef]
37. Grune, T.; Lietz, G.; Palou, A.; Ross, A.C.; Stahl, W.; Tang, G.; Thurnham, D.; Yin, S.-A.; Biesalski, H.K. β-Carotene is an important vitamin A source for humans. *J. Nutr.* **2010**, *140*, 2268S–2285S. [CrossRef]
38. Gaffney, M.; O'Rourke, R.; Taylor-Pickard, J.; Murphy, R. A comparative assessment of the fatty acid profiles and antioxidant status of supermarket eggs. *J. Appl. Anim. Nutr.* **2015**, *3*, e9. [CrossRef]
39. Bo, C.D.; Bernardi, S.; Marino, M.; Porrini, M.; Tucci, M.; Guglielmetti, S.; Cherubini, A.; Carrieri, B.; Kirkup, B.; Kroon, P.; et al. Systematic review on polyphenol intake and health outcomes: Is there sufficient evidence to define a health-promoting polyphenol-rich dietary pattern? *Nutrients* **2019**, *11*, 1355.
40. Islam, Z.; Sultan, A.; Khan, S.; Alhidary, I.A.; Abdelrahman, M.M.; Khan, R.U. Impact of varying housing systems on egg quality characteristics, fatty acid profile, and cholesterol content of Rhode Island Red × Fyoumi laying hens. *Trop. Anim. Health Prod.* **2021**, *53*, 456. [CrossRef]
41. Jiang, Z.R.; Ahn, D.U.; Sim, J.S. Effects of feeding flax and two types of sunflower seeds on fatty acid compositions of yolk lipid classes. *Poult. Sci.* **1991**, *70*, 2467–2475. [CrossRef]
42. Whelan, J.; Fritsche, K. Linoleic acid. *Adv. Nutr.* **2013**, *4*, 311–312. [CrossRef]
43. Mancinelli, A.C.; Veroli, A.D.; Mattioli, S.; Cruciani, G.; Bosco, A.D.; Castellini, C. Lipid metabolism analysis in liver of different chicken genotypes and impact on nutritionally relevant polyunsaturated fatty acids of meat. *Sci. Rep.* **2022**, *12*, 1888. [CrossRef]
44. Mancinelli, A.C.; Mattioli, S.; Twining, C.; Bosco, A.D.; Donoghue, A.M.; Arsi, K.; Angelucci, E.; Chiattelli, D.; Castellini, C. Poultry meat and eggs as an alternative source of n-3 long-chain polyunsaturated fatty acids for human nutrition. *Nutrients* **2022**, *14*, 1969. [CrossRef]
45. Fraeye, I.; Bruneel, C.; Lemahieu, C.; Buyse, J.; Muylaert, K.; Foubert, I. Dietary enrichment of eggs with omega-3 fatty acids: A review. *Food Res. Int.* **2012**, *48*, 961–969. [CrossRef]
46. Mugnai, C.; Sossidou, E.N.; Bosco, A.D.; Ruggeri, S.; Mattioli, S.; Castellini, C. The effects of husbandry system on the grass intake and egg nutritive characteristics of laying hens. *J. Sci. Food Agric.* **2014**, *94*, 459–467. [CrossRef]
47. Simopoulos, A.P. The importance of the omega-6/omega-3 fatty acid ratio in cardiovascular disease and other chronic diseases. *Exp. Biol. Med.* **2008**, *233*, 674–688. [CrossRef]
48. Jain, A.P.; Aggarwal, K.K.; Zhang, P.Y. Omega-3 fatty acids and cardiovascular disease. *Eur. Rev. Med. Pharmacol. Sci.* **2015**, *19*, 441–445.
49. Vannice, G.; Rasmussen, H. Position of the Academy of Nutrition and Dietetics: Dietary fatty acids for healthy adults. *J. Acad. Nutr. Diet.* **2014**, *114*, 136–153. [CrossRef]

50. Oleszczuk, J.; Oleszczuk, L.; Siwicki, A.; Skopińska-Różewska, E. Biological effects of conjugated linoleic acids supplementation. *Pol. J. Vet. Sci.* **2012**, *15*, 403–408. [CrossRef]
51. den Hartigh, L.J. Conjugated linoleic acid effects on cancer, obesity, and atherosclerosis: A review of pre-clinical and human trials with current perspectives. *Nutrients* **2019**, *11*, 370. [CrossRef]
52. Shinn, S.; Liyanage, R.; Lay, J.; Proctor, A. Improved fatty acid analysis of conjugated linoleic acid rich egg yolk triacylglycerols and phospholipid species. *J. Agric. Food Chem.* **2014**, *62*, 6608–6615. [CrossRef]
53. Raes, K.; Huyghebaert, G.; Smet, S.D.; Nollet, L.; Arnouts, S.; Demeyer, D. The deposition of conjugated linoleic acids in eggs of laying hens fed diets varying in fat level and fatty acid profile. *J. Nutr.* **2002**, *132*, 182–189. [CrossRef]
54. Ritzenthaler, K.L.; McGuire, M.K.; Falen, R.; Shultz, T.D.; Dasgupta, N.; McGuire, M.A. Estimation of conjugated linoleic acid intake by written dietary assessment methodologies underestimates actual intake evaluated by food duplicate methodology. *J. Nutr.* **2001**, *131*, 1548–1554. [CrossRef] [PubMed]
55. Benjamin, S.; Prakasan, P.; Sreedharan, S.; Wright, A.D.; Spener, F. Pros and cons of CLA consumption: An insight from clinical evidences. *Nutr. Metab.* **2015**, *12*, 4. [CrossRef] [PubMed]
56. Venn-Watson, S.; Lumpkin, R.; Dennis, E.A. Efficacy of dietary odd-chain saturated fatty acid pentadecanoic acid parallels broad associated health benefits in humans: Could it be essential? *Sci. Rep.* **2020**, *10*, 8161. [CrossRef] [PubMed]
57. Huang, L.; Lin, J.-S.; Aris, I.M.; Yang, G.; Chen, W.-Q.; Li, L.-J. Circulating saturated fatty acids and incident type 2 diabetes: A systematic review and meta-analysis. *Nutrients* **2019**, *11*, 998. [CrossRef] [PubMed]
58. Zhu, Y.; Tsai, M.Y.; Sun, Q.; Hinkle, S.N.; Rawal, S.; Mendola, P.; Ferrara, A.; Albert, P.S.; Zhang, C. A prospective and longitudinal study of plasma phospholipid saturated fatty acid profile in relation to cardiometabolic biomarkers and the risk of gestational diabetes. *Am. J. Clin. Nutr.* **2018**, *107*, 1017–1026. [CrossRef]
59. Forouhi, N.G.; Koulman, A.; Sharp, S.J.; Imamura, F.; Kröger, J.; Schulze, M.B.; Crowe, F.L.; Huerta, J.M.; Guevara, M.; Beulens, J.W.; et al. Differences in the prospective association between individual plasma phospholipid saturated fatty acids and incident type 2 diabetes: The EPIC-InterAct case-cohort study. *Lancet Diabetes Endocrinol.* **2014**, *2*, 810–818. [CrossRef]
60. Jiménez-Cepeda, A.; Dávila-Said, G.; Orea-Tejeda, A.; González-Islas, D.; Elizondo-Montes, M.; Pérez-Cortes, G.; Keirns-Davies, C.; Castillo-Aguilar, L.F.; Verdeja-Vendrell, L.; Peláez-Hernández, V.; et al. Dietary intake of fatty acids and its relationship with FEV(1)/FVC in patients with chronic obstructive pulmonary disease. *Clin. Nutr. ESPEN* **2019**, *29*, 92–96. [CrossRef]
61. Zhuang, P.; Cheng, L.; Wang, J.; Zhang, Y.; Jiao, J. Saturated fatty acid intake is associated with total mortality in a nationwide cohort study. *J. Nutr.* **2019**, *149*, 68–77. [CrossRef] [PubMed]
62. Tomaszewska, E.; Muszyński, S.; Arczewska-Włosek, A.; Domaradzki, P.; Pyz-Łukasik, R.; Donaldson, J.; Świątkiewicz, S. Cholesterol content, fatty acid profile and health lipid indices in the egg yolk of eggs from hens at the end of the laying cycle, following alpha-ketoglutarate supplementation. *Foods* **2021**, *10*, 596. [CrossRef]
63. Evans, J.; LaMont, M. Compositions and Methods for Introduction of Odd-Chain Fatty Acids into Poultry Eggs. U.S. Patent 10842173 B2, 24 November 2020.
64. Ran-Ressler, R.R.; Bae, S.; Lawrence, P.; Wang, D.H.; Brenna, J.T. Branched-chain fatty acid content of foods and estimated intake in the USA. *Br. J. Nutr.* **2014**, *112*, 565–572. [CrossRef]
65. Ran-Ressler, R.R.; Khailova, L.; Arganbright, K.M.; Adkins-Rieck, C.K.; Jouni, Z.E.; Koren, O.; Ley, R.E.; Brenna, J.T.; Dvorak, B. Branched chain fatty acids reduce the incidence of necrotizing enterocolitis and alter gastrointestinal microbial ecology in a neonatal rat model. *PLoS ONE* **2011**, *6*, e29032. [CrossRef] [PubMed]
66. Vahmani, P.; Salazar, V.; Rolland, D.C.; Gzyl, K.E.; Dugan, M.E.R. Iso- but not anteiso-branched chain fatty acids exert growth-inhibiting and apoptosis-inducing effects in MCF-7 cells. *J. Agric. Food Chem.* **2019**, *67*, 10042–10047. [CrossRef] [PubMed]
67. Dąbrowski, G.; Konopka, I. Update on food sources and biological activity of odd-chain, branched and cyclic fatty acids—A review. *Trends Food Sci. Technol.* **2022**, *119*, 514–529. [CrossRef]
68. Wang, D.H.; Wang, Z.; Brenna, J.T. Gas chromatography chemical ionization mass spectrometry and tandem mass spectrometry for identification and straightforward quantification of branched chain fatty acids in foods. *J. Agric. Food Chem.* **2020**, *68*, 4973–4980. [CrossRef]
69. Anderson, K.E. Comparison of fatty acid, cholesterol, and vitamin A and E composition in eggs from hens housed in conventional cage and range production facilities. *Poult. Sci.* **2011**, *90*, 1600–1608. [CrossRef] [PubMed]
70. National Institutes of Health. Vitamin A and Carotenoids. Available online: https://ods.od.nih.gov/factsheets/VitaminA-HealthProfessional/ (accessed on 12 May 2022).
71. Lima, H.J.D.; Souza, L.A.Z. Vitamin A in the diet of laying hens: Enrichment of table eggs to prevent nutritional deficiencies in humans. *World Poult. Sci. J.* **2018**, *74*, 619–626. [CrossRef]
72. Zhao, H.; Chen, Y.; Wang, S.; Wen, C.; Zhou, Y. Effects of dietary natural vitamin E supplementation on laying performance, egg quality, serum biochemical indices, tocopherol deposition and antioxidant capacity of laying hens. *Ital. J. Anim. Sci.* **2021**, *20*, 2254–2262. [CrossRef]
73. Areerob, P.; Dahlan, W.; Angkanaporn, K. Dietary crude palm oil supplementation improves egg quality and modulates tissue and yolk vitamin E concentrations of laying hen. *Anim. Prod. Sci.* **2019**, *59*, 1491–1500. [CrossRef]
74. Browning, L.C.; Cowieson, A.J. Vitamin D fortification of eggs for human health. *J. Sci. Food Agric.* **2014**, *94*, 1389–1396. [CrossRef]
75. Lee, M.R.F.; McAuliffe, G.A.; Tweed, J.K.S.; Griffith, B.A.; Morgan, S.A.; Rivero, M.J.; Harris, P.; Takahashi, T.; Cardenas, L. Nutritional value of suckler beef from temperate pasture systems. *Animal* **2021**, *15*, 100257. [CrossRef]

76. Hrubša, M.; Siatka, T.; Nejmanová, I.; Vopršalová, M.; Krčmová, L.K.; Matoušová, K.; Javorská, L.; Macáková, K.; Mercolini, L.; Remião, F.; et al. Biological properties of vitamins of the B-complex, part 1: Vitamins B1, B2, B3, and B5. *Nutrients* **2022**, *14*, 484. [CrossRef] [PubMed]
77. National Institutes of Health. Pantothenic Acid. Available online: https://ods.od.nih.gov/factsheets/PantothenicAcid-HealthProfessional/ (accessed on 12 May 2022).
78. Lourenco, J.M.; Rothrock, M.J., Jr.; Fluharty, F.L.; Callaway, T.R. The successional changes in the gut microbiome of pasture-raised chickens fed soy-containing and soy-free diets. *Front. Sustain. Food Syst.* **2019**, *3*, 35. [CrossRef]
79. United States Department of Agriculture. Beef, Variety Meats and By-products, Liver, Raw. Available online: https://fdc.nal.usda.gov/fdc-app.html#/food-details/169451/nutrients (accessed on 12 May 2022).

foods

MDPI

Article

Utilization of By-Products from Livestock: Study on the Mechanical and Thermal Properties of Biodegradable Containers Made with Pork Skin Gelatin Polymer

Sol-Hee Lee and Hack-Youn Kim *

Department of Animal Resources Science, Kongju National University, Yesan 32439, Chungnam, Korea
* Correspondence: kimhy@kongju.ac.kr

Abstract: This study aimed to develop a biodegradable container made of pork gelatin. Gelatin was extracted from pork skin by hot water at 80 °C, and containers were prepared by adding eggshell powder (20%) as a pore agent, and walnut powder (0.08 wt%; PEW1, 0.14 wt%; PEW2) to improve hardness. The blends were molded for each experiment and dried at 30 °C for 24 h, at 40 °C for 16 h, and at 121 °C for 16 h. The containers were analyzed with respect to morphological (SEM; scanning electron microscope), mechanical (tensile strain and stress), and thermal (DSA; differential scanning calorimetry and TGA; thermogravimetric analysis) properties, as well as biodegradability. SEM investigation showed a smoother surface for PEW1 than for PEW2. The tensile stress of PEW2 (37.86 MPa) was significantly higher than that of PEW1 (28.40 MPa), and the melting enthalpies were 137.60 J/g (PEW1) and 309.60 J/g (PEW2). TGA showed similar properties, but PEW2 contained more lignin; therefore, its decomposition temperature was higher. The PEW1 and PEW2 containers were completely biodegraded after approximately 7 and 11 weeks, respectively. Walnut shell powder increased the hardness, but slowed the biodegradation process. The applications of this biodegradable container are short-lived products such as food packaging.

Keywords: biodegradable container; gelatin polymer; walnut shell; by-products; sustainable goals

1. Introduction

Home meal replacement (HMR) refers to food that can be easily consumed, replacing meals prepared at home, and the global HMR market is on the rise [1]. According to Research and Markets [2], global ready-meal products are projected to grow at a compound annual growth rate of 6.83% from 2019 to 2025. Currently, as the proportion of people eating at home has increased owing to the COVID-19 pandemic, the use of plastic as a packaging material for HMR products has continued to increase [3]. As of 2020, approximately 9% of the 8.3 billion tonnes of plastics was recycled, 12% was incinerated, and the rest was mostly disposed of in landfills or as environmental waste [4]. Accordingly, the importance of eco-friendly containers is emerging worldwide, and their continuous development is in progress.

Eco-friendly packaging materials include biodegradable and biomass products. Biodegradability refers to decomposition by microbes when the material is buried in soil [5]. The material is converted into environmentally friendly carbon compounds, water and inorganic compounds by insects and microorganisms [6]. According to Zurier and Goddard [7], as biodegradable materials are decomposed, the possibility of improving the environment by biocatalysis can be expected. Biodegradable containers are relatively easy to remove compared to non-degradable containers, but they have the disadvantage that the storage period for food is shorter [8]. Therefore, it is necessary to study the manufacture of biodegradable and environmentally friendly containers with enhanced physical properties.

According to the research results of Yang and Park [9], collectivism, ecology, and health affect the consumption value of food, and ethical consumerism and food safety act as major variables. It is important that eco-friendly containers do not affect the preservation of food, but factors such as durability, low toxicity, recyclability, and biodegradation also play an important role [10]. In light of this consumer psychology, patents are being developed and studies on product development using plant by-products are being conducted in Korea [11,12]. Cellulose extracted from plant by-products is the most abundant polymer found in nature, and has been used as a major polymer in the packaging field until now [13]. Cellulose has also been processed and used in the form of hydroxypropyl-methylcellulose (HPMC) and carboxymethyl cellulose nitrate (CMCN). However, these materials have weak resistance to water absorption; therefore, a supplementary point is needed [14].

Gelatin is mainly obtained from pork skins (46%) and pork or cattle bones (23.1%), and is used industrially [15]. Gelatin used for commercial purposes is swollen and heated to be processed and used in various forms, and the chain shape changes depending on the process used: single chain, two single chains/single chain with a loop, and three single chains with one loop or two single chains [16]. Gelatin is found in 27 different forms depending on pH and temperature conditions [17], and has high adhesion when converted to a collagen state by hydrothermal treatment [18]. It has been reported that the strength of gelatin increases after gelation is completed, and the higher the purity of the gelatin, the higher the strength maintained [19]. Owing to these properties, gelatin is used not only as a gel emulsion, an adhesive, and in medicinal capsules, but also as a medical tissue adhesive [20,21].

By-products such as gelatin are substances that are produced during the manufacturing and processing of food. These by-products are thrown away, because they are difficult to ingest for humans or require processing before they can be used. Therefore, the consumed pork skin, eggshell and walnut shell were used as a polymer, a pore agent, and a flavoring agent, respectively, to prepare a container. As interest in fields requiring biodegradable substances increases, we developed a biodegradable container that can be used in the food industry. To find out whether this biodegradable container can be used in industry, we studied biodegradable containers' morphology, physics and biodegradability properties using a pork gelatin polymer.

2. Materials and Methods

2.1. Material and Preparation of Pork Gelatin Powder

Pork skin purchased from local markets (Yesan, Chungnam, Korea) was used for the production of gelatin [22]. Excess fat and impurities were removed from the pork skin by washing with water. The prepared material was allowed to swell for 24 h by diluting by a factor of ten (v/w) using 0.1 N HCl (purity 35–37%, pH 2). The swollen sample was placed in flowing water and neutralized until the pH reached 5.5. The samples were vacuum packed and extracted in a water bath (JSWB-30T, JSR, Gongju, Korea) at 80 °C for 2 h. The extracted gelatin was dried with hot air in an atmospheric dryer (C-F03, Vision Scientific, Daejeon, Korea) set at 105 °C, and walnut powder and eggshell powder were purchased from Ganong Bio (Incheon, Korea) and The witch company (Seoul, Korea), respectively; the particle size was 180 mesh.

2.2. Manufacture of Biodegradable Containers

Biodegradable containers were manufactured by mixing eggshell powder (purity 100%) and walnut powder (purity 100%) with the prepared gelatin [23] using the following method for establishing mixing ratios. The prepared gelatin powder was hydrated by adding three volumes of distilled water to a water bath (JSWB-30T, Korea) at 100 °C for 3 h. Then, 20% eggshell powder was added to the liquid gelatin. Two mixtures were prepared using this mixture: 0.08 wt% walnut powder (pork gelatin, eggshell and 0.08 wt% walnut shell; PEW1) and 0.14 wt% walnut powder (pork gelatin, eggshell and 0.14 wt% walnut shell: PEW2). The eggshell and walnut powder were stirred at 80 °C for 3 h until it was completely dissolved in the pig skin gelatin solution for every step. The final mixtures were prepared by stirring at 80 °C for 2 h in a water bath (JSWB-30T, Korea). The prepared mixture was poured into a 30 × 20 cm plate, cooled, and cut into smaller pieces for each experiment. The drying process was carried out at 30 °C for 24 h, at 40 °C for 16 h, and at 121 °C for 16 h in an atmospheric dryer (C-F03, Korea). The dried containers were sprayed with a 1% sodium alginate dilution solution and 1% calcium lactate for waterproofing [24], and dried at 30 °C for 24 h in an atmospheric dryer (C-F03, Korea). The biodegradable container is shown in Figure 1.

2.3. *Morphological Properties*

The mechanisms of surface fracture were determined using a Scanning electron microscope (SEM; ECLIPSE Ci-L, Nikon, Tokyo, Japan) and images were used to analyze the morphology of the samples. The SEM sample was prepared by cutting to 35 mm × 35 mm × 10 mm. The images were taken at a voltage of 15 kV and magnifications of 300, 500 and 1000×.

2.4. *Mechanical Properties*

The samples for ultimate tensile stress and ultimate tensile strain were cut to 10 mm × 10 mm × 20 mm and 30 mm × 40 mm × 35 mm, respectively. The ultimate tensile stress and ultimate tensile strain of the samples were measured using a universal testing machine (AGS-X, Shimadzu Corporation, Kyoto, Japan) at a crosshead speed of 1 mm/min. The test conditions were 20 °C and 45% relative humidity. The tensile stress and strain were calculated based on the obtained tensile stress-strain curves. The experiments were repeated three or more times in total.

2.5. *Thermal Properties*

DSC and TGA were carried out on powder with a particle size of 32 mesh. DSC (differential scanning calorimetry) was used to measure the behavior of particles on a DSC-Q20 (TA Instruments, New Castle, DE, USA). The sample weight was 3–3.8 mg and analysis was carried out at a heating rate of 10 °C/min in the range of 0 to 250 °C. The experiment was conducted as follows: 1st heating → 1st cooling → 2nd heating → 2nd cooling. The indium (99.99%) was used for the temperature and enthalpy calibration in the experiment. Tg was measured as the midpoint of the transition during the scan, which was obtained by repeating the experiment twice. TGA (thermos-gravimetric analysis) was performed using TA-Q500 (TA Instruments, New Castle, DE, USA). Approximately 7–10 mg were weighed into a platinum pan and heated over the temperature range of 0 to 800 °C at a heating rate of 10 °C/min under a nitrogen purge [25].

2.6. *Biodegradability Performed Indoors*

To exclude variables, biodegradability was determined indoors in a simulated natural environment [26]. The soil was collected from the surface layer and placed in a 36 L plastic tray at a thickness of approximately 23 cm. After planting crops to create a natural environment, an analysis was performed. Uniform samples weighing 10 g were placed on the soil between plants where the roots of the crops were not in contact with the samples. A total of 4 or more samples were used per treatment group, and this was repeated twice. The test was performed for a total of 4 weeks, and the weight was checked every week. Water was sprayed once every four days to maintain moisture, and the decomposed sample was gently washed with distilled water and dried at 37 °C. After drying, the weight of each sample was determined. Based on the weight of the dried samples, the decomposition curve was plotted:

$$D_1\ [\text{Biodegradability}\ (\%)] = (m_1 - m_2)/m_1 \times 100(\%)$$

where m_1 = amount of the injected test material (g), and m_2 = amount of the recovered test material (g).

2.7. Statistical Analysis

The biodegradation curve was duplicated by performing at least three or more repeated experiments. Statistical analysis was performed using the statistical processing program SAS (Windows version 9.4, SAS Institute, Cary, NC, USA), and ANOVA, and Duncan's multiple range test were used to determine if there was a significant difference for each characteristic.

3. Results

3.1. Scanning Electron Microscopy

SEM images of the outer surface of the biodegradable container made of pork skin gelatin polymers, eggshell, and walnut powder are shown in Figure 1. External observation of the biodegradable container with 10% walnut powder showed a smooth surface without any convex parts, but the biodegradable container with 20% walnut powder appeared rougher (green arrow). This phenomenon is likely due to the excess addition of walnut powder compared to the amount of gelatin. The black spots seen in Figure 1a,b appear to be a walnut powder (white arrow). In the SEM images, the presence of holes in the biodegradable surface is indicated by white dots, indicated by green arrows. This also appears in the biodegradable container with 20% walnut powder and seems to be due to the agglomeration of walnut powder. Katekhong et al. [27] reported that an inaccurate polymer blending ratio may lead to matrix breakdown. Therefore, it was decided that the addition of only 10% walnut powder was better for binding with gelatin.

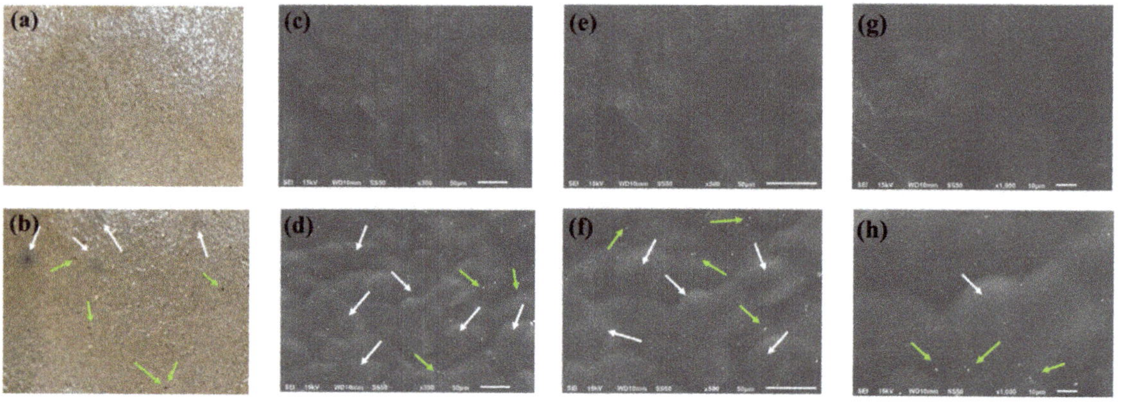

Figure 1. Images of biodegradable containers made from pork skin gelatin, walnut shells, and eggshells. (**a**) Surface of PEW1: pork gelatin, eggshell and 0.08 wt% walnut shell; (**b**) surface of PEW2: pork gelatin, eggshell and 0.14 wt% walnut shell; (**c**,**e**,**g**) representative SEM photographs of PEW1 surfaces at magnifications of 300, 500 and 1000×, respectively; (**d**,**f**,**h**) representative SEM photographs of PEW2 surfaces at magnifications of 300, 500 and 1000×, respectively.

3.2. Mechanical Properties

A universal testing machine (UTM) was used to analyze the mechanical properties of the materials. Analysis of the results can help in determining the best mixing ratio of materials and processing methods [28]. Table 1 shows the ultimate tensile stress (MPa) and ultimate tensile strain (%) of the biodegradable containers prepared with pork skin gelatin, eggshell and walnut powder. The tensile stress was 28.40 MPa in the container with 10% walnut powder (PEW1), significantly lower than that of the container with 20% walnut powder (PEW2) (37.86 MPa) ($p < 0.05$). Tensile strength and compressive are reported to have a positive correlation at an appropriate mixing ratio, and Nataraja and Dhang (2001) reported that compression tensile was 0.09 times of its compressive strength [29,30]. Although the tensile strain value of the PEW2 was higher than that of PEW1, the difference was found not to be significant ($p > 0.05$). Therefore, it can be concluded that PEW2, with a higher tensile strain, will have a higher compressive strength than PEW1. It was determined that the biodegradable container containing more walnut powder had a higher density. The addition of an excess of walnut powder to the polymer could interfere with the uniformity of the mixture [31]; however, no negative effects on uniformity were seen with the addition of 20%. Phothisarattana et al. [32] reported that solid particles in a polymer matrix can reinforce strength, and showed similar results to this study. It has been reported that tensile strength decreases as the particle size of agents used for biodegradable containers increases [33]. Therefore, if it is necessary to develop a container with high tensile strength, which may be achievable by reducing the amount of eggshell and walnut powder particles.

Table 1. Mechanical properties of biodegradable containers made from pork skin gelatin, walnut shells, and eggshells.

Treatments [1]	Ultimate Tensile Stress (MPa)	Ultimate Tensile Strain (%)
PEW1	28.40 ± 0.72^{b}	26.23 ± 0.68^{NS}
PEW2	37.86 ± 1.38^{a}	33.37 ± 1.26

[1] PEW1: pork gelatin, eggshell and 0.08 wt% walnut shell; PEW2: pork gelatin, eggshell and 0.14 wt% walnut shell. All values are mean ± SE. [a,b] Means in the same column with different letters are significantly different ($p < 0.05$). [NS]: non-significant.

3.3. DSC (Differential Scanning Calorimetry)

The results of thermal analysis of the biodegradable containers made of gelatin polymer material are shown in Figure 2 and Table 2. There are several factors that affect DSC results, including polymer quality, heat capacity and thermal stability. In this study, DSC was conducted to analyze the characteristics of the unknown polymers [34]. Because both PEW1 and PEW2 show two transition points, the biodegradable containers made of gelatin, eggshell and walnut powder feature both a Tg (glass transition temperature) and a Tm (melting temperature), so it is considered a crystalline polymer. Xi Wang et al. [35] reported that the Tg is due to the Brownian motion of the amorphous polymer and Tm is the melting point of materials that appear only in crystalline polymers. The Tg values of PEW1 and PEW2 were 38.30 and 39.78, respectively, and the Tm values were 89.59 and 88.68, respectively. Rebers et al. [36] reported that the Tm value of gelatin was 28.0, which is much lower than in this study. This is because gelatin is a triple helical structured protein that exhibits the behavior of an upper critical solution temperature, and thus has high binding ability with other materials [37]. In contrast, the amount of change in the melting enthalpy of the sample during the heat scan was 137.60 J/g for PEW1 and 309.60 J/g for PEW2, indicating a large difference. This is considered to be due to the amount of hydrogel bonding in the helical structure of gelatin. Bigi et al. [38] reported different crosslinking according to the binding agents used for the gelatin polymer. In the 2nd heating, since the DSC of the mass excluding the mass removed in 1st was measured, it showed a lower value than in 1st. The PEW1 showed overall lower values of Tg, Tc and Hf compared to those of PEW2, this can be understood in the same context as the tensile strain (%), because PEW2 is a denser material.

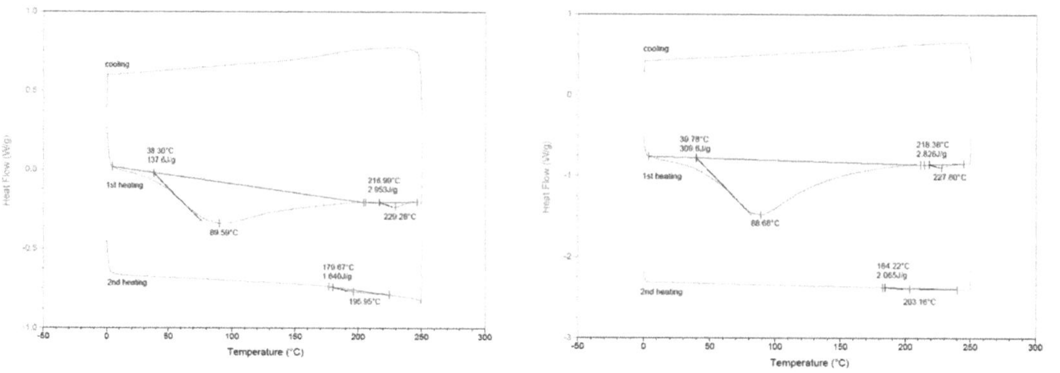

Figure 2. DSC scans of biodegradable containers made from pork skin gelatin, walnut shells, and eggshells.

Table 2. Analysis of the effect of seal integrity of DSC sample pan on thermal transitions of biodegradable container of pork skin gelatin, walnut shells, and eggshells.

Treatments [1]	Sample mass/mg	1st Heating			2nd Heating		
		T_g/°C	T_m/°C	ΔH/J_g	T_g/°C	T_m/°C	ΔH/J_g
PEW1	3.04	38.30	89.59	137.60	179.67	195.95	1.64
PEW2	3.84	39.78	88.68	309.60	184.22	203.16	2.07

[1] PEW1: pork gelatin, eggshell and 0.08 wt% walnut shell; PEW2: pork gelatin, eggshell and 0.14 wt% walnut shell. All values are mean ± SE. Glass transition temperature (Tg), melting temperature (Tm) and enthalpy (ΔHc) of individual and multilayer films were obtained by differential scanning calorimetry.

3.4. TGA (Thermogravimetry Analysis)

TGA is not only a good indicator of the thermal stability of polymer materials, it is also a good method for measuring the decomposition temperature of biodegradable polymer materials [39]. Accordingly, Figure 3 shows the weight loss (%) curve and derivative weight (%/°C) plotted against the degradation temperature, with different amounts of walnut powder. In the TGA curves of both materials, three different weight losses are observed. The weight loss at 1–200 °C is considered to be due to the decomposition of the eggshell powder and the low molecular weight sodium alginate sprayed as a coating [40,41]. Thereafter, the weight of the PEW1 biodegradation container was reduced to 49.20% at 533.56 °C, and that of PEW2 was reduced to 53.94% at 576.99 °C, showing significant weight loss. This is similar to the decomposition temperature of lignin present in lignocellulosic biomass [42]. The two samples show the same result because they are similar materials, but PEW2 contains a higher amount of walnut powder per gram. In the third loss section, the two samples show similar final temperature values of 800.04 °C (PEW1) and 800.03 °C (PEW2). PEW1 shows initial decomposition at 225 °C and peak thermal decomposition at 301.20 °C. A second decomposition occurs at 620 °C, and the third decomposition at 696.08 °C. PEW2 shows double decomposition at 210 and 670 °C, and peak thermal decomposition at 317.94 and 721.59 °C. Overall, PEW1 and PEW2 show similar behavior and derivative weight, but it can be seen that PEW2 has slightly higher decomposition temperatures. This is thought to be because PEW2 contains more substances such as lignin, thus raising the decomposition temperature [43].

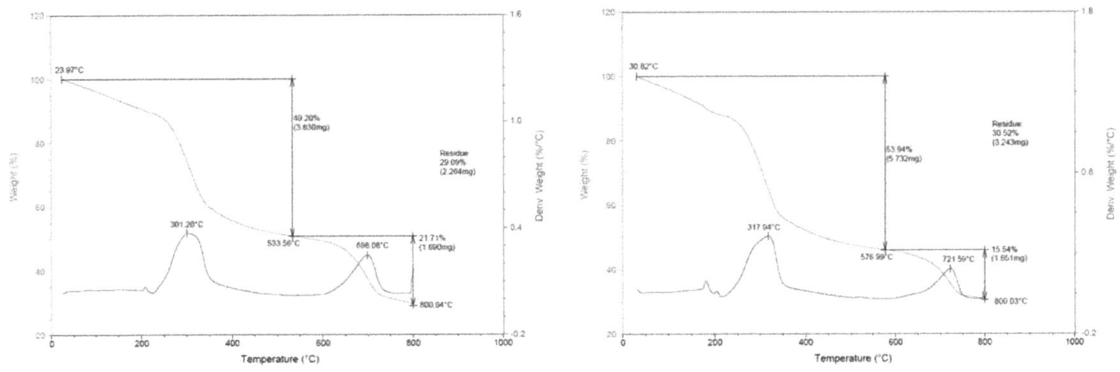

Figure 3. TGA scans of biodegradable containers made from pork skin gelatin, walnut shells, and eggshells.

3.5. Biodegradability

The biodegradability of the containers made of gelatin polymer material was analyzed using a soil burial method similar to that of the external environment (Figure 4). Biodegradable packaging materials not only decompose naturally due to microorganisms in the soil and have no effect on the environment, they are also useful because they provide nutrients to the soil [44,45]. In this study, based on experimental results over up to four weeks, during which time the shape of the sample can still be distinguished, a prediction of the time at which 100% decomposition will occur is shown. Evidently, as decomposition was estimated by measuring the rate of decrease in weight over time, it cannot be completely excluded that damage to the sample had no effect [46,47]. Nevertheless, biodegradability can be inferred based on the results of this study. In the first week, PEW1 showed a degradation rate of 47.70%, and PEW2 showed a degradation rate of 28.13%. It was later confirmed that PEW1 was completely biodegraded at approximately 7 weeks and PEW2 was completely biodegraded at approximately 11 weeks; therefore, the rate of degradation of PEW1 was much faster ($p < 0.05$). Oyeoka et al. [48] reported that incompletely dispersed composite materials are easier to physically damage than completely dispersed composites. Therefore, it can be inferred from the experimental results that PEW2 has safer dispersion than PEW1. In addition, PEW1 had a larger surface area and decomposed faster than PEW2 (Figure 4). This result is attributed to the fact that PEW2 is a denser material. In addition, from the fifth week onwards, the amount of decomposition was difficult to measure. It is reported that the packaging material of food is affected by the sorption and diffusion between water and hydrophilic matrix [49]. The higher the hydrophilicity, the faster the biodegradation, therefore the actual decomposition rate may be faster than the predicted decomposition curve shown in Figure 4 [50].

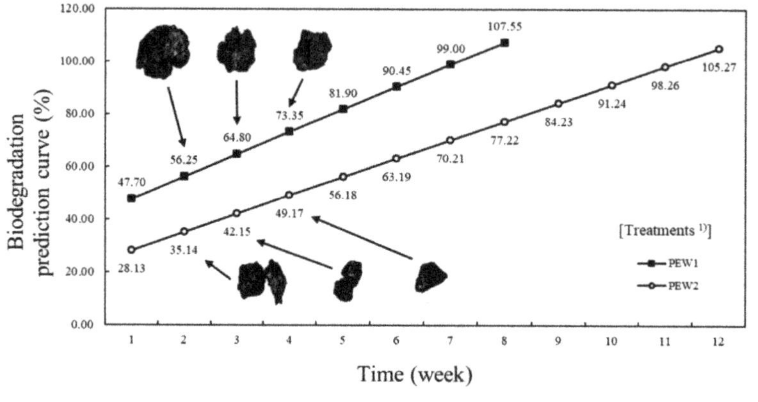

Figure 4. Biodegradation prediction curve (%) for biodegradable containers made from pork skin gelatin, walnut shells, and eggshells. [1] PEW1: pork gelatin, eggshell and 0.08 wt% walnut shell; PEW2: pork gelatin, eggshell and 0.14 wt% walnut shell.

4. Conclusions

With the utilization of by-products from livestock, the development of biodegradable containers made with pork gelatin showed the following results. Biodegradable containers with less walnut powder (PEW1) had a smoother appearance than PEW2. Additionally, PEW1 had a shorter biodegradation time compared to PEW2. Biodegradable containers with more walnut shell powder (PEW2) had higher tensile strain, stress and melting enthalpy. PEW1 could be useful in consideration of its clean appearance and biodegradation period, but when product strength is required, PEW2 showed better results. Therefore, it seems necessary to select the amount of walnut powder in biodegradable containers made of gelatin, varying its value according to the use, thickness and shape of the product being stored in such a container.

Author Contributions: Conceptualization, S.-H.L.; methodology, S.-H.L. and H.-Y.K.; software, S.-H.L.; validation, S.-H.L. and H.-Y.K.; formal analysis, S.-H.L.; investigation, S.-H.L. and H.-Y.K.; resources, H.-Y.K.; data curation, S.-H.L.; writing—original draft preparation, S.-H.L.; writing—review and editing, S.-H.L. and H.-Y.K.; visualization, S.-H.L.; supervision, H.-Y.K.; project administration, S.-H.L. and H.-Y.K.; funding acquisition, S.-H.L. and H.-Y.K. All authors have read and agreed to the published version of the manuscript.

Funding: This work was carried out with the support of the "Cooperative Research Program for Agriculture Science and Technology Development" (Project No. PJ016217), Rural Development Administration, Republic of Korea.

Institutional Review Board Statement: Not applicable.

Informed Consent Statement: Not applicable.

Data Availability Statement: The date are available from the corresponding author.

Conflicts of Interest: The authors declare no conflict of interest.

References

1. Bumbudsanpharoke, N.; Ko, S. Packaging technology for home meal replacement: Innovations and future prospective. *Food Control* **2022**, *132*, 108470. [CrossRef]
2. Global $156+ Billion Ready Meals Market to 2025, Research and Markets; 2020. Available online: https://www.globenewswire.com/news-release/2020/12/02/2138109/28124/en/Global-156-Billion-Ready-Meals-Market-to-2025.html#:~{}:text=The%20global%20ready%20meals%20market,US%24105.489%20billion%20in%202019. (accessed on 8 June 2021).
3. Lee, S.; Ham, S. Food service industry in the era of COVID-19: Trends and research implications. *Nutr. Res. Pract.* **2021**, *15*, S22–S31. [CrossRef] [PubMed]

4. Lee, A.; Liew, M.S. Tertiary recycling of plastics waste: An analysis of feedstock, chemical and biological degradation methods. *J. Mater. Cycles Waste Manag.* **2021**, *23*, 32–43. [CrossRef]
5. Sintim, H.Y.; Flury, M. Is biodegradable plastic mulch the solution to agriculture's plastic problem? *Environ. Sci. Technol.* **2017**, *51*, 1068–1069. [CrossRef] [PubMed]
6. Amobonye, A.; Bhagwat, P.; Singh, S.; Pillai, S. Plastic biodegradation: Frontline microbes and their enzymes. *Sci. Total Environ.* **2021**, *759*, 143536. [CrossRef]
7. Zurier, H.S.; Goddard, J.M. Biodegradation of microplastics in food and agriculture. *Curr. Opin. Food Sci.* **2021**, *15*, 37–44. [CrossRef]
8. Davis, G.; Song, J.H. Biodegradable packaging based on raw materials from crops and their impact on waste management. *Ind. Crops Prod.* **2006**, *23*, 47–161. [CrossRef]
9. Yang, I.M.; Park, C. Factors influencing consumer's attitude and buying behavior on the green products. *Asia Pac. J. Bus. Ventur. Entrep.* **2012**, *7*, 55–64. [CrossRef]
10. Cenci, M.P.; Scarazzato, T.; Munchen, D.D.; Dartora, P.C.; Veit, H.M.; Bernardes, A.M.; Dias, P.R. Eco-friendly electronics—A comprehensive review. *Adv. Mater. Technol.* **2022**, *7*, 2001263. [CrossRef]
11. Jafarzadeh, S.; Jafari, S.M.; Salehabadi, A.; Nafchi, A.M.; Uthaya Kumar, U.S.; Khalil, H.P.S.A. Biodegradable green packaging with antimicrobial functions based on the bioactive compounds from tropical plants and their by-products. *Trends Food Sci. Technol.* **2020**, *100*, 262–277. [CrossRef]
12. Abdul Khalil, H.P.S.; Lai, T.K.; Tye, Y.Y.; Paridah, M.T.; Fazita, M.R.N.; Azniwati, A.A.; Dungani, R.; Rizal, S. Preparation and characterization of microcrystalline cellulose from sacred bali bamboo as reinforcing filler in seaweed-based composite film. *Fibers Polym.* **2018**, *19*, 423–434. [CrossRef]
13. Weng, J.K.; Li, X.; Bonawitz, N.D.; Chapple, C. Emerging strategies of lignin engineering and degradation for cellulosic biofuel production. *Curr. Opin. Biotechnol.* **2008**, *19*, 166–172. [CrossRef]
14. Di Filippo, M.F.; Dolci, L.S.; Liccardo, L.; Bigi, A.; Bonvicini, F.; Gentilomi, G.A.; Passerini, N.; Panzavolta, S.; Albertini, B. Cellulose derivatives-snail slime films: New disposable eco-friendly materials for food packaging. *Food Hydrocoll.* **2021**, *111*, 106247. [CrossRef]
15. Thakur, S.; Govender, P.P.; Mamo, M.A.; Tamulevicius, S.; Thakur, V.K. Recent progress in gelatin hydrogel nanocomposites for water purification and beyond. *Vacuum* **2017**, *146*, 396–408. [CrossRef]
16. Guo, L.; Colby, R.H.; Lusignan, C.P.; Whitesides, T.H. Kinetics of triple helix formation in semidilute gelatin solutions. *Macromolecules* **2003**, *36*, 9999–10008. [CrossRef]
17. Gómez-Guillén, M.C.; Giménez, B.; López-Caballero, M.E.; Montero, M.P. Functional and bioactive properties of collagen and gelatin from alternative sources: A review. *Food Hydrocoll.* **2011**, *25*, 1813–1827. [CrossRef]
18. Farris, S.; Song, J.; Huang, Q. Alternative reaction mechanism for the cross-linking of gelatin with glutaraldehyde. *J. Agric. Food Chem.* **2010**, *58*, 998–1003. [CrossRef]
19. Fukunaga, S.; Karck, M.; Harringer, W.; Cremer, J.; Rhein, C.; Haverich, A. The use of gelatin-resorcin-formalin glue in acute aortic dissection type A. *Eur. J. Cardio Thorac. Surg.* **1999**, *15*, 564–570. [CrossRef]
20. Siburian, W.Z.; Rochima, E.; Andriani, Y.; Praseptiangga, D. Fish gelatin (definition, manufacture, analysis of quality characteristics, and application): A review. *Int. J. Fish. Aquat. Stud.* **2020**, *8*, 90–95.
21. Mo, X.; Iwata, H.; Matsuda, S.; Ikada, Y. Soft tissue adhesive composed of modified gelatin and polysaccharides. *J. Biomater. Sci. Polym. Ed.* **2000**, *11*, 341–351. [CrossRef]
22. Choe, J.; Kim, H.Y. Effects of chicken feet gelatin extracted at different temperatures and wheat fiber with different particle sizes on the physicochemical properties of gels. *Poult. Sci.* **2018**, *97*, 1082–1088. [CrossRef] [PubMed]
23. Lee, S.H.; Jeong, C.H.; Oh, Y.; Kim, H.Y. Analysis of quality properties, microbial stability, and biodegradability of biodegradable containers made of pork skin gelatin. *Food Eng. Prog.* **2021**, *25*, 347–353. [CrossRef]
24. Han, J.M. Eco Friendly Bowl Made by Agar, Sodium Alginate and Calcium Lactate. Korea Patent 10-2021-0029958, 29 March 2021.
25. Naresh, G.; Rajasekhar, A.; Bharali, J.; Ramesh, K. Homogeneous molten salt formulations as thermal energy storage media and heat transfer fluid. *J. Energy Storage* **2022**, *50*, 104200. [CrossRef]
26. Fernando, T.N.; Ariadurai, S.A.; Disanayaka, C.K.; Kulathunge, S.; Aruggoda, A.G.B. Development of radiation grafted super absorbent polymers for agricultural applications. *Energy Procedia* **2017**, *127*, 163–177. [CrossRef]
27. Katekhong, W.; Wongphan, P.; Klinmalai, P.; Harnkarnsujarit, N. Thermoplastic starch blown films functionalized by plasticized nitrite blended with PBAT for superior oxygen barrier and active biodegradable meat packaging. *Food Chem.* **2022**, *374*, 131709. [CrossRef] [PubMed]
28. Kweon, H.; Choi, S.; Kim, Y.; Nam, K. Development of a new UTM (universal testing machine) system for the nano/MICRO in-process measurement. *Int. J. Mod. Phys. B* **2006**, *20*, 4432–4438. [CrossRef]
29. Choi, Y.; Yuan, R.L. Experimental relationship between splitting tensile strength and compressive strength of GFRC and PFRC. *Cem. Concr. Res.* **2005**, *35*, 1587–1591. [CrossRef]
30. Nataraja, M.C.; Dhang, N.; Gupta, A.P. Splitting tensile strength of SFRC. *Indian Concr. J.* **2001**, *75*, 287–290.
31. Lu, J.Z.; Wu, Q.; McNabb, H.S. Chemical coupling in wood fiber and polymer composites: A review of coupling agents and treatments. *Wood Fiber Sci.* **2000**, *32*, 88–104.

32. Phothisarattana, D.; Wongphan, P.; Promhuad, K.; Promsorn, J.; Harnkarnsujarit, N. Biodegradable poly (butylene adipate-co-terephthalate) and thermoplastic starch-blended TiO2 nanocomposite blown films as functional active packaging of fresh fruit. *Polymers* **2021**, *13*, 4192. [CrossRef]
33. Petchwattana, N.; Covavisaruch, S. Effects of rice hull particle size and content on the mechanical properties and visual appearance of wood plastic composites prepared from poly (vinyl chloride). *J. Bionic Eng.* **2013**, *10*, 110–117. [CrossRef]
34. Crighton, J.S.; Wilburn, F.W. The role of heat transfer in the production of DSC curves. *Thermochim. Acta* **1992**, *203*, 1–5. [CrossRef]
35. Wang, T.X.; Huang, W.M.; Chen, H.; Xiao, R.; Lu, H.B.; Kang, S.F. Temperature memory effect and its stability revealed via differential scanning calorimetry in ethylene-vinyl acetate within glass transition range. *J. Polym. Sci. B* **2016**, *54*, 1731–1737. [CrossRef]
36. Rebers, L.; Reichsöllner, R.; Regett, S.; Tovar, G.E.M.; Borchers, K.; Baudis, S.; Southan, A. Differentiation of physical and chemical cross-linking in gelatin methacryloyl hydrogels. *Sci. Rep.* **2021**, *11*, 3256. [CrossRef] [PubMed]
37. Van Hoorick, J.; Gruber, P.; Markovic, M.; Tromayer, M.; Van Erps, J.; Thienpont, H.; Liska, R.; Ovsianikov, A.; Dubruel, P.; Van Vlierberghe, S. Cross-linkable gelatins with superior mechanical properties through carboxylic acid modification: Increasing the two-photon polymerization potential. *Biomacromolecules* **2017**, *18*, 3260–3272. [CrossRef]
38. Bigi, A.; Cojazzi, G.; Panzavolta, S.; Rubini, K.; Roveri, N. Mechanical and thermal properties of gelatin films at different degrees of glutaraldehyde crosslinking. *Biomaterials* **2001**, *22*, 763–768. [CrossRef]
39. Saldarriaga, J.F.; Aguado, R.; Pablos, A.; Amutio, M.; Olazar, M.; Bilbao, J. Fast characterization of biomass fuels by thermogravimetric analysis (TGA). *Fuel* **2015**, *140*, 744–751. [CrossRef]
40. Aida, T.M.; Yamagata, T.; Watanabe, M.; Smith, R.L., Jr. Depolymerization of sodium alginate under hydrothermal conditions. *Carbohydr. Polym.* **2010**, *80*, 296–302. [CrossRef]
41. Yu, H.; Tang, Q.; Wu, J.; Lin, Y.; Fan, L.; Huang, M.; Lin, J.; Li, Y.; Yu, F. Using eggshell membrane as a separator in supercapacitor. *J. Power Sources* **2012**, *206*, 463–468. [CrossRef]
42. El-Sayed, S.A.; Mostafa, M.E. Pyrolysis characteristics and kinetic parameters determination of biomass fuel powders by differential thermal gravimetric analysis (TGA/DTG). *Energy Convers. Manag.* **2014**, *85*, 165–172. [CrossRef]
43. Ago, M.; Okajima, K.; Jakes, J.E.; Park, S.; Rojas, O.J. Lignin-based electrospun nanofibers reinforced with cellulose nanocrystals. *Biomacromolecules* **2012**, *13*, 918–926. [CrossRef] [PubMed]
44. Arvanitoyannis, I.S. Totally and partially biodegradable polymer blends based on natural and synthetic macromolecules: Preparation, physical properties, and potential as food packaging materials. *J. Macromol. Sci. Part C Polym. Rev.* **1999**, *39*, 205–271. [CrossRef]
45. Siracusa, V.; Rocculi, P.; Romani, S.; Dalla Rosa, M.D. Biodegradable polymers for food packaging: A review. *Trends Food Sci. Technol.* **2008**, *19*, 634–643. [CrossRef]
46. Martucci, J.F.; Ruseckaite, R.A. Biodegradation behavior of three-layer sheets based on gelatin and poly (lactic acid) buried under indoor soil conditions. *Polym. Degrad. Stab.* **2015**, *116*, 36–44. [CrossRef]
47. Di Franco, C.R.; Cyras, V.P.; Busalmen, J.P.; Ruseckaite, R.A.; Vázquez, A. Degradation of polycaprolactone/starch blends and composites with sisal fibre. *Polym. Degrad. Stab.* **2004**, *86*, 95–103. [CrossRef]
48. Oyeoka, H.C.; Ewulonu, C.M.; Nwuzor, I.C.; Obele, C.M.; Nwabanne, J.T. Packaging and degradability properties of polyvinyl alcohol/gelatin nanocomposite films filled water hyacinth cellulose nanocrystals. *J. Bioresour. Bioprod.* **2021**, *6*, 168–185. [CrossRef]
49. Kimbuathong, N.; Leelaphiwat, P.; Harnkarnsujarit, N. Inhibition of melanosis and microbial growth in Pacific white shrimp (Litopenaeus vannamei) using high CO_2 modified atmosphere packaging. *Food Chem.* **2020**, *312*, 126114. [CrossRef]
50. Dalev, P.G.; Patil, R.D.; Mark, J.E.; Vassileva, E.; Fakirov, S. Biodegradation of chemically modified gelatin films in soil. *J. Appl. Polym. Sci.* **2000**, *78*, 1341–1347. [CrossRef]

foods

MDPI

Article

Nutritional Properties of Milk from Dairy Ewes Fed with a Diet Containing Grape Pomace

Francesca Bennato [1], Andrea Ianni [1], Marco Florio [1], Lisa Grotta [1], Francesco Pomilio [2], Maria Antonietta Saletti [2] and Giuseppe Martino [1,*]

[1] Faculty of BioScience and Technology for Food, Agriculture, and Environment, University of Teramo, 64100 Teramo, Italy; fbennato@unite.it (F.B.); aianni@unite.it (A.I.); mflorio@unite.it (M.F.); lgrotta@unite.it (L.G.)

[2] Istituto Zooprofilattico Sperimentale Dell'Abruzzo e del Molise "G. Caporale", 64100 Teramo, Italy; f.pomilio@izs.it (F.P.); m.saletti@izs.it (M.A.S.)

* Correspondence: gmartino@unite.it; Tel.: +39-0861-266-950

Abstract: The aim of the present study was to evaluate the effects of a diet containing a 10% of grape pomace (GP) on the milk yield, chemical-nutritional characteristics, total phenolic compounds (TPCs), antioxidant activity (AOA), fatty acids and proteins profile of dairy ewe's milk. Forty-six ewes were dived into two groups: a control group (Ctrl), fed a standard diet, and an experimental group (GP+), whose diet was supplemented with 10% of GP on dry matter. The trial lasted 60 days and milk samples were collected and analyzed at the beginning (T0) and after 60 (T60) days. Dietary enrichment with GP did not affect the yield and the chemical composition of the milk. TPCs and AOA were not affected by the diet. After 60 days, the diet induced an increase in monounsaturated fatty acids (MUFA) and a decrease in medium chain saturated fatty acids (MCSFA), but the total saturated fatty acids (SFA), polyunsaturated fatty acids (PUFA), short chain saturated fatty acids (SCSFA) and long chain saturated fatty acids (LCSFA) were not modified. A decrease in the C14 desaturation index and an increase in the C18 index were also detected. Total caseins and whey protein were not affected by GP, even if a lower content of k-casein in GP+ milk compared to Ctrl milk was observed on the 60th day. The results of the present study suggest that 10% of GP can be included in the diet of lactating ewes without modifying milk gross composition but inducing significantly changes the fatty acid profile.

Keywords: grape pomace; ewes; milk; fatty acids; caseins

1. Introduction

The use of agri-food by-products in animal nutrition represents a fairly consolidated strategy that has its foundations in two main aspects. First of all, the valorization of these by-products represents a sustainable method both from an environmental and an economic point of view, since the industries that accumulate these wastes incur costs for their disposal. In addition to this, it should be emphasized that these matrices are rich in compounds with high biological value, and this generally has positive effects on the qualitative traits of products of animal origin, as well as on the animal welfare [1–3]. For instance, it has been demonstrated that the use of olive crude phenolic concentrate obtained from olive oil wastewater in the diet of dairy ewes induced an increase in milk polyunsaturated fatty acids (PUFA) [4]. The inclusion of pomegranate pulp in the diet of sheep reduced the concentration of saturated fatty acids (SFA) and increased that of PUFA [5]. In addition, in the milk of cows fed a diet containing dried grape pomace, there has been observed a significant increase in lactose and β-lactoglobulin but not in α-lactalbumin, albumin and caseins [6].

However, the utilization of agro-industry by-products in animal feeding has limits due to the presence of antinutritional factors such as lectin, tannin and non-starch polysaccharide, which can influence feed palatability, digestibility, and animal performance [7,8].

One of the by-products that has attracted more attention is certainly grape pomace, derived from grape processing in the wine industry, a particularly developed sector in Europe, with Italy and France representing the major producers [9]. This waste is characterized by the richness of tannins, phenolic compounds credited with antioxidant and anti-inflammatory functions and by the presence of PUFA, the consumption of which, as known, is associated with a lower risk of chronic diseases of the cardiovascular system [3,10].

The introduction of grape pomace in the diet of dairy ruminants has shown several advantages, particularly related to the fatty acids profile of milk and cheeses. Most of the studies were specifically conducted on dairy cows, evidencing the ability of this ingredient to influence rumen function, slowing down the last steps of biohydrogenation, thus favoring the accumulation of trans-vaccenic acid (C18:1, t11) [11]. This compound represents the precursor of rumenic acid (C18:2, c9, t11), a conjugate of linoleic acid (CLA) to which is associated high bioactive potential [12]. With particular regard to cattle, the introduction of grape pomace in their diet was also effective in reducing the production of ruminal methane, especially by inducing changes in the rumen microbiota [13]. With specific regard to the ruminal function, various considerations have also been reported for small ruminants [14], even with regard to the mechanisms associated with the biohydrogenation process [15].

There is still little knowledge on the dietary effect of grape pomace on the quality of ewes' milk and deriving dairy products. Conversely to reports for dairy cows [3], the marked ability of this ingredient to modify some parameters of interest such as, for example, the fatty acids profile of ewes' milk, has not been highlighted [16]. Therefore, the aim of the present study was to evaluate if a diet containing 10% of grape pomace (GP) was able to induce changes in milk yield, chemical-nutritional characteristics, total phenolic compounds (TPCs), antioxidant activity (AOA), fatty acids and proteins profile of dairy ewes milk.

2. Materials and Methods

2.1. Experimental Design and Animal Management

The animals were managed according to Directive 2010/63/EU of the European Parliament (European Union, 2010) and Directive 86/609/EEC (European Economic Community, 1986), which deals with the protection of animals used for experimentation or other scientific purposes [17,18].

Forty-six crossbreed dairy ewes were involved in this study. All the animals were in the first part of the lactation (<45 days in milk) and, as a result, were homogeneous for milk production, body weight, and number of lactations. Ewes were randomly assigned to two groups of 23 animals each and confined into two separated areas: a control group (Ctrl) and an experimental group (GP+). The trial lasted for 70 days, preceded by 10 days adaptation period to the experimental diet. During the adaptation period, in both groups, the concentrate was gradually increased up to 1 kg of DM. This was achieved on day 10, which was considered time zero. Animals received, for 60 days, alfalfa hay ab libitum and a custom-formulated concentrate, the ingredients and chemical composition of which is reported in Table 1. The concentrate ration (1 kg/day per head) was administered in correlation with two daily milkings (8:00 and 18:00).

Table 1. Nutrient and chemical composition of the diet administered to the control group (Ctrl) and experimental group (GP+).

	Ctrl	GP+
Ingredients (%)		
Triticum	20.20	19.70
Corn	18.00	19.00
Barley	14.00	15.00
Sunflower	10.00	7.00
Beet pulps	10.00	-
Grape Pomace	-	10.00
Soy	10.00	11.50
Wheat	8.00	8.00
White sorghum	5.00	5.00
Calcium carbonate	1.50	1.50
Sodium bicarbonate	1.00	1.00
Molasses	1.00	1.00
Dicalcium phosphate	0.80	0.80
Vitamins and microelements	0.50	0.50
Chemical composition (%)		
Dry Matter	89.00	88.00
Crude protein [1]	18.76	19.09
Ether extract [1]	3.03	3.30
Crude fiber [1]	6.97	7.27
Ash [1]	6.63	6.70
ME [1], MJ/kg	7.67	7.35

ME = Metabolizable Energy. [1] Data are reported on dry matter basis.

The control group (Ctrl) received a standard diet according to the common nutritional needs of dairy ewes in lactation [19], while, in the experimental group (GP+), a 10% (*w*/*w*) of grape pomace (on a dry matter basis) was used as a substitute for beet pulp. Specifically, red pomace, derived from the processing of grapes obtained from vines of Montepulciano d'Abruzzo (Abruzzo Region, Italy), was used; grape pomace before its use was treated as previously described by Ianni et al. [20]. Briefly, GP was subjected to a fermentation phase and treatment with steam to eliminate ethyl alcohol and, subsequently, it was treated with water at 90 °C in order to recover the tartaric acid. Finally, the GP was dried for the production of flour, incorporated in concentrate formulation.

2.2. Sampling of Feed and Milk

The feeds of each group were collected at the beginning and at the end of the trial. The total amount collected at each time and for each group was mixed, and a final sample was taken. Samples of feed were taken and analyzed for chemical composition.

At the end of the adaptation period and on 60th day, the daily milk yield, determined as the sum of morning and evening milkings, was measured by using lactometers and milk samples of the morning milking were collected separately from each ewe. Fresh milk was immediately used for the chemical-centesimal analysis and the remaining part was suitably aliquoted and frozen at −20 °C before carrying out further analysis.

2.3. Chemical Analysis of Feed and Milk

The samples of concentrate were analyzed for dry matter (method 930.15), crude protein (method 954.01), ether extract (method 920.39), crude fiber (method 962.09) and ash (method 942.05) according to AOAC International [21]. Milk fat, protein, casein, lactose, and urea contents were determined using a MilkoScan FT 6000 (Foss Integrator IMT: Foss Analytics, Hiller'ød, Denmark); pH values were determined by using a portable pH meter equipped with an electrode (Crison, Barcelona, Spain).

2.4. Feed and Milk Phenolic Content and Antioxidant Capacity

The content of TPCs in feed and milk was determined according to the protocol reported by Singleton and Rossi [22]. Both in the feeds and in the defatted milk, the phenolic compounds were previously extracted by a hydroalcoholic solution (ethanol:water 70:30, *v*/*v*). The samples were agitated for 40 min in the dark, centrifugated and the supernatants were recovered, filtered and analyzed. The content of TPCs was determined according to Ianni et al. [23]. Briefly, 200 μL of extract was mixed with 1 mL of 0.2 N Folin–Ciocalteu reagent and 800 μL of sodium carbonate 7.5% in water. The samples were incubated at room temperature (20–22 °C) in the dark and after 30 min the absorbance at 765 nm (Jenway 6305 UV/Vis spectrophotometer, Jenway, Barloworld Scientific, Dunmow, United Kingdom) was evaluated. The results were reported as mg of gallic acid equivalent (GAE)/g of sample on a dry matter basis for the feeds and as μg GAE/mL for the milk. The AOA was determined as reported by Ianni et al. [23], mixing 100 μL of extract with 1 mL of opportunely diluted ABTS solution. The colorimetric evaluations were performed at 734 nm after 4 min and the results were expressed as μmol equivalent antioxidant capacity (TEAC)/g of sample on a dry matter basis for the feeds and as μg GAE/mL for the milk.

2.5. Milk Protein Extraction and SDS-PAGE

Fifteen mL of each milk sample were cool centrifugated at 4000× *g* rpm for 15 min. The fat was removed, the supernatant was recovered, filtered and quantified using Bradford method [24] and Bovine Serum Albumine (BSA) as standard. Protein samples (7.5 μg) were diluted in sample buffer (0.5 M Tris-HCl, pH 6.8; 2% (*w*/*v*) SDS; 7% (*v*/*v*) glycerol; 4.3% (*v*/*v*) β-mercaptoethanol; 0.0025% (*w*/*v*) bromophenol blue), boiled for 5 min and loaded into a 12% polyacrylamide gel. After the run, the gels were placed for 30 min in a staining solution containing 40% (*v*/*v*) methanol, 10% (*v*/*v*) acetic acid and 0.1% Comassie Brillant Blue G-250. Thereafter, gels were de-stained by two washings in distilled water containing 40% (*v*/*v*) methanol and 10% (*v*/*v*) acetic acid. The densitometric values of caseins and whey proteins were analyzed by using ImageLab 6.0.1 software.

2.6. Feed and Milk Fatty Acid Profile

Feed fatty acids were extracted using a mix of chloroform and methanol (2:1, vol/vol) and analyzed by a gas chromatography (GC) coupled with a flame ionization detector (FID) equipped with a capillary column (Restek rt-2560 Column, fused silica 100 m × 0.25 mm highly polar phase; Restek Corporation, Bellefonte, PA, USA). Hydrogen was used as a carrier gas a flow rate of 1 mL/min. For the separation of methyl esters of fatty acids (FAMEs) the temperature program previously reported by Bennato et al. was used [25]. The milk lipid fraction was analyzed according to the AOAC official method [26]. The detection of FAMEs was performed by GC-FID. The injector temperature was 280 °C. The oven temperature was initially settled at 80° C and held for 10 min, then increased up to 172 °C at 4 °C/min for 30 min, finally increased up to 190 °C at 4 °C/min and held for 10 min. The FAMEs were identified by a comparison of the FAME retention times with different FAME analytical standards for feed (FAME Mix C8-C24 Supelco, Bellefonte, PA, USA) and milk (FIM-FAME-7-Mix; Matreya LLC, Pleasant Gap, PA, USA). Peak areas were quantified by Chrome Card software and the relative value of each individual fatty acid was expressed as a relative percentage of the total FAMEs identified. The relative percentage of each fatty acid was also used to calculate the sum of SFA, MUFA = monounsaturated fatty acids (MUFA), PUFA, short chain saturated fatty acids (SCSFA), medium chain saturated fatty acids (MCSFA) and long chain saturated fatty acids (LCSFA). Atherogenic Index (AI), Thrombogenic Index (TI) and Desaturation (C14, C16, C18 and CLA) Indices were calculated as previously reported by Innosa et al. [27].

2.7. Statistical Analysis

Statistical data analysis was performed using the JMP Pro 14 program (SAS Institute, Cary, NC, USA). All data were processed with ANOVA (Analysis of Variance) to analyze the impact of the diet and time. Sample means were assessed by HSD Tukey's test and differences were considered significant for $p < 0.05$. Results were reported as means ± standard error (SE).

3. Results

3.1. Diet Fatty Acid Profile, Polyphenol Content and Antioxidant Activity

The fatty acid composition of the diet administrated to the animals is reported in Table 2. In both groups, the most abundant fatty acid was linoleic acid (C18:2, c9, c12). Oleic acid (C18:1, c9) and palmitic acid (C16:0) were also presented, whereas stearic (C18:0) and linolenic (C18:3, c9, c12, c15) acids were the fatty acids less present. No significant differences in TPCs and AOA were found between the two diets.

Table 2. Fatty acid profile, total phenolic content and antioxidant activity of the diet administered to the control group (Ctrl) and experimental group (GP+).

	Ctrl	GP+
Fatty acids [1] (%)		
C14:0	0.35 ± 0.19	0.19 ± 0.01
C16:0	16.84 ± 1.67	16.67 ± 1.41
C18:0	1.81 ± 0.05	1.73 ± 0.25
C16:1, c9	0.17 ± 0.03	0.16 ± 0.03
C18:1, c9	18.94 ± 0.65	18.68 ± 0.31
C18:1, c11	0.18 ± 0.09	0.20 ± 0.06
C18:2, c9, c12	56.56 ± 1.59	57.75 ± 1.53
C18:3, c9, c12, c15	0.48 ± 0.09	0.40 ± 0.02
C22:0	0.06 ± 0.01	0.05 ± 0.01
C22:1	0.12 ± 0.05	0.11 ± 0.05
Others	4.49 ± 0.20	4.13 ± 0.27
SFA	19.06 ± 1.79	18.64 ± 1.18
MUFA	19.41 ± 0.75	18.71 ± 1.02
PUFA	57.04 ± 0.20	58.14 ± 1.53
TPCs [2] (mg GAE/g)	2.19 ± 0.07	2.24 ± 0.13
AOA [2] (μmol TEAC/g)	461.90 ± 31.49	502.59 ± 29.20

SFA = Saturated Fatty Acid, MUFA = Monounsaturated Fatty Acid, PUFA = Polyunsaturated Fatty Acid, TPCs = Total Phenolic Compounds, AOA = Antioxidant Activity. [1] Data are reported as mean relative percentage of total FAMEs ± SE. [2] Data are reported on dry matter basis. Differences in TPCs, AOA and fatty acid profile between Ctrl and GP+ were not significant ($p > 0.05$).

3.2. Milk Yield, Chemical Composition, Polyphenols Content and Antioxidant Capacity

The average daily milk yield at the beginning of the study was 1.58 ± 0.09 kg/head and 1.62 ± 0.11 for Ctrl and GP+, respectively. After 60 days of experimental period, the daily milk yield decreased significantly, both in Ctrl ($p < 0.05$) and GP+ ($p < 0.001$), reaching values of 1.28 ± 0.07 and 1.18 ± 0.10 Ctrl and GP+, respectively. Moreover, there was no significant difference in milk yield between Ctrl and GP+, either at the beginning or after 60 days.

As reported in Table 3, during the entire period of the trial, the chemical composition of the milk samples was not affected by diet, with the exception of lactose, whose values were lower ($p < 0.01$) in GP+ compared to Ctrl on the 60th day. In Ctrl, the milk lactose percentage increased significantly at 60 days compared to T0; conversely, in GP+, the content remained constant throughout the entire period. Both in Ctrl and GP+, the lipid percentage increased significantly during the trial and significant differences were observed in both groups between the beginning and after 60 days of experiment. In both groups, urea content showed a decrease after 60 days. No significant differences within treatments and between treatments were observed in casein and protein content and pH values.

Table 3. Chemical composition of at the beginning (T0) and after 60 (T60) days of trial of raw milk samples obtained from ewes fed a standard diet (Ctrl) and grape pomace diet (GP+).

	T0			T60		
	Ctrl (n = 23)	**GP+ (n =23)**	p	**Ctrl (n = 23)**	**GP+ (n = 23)**	p
Lipid (%)	5.29 ± 0.25 [b]	5.04 ± 0.22 [b]	ns	6.38 ± 0.27 [a]	6.12 ± 0.21 [a]	ns
Casein (%)	4.48 ± 0.09	4.47 ± 0.10	ns	4.18 ± 0.08	4.35 ± 0.08	ns
Urea (mg/dL)	66.40 ± 1.38 [a]	65.42 ± 1.40 [a]	ns	58.53 ± 1.42 [b]	58.10 ± 1.33 [b]	ns
Lactose (%)	5.04 ± 0.05 [b]	5.03 ± 0.05	ns	5.24 ± 0.06 [a]	4.98 ± 0.06	<0.01
Protein (%)	5.68 ± 0.10	5.82 ± 0.09	ns	5.43 ± 0.09	5.56 ± 0.10	ns
pH	6.69 ± 0.20	6.75 ± 0.20	ns	6.85 ± 0.25	6.85 ± 0.30	ns

Data are reported as mean ± SE. [a,b] Means in the same row with different uppercase superscript letters are significantly different by time ($p < 0.05$). ns = not significant ($p > 0.05$).

The TPCs and AOA in ewes' milk were not affected by the diet (Figure 1). No significant differences in TPCs and AOA both in Ctrl and GP+ were observed during the trial.

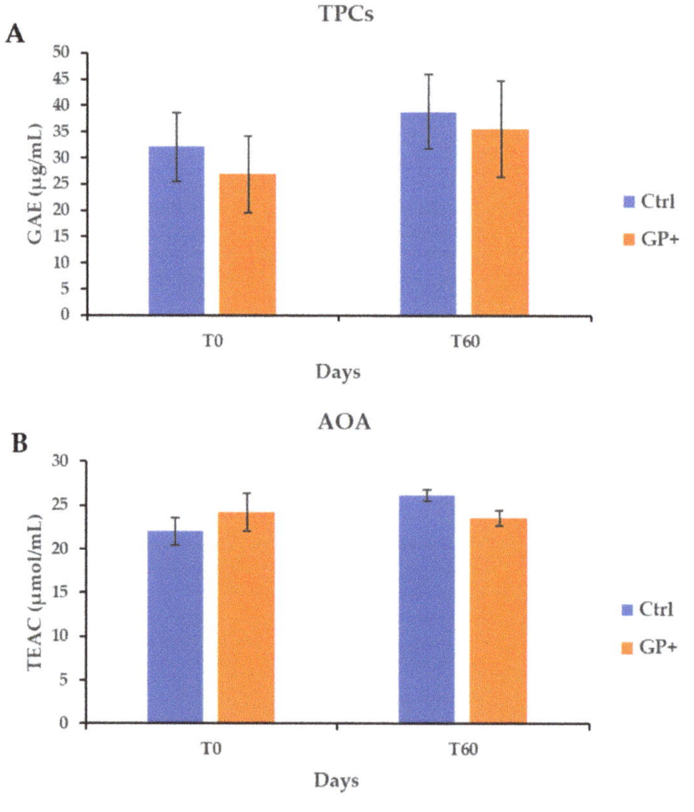

Figure 1. Total Phenolic Compounds (TPCs) (**A**) and Antioxidant Activity (AOA) (**B**) at the beginning (T0) and after 60 (T60) days of trial in raw milk samples obtained from ewes fed a standard diet (Ctrl) and grape pomace diet (GP+). GAE = gallic acid equivalent; TEAC = trolox equivalent antioxidant capacity. Data are reported as mean ± SE for n = 23. Differences between Ctrl and GP+ were not significant ($p > 0.05$).

3.3. SDS-PAGE Milk Protein Profile

The sodium dodecyl sulfate polyacrylamide gel electrophoresis (SDS-PAGE) allowed us to separate the major proteins in ewes milk (α-casein, β-casein, k-casein, β-lactoglobulin and α-lactalbumin). Attention was paid particularly to the determination of the total caseins (α-casein, β-casein, k-casein) and whey proteins (β-lactoglobulin and α-lactalbumin). As reported in Figure 2, no significant differences between the two groups were found both in caseins and whey proteins at the beginning and after 60 days. However, in Ctrl milk, lower ($p < 0.05$) levels of caseins (Figure 2A) and higher ($p < 0.01$) whey proteins (Figure 2B) content were found at 60 days compared to T0 samples. Conversely, in GP milk, no significant differences in caseins and whey proteins content between T0 and T60 were detected.

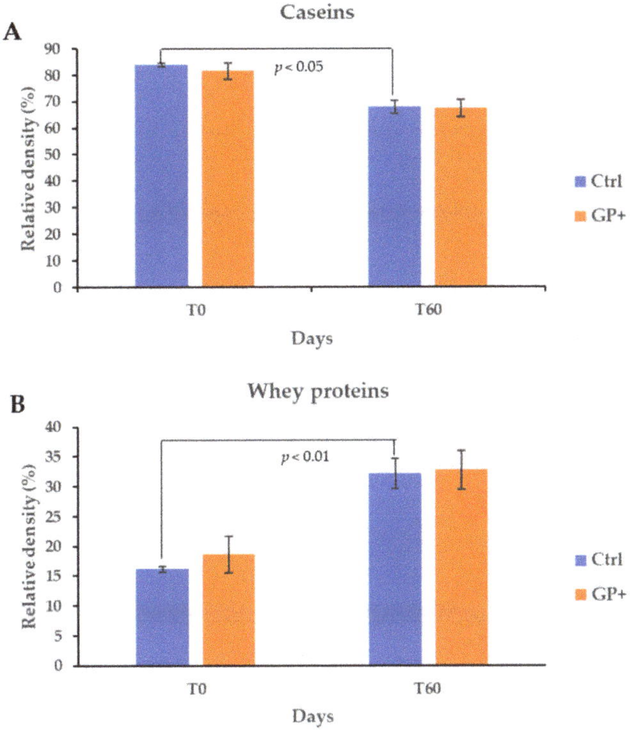

Figure 2. Total caseins (**A**) and whey proteins (**B**) at the beginning (T0) and after 60 (T60) days of trial in raw milk samples obtained from ewes fed a standard diet (Ctrl) and grape pomace diet (GP+). Data are reported as mean percentage ± SE for n = 23. Differences between Ctrl and GP+ were not significant ($p > 0.05$).

Specifically, the main changes, both in Ctrl and GP+ proteins profile, were relative to a significant decrease in β-casein between T0 and T60 (Table 4). Conversely, after 60 days, the values of k-casein and β-lactoglobulin increased, both in Ctrl and Gp+ milk. At 60 days, the values of k-casein in GP+ milk were significantly lower ($p < 0.05$) compared to Ctrl. No significant differences between Ctrl and GP+ were detected in all the other protein fractions at T0 and T60.

Table 4. Densitometric analysis of SDS-PAGE protein bands at the beginning (T0) and after 60 (T60) days of trial in raw milk samples obtained from ewes fed a standard diet (Ctrl) and grape pomace diet (GP+).

	T0			T60		
	Ctrl (n = 23)	GP+ (n = 23)	p	Ctrl (n = 23)	GP+ (n = 23)	p
α-casein (%)	12.43 ± 0.91	12.79 ± 2.35	ns	8.01 ± 1.10	9.60 ± 0.69	ns
β-casein (%)	61.85 ± 2.52 [a]	60.61 ± 2.26 [a]	ns	37.77 ± 2.36 [b]	43.27 ± 2.60 [b]	ns
κ-casein (%)	9.56 ± 0.62 [b]	8.01 ± 1.67 [b]	ns	22.13 ± 1.54 [a]	14.45 ± 1.43 [a]	<0.05
β-lactoglobulin (%)	10.12 ± 0.77 [b]	11.82 ± 2.12 [b]	ns	22.23 ± 2.26 [a]	23.92 ± 2.30 [a]	ns
α-lactalbumin (%)	6.04 ± 0.35	6.74 ± 1.14	ns	8.89 ± 0.92	8.77 ± 4.66	ns

Data are reported as mean percentage ± SE of the main proteins found in the electrophoretic profile of each sample. [a,b] Means in the same row with different uppercase superscript letters are significantly different by time ($p < 0.05$). ns = not significant ($p > 0.05$).

3.4. Milk Fatty Acids Profile

The composition of milk fatty acids in Ctrl and GP+ at the beginning and at the end of the study is shown in Table 5. The inclusion of GP modified the milk fatty acids profile. After 60 days, significantly higher values of MUFA ($p < 0.05$) and lower levels of MCSFA ($p < 0.01$) were detected in GP+ milk compared to the Ctrl, but total SFA, PUFA, SCSFA and LCSFA content was not modified. In addition, at the end of the trial compared to the beginning of the experimental period, GP+ milk samples contained a lower percentage of C15:0 ($p < 0.01$) and C14:1, c9 ($p < 0.001$). After 60 days, a lower content of C12:0 ($p < 0.05$), C14:0 ($p < 0.05$) and C14:1, c9 ($p < 0.01$) was highlighted in GP+ milk samples compared to Ctrl. A lower ($p < 0.05$) C14 desaturation index and higher C18 desaturation index were found in T60 GP+ samples. No significant differences in C16 and CLA desaturation indices and TI and AI were found between the two groups.

Table 5. Fatty acid profile at the beginning (T0) and after 60 (T60) days of trial in raw milk samples obtained from ewes fed a standard diet (Ctrl) and grape pomace diet (GP+).

	T0			T60		
	Ctrl (n = 23)	GP+ (n = 23)	p	Ctrl (n = 23)	GP+ (n = 23)	p
Fatty Acids (%)						
C4:0	2.69 ± 0.37	2.60 ± 0.33	ns	3.13 ± 0.20	3.41 ± 0.28	ns
C6:0	3.04 ± 0.35	2.85 ± 0.33	ns	3.31 ± 0.15	3.24 ± 0.24	ns
C8:0	3.63 ± 0.33	3.39 ± 0.35	ns	3.73 ± 0.10	3.38 ± 0.24	ns
C10:0	13.14 ± 0.86	12.12 ± 0.95	ns	13.20 ± 0.27	11.26 ± 0.64	ns
C12:0	7.98 ± 0.36	7.17 ± 0.39	ns	7.24 ± 0.26	6.12 ± 0.34	<0.05
C14:0	13.17 ± 0.28	13.14 ± 0.34	ns	14.21 ± 0.27	13.38 ± 0.49	<0.05
C15:0	1.12 ± 0.05	1.13 ± 0.04 [a]	ns	1.08 ± 0.02	0.92 ± 0.04 [b]	<0.01
C16:0	26.22 ± 0.86	27.69 ± 0.95	ns	26.47 ± 0.60	27.89 ± 1.19	ns
C17:0	0.44 ± 0.03	0.50 ± 0.04	ns	0.46 ± 0.02	0.42 ± 0.03	ns
C18:0	5.50 ± 0.43	6.19 ± 0.42	ns	6.23 ± 0.24	6.06 ± 0.63	ns
C14:1, c9	0.44 ± 0.02	0.48 ± 0.02 [a]	ns	0.51 ± 0.02	0.38 ± 0.02 [b]	<0.001
C16:1, c9	1.21 ± 0.05	1.27 ± 0.05	ns	1.22 ± 0.04	1.38 ± 0.10	ns
C18:1, t11	1.01 ± 0.60	0.42 ± 0.08	ns	0.45 ± 0.05	1.76 ± 0.92	ns
C18:1, c9	11.69 ± 0.77	12.79 ± 0.92	ns	11.34 ± 0.49	12.30 ± 0.79	ns
C18:1, c11	0.13 ± 0.03	0.10 ± 0.01	ns	0.08 ± 0.01	0.09 ± 0.02	ns
CLA	1.13 ± 0.08	1.18 ± 0.10	ns	1.10 ± 0.05	1.20 ± 0.11	ns
C 18:2, c9, c11	3.37 ± 0.44	2.97 ± 0.21	ns	2.42 ± 0.14	3.23 ± 0.46	ns
C 18:3, c9, c12, c15	0.77 ± 0.06	0.79 ± 0.06	ns	0.57 ± 0.03	0.55 ± 0.05	ns
Others	3.31 ± 0.08	3.22 ± 0.05	ns	3.26 ± 0.08	3.02 ± 0.11	ns

Table 5. *Cont.*

	T0			T60		
	Ctrl (n = 23)	GP+ (n = 23)	p	Ctrl (n = 23)	GP+ (n = 23)	p
SFA	76.94 ± 1.26	76.78 ± 1.23	ns	79.06 ± 0.75	76.09 ± 1.44	ns
MUFA	14.48 ± 0.79	15.06 ± 0.91	ns	13.59 ± 0.55	15.91 ± 0.93	<0.05
PUFA	5.27 ± 0.53	4.93 ± 0.35	ns	4.09 ± 0.53	4.98 ± 1.74	ns
SCSFA	5.73 ± 0.72	5.45 ± 0.66	ns	6.43 ± 0.34	6.65 ± 0.49	ns
MCSFA	39.05 ± 1.59	36.96 ± 1.79	ns	39.45 ± 0.64	35.06 ± 1.38	<0.01
LCSFA	32.16 ± 1.26	34.37 ± 1.30	ns	33.17 ± 0.52	34.37 ± 1.35	ns
C14 Index	32.38 ± 0.17	31.81 ± 0.19	ns	35.48 ± 0.17	28.17 ± 0.24	<0.05
C16 Index	44.56 ± 0.21	44.01 ± 0.13	ns	44.30 ± 0.16	46.62 ± 0.19	ns
C18 Index	68.24 ± 0.66 ^{a}	67.32 ± 0.63	ns	64.45 ± 0.56 ^{b}	67.71 ± 1.41	<0.05
CLA Index	68.49 ± 0.59	74.46 ± 3.67	ns	71.83 ± 2.63	63.21 ± 7.82	ns
AI	4.64 ± 0.37	4.59 ± 0.33	ns	5.25 ± 0.31	4.45 ± 0.44	ns
TI	8.30 ± 0.36	8.04 ± 0.08	ns	7.53 ± 0.15	7.92 ± 0.29	ns

Data are reported as mean relative percentage of total FAMEs ± SE. CLA = Conjugated Linoleic Acids; SFA = Saturated Fatty Acids; MUFA = Monounsaturated Fatty Acids; PUFA = Polyunsaturated Fatty Acids; SCSFA = Short Chain Saturated Fatty Acids (C4:0-C6:0), MCSFA = Medium Chain Saturated Fatty Acids (C8:0-C15:0); LCSFA = Long Chain Saturated Fatty Acids (C16:0-C18:0); C14 Index = [(C14:1, c9)/(C14:0 + C14:1, c9)]*100; C16 Index = [(C16:1, c9)/(C16:0 + C16:1, c9)]*100; C18 Index = [(C18:1, c9)/(C18:0 + C18:1, c9)]*100; CLA Index = (CLA/(C18:1, t11 + CLA)]*100; AI = Atherogenic Index; TI = Thrombogenic Index. ^{a,b} Means in the same row with different uppercase superscript letters are significantly different by time ($p < 0.05$). ns = not significant ($p > 0.05$).

4. Discussion

The results of the present study revealed that GP supplementation in the diet of lactating ewes did not influence milk yield during the whole study. Both Ctrl and GP+ follow the same trend characterized by a decrease in milk yield. This trend follows the general pattern of the lactation curve in dairy sheep, characterized by an increase from lambing to the peak of lactation and a gradual decrease toward the end of lactation with different persistence, depending on rearing conditions, breed, and individual animal [28,29]. Similarly, studies conducted on dairy cows fed a diet containing GP showed no significant effects on milk yield [3,30].

The stage of lactation, as well as genic and physiological factors, breed, milking interval and techniques, breeding strategies and nutrition, affect milk composition [29–32]. Ewes milk is characterized by a low quantity of fat and proteins at the beginning of the lactation and highest content at the end of lactation stage [33]. It is widely reported that a negative correlation between milk yield and milk composition at a higher milk production is usually associated with a lower fat and protein concentration [34,35]. In our study, lipid content increased in both groups on the 60th day. Conversely, protein content showed higher stability, probably reflecting the lower variation in protein content throughout the lactation. As reported in other studies, the use of GP in the diet did not affect fat and protein concentration [16]. Conversely to a report by Manso et al. [16], the dietary inclusion with GP did not affect lactose percentage during the period; in fact, in GP+, the values of lactose remained constant up to sixty days. On the contrary, on the 60th day, in Ctrl, the amount of lactose increased. This finding disagrees with data in literature that reported a low content of lactose in sheep milk as in other ruminants at the beginning of lactation toward the end of lactation, contrary to the behavior of fat and protein contents in the milk [36,37]. The mechanism governing this process is very difficult to understand and little is known about its variation in ewes. It is known that lactose synthesis is predominantly subjected to hormonal control with very few dietary factors influencing its production [38–40]. However, the significant increase in lactose in Ctrl milk was not associated with an increase in milk volume production, of which the lactose is the major determinant, since the entry of water into the vesicle formed in the trans-Golgi is linked with lactose synthesis to maintain osmotic equilibrium with surrounding fluids. Thus, the rate of lactose synthesis regulates water secretion and, consequently, milk yield. Lactose is the major sugar of milk, and the glucose is the primary substrate for its synthesis. Lactose synthesis is initiated in the Golgi apparatus and continues in the vesicles with an influx of water and ionic constituents that causes the vesicles to swell as they pass toward the cell surface. Glucose and uridine diphosphate (UDP)-galactose combine to form lactose under the action of the enzyme lactose synthetase. The milk protein α-lactalbumin must be present for glucose and UDP-galactose to combine. Thus, α-lactalbumin appears to be a prime regulator of lactose synthesis [41,42].

However, contrary to what is expected, electrophoretic profile analysis did not show a significant increase in α-lactalbumin in Ctrl. GP supplementation did not affect the α-lactalbumin quantity in accordance with a study by Chedea et al. [6] in dairy cows fed a diet containing 15% GP. Conversely to a report by Chedea and colleagues [6], in our study, GP did not affect β-lactoglobulin percentage. Concerning the caseins fractions, GP did not modify α-casein and β-casein amount, but a significant increase in k-casein was observed, even if the percentage of k-casein in GP+ was, however, lower compared to Ctrl on 60 days.

Caseins are the main protein in sheep milk, and they are responsible for the enzymatic coagulation of milk and their hydrolysis pattern is an important characteristic of cheese ripening, so they are positively correlated with cheese production [43]. The different protein profile found in the milk of the two groups may induce changes in the cheese-making property of the milk, influencing the quality and the yield of the cheeses.

The amount of phenolic compound in the milk depends mainly on the type of diet. In our experiment, the diets administered to animals were characterized by similar polyphenols concentrations, so the TPCs and AOA in ewes' milk did not differ between the two groups.

The fatty acids composition of the milk varies according to the stage of lactation, which can influence the milk fatty acids profile as result of the balance between body fat mobilization and de novo synthesis in the mammary gland. However, feeds composition and feeding strategies have a decisive influence on the milk composition, especially fat LCSFA. In the milk of ewes fed GP, lower values of C15:0 and C14:1, c9 were highlighted in T60 samples compared to T0 milk. After 60 days of 10% GP inclusion in the diet, GP+ milk compared to Ctrl had a lower percentage of MCSFA and lower levels of C12:0, C14:0 and C15:0. A decrease of C10:0 and C12:0 was also observed in Churra ewes fed a diet containing 10% of GP [11] and in dairy cows [13]. Saturated fatty acids from C6:0 to C14:0 are almost exclusively synthesized ex novo by the mammary gland, and it has been demonstrated that their synthesis is related to the dietary intake of fermentable carbohydrate [44]. As reported by those authors, the high amount of lignin in GP, whose rumen digestion is limited, may interfere with de novo synthesis of fatty acids in the mammary gland. In our experiment, dietary supplementation with GP resulted in increased MUFA concentrations in the milk, suggesting that ruminal biohydrogenation was indeed inhibited.

Saturated fat intake is usually associated with an increased risk of coronary artery disease or cardiovascular disease. However, studies on the effects of C12:0 and C14:0 on human health are often contradictory. Lauric (C12:0), myristic (C14:0) and palmitic (C16:0) acids seem to exert atherogenic effects by increasing the synthesis of (low-density lipoprotein) LDL cholesterol and the level of total cholesterol [45] and the potential of C14:0 to increase total serum cholesterol would seem fourfold or even sixfold higher than that of C16:0. However, several studies have demonstrated that lauric acid increases the level of high-density lipoprotein (HDL) [45]. The low amount of MCFA, as C12:0 and C14:0, detected in GP+ milk may represent a benefit for the consumer.

In order to better evaluate the nutritional and health aspects of the milk, IA and TI were also calculated. These indices are used to value the nutritional and animal fat for consumers. In this study, we did not find differences related to the diet in AI and TI, whose values remained constant during the period in both groups. To study the effect of the diet on the capacity of stearoyl-CoA desaturase (SCD) to desaturate SFA to Δ^9-unsaturated fatty acids, the desaturation indices for C16:1, c9 C18:1, c9, C14:1, c9, and CLA were the estimated [15]. On the 60th day, the lower C18 index found in Ctrl compared to T0 samples, suggests that a low quantity of the C18:0 that is transferred from the blood to the mammary gland is desaturated in the mammary gland to produce C18:1, c9. This trend was not observed in GP+, in which a higher value was detected at 60 days. This finding suggests that milk fluidity was maintained, since the conversion of 18:0, absorbed by the mammary gland to C18:1, c9 is essential to maintain the fluidity of the milk. Conversely, to 60 days, the higher C14:1, c9 level in GP+ milk compared to T0 milk, might be the result of an increase in the desaturation of C14:0 occurring in the mammary gland by SCD, a finding also supported by the higher values of the relative C14 index. The only source of C14:1, c9 in the milk, is the desaturation of C14:0; fatty acid was produced almost entirely in the mammary gland, so the C14 index is the main representative of the SCD compared to the other indices [46]. In cow milk somatic cells, a positive correlation between C14 index and SCD mRNA has been observed [47]. The lower C14 index in GP+ milk could be due to the stimulatory effect of GP compounds on the expression, or on the activity of SCD.

5. Conclusions

The results of the present study suggest that grape pomace can be included in the diet of lactating ewes at a level of 10% without adverse effects on milk production and chemical composition. The different protein profile found in the milk of ewes fed grape pomace may affect the cheese-making property of the milk, influencing the quality and the yield of the cheeses, an aspect that needs further investigation. In addition, grape pomace is able to modify the milk fatty acid profile, whose composition is one of the factors that affects the formation and the development of cheese aroma. The low content of saturated fatty acids, such as lauric and myristic in the milk of ewes fed with grape pomace, also represents a benefit for human health.

Author Contributions: Conceptualization, G.M.; methodology, A.I. and L.G.; software, M.F.; formal analysis, F.B., M.F., F.P. and M.A.S.; investigation, F.B. and A.I.; resources, G.M. and L.G.; data curation, F.B.; writing—original draft preparation, F.B.; writing—review and editing, A.I.; visualization, F.B.; supervision, G.M. All authors have read and agreed to the published version of the manuscript.

Funding: This research is part a project funded by the Italy's Ministry of Economic Development (MISE), entitled "Sviluppo di tecnologie innovative nella filiera latte per la riduzione dell'emissione di gas serra, miglioramento dei parametri salutistici nei prodotti caseari e sicurezza alimentare (INNOLAT)"; MISE-Fondo Crescita Sostenibile "Agrifood" Accordo Innovazione – DM 5 marzo 2018, Capo II-Posizione n.11, MISE F/170011/01-05/X42 (Scientific coordinator: Prof. Giuseppe Martino).

Institutional Review Board Statement: Ethical review and approval were not necessary for this study, due to the fact that no breeding practices other than those normally adopted by the farm were introduced.

Informed Consent Statement: Not applicable.

Data Availability Statement: The data presented in this study are available on request from the corresponding author.

Conflicts of Interest: The authors declare no conflict of interest.

References

1. Dou, Z.; Toth, J.D.; Westendorf, M.L. Food waste for livestock feeding: Feasibility, safety, and sustainability implications. *Glob. Food Sec.* **2018**, *17*, 154–161. [CrossRef]
2. Salami, S.A.; Luciano, G.; O'Grady, M.N.; Biondi, L.; Newbold, C.J.; Kerry, J.P.; Priolo, A. Sustainability of feeding plant by-products: A review of the implications for ruminant meat production. *Anim. Feed Sci. Technol.* **2019**, *251*, 37–55. [CrossRef]
3. Ianni, A.; Martino, G. Dietary grape pomace supplementation in dairy cows: Effect on nutritional quality of milk and its derived dairy products. *Foods* **2020**, *9*, 168. [CrossRef] [PubMed]
4. Cappucci, A.; Alves, S.P.; Bessa, R.J.; Buccioni, A.; Mannelli, F.; Pauselli, M.; Viti, C.; Pastorelli, R.; Roscini, V.; Serra, A.; et al. Effect of increasing amounts of olive crude phenolic concentrate in the diet of dairy ewes on rumen liquor and milk fatty acid composition. *J. Dairy Sci.* **2018**, *101*, 4992–5005. [CrossRef]
5. Valenti, B.; Luciano, G.; Morbidini, L.; Rossetti, U.; Codini, M.; Avondo, M.; Priolo, A.; Bella, M.; Natalello, A.; Pauselli, M. Dietary pomegranate pulp: Effect on ewe milk quality during late lactation. *Animals* **2019**, *9*, 283. [CrossRef]

6. Chedea, V.S.; Pelmus, R.S.; Lazar, C.; Pistol, G.C.; Calin, L.G.; Toma, S.M.; Dragomir, C.; Taranu, I. Effects of a diet containing dried grape pomace on blood metabolites and milk composition of dairy cows. *J. Sci. Food Agric.* **2017**, *97*, 2516–2523. [CrossRef]
7. Abarghuei, M.J.; Rouzbehan, Y.; Alipour, D. The influence of the grape pomace on the ruminal parameters of sheep. *Livest. Sci.* **2010**, *132*, 73–79. [CrossRef]
8. Bahrami, Y.; Foroozandeh, A.D.; Zamani, F.; Modarresi, M.; Eghbal-Saeid, S.; Chekani-Azar, S. Effect of diet with varying levels of dried grape pomace on dry matter digestibility and growth performance of male lambs. *J. Anim. Plant Sci.* **2010**, *6*, 605–610.
9. Food and Agriculture Organization of the United Nations (FAO). 2019. Available online: http://faostat.fao.org/site339/default.aspx (accessed on 11 January 2022).
10. Marventano, S.; Kolacz, P.; Castellano, S.; Galvano, F.; Buscemi, S.; Mistretta, A.; Grosso, G. A review of recent evidence in human studies of n-3 and n-6 PUFA intake on cardiovascular disease, cancer, and depressive disorders: Does the ratio really matter? *Int. J. Food Sci. Nutr.* **2015**, *66*, 611–622. [CrossRef]
11. Szczechowiak, J.; Szkudelska, K.; Szumacher-Strabel, M.; Sadkowski, S.; Gwozdz, K.; El-Sherbiny, M.; Kozłowska, M.; Rodrigue, V.; Cieslak, A. Blood hormones, metabolic parameters and fatty acid proportion in dairy cows fed condensed tannins and oils blend. *Ann. Anim. Sci.* **2018**, *18*, 155. [CrossRef]
12. Mcguire, M.A.; Mcguire, M.K. Conjugated linoleic acid (CLA): A ruminant fatty acid with beneficial effects on human health. *J. Anim. Sci.* **2000**, *77*, 3. [CrossRef]
13. Moate, P.J.; Williams, S.R.O.; Torok, V.A.; Hannah, M.C.; Ribaux, B.E.; Tavendale, M.H.; Eckard, R.J.; Jacobs, J.L.; Auldist, M.J.; Wales, W.J. Grape marc reduces methane emissions when fed to dairy cows. *J. Dairy Sci.* **2014**, *97*, 5073–5087. [CrossRef]
14. Abarghuei, M.J.; Rouzbahan, Y. Influence of grape pomace extract on in vitro gas production kinetics and on ruminal unicellular population of inoculum in sheep. *Iran J. Anim. Sci.* **2013**, *44*, 375–384. [CrossRef]
15. Correddu, F.; Nudda, A.; Battacone, G.; Boe, R.; Francesconi, A.H.D.; Pulina, G. Effects of grape seed supplementation, alone or associated with linseed, on ruminal metabolism in Sarda dairy sheep. *Anim. Feed Sci. Technol.* **2015**, *199*, 61–72. [CrossRef]
16. Manso, T.; Gallardo, B.; Salvá, A.; Guerra-Rivas, C.; Mantecón, A.R.; Lavín, P.; De la Fuente, M.A. Influence of dietary grape pomace combined with linseed oil on fatty acid profile and milk composition. *J. Dairy Sci.* **2016**, *99*, 1111–1120. [CrossRef]
17. European Economic Community. EEC Council Directive 86/609/EEC of 24 November 1986 on the approximation of laws, regulations and administrative provisions of the Member States regarding the protection of animals used for experimental and other scientific purposes. *Off. J. Eur. Commun.* **1986**, *358*, 1–28.
18. European Union. Directive 2010/63/EU of the European Parliament and of the Council of 22 September 2010 on the Protection of Animals Used for Scientific Purposes. 2010. Available online: http://eur-lex.europa.eu/legal-content/EN/TXT/PDF/?uri=CELEX:32010L0063&from=EN (accessed on 11 January 2022).
19. NRC National Research Council (US); Committee on Nutrient Requirements of Small Ruminants; Board on Agriculture, Division on Earth, & Life Studies. *Nutrient Requirements of Small Ruminants: Sheep, Goats, Cervids, and New World Camelids*; China Legal Publishing House: Beijing, China, 2007.
20. Ianni, A.; Di Luca, A.; Martino, C.; Bennato, F.; Marone, E.; Grotta, L.; Cichelli, A.; Martino, G. Dietary supplementation of dried grape pomace increases the amount of linoleic acid in beef, reduces the lipid oxidation and modifies the volatile profile. *Animals* **2019**, *9*, 578. [CrossRef]
21. AOAC International. *Official Methods of Analysis*, 15th ed.; AOAC International: Washington, DC, USA, 1990.
22. Bradford, M.M. A rapid and sensitive method for the quantitation of microgram quantities of protein utilizing the principle of protein dye binding. *Anal. Biochem.* **1976**, *72*, 248–254. [CrossRef]
23. Singleton, V.L.; Rossi, J.R. Colorimetry of total phenolics with phosphomolybdic phosphotungstic acid. *Am. J. Enol. Vitic.* **1965**, *16*, 144–158.
24. Ianni, A.; Innosa, D.; Oliva, E.; Bennato, F.; Grotta, L.; Saletti, M.A.; Pomilio, F.; Sergi, M.; Martino, G. Effect of olive leaves feeding on phenolic composition and lipolytic volatile profile in goat milk. *J. Dairy Sci.* **2021**, *104*, 8835–8845. [CrossRef]
25. Bennato, F.; Ianni, A.; Martino, C.; Grotta, L.; Martino, G. Evaluation of Chemical Composition and Meat Quality of Breast Muscle in Broilers Reared under Light-Emitting Diode. *Animals* **2021**, *11*, 1505. [CrossRef]
26. AOAC International. *Official Methods of Analysis*, 17th ed.; AOAC International: Washington, DC, USA, 2000.
27. Innosa, D.; Bennato, F.; Ianni, A.; Martino, C.; Grotta, L.; Pomilio, F.; Martino, G. Influence of olive leaves feeding on chemical-nutritional quality of goat ricotta cheese. *Eur. Food Res. Technol.* **2020**, *246*, 923–930. [CrossRef]
28. Ringdorfer, F.; Pockl, E. Feed intake and milk performance of Austrian mountain sheep depending on roughage quality and feeding intensity. *Arch. Tierzucht.* **2004**, *47*, 97–102.
29. Oravcova, M.; Margetin, M.; Peškovičova, D.; Daňo, J.; Milerski, M.; Hetenyi, L.; Polak, P. Factors affecting milk yield and ewe's lactation curves estimated with test-day models. *Czech J. Anim. Sci.* **2006**, *51*, 483–490. [CrossRef]
30. Haenlein, G. *Nutritional Value of Dairy Products of Ewe and Goat Milk*; Cooperat Extens Dairy Spec Univ Delaware: Newark, DE, USA, 2002.
31. Kuchtik, J.; Šustova, K.; Urban, T.; Zapletal, D. Effect of the stage of lactation on milk composition, its properties and the quality of rennet curdling in East Friesian ewes. *Czech J. Anim. Sci.* **2008**, *53*, 55–63. [CrossRef]
32. Soják, L.; Blasko, J.; Kubinec, R.; Górová, R.; Addová, G.; Ostrovský, I.; Margetín, M. Variation among individuals, breeds, parities and milk fatty acid profile and milk yield of ewes grazed on pasture. *Small Rumin. Res.* **2013**, *109*, 173–181. [CrossRef]

33. Yilmaz, O.; Çak, B.; Bolacali, M. Effects of Lactation Stage, Age, Birth Type and Body Weight on Chemical composition of Red Karaman Sheep Milk. *Kafkas Univ. Vet. Fak. Derg.* **2011**, *17*, 383–386.
34. Barillet, F.; Elsen, J.M.; Roussely, M. Optimization of a selection scheme for milk composition and yield in milking ewes: Example of the Lacaune breed. In Proceedings of the 3rd World Congress on Genetics Applied to Livestock Production, Lincoln, NE, USA, 16–22 July 1986.
35. Casu, S.; Sanna, S. Aspects and problems of genetic improvement of the composition of milk from sheep and goats. In Proceedings of the Second International Symposium—Nuove Prospettive Della Ricerca Sugli Ovicaprini, Varese, Italy, 1990.
36. Ploumi, K.; Belibasaki, S.; Triantaphyllidis, G. Some factors affecting daily milk yield and composition in a flock of Chios ewes. *Small Rumin. Res.* **1998**, *28*, 89–92. [CrossRef]
37. Pavič, V.; Antunac, N.; Mioč, B.; Ivankovič, A.; Havranek, J.I. Influence of stage of lactation on the chemical composition and physical properties of sheep milk. *Czech J. Anim. Sci.* **2002**, *47*, 80–84.
38. Bauman, D.E.; Mather, I.H.; Wall, R.J.; Lock, A.L. Major advances associated with the biosynthesis of milk. *J. Dairy Sci.* **2006**, *89*, 1235–1243. [CrossRef]
39. Jenkins, T.C.; McGuire, M.A. Major advances in nutrition: Impact on milk composition. *J. Dairy Sci.* **2006**, *89*, 1302–1310. [CrossRef]
40. Neville, M.C.; McFadden, T.B.; Forsyth, I. Hormonal regulation of mammary differentiation and milk secretion. *J. Mammary Gland Biol. Neoplasia* **2002**, *7*, 49–66. [CrossRef]
41. Kuhn, N.J. The biosynthesis of lactose. In *Biochemistry of Lactation*; Mepham, T.B., Ed.; Elsevier Science Publishers: Amsterdam, The Netherlands, 1983; pp. 159–176.
42. Larson, B.L. Biosynthesis and cellular secretion of milk. In *Lactation*; Larson, B.L., Ed.; Iowa State University Press: Ames, IA, USA, 1985; pp. 129–163.
43. Park, Y.W.; Juarez, M.; Ramos, M.; Haenlein, G.F.W. Physico-chemical characteristics of goat and sheep milk. *Small Rumin. Res.* **2007**, *68*, 88–113. [CrossRef]
44. Moate, P.J.; Chalupa, W.; Boston, R.C.; Lean, I.J. Milk Fatty Acids II: Prediction of the Production of Individual Fatty Acids in Bovine Milk. *J. Dairy Sci.* **2008**, *91*, 1175–1188. [CrossRef]
45. Dayrit, F.M. The Properties of Lauric Acid and Their Significance in Coconut Oil. *J. Am. Oil Chem. Soc.* **2014**, *92*, 1–15. [CrossRef]
46. Palmquist, D.L.; Davis, C.L.; Brown, R.E.; Sachan, D.S. Availability and metabolism of various substrates in ruminants. V. Entry rate into the body and incorporation into milk fat of D(-) betahydroxybutyrate. *J. Dairy Sci.* **1969**, *52*, 633–638. [CrossRef]
47. Feng, S.; Salter, A.M.; Parr, T.; Garnsworthy, P.C. Extraction and quantitative analysis of stearoyl-coenzyme A desaturase mRNA from dairy cow milk somatic cells. *J. Dairy Sci.* **2007**, *90*, 4128–4136. [CrossRef]

foods

Article

Optimization of a Simultaneous Enzymatic Hydrolysis to Obtain a High-Glucose Slurry from Bread Waste

Teresa Sigüenza-Andrés [1,*], Valentín Pando [2], Manuel Gómez [1] and José M. Rodríguez-Nogales [1]

1 Food Technology Area, College of Agricultural Engineering, University of Valladolid, 34004 Palencia, Spain; mgpallares@uva.es (M.G.); josemanuel.rodriguez@uva.es (J.M.R.-N.)
2 Department of Statistics and Operational Investigation, College of Agricultural Engineering, University of Valladolid, 34004 Palencia, Spain; vpando@uva.es
* Correspondence: teresa.siguenza@uva.es

Abstract: Bread and bakery products are among the most discarded food products in the world. This work aims to investigate the potential use of wasted bread to obtain a high-glucose slurry. Simultaneous hydrolysis of wasted bread using α-amylase and glucoamylase was carried out performing liquefaction and saccharification at the same time. This process was compared with a traditional sequential hydrolysis. Temperature and pH conditions were optimized using a response surface design determining viscosity, reducing sugars and glucose concentration during the enzymatic processes. The optimal conditions of pH and temperature in the saccharification stage and the simultaneous hydrolysis were pretty similar. Results show that the slurry produced with simultaneous process had a similar glucose yield at 2 h, and at 4 h a yield higher than that obtained by the sequential method of 4 h and could reduce time and energy.

Keywords: discarded bread; liquefaction; saccharification reducing sugars; glucose; α-amylase; glucoamylase

1. Introduction

It is estimated that one-third of the food produced in the world is wasted. In retail shops, bread and bakery products are usually the most discarded, after fruits and vegetables [1]. Melikoglu et al. [2] estimated that 10% of the world production of around 100 million bakery products are wasted annually. Moreover, the European Union recommends reducing food waste, but when not possible, the institution advises the redirection of this waste for human consumption before other uses [3].

Currently, despite these recommendations, discarded bread is mostly used for the production of animal feed, while studies focused on the use of bread waste in human consumption are scarce. Gélinas et al. [4] suggested the use of discarded bread for the production of sourdoughs More recently, wasted bread has been studied as a potential source to generate extruded products [5,6].

Another frequent use of wasted bread is its hydrolysis and fermentation as bread is an ideal substrate for enzymes and microorganisms. This amorphous matrix, consisting mainly of gelatinized starch, will be further modified during cooling as the starch retrogrades [7]. Besides, gelatinized starch can be attacked by amylases, owned or added, to generate sugars [8]. Immonen et al. [9] worked on the effects of bread slurry addition on bread quality and Rosa-Sibakov et al. [10] replaced sucrose with bread hydrolysate. The use of bread waste for the production of a growth medium for starters cultivation [11] or a new antifungal ingredient [12] have been also studied. Many studies took advantage of these changes and used bakery products as good feedstocks for ethanol production with high fermentation efficiencies through starch enzymatic hydrolysis [2,13,14]. For alcohol production, a first phase of starch hydrolysis is necessary to obtain a glucose syrup. This hydrolysis usually consists of two different stages, liquefaction (LIQ), carried out with α-amylase, and saccharification (SAC).

Several studies have optimized the LIQ and SAC processes of wasted bread separately to obtain a glucose-fructose syrup changing different parameters such as substrate, water and enzyme ratio [15,16]. As each enzyme has an optimum pH and temperature, performing the hydrolysis in two steps may complicate the process and its control may lead to a major source of error.

Only few studies did combine both stages to hydrolyze different substrates such as cassava [17], potato starch [18] and wheat starch [19]. All these studies showed the convenience of the single-step method for starch hydrolysis since it could facilitate a more efficient use of energy and time, as well as a higher efficiency of starch conversion. However, there is no research that considers the potential use of this shorter, economical and simultaneous hydrolysis of wasted bread taking advantage of the fact that this substrate could also avoid the cost of starch isolation, purification and gelatinization The statistical optimization of this simultaneous enzymatic hydrolysis of waste bread may give an idea to adapt this procedure more easily to a large or industrial scale and the possibility of getting a glucose syrup in only one stage.

This study aims to analyze the possibility of performing hydrolysis of wasted bread starch in one step and to compare it to the traditional two-step process. We hypothesized to reduce costs and time and improve the yield with the new alternative process. Two different and optimized hydrolysis processes were compared and evaluated to get as much glucose content as possible. One was the classic sequential hydrolysis where each phase was carried out separately. The other was carried out simultaneously, i.e., both steps were implemented under the same working conditions at the same time. We optimized the pH and temperature conditions of each enzyme in both processes with a robust response surface design, during 2 and 4 h for the one-step hydrolysis, and 4 h for the two-step hydrolysis. The solution viscosity and the reducing sugar and glucose contents were analysed. We also modelled the evolution of these parameters over time.

2. Materials and Methods

2.1. Materials

Bread

La Tahona de Sahagún (Sahagún, Spain), a local bakery, provided the wasted wheat bread. Each bread weighted ~250 g and its formulation (per 100 g of wheat flour) was 1.8% salt, 4% fresh yeast, 55% water, and 2% improver. The bread was cut into small pieces, dried at room temperature for three days, and milled in an LM 3100 hammer mill (Perten Instruments, Huddinge, Sweden) with a sieve of 1 mm. The bread flour was stored in a plastic box at ambient temperature (22 to 25 °C) for a maximum of 3 months. The protein, fat, ash, moisture content was analysed in the discarded bread (fat: 0.36 ± 0.03%; protein: 10.65 ± 1.36%; moisture: 11.10 ± 0.21% and ash: 1.95 ± 0.06%).

2.2. Enzymatic Hydrolysis

2.2.1. Enzymatic Sequential Hydrolysis: Liquefaction (LIQ)

To optimize the LIQ conditions a response surface design was used (Table 1) with two independent variables (pH (4.09–6.91) and temperature (43.8–86.2 °C)). These ranges were chosen as they encompassed the optimal values provided by Novozyme JSC (Bagsværd, Denmark). Bread flour was mixed with distilled water (20 g flour/80 mL of water) in a 100-mL Erlenmeyer flask. 1 M NaOH or HCl were used to adjust each flask to a specific pH at the beginning of the hydrolysis. LIQ was conducted by adding the maximum standard dose of α-amylase (0.0179 mL/100 g flour, Liquoflow® Yield, Novozyme JSC, Bagsværd, Denmark) according to its datasheet. Immediately after, the slurry was left for 120 min in a shaking water bath (70 rpm) at the temperature indicated in Table 1. A sample of the mixture was analysed before adding the enzyme (control sample) and during LIQ (30–60–90–120 min).

Table 1. Central composite design and responses for the study of pH and temperature in the sequential enzymatic hydrolysis.

Run	Experimental Factors		Responses						
			Liquefaction				Saccharification		
	pH	T	RS_{E120}	RS_{T120}	RS_{mi}	V_{mi}	GL_{E120}	GL_{T120}	GL_{mi}
1	4.50	50.0	76.34	72.67	0.89	122.65	81.80	83.06	0.81
2	6.50	50.0	62.74	62.19	0.67	249.75	57.02	56.72	0.48
3	4.50	80.0	98.58	98.20	1.08	937.05	89.72	88.34	2.69
4	6.50	80.0	91.33	85.95	1.70	626.06	35.76	37.47	0.58
5	4.09	65.0	64.68	64.40	0.67	314.21	141.20	137.30	1.57
6	6.91	65.0	88.62	88.97	1.03	1783.97	41.99	38.59	0.56
7	5.50	43.8	56.66	56.15	0.48	151.14	63.16	63.00	0.53
8	5.50	86.2	102.52	99.69	1.34	515.48	29.37	30.26	0.47
9	5.50	65.0	98.48	95.03	1.24	364.82	108.95	104.94	1.08
10	5.50	65.0	84.05	85.27	0.96	340.56	124.75	121.64	1.10
11	5.50	65.0	88.09	86.79	1.07	395.04	114.62	113.37	1.32
12	5.50	65.0	82.86	80.32	0.94	609.25	109.72	105.55	1.26
13	5.50	65.0	84.50	81.29	1.00	515.81	104.76	101.15	1.11
14	5.50	65.0	85.28	80.55	0.94	346.10	126.72	123.03	1.05
15	5.50	65.0	88.72	84.72	1.08	577.47	114.35	112.26	1.38
16	5.50	65.0	89.08	85.56	1.21	368.40	116.24	112.50	1.25

T: Incubation temperature (°C); RS_{E120}: Experimental reducing sugars concentration after 120 min (g/L); RS_{T120}: Theoretical reducing sugars concentration after 120 min (g/L); RS_{mi}: Rate of reducing sugars production (g/L·min); GL_{E120}: Experimental glucose concentration after 120 min (g/L); GL_{T120}: Theoretical glucose concentration after 120 min (g/L); GL_{mi}: Rate of glucose production (g/L·min); V_{mi}: Slope of the viscosity curve (cP/min).

2.2.2. Enzymatic Sequential Hydrolysis: Saccharification (SAC)

This phase was carried out after LIQ. To optimize this phase, the response surface design of Table 1 was used. Previously, LIQ was carried out at the optimum conditions obtained in Section 3.1 (pH 5.75 and 86.2 °C) in a 100-mL Erlenmeyer flask. All the Erlenmeyer flasks were frozen (−20 °C) after LIQ. The day of the experiment, the flasks were defrosted in a water bath at 35 °C during 20 min and each flash was adjusted to a specific pH. Then, the maximum dose estimated by the glucoamylase datasheet was added to the flask (0.029 mL/100 g flour, Saczyme® Go, Novozyme JSC, Bagsværd, Denmark) and it was incubated at the temperature indicated in Table 1 at 70 rpm for 120 min. A sample of the slurry was analysed immediately after LIQ and at different times during SAC (30–60–90–120 min).

2.3. Enzymatic Simultaneous Hydrolysis (SH)

LIQ and SAC were carried out simultaneously using the same response surface design (Table 2). All the process was performed as in LIQ except that both enzymes were added simultaneously at the same enzyme concentrations used in LIQ and SAC.

Table 2. Central composite design and responses for the study of pH and temperature in the simultaneous enzymatic hydrolysis.

Run	Experimental Factors		Responses								
			Simultaneous Hydrolysis								
	pH	T	RS_{E120}	RS_{T120}	RS_{mi}	GL_{E120}	GL_{T120}	GL_{mi}	GL_{E240}	GL_{T240}	V_{mi}
1	4.50	50.0	127.85	125.46	1.33	80.66	80.74	1.13	138.19	127.27	140.77
2	6.50	50.0	106.37	105.97	1.16	40.77	39.80	0.46	70.01	69.28	157.54
3	4.50	80.0	143.49	135.89	3.01	78.28	79.11	1.61	90.42	85.45	827.42
4	6.50	80.0	86.59	85.30	1.17	16.19	15.94	0.22	15.65	15.43	634.58
5	4.09	65.0	204.58	197.17	2.00	115.46	113.18	1.40	154.77	145.68	621.11
6	6.91	65.0	113.87	113.78	1.46	30.93	30.22	0.30	35.29	34.95	6173.47
7	5.50	43.8	110.20	108.77	1.30	47.69	49.32	0.62	87.58	85.88	53.15
8	5.50	86.2	122.89	118.45	2.05	27.43	28.02	0.42	37.53	35.49	597.52
9	5.50	65.0	219.88	213.48	2.64	96.06	93.30	1.31	135.71	126.86	1.59×10^6
10	5.50	65.0	226.33	220.43	2.50	115.97	108.44	1.31	132.97	127.05	2246.65
11	5.50	65.0	225.44	227.54	2.46	101.51	97.73	1.17	130.87	129.83	391.60
12	5.50	65.0	206.92	204.93	2.82	100.92	96.12	1.33	143.48	127.58	862.77
13	5.50	65.0	206.11	205.13	2.58	96.74	92.04	1.46	148.40	130.38	489.84
14	5.50	65.0	200.12	203.61	2.32	108.73	105.86	1.54	133.27	126.59	370.87
15	5.50	65.0	201.16	198.65	2.30	84.24	93.28	1.18	138.72	132.11	146,197.85
16	5.50	65.0	228.73	227.13	2.44	99.54	95.10	1.53	140.49	121.89	724.20

T: Incubation temperature (°C); RS_{E120}: Experimental reducing sugars concentration after 120 min (g/L); RS_{T120}: Theoretical reducing sugars concentration after 120 min (g/L); RS_{mi}: Rate of reducing sugars production (g/L·min); GL_{E120}: Experimental glucose concentration after 120 min (g/L); GL_{T120}: Theoretical glucose concentration after 120 min (g/L); GL_{mi}: Rate of glucose production (g/L·min); GL_{E240}: Experimental glucose concentration after 240 min (g/L); GL_{T240}: Theoretical glucose concentration after 240 min (g/L)

2.4. Analytical Methods

2.4.1. Determination of Reducing Sugars (RS)

The concentration of RS during LIQ and SH was determined in triplicate according to the DNS method [20].

2.4.2. Viscosity Analysis

A mixture of bread flour and distilled water (6.25 g flour/25 mL of water) was prepared and the enzymes were added at the same concentration used in LIQ (α-amylase) and SH (α-amylase and glucoamylase). The solution viscosity was measured using a Rapid Viscoanalyser (RVA-4) (Newport Scientific model 4-SA, Warriewood, Australia). The mixture of bread flour was submitted to the viscosity analysis imitating the conditions tested (pH and temperature) in the sixteen trials of LIQ (160 rpm for 60 min) and SH (160 rpm for 120 min). The equipment provides a curve showing the viscosity of the solution over time.

2.4.3. Glucose Analysis (GL)

The concentration of GL was determined using a D-glucose assay kit (K-FRUGL 04-18, Megazyme, Bray, Ireland) and a 96-well microplate spectrophotometer (Multiskan Go, Thermo Scientific, Waltham, MA, USA). Samples of slurry during SAC and SH were analysed in triplicate. GL was also measured in samples at 180 and 240 min after adding the enzymes in SH process.

2.5. Modelling and Experimental Design

The kinetics of concentration of RS and GL, and viscosity were modelled using a non-linear regression with a parameter estimation by the Marquardt method. The following logistic growth model was used to fit the kinetics of RS and GL concentration:

$$y = \frac{a}{1 + be^{-ct}} \tag{1}$$

where y is the dependent variable (concentration of RS or GL) to be modelled; a, b, and c are the regression coefficients; and t is the incubation time. RS_{T120} and GL_{T120} (Tables 1 and 2) were calculated substituting 120 min for t in the above equation. The production rate of RS (RS_{mi}) and GL (GL_{mi}) (Tables 1 and 2) were estimated with this equivalence:

$$mi = f(t_i) = \frac{ac}{4} \tag{2}$$

The viscosity was fitted using this formula:

$$y = \frac{a}{1 + bt^c} \tag{3}$$

The value of the maximum slope (V_{mi}) (Tables 1 and 2), calculated as the derivative of the function at the inflexion point, is:

$$mi = f(t_i) = \left(\frac{-ab(c+1)^2}{2c}\right)\left(\frac{c-1}{b(c+1)}\right)^{c-1/c} \tag{4}$$

where the value of the function at 120 min was determined substituting 120 min for t in the above equation. Statistical package SAS 9.4 (Statistical Analysis System, Cary, NC, USA) was used for modelling. The model fits were very good in all cases with a R^2 value between 0.88–0.99.

A central composite design (2^2 + star) of sixteen randomized assays (Tables 1 and 2) was used to optimize the starch hydrolysis. Each factor (temperature and pH) varied at five levels, and eight centre points for replication were included. The ranges of pH and temperature were chosen following the enzyme technical sheets. The response variables measured in LIQ were the experimental and theoretical RS concentration at 120 min (RS_{E120} and RS_{T120}, respectively), the rate of RS production (RS_{mi}) and the slope of the viscosity curve (V_{mi}); and in the SAC phase, the experimental and theoretical GL concentration at 120 min (GL_{E120} and GL_{T120}, respectively) and the rate of GL production (GL_{mi}). These seven variables and the experimental and theoretical GL values at 240 min (GL_{E240} and GL_{T240}) were measured in SH. The theoretical maximum values were calculated by substituting the optimal pH and temperature values in the mathematical equations obtained from the design of experiments. The optimization was performed using Statgraphics Plus V5.1 software (StatPoint Technologies, Warrenton, VA, USA).

3. Results and Discussion

3.1. Optimization of Liquefaction (LIQ)

Evolution curves of the RS concentration and viscosity during liquefaction (LIQ) were modelled as described in Section 2.5. The experimental design with the different experimental and theoretical RS concentrations is presented in Table 1. There is a strong correlation between both responses so all results discussed will be referred to the theoretical data. Despite the pH and the temperature of the LIQ process, the viscosity of the sixteen mixtures drops rapidly during the first 5 min and, in less than 30 min, the curves were at viscosities close to 1 cP (data not shown). These results are very significant for the effective transport of slurry on an industrial scale and also for the improvement of the substrate-enzyme interactions [21]. These authors also observed a rapid drop in viscosity in the first 10 min during LIQ of raw wheat flour.

The viscosity reduction of wheat flour mixture due to the action of amylases is also in line with observations made with different substrates by other authors [15,22]. This effect is due to the rapid breakdown of starch chains and the creation of maltodextrins, which increase the dextrose equivalent [23,24]. As an example, Figure 1a shows this behaviour in experiment 16, as well as the modelling of the viscosity curve.

The evolution of the viscosity and the value of maximum rate in viscosity reduction (V_{mi}) were considered to observe the effect of pH and temperature on the LIQ of the bread flour mixture. The optimum values to obtain the maximum rate in V_{mi} were pH 6.91 and 69.2 °C (Table 3). The effect of temperature in V_{mi} was higher at low pH than at high pH (Figure 2a). We are not aware of studies that optimize the conditions for viscosity reduction; it is to be expected that these conditions are specific to each enzyme and substrate [25]. According to the manufacturer information, the α-amylase used in this study works best at pHs above 5.3, which coincides with that observed in the optimization. However, the optimum temperature is somewhat higher (80–95 °C). Note the smaller effect of temperature on LIQ at high pH than at low pH, and the fact that viscosity reduction is very fast in all cases.

a

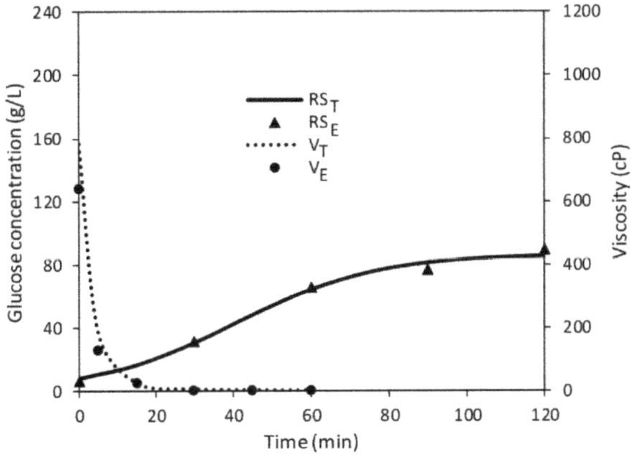

b

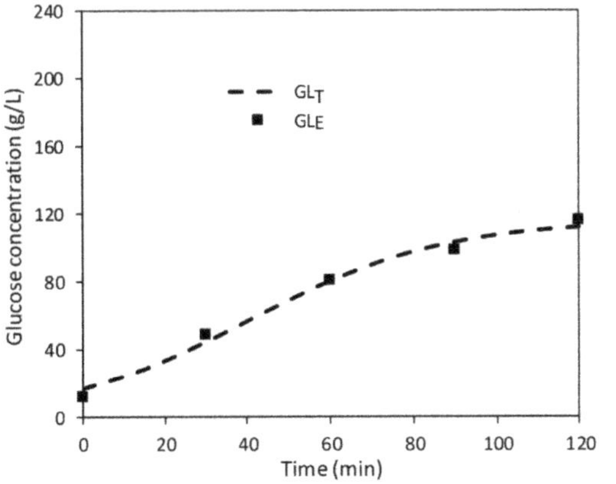

Figure 1. *Cont.*

c

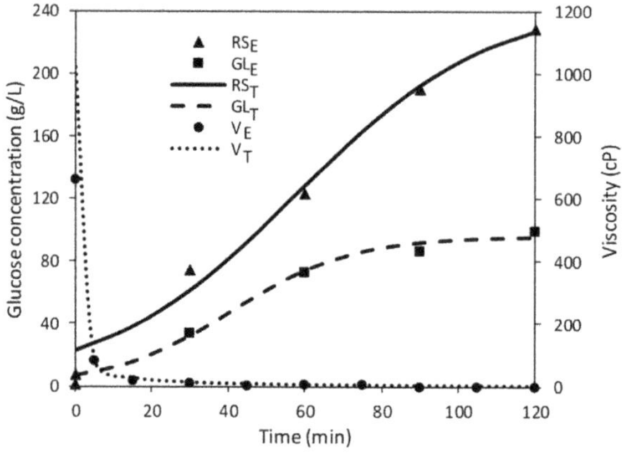

Figure 1. (**a**) Evolution of experimental and theoretical reducing sugar concentration (RS_E and RS_T, respectively) and viscosity curve (V_E and V_T) during liquefaction of trial No. 16. (**b**) Evolution of experimental and theoretical glucose concentration (GL_E and GL_T) during saccharification of trial No. 16. (**c**) Evolution of RS_E, RS_T, GL_E, GL_T and V_T and V_E during simultaneous hydrolysis of the trial No. 16.

Table 3. Optimal pH and temperature for each response in the sequential and simultaneous enzymatic hydrolysis.

Responses	Optimal Conditions			Maximum Theoretical Response
	pH	T	R^2	
Liquefaction				
RS_{E120}	5.49	86.2	88.75	103.00
RS_{T120}	5.24	86.2	85.33	99.73
RS_{mi}	6.91	86.2	85.12	1.92
V_{mi}	6.91	69.2	62.73	1288.46
Saccharification				
GL_{E120}	4.39	65.1	94.48	130.10
GL_{T120}	4.38	64.8	93.97	126.92
GL_{mi}	4.09	86.2	78.09	2.63
Simultaneous hydrolysis				
RS_{E120}	5.15	65.4	94.23	219.52
RS_{T120}	5.10	65.7	94.61	216.73
RS_{mi}	4.63	77.5	92.30	2.80
GL_{E120}	4.47	64.7	96.16	114.19
GL_{T120}	4.51	64.7	98.13	112.11
GL_{mi}	4.28	69.3	92.51	1.63
GL_{E240}	4.59	61.0	98.55	158.32
GL_{T240}	4.61	60.9	99.05	146.84
V_{mi}	6.91	67.04	54.56	4055.02

R^2: Coefficient of determination of the response surface design; T: Incubation temperature (°C); RS_{E120}: Experimental reducing sugars concentration after 120 min (g/L); RS_{T120}: Theoretical reducing sugars concentration after 120 min (g/L); RS_{mi}: Rate of reducing sugars production (g/L·min); GL_{E120}: Experimental glucose concentration after 120 min (g/L); GL_{T120}: Theoretical glucose concentration after 120 min (g/L); GL_{mi}: Rate of glucose concentration (g/L·min); GL_{E240}: Experimental glucose concentration after 240 min (g/L); GL_{T240}: Theoretical glucose concentration after 240 min (g/L); V_{mi}: Slope of the viscosity curve (cP/min).

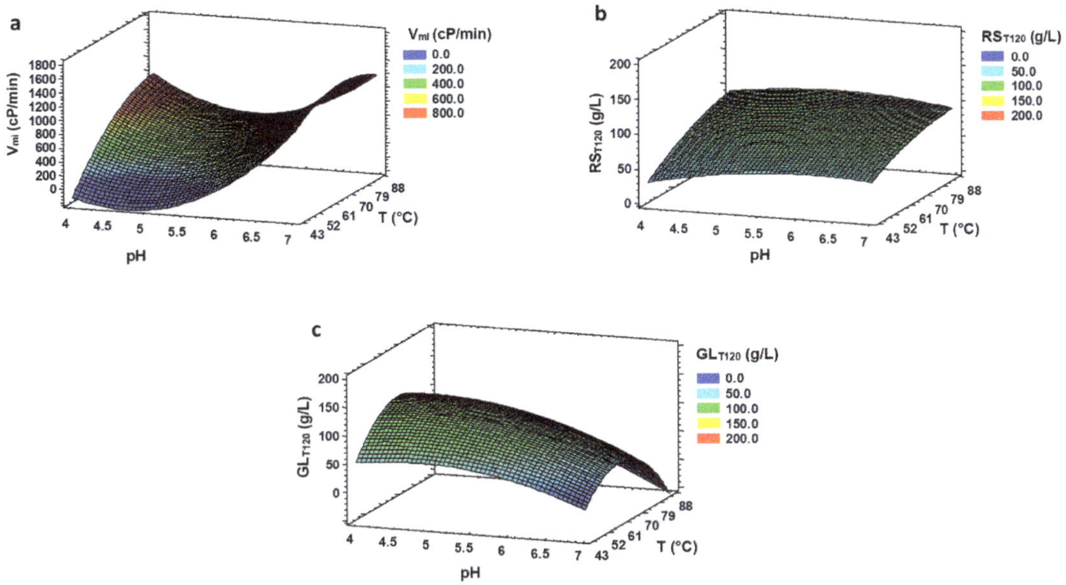

Figure 2. Response surface plots for the sequential hydrolysis. LIQ and SAC phases as function of temperature and pH for (**a**) slope of the viscosity curve (V_{mi}), (**b**) theoretical reducing sugar concentration at 120 min (RS_{T120}) and (**c**) theoretical glucose concentration at 120 min (GL_{T120}).

The maximum concentration of RS achieved was 102.52 g/L at intermediate values of pH (5.5) and the highest temperature (86.2 °C), as shown in Table 1. This value of RS is a bit higher than that observed in other similar studies [16,26]. These differences could be explained by the different hydrolysis conditions (enzyme dose and substrate concentration) or the type of substrate used. In assessing this response, we also considered the RS production rate (RS_{mi}) obtained with the model. The highest rate (1.70 g/L·min) was reached at high temperatures (80.0 °C) and pHs (6.50). However, our main aim was to have the largest amount of RS in a specific time in our slurry. Thus, we contemplated the optimum conditions to obtained RS theoretically after 2 h (RS_{T120}) as the optimal LIQ conditions, achieving R^2 values higher than 85.3 (Table 3). Figure 2b shows that low pH and temperature values harmed RS production. According to Ebrahimi et al. [13] also reported that lower LIQ temperature caused a decrease in the hydrolysis. These results are closely linked to the optimal enzyme conditions that depend on each enzyme. In general, amylase enzymes need high temperatures as they are designed to act on starches that have been previously gelatinized by heating. In the case of RS_{mi}, the optimum values are also obtained at the highest temperature but with higher pH values, which agrees with results obtained for V_{mi} (Table 3).

Overall, evaluating the best conditions provided by viscosity and RS, we decided to take the optimal conditions from the RS response. Thus, these conditions are set in pH 5.24 and a temperature of 86.2 °C, where maximum production of RS and excellent viscosity reduction are obtained. According to Demirci et al. [15] a high optimal LIQ temperature could be explained because this temperature makes it easier for enzymes to attack starch granules. However, we must take into consideration that we are working with bread flour when the starch is pre-gelatinized (almost) completely [27–29]. Moreover, this temperature may be due to the chosen enzyme, whose recommended range of action is between 80–85 °C (data provided by Novozyme). The optimum value of pH (5.24) was very close to the natural pH of mixture (5.75 ± 0.20), observing no statistically significant differences ($p > 0.05$) between RS_{T120} at pH 5.24 and 5.75 at 86.2 °C (99.72 ± 11.40 g/L and 98.60 ± 11.44 g/L, respectively). Hence, we decided not to modify the pH of mixture in order to save energy, reactive and time. These optimal conditions (natural pH and 86.2 °C) were used in the process of LIQ before all SAC experiments with glucoamylase were carried out.

3.2. Optimization of Saccharification (SAC)

The initial concentration of GL after LIQ was 12.31 ± 1.6 g/L (Figure 1b), whereas the final GL production ranged from 29.37 g/L for trial 8 (86.2 °C, pH 5.50) to 141.20 g/L, according to the conditions of trial 5 (65 °C, pH 4.09) (Table 1). Our maximum is much higher than that obtained in studies with waste bread [16,30] although in this study a slightly higher bread:liquid ratio (20%, *w*/*v*) was used. The use of the predictive models enabled the calculation of optimal sets of conditions under which the theoretical maximal values could be attained as shown in Table 3.

As in the previous step, we decided to consider GL_{T120} for both factors (pH and temperature), with optimal values of 4.38 and 64.8 °C (Table 3). These values are in line with those used in similar works [22,31,32]. As shown in Figure 2c, an increase in pH from the lowest (4.09) to the highest pH studied (6.91) at 65 °C (optimal) causes a significant decrease in GL concentration of 61%. Amaral-Fonseca et al. [33] also found that a pH equal or greater than 7.0 reduced the action of glucoamylase. In terms of temperature, higher GL values were obtained at intermediate temperatures within the pH range studied. At pH 4.38 (optimum), the GL concentration was 50% lower when decreasing from 65 °C to 43.8 °C and when increasing from 65 °C to 86.2 °C. On the one hand, lowering the temperature reduces enzyme activity, as enzymes are activated with increasing temperature. On the other hand, as the temperature increases, the enzyme starts to denature due to the effect of the temperature and, consequently, its activity is reduced [34]. In general, both pH and temperature conditions depend on the type of glucoamylase used.

Maximum rate of GL production (GL_{mi}) was observed at pH 4.09, value very similar to that found for RS concentration (4.38). However, the optimum temperature (86.2 °C) was significantly higher than that observed for RS concentration (64.8 °C) (Table 3). These results agree with those observed for RS concentration where the use of high temperatures for a long time could cause a denaturalization of the enzyme. It is important to note that the determination of GL_{mi} takes place at the beginning of the reaction where the protein denaturalization process is low.

3.3. Optimization of Simultaneous Hydrolysis (SH)

Viscosity drop and RS concentration were analysed after 2 h of enzymatic hydrolysis, and GL was performed at 4 h to compare the results with the sequential process of LIQ (2 h) and SAC (2 h).

The optimum conditions for viscosity descent rate (V_{mi}) were observed at pH 6.91 and 67 °C (Table 3 and Figure 3a). These conditions are similar to those obtained in LIQ, which shows that V_{mi} was not influenced by the presence of glucoamylase. All the viscosity curves dropped rapidly, reaching values below 70 cP from the initial ones (between 400–950 cP) in 15 min. Figure 1c illustrates this trend in trial No. 16. It is therefore not a problem in any of the pH or temperature conditions studied, and V_{mi} was not used for the optimization process.

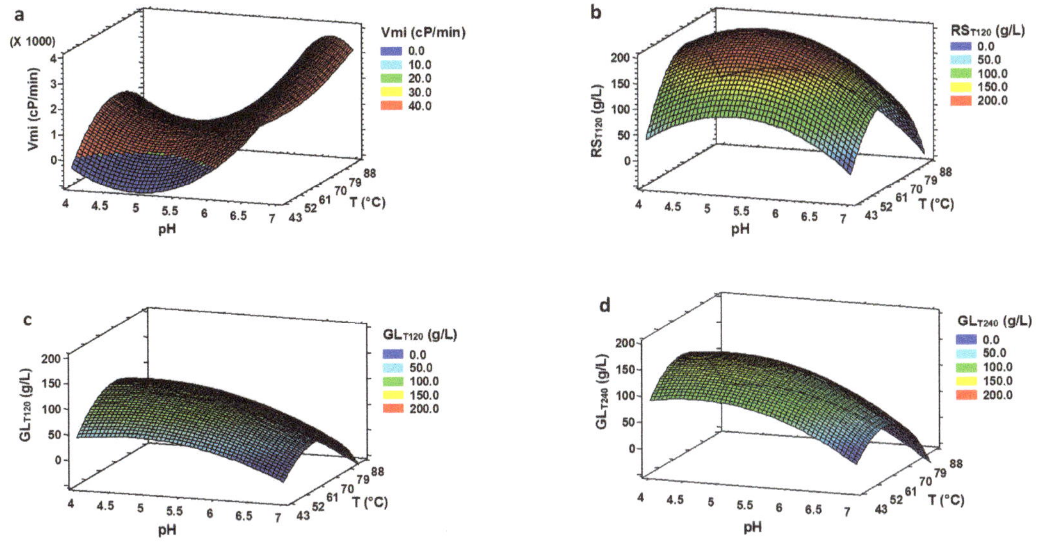

Figure 3. Response surface plots for simultaneous hydrolysis phase as function of temperature and pH for (**a**) slope of viscosity curve (V_{mi}), (**b**) theoretical RS concentration at 120 min (RS_{T120}), (**c**) theoretical glucose concentration at 120 min (GL_{T120}) and (**d**) theoretical glucose concentration at 240 min (GL_{T240}).

Up to 228 g/L of RS were obtained in SH, 81% more than in the LIQ process during the same time (Table 2). Similar results was found by Fujii et al. [35] in the SAC process from starch granules because the RS measurement of the SH includes the GL generated by glucoamylase. Some authors observed that there was a synergistic action of α-amylase and glucoamylase enzymes during hydrolysis [35–38]. However, other authors remark a negative effect when both enzymes are added together in high concentrations [15] or an interaction through complex kinetics more than a simple synergy [39]. The effect of pH and temperature on theoretical RS concentration can be seen in Figure 3b. RS production was optimized at pH 5.1 and 65.7 °C. The optimum temperature has changed from the LIQ phase (86.2 °C) but is similar to the optimal of the SAC (64.8 °C) (Table 3). For RS_{mi}, the optimum pH value obtained (4.63) is much lower than that obtained with LIQ (6.91), and similar to the optimum value of GL_{mi} for SAC (4.09). These behaviours could be caused by the presence of glucoamylase in the SH process which is more active at temperatures of ~65 °C and pH of ~4.3 (Table 3). There is also a decrease in the optimum temperature (77.5 °C) versus that obtained in LIQ (86.2 °C). In general, it seems that the optimal conditions for viscosity reduction and RS increase resemble more those of SAC than to those of LIQ. This may indicate a lesser influence of the optimal conditions on the action of α-amylases than on those of glucoamylases.

Regarding GL_{T120}, the maximum amount of GL was 113 g/L in experiment 5 (pH 4.09, 65 °C) (Table 2). Sarbatly [13] only got 65 g/L of GL at 60 °C during 30 h but with less substrate loading of cassava starch. A difference of only 15% of production was found between the theoretical maximum response of SAC (126.92 g/L) and SH (112.11 g/L) under optimal conditions (Table 3), emphasizing the suitability of the SH to obtain a high-glucose slurry. Słomińska et al. [14], using potato starch, obtained a slightly higher yield of GL using a single-stage method than the two-stage one during 48 h. Differences with our study may originate in the substrate used and the times analysed. The optimum pH and temperature conditions for the theoretical GL concentration obtained were pH 4.51 at 64.7 °C. These values are similar to the optimum conditions for the SAC process (pH 4.38, 64.8 °C). Słomińska et al. [14] achieved the maximum amount of GL at the optimal temperature of

60 °C, like our optimal, for both SH and SAC in the two-stage method. Thus, although the optimum pH and temperature conditions are different for α-amylase (86.2 °C, pH 5.24), this enzyme can cause the starch depolymerisation required by glucoamylase. Both α-amylase and glucoamylase can work together in balance at 64.7 °C and pH 4.5 without significantly decreasing the efficiency of the process. The wide temperature range of thermostable α-amylase allows it to remain active at 20 °C below its optimum temperature in LIQ [36].

GL concentration at 4 h was also measured to compare sequential hydrolysis with SH during the same time. The optimal conditions after extending the HS time were pH 4.61 at 61.0 °C for both experimental and theoretical responses. These results are slightly different from that obtained in SAC or HS at 2 h. However, all three surface plots show the same tendency. Figures 2c and 3c,d show that higher pHs and temperatures reduced GL production. The decrease in GL production was more intense at the highest temperature (86.2 °C) at 4 h (65.0%) than at 2 h (54.9%) of HS and SAC (53.3%), compared to the optimal temperature for each response. The GL production (GL_{mi}) at the highest pH (6.91) was only ~30.0% in HS at 2 and 4 h and 39.0% in SAC, compared to the optimal pH for each response.

Comparing GL concentration in SH at 4 h (Table 2) with GL concentration in SAC (Table 1), we observed that in almost all the experiments a higher amount of GL was achieved at 4 h. Under optimal conditions of pH and temperature, 15.7% more GL concentration was obtained in SH at 4 h (146.84 g/L) than in sequential process (126.92 g/L). The evolution of GL concentration in SH from 2 to 4 h was very irregular, remaining stable in some curves and increasing more than 70% in others (Table 2) with a light increase from 112.1 g/L (at 2 h) to 146.8 g/L (at 4 h) under optimal conditions (Table 3). To optimize the process parameters, the objective was fixed to maximize the GL concentration and minimize the hydrolysis time. Thus, we chose pH 4.51 and 64.7 °C as the final optimal conditions for the HS.

4. Conclusions

This work aims to propose a simultaneous LIQ and SAC method for the enzymatic hydrolysis of bread flour to produce a glucose-rich product. Under optimal conditions of pH and temperature, a slightly lower concentration of GL was obtained using the simultaneous process at 2 h (112.1 g/L) in comparison with the traditional enzymatic hydrolysis that takes 4 h of operation (126.9 g/L). However, using the simultaneous process at 4 h, the GL concentration obtained (146.8 g/L) was substantially higher than the classic method. These results confirm the potential and the advantages of the simultaneous process, namely, the reduction of time, labour, energy, equipment and other costs required for starch isolation, purification and gelatinization. Moreover, bread flour is a perfect substrate for this process since the starch is almost gelatinized and, no pre-treatment is needed.

Author Contributions: T.S.-A.: Data curation, Investigation, Formal analysis, Methodology, Resources, Writing—original draft, Writing—review & editing. V.P.: Methodology, Validation, Data curation and Supervision. M.G.: Conceptualization, Validation, Funding acquisition, Investigation, Methodology, Resources, Supervision, Writing—original draft, Writing—review & editing. J.M.R.-N.: Conceptualization, Validation, Investigation, Methodology, Resources, Supervision, Writing—original draft, Writing—review & editing. All authors have read and agreed to the published version of the manuscript.

Funding: The authors are grateful to Junta de Castilla y León (VA177P20), Spain and, FEDER-Interreg Spain-Portugal programme (TRANSCOLAB, 0612_TRANS_CO_LAB_2_P).

Institutional Review Board Statement: Not applicable.

Informed Consent Statement: Not applicable.

Data Availability Statement: The data presented in this study are available on request from the corresponding author.

Conflicts of Interest: The authors declare no conflict of interest with respect to the work described in this manuscript.

References

1. FAO Food Wastage Footprint: Impacts on Natural Resources—Summary Report. Available online: http://www.fao.org/3/a-i3347e.pdf (accessed on 22 April 2021).
2. Melikoglu, M.; Webb, C. Use of waste bread to produce fermentation products. In *Food Industry Wastes*; Maria Kosseva, C.W., Ed.; Elsevier: Amsterdam, The Netherlands, 2013; pp. 63–76, ISBN 9780123919212.
3. European Comission Directive 2008/98/EC of the European Parliament and of the Council of 19 November 2008 on Waste and Repealing Certain Directives (Text with EEA Relevance). Available online: http://data.europa.eu/eli/dir/2008/98/oj (accessed on 17 May 2021).
4. Gélinas, P.; McKinnon, C.M.; Pelletier, M. Sourdough-type bread from waste bread crumb. *Food Microbiol.* **1999**, *16*, 37–43. [CrossRef]
5. Luo, S.; Koksel, F. Physical and technofunctional properties of yellow pea flour and bread crumb mixtures processed with low moisture extrusion cooking. *J. Food Sci.* **2020**, *85*, 2688–2698. [CrossRef]
6. Samray, M.N.; Masatcioglu, T.M.; Koksel, H. Bread crumbs extrudates: A new approach for reducing bread waste. *J. Cereal Sci.* **2019**, *85*, 130–136. [CrossRef]
7. Martínez, M.M.; Gómez, M. Rheological and microstructural evolution of the most common gluten-free flours and starches during bread fermentation and baking. *J. Food Eng.* **2017**, *197*, 78–86. [CrossRef]
8. Dura, A.; Rosell, C.M. Enzymes in Baking. *Microb. Enzym. Technol. Food Appl.* **2017**, 295–314. [CrossRef]
9. Immonen, M.; Maina, N.H.; Wang, Y.; Coda, R.; Katina, K. Waste bread recycling as a baking ingredient by tailored lactic acid fermentation. *Int. J. Food Microbiol.* **2020**, *327*, 108652. [CrossRef]
10. Rosa-Sibakov, N.; Sorsamäki, L.; Immonen, M.; Nihtilä, H.; Maina, N.H.; Siika-aho, M.; Katina, K.; Nordlund, E. Functionality and economic feasibility of enzymatically hydrolyzed waste bread as a sugar replacer in wheat bread making. *J. Food Process. Preserv.* **2022**, *46*, e16378. [CrossRef]
11. Verni, M.; Minisci, A.; Convertino, S.; Nionelli, L.; Rizzello, C.G. Wasted bread as substrate for the cultivation of starters for the food industry. *Front. Microbiol.* **2020**, *11*. [CrossRef]
12. Nionelli, L.; Wang, Y.; Pontonio, E.; Immonen, M.; Rizzello, C.G.; Maina, H.N.; Katina, K.; Coda, R. Antifungal effect of bioprocessed surplus bread as ingredient for bread-making: Identification of active compounds and impact on shelf-life. *Food Control* **2020**, *118*. [CrossRef]
13. Ebrahimi, F.; Khanahmadi, M.; Roodpeyma, S.; Taherzadeh, M. Ethanol production from bread residues. *Biomass Bioenergy* **2008**, *32*, 333–337. [CrossRef]
14. Torabi, S.; Satari, B.; Hassan-Beygi, S.R. Process optimization for dilute acid and enzymatic hydrolysis of waste wheat bread and its effect on aflatoxin fate and ethanol production. *Biomass Conv. Bioref.* **2020**, *11*, 2617–2625. [CrossRef]
15. Sükrü Demirci, A.; Palabıyık, I.; Gümüs, T.; Özalp, Ş. Waste bread as a biomass source: Optimization of enzymatic hydrolysis and relation between rheological behavior and glucose yield. *Waste Biomass Valorization* **2017**, *8*, 775–782. [CrossRef]
16. Hudečková, H.; Šupinová, P.; Babák, L. Optimization of enzymatic hydrolysis of waste bread before fermentation. *Acta Univ. Agric. Silvic. Mendel. Brun.* **2017**, *65*, 35–40. [CrossRef]
17. Sarbatly, R. The simultineous enzymatic hydrolysis of tapioca starch for instant formation of glucose. *J. Appl. Sci.* **2007**, *15*, 2057–2062. [CrossRef]
18. Słomińska, L.; Zielonka, R.; Jarosławski, L. The unconventional single stage hydrolysis of potato starch. *Pol. J. Chem. Technol.* **2013**, *15*, 7–14. [CrossRef]
19. Legin, E.; Copinet, A.; Duchiron, F. A single step high temperature hydrolysis of wheat starch. *Starch Stärke* **1998**, *50*, 84–89. [CrossRef]
20. Miller, G.L. Use of dinitrosalicylic acid reagent for determination of reducing sugar. *Anal. Chem.* **1959**, *31*, 426–428. [CrossRef]
21. Montesinos, T.; Navarro, J.M. Production of alcohol from raw wheat flour by amyloglucosidase and Saccharomyces cerevisiae. *Enzyme Microb. Technol.* **2000**, *27*, 362–370. [CrossRef]
22. Zhang, C.; Kang, X.; Wang, F.; Tian, Y.; Liu, T.; Su, Y.; Qian, T.; Zhang, Y. Valorization of food waste for cost-effective reducing sugar recovery in a two-stage enzymatic hydrolysis platform. *Energy* **2020**, *208*, 118379. [CrossRef]
23. Leman, P.; Goesaert, H.; Delcour, J.A. Residual amylopectin structures of amylase-treated wheat starch slurries reflect amylase mode of action. *Food Hydrocoll.* **2009**, *23*, 153–164. [CrossRef]
24. Martinez, M.M.; Gómez, M. Insight of the α-amylase family of enzymes. *Microb. Enzym. Technol. Food Appl.* **2020**, 21–45. [CrossRef]
25. Robyt, J.F. Enzymes and Their Action on Starch. In *Starch*; BeMiller, J., Whistler, R., Eds.; Elsevier: Amsterdam, The Netherlands, 2009; pp. 237–292, ISBN 9780127462752.
26. Sirohi, R.; Pandey, J.P.; Goel, R.; Singh, A.; Lohani, U.C.; Kumar, A. Two-stage enzymatic hydrolysis for fermentable sugars production from damaged wheat grain starch with sequential process optimization and reaction kinetics. *Starch Stärke* **2021**, *73*, 2000082. [CrossRef]

27. Fernández-Peláez, J.; Guerra, P.; Gallego, C.; Gomez, M. Physical properties of flours obtained from wasted bread crusts and crumbs. *Foods* **2021**, *10*, 282. [CrossRef]
28. Martínez, M.M.; Román, L.; Gómez, M. Implications of hydration depletion in the in vitro starch digestibility of white bread crumb and crust. *Food Chem.* **2018**, *239*, 295–303. [CrossRef]
29. Primo-Martín, C.; van Nieuwenhuijzen, N.H.; Hamer, R.J.; van Vliet, T. Crystallinity changes in wheat starch during the bread-making process: Starch crystallinity in the bread crust. *J. Cereal Sci.* **2007**, *45*, 219–226. [CrossRef]
30. Han, W.; Liu, W.-X.; Yu, C.-M.; Huang, J.-G.; Tang, J.-H.; Li, Y.-F. BioH2 production from waste bread using a two-stage process of enzymatic hydrolysis and dark fermentation. *Int. J. Hydrogen Energy* **2017**, *42*, 29929–29934. [CrossRef]
31. Govindasamy, S.; Campanella, O.H.; Oates, C.G. Influence of extrusion variables on subsequent saccharification behaviour of sago starch. *Food Chem.* **1995**, *54*, 289–296. [CrossRef]
32. Rendón-Villalobos, J.R.; Solorza-Feria, J.; Aguilar-Sandoval, A. Optimisation of conditions for glucose syrup production from banana (Musa paradisiaca L.) pulp using response surface methodology. *Int. J. Food Sci. Technol.* **2011**, *46*, 739–745. [CrossRef]
33. Amaral-Fonseca, M.; Morellon-Sterling, R.; Fernández-Lafuente, R.; Tardioli, P.W. Optimization of simultaneous saccharification and isomerization of dextrin to high fructose syrup using a mixture of immobilized amyloglucosidase and glucose isomerase. *Catal. Today* **2021**, *362*, 175–183. [CrossRef]
34. Brandam, C.; Meyer, X.M.; Proth, J.; Strehaiano, P.; Pingaud, H. An original kinetic model for the enzymatic hydrolysis of starch during mashing. *Biochem. Eng. J.* **2003**, *13*, 43–52. [CrossRef]
35. Fujii, M.; Homma, T.; Taniguchi, M. Synergism of α-amylase and glucoamylase on hydrolysis of native starch granules. *Biotechnol. Bioeng.* **1988**, *32*, 910–915. [CrossRef]
36. Guzmán-Maldonado, H.; Paredes-López, O.; Biliaderis, C.G.; Maldonado-Guzmán, H.; López-Paredes, O.; Biliaderis, G.C. Amylolytic enzymes and products derived from starch: A review. *Crit. Rev. Food Sci. Nutr.* **1995**, *35*, 373–403. [CrossRef]
37. Wong, D.W.S.; Robertson, G.H.; Lee, C.C.; Wagschal, K. Synergistic action of recombinant α-amylase and glucoamylase on the hydrolysis of starch granules. *Protein J.* **2007**, *26*, 159–164. [CrossRef]
38. Zhang, B.; Dhital, S.; Gidley, M.J. Synergistic and antagonistic effects of α-amylase and amyloglucosidase on starch digestion. *Biomacromolecules* **2013**, *14*, 1945–1954. [CrossRef]
39. Liu, D.; Zhang, H.; Xu, B.; Tan, J. Development of kinetic model structures for glutinous rice saccharification by different enzymes. *J. Food Process Eng.* **2014**, *37*, 317–328. [CrossRef]

foods

Article

Coffee Leaf Tea from El Salvador: On-Site Production Considering Influences of Processing on Chemical Composition

Marc C. Steger [1,2], Marina Rigling [1], Patrik Blumenthal [2], Valerie Segatz [3,4], Andrès Quintanilla-Belucci [5], Julia M. Beisel [3], Jörg Rieke-Zapp [6], Steffen Schwarz [2], Dirk W. Lachenmeier [3,*] and Yanyan Zhang [1,*]

1 Department of Flavor Chemistry, University of Hohenheim, Fruwirthstr. 12, Verfügungsgebäude 221, 70599 Stuttgart, Germany
2 Coffee Consulate, Hans-Thoma-Strasse 20, 68163 Mannheim, Germany
3 Chemisches und Veterinäruntersuchungsamt (CVUA), Weissenburger Strasse 3, 76187 Karlsruhe, Germany
4 Hochschule für Angewandte Wissenschaften Coburg, Friedrich-Streib-Strasse 2, 96450 Coburg, Germany
5 Finca La Buena Esperanza, Pasaje Senda Florida Norte 124, San Salvador, El Salvador
6 Rubiacea Research and Development GmbH, Hans-Thoma-Strasse 20, 68163 Mannheim, Germany
* Correspondence: lachenmeier@web.de (D.W.L.); yanyan.zhang@uni-hohenheim.de (Y.Z.); Tel.: +49-721-926-5434 (D.W.L.); +49-711-459-24871 (Y.Z.)

Abstract: The production of coffee leaf tea (*Coffea arabica*) in El Salvador and the influences of processing steps on non-volatile compounds and volatile aroma-active compounds were investigated. The tea was produced according to the process steps of conventional tea (*Camellia sinensis*) with the available possibilities on the farm. Influencing factors were the leaf type (old, young, yellow, shoots), processing (blending, cutting, rolling, freezing, steaming), drying (sun drying, oven drying, roasting) and fermentation (wild, yeast, *Lactobacillus*). Subsequently, the samples were analysed for the maximum levels of caffeine, chlorogenic acid, and epigallocatechin gallate permitted by the European Commission. The caffeine content ranged between 0.37–1.33 g/100 g dry mass (DM), the chlorogenic acid was between not detectable and 9.35 g/100 g DM and epigallocatechin gallate could not be detected at all. Furthermore, water content, essential oil, ash content, total polyphenols, total catechins, organic acids, and trigonelline were determined. Gas chromatography—mass spectrometry—olfactometry and calculation of the odour activity values (OAVs) were carried out to determine the main aroma-active compounds, which are β-ionone (honey-like, OAV 132-927), decanal (citrus-like, floral, OAV 14-301), α-ionone (floral, OAV 30-100), (*E*,*Z*)-2,6-nonadienal (cucumber-like, OAV 18-256), 2,4-nonadienal (melon-like, OAV 2-18), octanal (fruity, OAV 7-23), (*E*)-2 nonenal (citrus-like, OAV 1-11), hexanal (grassy, OAV 1-10), and 4-heptenal (green, OAV 1-9). The data obtained in this study may help to adjust process parameters directly to consumer preferences and allow coffee farmers to earn an extra income from this by-product.

Keywords: coffee leaf tea; novel food; coffee by-products; *Coffea arabica*; caffeine; epigallocatechin gallate

1. Introduction

Coffee is one of the most consumed beverages in the world. Worldwide, more than 10 million tons are produced each year with a turnover of more than 20 billion US dollars [1]. However, the volatile world coffee market and climate change put pressure on the farmers to keep their farms in good condition. The use of coffee by-products, i.e., products that are created during the production of coffee, could help to create increased income along with increased sustainability [2]. One of these coffee by-products is the coffee leaf, which has been approved as a tea beverage in the EU since July 2020, with maximum permitted quantities of 80 mg/L for caffeine, of 100 mg/L for chlorogenic acid, and of 700 mg/L for epigallocatechin gallate [3]. Coffee leaves are usually produced as a by-product during the pruning of the plants [4].

Coffee leaves from the coffee plant are typically light green (buds and young leaves) to dark green (matured leaves) with a size range of 15 cm (*Coffea arabica*) up to 50 cm (*Coffea liberica*). The lifetime of a leaf is about 7–10 months [5]. Considerable evidence suggests that the leaves of the coffee plant have long been used as a traditional food in the countries where it is grown. Von Pröpper observed in 1882 [6] that "The leaves of the coffee plant, roasted and poured over with hot water, make an excellent tea, which has long been one of the staple foods of the entire Indian archipelago, and is said to be not inferior in effect to the true Chinese tea, but apparently has not yet come into commerce" (authors' translation from German). Further mentioned in the literature are the countries Ethiopia, West Sumatra, Jamaica, Java, and South Sudan [7,8]. Novita et al. [9] described the traditional production of "Kahwa daun", a herbal tea from coffee leaves produced in West Sumatra. Herby branches with leaves were clasped on a bamboo stick and then smoked or dried over a cooking fire. In Indonesia, the infusion of coffee leaves is called "copi daon" or "leaf coffee" and in Ethiopia it is called "Quti"; in both countries the leaves are sun dried [10]. Consumption of teas in general may be associated with beneficial health effects, but in most countries, it is consumed just for the taste or the effects of caffeine [11–13]. Coffee leaves also contain nutritionally interesting compounds, including carbohydrates, amino acids, organic acids, alkaloids, phenolic compounds, terpenes, carotenoids, phytosterols and flavour compounds such as aldehydes, alcohols, ketones, and esters [14–17]. The chemical composition of a leaf is highly influenced by the light intensity, nitrogen concentration of the soil, age of the plant and the leaf, growing region and the coffee species [16].

Novel foods or traditional foods from third countries require an authorization to be placed on the market in the European Union (EU) [18]. The application to authorise the placing on the market of an infusion of coffee leaves of the species *Coffea arabica* and/or *Coffea canephora* as a traditional food was approved by the EU Commission on 1 July 2020. Since the applicant could not provide evidence of the use of coffee leaves as an ingredient in other beverages, only the infusion of coffee leaves was approved as a novel food. Critical values have been set for the substances caffeine, chlorogenic acid and epigallocatechin gallate [3].

In this study, the possibility of producing coffee leaf tea in a country of origin were investigated aiming to provide coffee farmers with easily applicable procedures. With the locally available resources, different coffee tea samples were produced. Furthermore, the effects of leaf types, coffee varieties, processing, and drying methods on consumer acceptance, the aroma profile and the product characteristics were studied. This understanding can later help to optimally adjust process and manufacturing parameters to the desired taste. For the analyses of non-volatiles, near-infrared spectroscopy (NIR), high-performance liquid chromatography (HPLC) and nuclear magnetic resonance (NMR) spectroscopy were used. The water content, essential oil, ash content, caffeine, polyphenol content, catechins, organic acids, trigonelline and lactic acid were determined. Afterwards, the samples were sensorically evaluated by a panel according to DIN 10,809 [19], followed by aroma analysis using gas chromatography-olfactometry.

2. Materials and Methods

2.1. Coffee Leaf Tea Production in El Salvador

The collected leaves are shown in Table 1. The leaf types, their variety, and their collection place are given. Finca La Palma is located in Chinameca, San Miguel and Finca La Quintanilla is located on the north side of Cacahuatique mountain in Morazan, El Salvador. Furthermore, different leaf types are shown in Figure 1.

The harvest of the leaves was conducted between February and March 2021. Old leaves were cut directly from the plant on the field. About five leaves were harvested per plant. All yellow leaves of a plant were picked directly from the branches. Shoots were completely cut off from the plants and then divided into old and young leaves. All leaves were cleaned with fresh water before further processing. For each sample, 600 g of fresh leaves were picked. Typically, about 5–10 leaves were picked per plant, so that about 1000 different plants were picked per variety. For determination of the water content of fresh leaves, one sample of each leaf type was dried in the oven at approx. 80 °C. The weight was determined before drying (approx. 2 g) and after drying with a precision balance. Each experiment was carried out in triplicates.

Table 1. Leaf types, their variety, and their collection place.

Leaf Type	*Coffea arabica* Variety	Collection Place
Old leaf	Pacamara	Finca La Palma
Yellow leaf	Pacamara	Finca La Palma
Old leaf	Bourbon Tekisic	Finca La Quintanilla
Young leaf	Bourbon Tekisic	Finca La Quintanilla
Shoot	Bourbon Tekisic	Finca La Quintanilla

Figure 1. From left to right: whole shoots, young leaves, old leaves (Bourbon Tekisic).

Different processing steps were applied to obtain a variety of different teas. These steps were related to the processing, drying and fermentation of the leaves. All leaves underwent a withering process where the leaves were stored on drying beds for 12 h overnight at 20 °C. Subsequently, the leaves were either dried whole or processed by various mechanical methods. The methods are shown in Table 2. Furthermore, the processing steps are shown in Figure 2.

Table 2. Processing steps and their explanation.

Processing	Explanation
None (whole leaf)	No further mechanical intervention.
Blending	The leaves were blended in a kitchen blender. For 100 g leaves, around 400 mL tap water was added.
Cutting	Leaves were cut with a kitchen knife to small strips (20 mm wide).
Rolling	The leaves were rolled by hand.
Freezing	The leaves were frozen in a freezer at −20 °C for 2 days.
Crumbling	Leaves were crumbled by hand.
Steaming	Leaves were steamed in a 50 L pot. A sieve was placed in the centre of the pot and approximately 2 L of tap water was boiled under the leaves. The temperature was measured at the lid of the pot. The process was stopped when the temperature reached around 100 °C.

(a) (b) (c) (d)

Figure 2. (**a**) Leaves after blending. (**b**) Leaves in the steaming pot. (**c**) Leaves directly after crumbling. (**d**) Crumbled, whole and cut leaves on the drying bed.

2.1.1. Drying Methods

Three different drying methods were carried out for the samples, namely sun/air drying, oven drying and roasting. For sun drying, the leaves were stored on a drying bed for at least 48 h until they were crispy. The oven drying was performed at 70 °C in a gas oven with circulating air for 4 h. The roasting was done on a gas stove until the leaves were crispy. The sun drying process is shown in Figure 3.

Figure 3. Sun drying of the different samples on the drying bed in Chinameca.

2.1.2. Fermentation

For pre-cultivation of the microorganisms (*Saccharomyces cerevisiae* var. bayanus, *Lactobacillus plantarum*), around 1 g of the dry culture was dissolved in 100 mL of water, 1 h before mixing it with the samples. Following that, 40 mL starter cultures were sprayed on the leaves. The leaves were then fermented by storing in closed plastic buckets (anaerobically) for 12 h at a temperature of around 25 °C (overnight). For the wild fermentation, the leaves were stored in the buckets without adding a starter culture.

2.1.3. Postprocessing and Packaging

Samples that had not been already mixed during processing (Table 2) were brought to approximately the same sheet size with the blender. Subsequently, all samples were packed into zip bags and vacuum sealed for transport.

2.2. Analysing of Non-Volatile Compounds

2.2.1. Sample Preparation for HPLC, NMR, NIR, and Photometry

The sample preparation of the ground tea sample was carried out according to the international standard ISO 1572 [20]. In accordance with this standard, the samples were prepared using a comminution mill so that the ground material subsequently fell completely through a test sieve with a mesh size of 500 µm. For each sample, a small portion of the sample was first ground in the mill and discarded. Subsequently, the amount of sample required for further testing was ground and packed in a separate, airtight package.

2.2.2. Standard Analytical Procedures

The HPLC analysis of catechins was conducted according to a procedure previously described [21]. For the determination of total phenols, the Folin method was applied using spectrophotometry with a Lambda 35 instrument (PerkinElmer, Rodgau, Germany). For near infrared (NIR) spectroscopic measurement of water, essential oil and ash, a layer of about 1 cm of the tea sample was spread on a Petri dish to cover the entire bottom. Once spread, the sample was pressed firmly with a stamp and placed in the measuring device (Büchi NIRFlex Solids, N-500, Büchi Labortechnik AG, Flawil, Switzerland). Each measurement was performed in triplicate.

2.2.3. Nuclear Magnetic Resonance (NMR) Spectrometry

200 mg of the ground coffee leaf tea was weighed into a 15 mL centrifuge tube and 8 mL of deionised water was added. The tubes were then placed on the combination shaker and shaken on level 8 for 20 min. 2 mL of the solution was membrane filtered into a 4 mL glass vial. Sodium dihydrogen phosphate buffer, pH 6.1, and trimethylsilylpropanoic acid (TSP) were brought to room temperature and 70 µL each of TSP and 100 µL of buffer were pipetted into an empty vial. Then, 600 µL of the membrane-filtered sample was pipetted into each of these prefilled vials. The solution was homogenised before 600 µL of each was pipetted into an NMR tube. The NMR tubes were finally sealed with a lid and a spinner for the NMR instrument. The measurements were performed according to a previously described procedure developed for cold brew coffee [22].

2.3. Preliminary Sensory Analysis

Due to COVID-19 pandemic-related contact restrictions during the research period in 2021, only a preliminary sensory analysis was possible using a limited number of tasters. The 24 tea samples from El Salvador of the Bourbon and Pacamara coffee varieties were tasted by a trained test panel composed of 7 people with experience in tea tasting. All respondents have consented to participation in the study. The tasting was done at room temperature with cupping spoons. The main questions of the tasting are: (i) Which teas exhibit the highest popularity among testers? (ii) What flavour profiles do each of the top eight tea samples exhibit?

The coffee leaf tea is prepared according to DIN 10,809 [19] in infusion vessels. After 5 min, the tea is poured off into the bowl and can be tasted after a short cooling period.

The individual samples were rated according to the personal preference of the tasters with values from 0 to 5 (0 = dislikes very badly 5 = likes very well). The 8 highest scoring samples are brewed for a simple descriptive test, a profile test and a rank order test.

The tasted "best eight" were then evaluated in a ranking test by all participants. The samples are ranked from 1 to 8 according to personal preference (1 = best/8 = worst). Multiple assignment of numbers is excluded in this test. A selection must be made even if there are only slight differences (forced-choice test).

The test material was tested for the perceptual attributes of colour, odour, and flavour. Participants were free to add other properties. The individual results were then shared. The terms were collected and either accepted or rejected by the testers. A minimum of 50% agreement was required to define a term.

Finally, the given characteristics (sweet, salty, sour, bitter, body and the dwell time of the taste (finish)) were described with values ranging from 0 (absent) to 5 (strongly expressed).

2.4. *Analysis of Volatile, Odour-Active Compounds*

2.4.1. Tea Preparation

To prepare the tea infusions, 2 g of tea were filled into cellulose bags. These were infused with 200 mL of boiling distilled water in a beaker and infused for 5 min. The tea bag was then removed and the samples were frozen in aluminium bottles at −20 °C.

2.4.2. Direct-Immersion Stir Bar Sorptive Extraction (DI-SBSE)

To extract the compounds from the tea, DI-SBSE technique was applied. Therefore, 10 mL of tea and 3.3 g of NaCl and 0.05 mL of thymol standard (4.2 mg/L) were transferred to a headspace vial (20 mL). The mixture was stirred by a Twister (PDMS) with 1000 rpm for 2 h at room temperature. The Twister was then taken out and rinsed with distilled water and dried off with a lint-free tissue. Afterwards, the Twister was placed in the autosampler of the gas chromatograph. Each Twister was conditioned for 1 h at 250 °C after use. Each measurement was done in triplicate.

2.4.3. Gas Chromatography

Gas chromatography (GC) was performed according to Rigling et al. [23]. In short, an Agilent 7890 B gas chromatograph connected to a 5977 B mass spectrometry detector (Agilent Technologies, Waldbronn, Germany) was equipped with thermal desorption unit (TDU), cooled injection system (CIS) as well as an olfactometry detection port (ODP 3, Gerstel, Mülheim an der Ruhr, Germany). An Agilent J&W DB-WAXms column (30 m × 0.25 mm ID × 0.25 μm film thickness) (Agilent Technologies) was installed. Helium (5.0) served as carrier gas with a constant flow rate of 1.62 mL/min. The gas flow was split 1:1 into the MS detector and the ODP using a μFlowManager Splitter (Gerstel) with a column outlet pressure of 20 kPa. The GC oven temperature was held at 40 °C (3 min), then ramped with 5 °C/min to 240 °C (10 min). The following parameters were applied: MS mode, scan; scan range, m/z 40–330; electron ionisation energy, 70 eV; source temperature, 230 °C; quadrupole temperature, 150 °C; ODP 3 transfer line temperature, 250 °C; ODP mixing chamber temperature, 150 °C; ODP 3 makeup gas, N_2 (5.0) (Westfalen). The data were collected using Gerstel ODP1 and Agilent Mass Hunter B07.06 combined with Gerstel Maestro.

Semi quantification was performed using the internal standard thymol (c = 4.2 mg/L) and the weighed-in standard solutions. The response factor of the respective substances could then be calculated using the peak areas of the standard.

2.4.4. Odour Activity Value (OAV)

To determine the odour activity value (OAV) (for details see [23]), the odour threshold of each substance was retrieved from the literature. Values above 1 indicate the possibility of sensory perception of the respective substance.

2.5. Statistical Analysis

Microsoft Excel was used to calculate means and standard deviation and for graphical illustrations. The statistical evaluation of the sensory test was performed using the Friedmann test. The calculations (One way ANOVA (confidence level $p < 0.05$) were applied using SPSS (IBM Corporation, Armonk, NY, USA). For the analytical data, the statistical evaluation was carried out using the programme Design Expert 12 (Stat-Ease, Inc., Minneapolis, MN, USA). Hereby an ANOVA for selected factorial model (confidence level $p < 0.05$) was applied. The results are presented as mean value ± standard deviation of the respective parameters.

3. Results and Discussion

3.1. Moisture Content of Fresh Leaves

The moisture content of the different types of fresh leaves is shown in Table 3. Young leaves and shoots showed the highest water content with 72.89 ± 0.99% and 72.74 ± 1.73%, respectively. Weatherley [24] described a correlation between leaf age and its water content. The water content decreased in all plants during the ageing process. Yellow leaves show the lowest water content with 56.28 ± 1.02%. This effect was to be expected as the plant tries to extract all nutrients and water from the dead leaf before dropping it [25]. To obtain sufficient quantities for the analyses, 600 g of fresh leaves were collected for each sample. Since the pruning of the plants was already done, it was important not to cut off too many leaves and especially not the fresh buds. This could lead to deterioration of the plant [26].

Table 3. Moisture content of the different coffee leaves directly after picking shown as mean ± standard deviation.

Leaf Type	Moisture Content [%]
Pacamara yellow	56.28 ± 1.02
Shoots Bourbon whole	72.74 ± 1.73
Bourbon old	62.83 ± 2.22
Bourbon young	72.89 ± 0.99

3.2. Preparation of Coffee Leaf Tea Samples

In total, 24 different samples were produced during this field study. All samples are shown in Table 4.

Table 4. Produced coffee leaf tea samples in this study including the test number, the variety, the leaf type, the processing, the drying and the fermentation process.

	Test No.	*Coffea arabica* Variety	Leaf Type	Processing	Drying	Fermentation
1	936	Pacamara	old	whole	air	none
2	324	Pacamara	yellow	whole	air	none
3	742	Pacamara	old	crumbled	air	none
4	183	Pacamara	yellow	crumbled	air	none
5	502	Pacamara	old	cutted	air	none
6	643	Pacamara	old	blended	air	none
7	842	Pacamara	old	crumbled	air	Yeast
8	234	Pacamara	old	whole	oven	none

Table 4. *Cont.*

	Test No.	*Coffea arabica* Variety	Leaf Type	Processing	Drying	Fermentation
9	238	Pacamara	old	blended	air	*Lactobacillus*
10	182	Pacamara	old	blended	air	Yeast
11	789	Pacamara	old	blended	air	Wild
12	156	Pacamara	yellow	blended	air	Wild
13	687	Pacamara	yellow	blended	air	Yeast
14	463	Pacamara	yellow	blended	air	*Lactobacillus*
15	289	Bourbon	shoot	blended	air	Wild
16	138	Bourbon	old	blended	air	Yeast
17	147	Bourbon	young	blended	air	Wild
18	392	Bourbon	young	steamed/rolled	air	none
19	305	Bourbon	young	rolled/fermented	air	Wild
20	930	Bourbon	shoot	blended/steamed	air	none
21	743	Bourbon	old	whole	air	Wild
22	901	Bourbon	old	whole/frozen	air	none
23	369	Bourbon	old	whole	roasted	none
24	220	Bourbon	shoot	whole	air	none

3.3. *Water Content, Essential Oil Content and Ash Content*

The results of the NIR analysis for water, essential oil and ash are shown in Table A1 in the Appendix A. Furthermore, the influences of the manufacturing methods are shown in Figure 4.

The water content of the different samples varied from 3.92 g/100 g to 17.42 g/100 g. Statistical analysis of the individual samples showed significant indication of an influence of the leaf type, the processing and the drying method. On average, yellow leaves show the highest water content (8.59 g/100 g), while shoots show the lowest water content (5.83 g/100 g). In processing, freezing (6.58 g/100 g), the use of the whole leaf (6.65 g/100 g) and blending (7.93 g/100 g) lead to lower water content. These values are up to 33% lower compared to the other processing steps. The different surface area plays a role here. Blended samples will dry much faster than less processed ones [27]. Furthermore, up to 46% lower water contents could be achieved during drying by roasting and by the oven. According to Arslan et al. [28], oven drying has a more than double drying rate compared to sun drying. However, the water content of sample 16 with the highest water content was significantly higher compared to the other samples, indicating an error during sun drying. Here, the sample was probably removed from the drying bed too early. Furthermore according to German guidelines [29], the water content of a tea or tea-like product must not exceed 8%. In future production, special care must be taken to ensure that the sun-drying process is not terminated too early.

The essential oil content of the leaves in this experiment varies from not detectable to 1.52 mL/100 g. The statistical analysis showed a significant influence of the variety and the drying method. Leaves from Pacamara have a significantly higher oil content compared to Bourbon. The effects of drying methods on the essential oil content have already been investigated in many previous studies for different plants. Here, gentle drying in the oven resulted in a higher oil content than drying under direct sunlight [30–32].

The ash content of the leaves showed no significant influencing factors. The values of all samples ranged between 7.81 g/100 g and 10.20 g/100 g regardless of the influences.

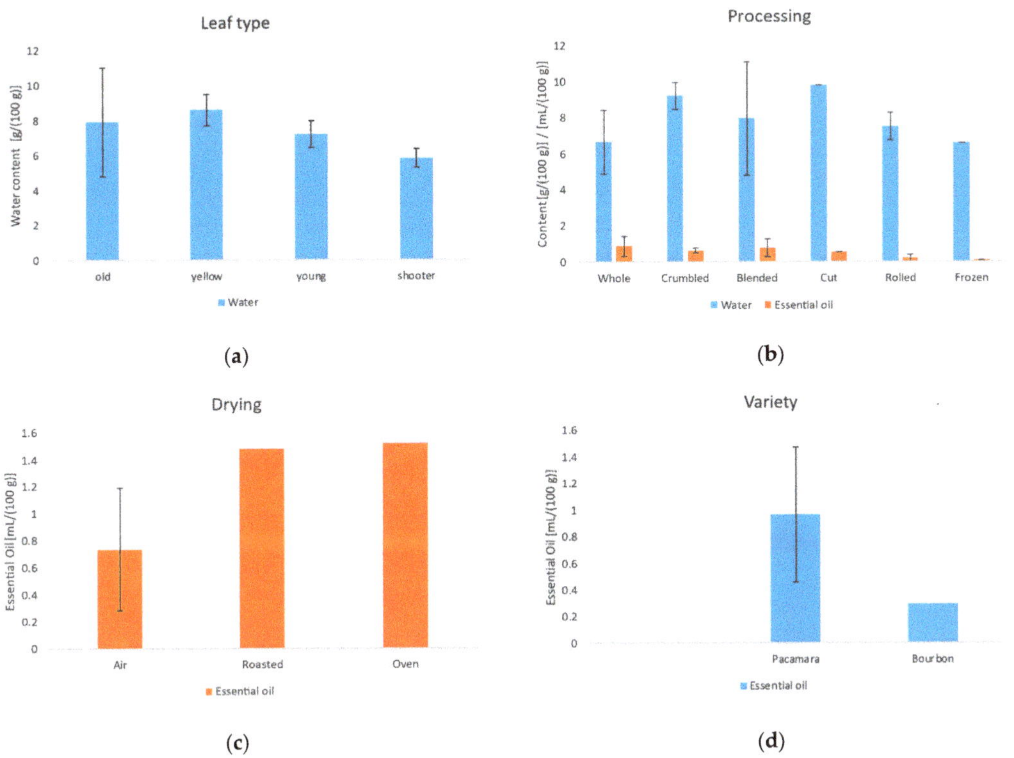

Figure 4. (**a**) Influence of the leaf type on the water content. (**b**) Influence of the processing methods on the water (g/100 g) and the essential oil content (mL/100 g). (**c**) Influence of the drying method on the essential oil content. (**d**) Influence of the variety on the essential oil content.

3.4. Content of Caffeine and Catechins

The results of the HPLC analysis for caffeine and catechins are shown in Table A2 in the Appendix A. Furthermore, the influence of the manufacturing methods on caffeine and the total catechins are shown in Figures 5 and 6.

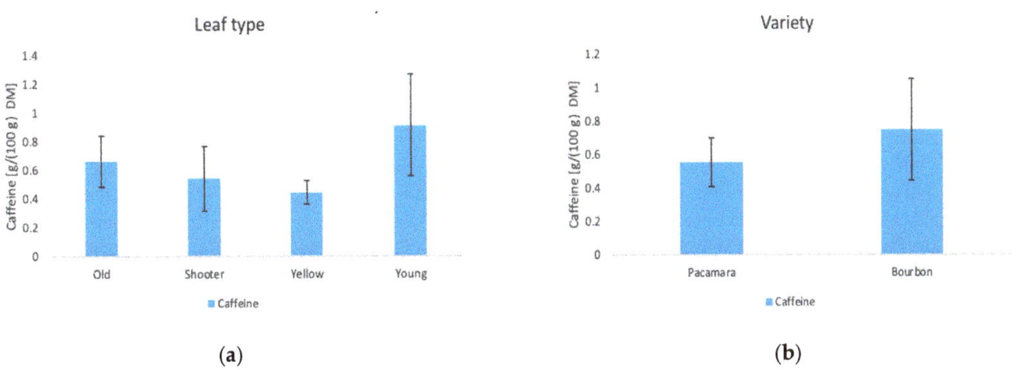

Figure 5. *Cont.*

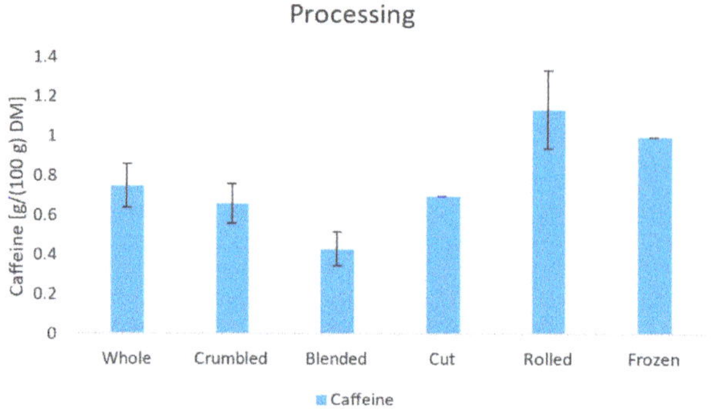

(c)

Figure 5. Influence of the leaf type (**a**), the variety (**b**), and the processing method (**c**) on the caffeine content.

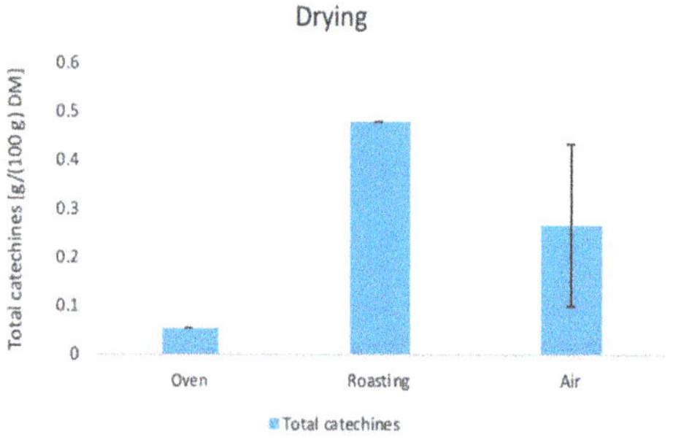

(a)

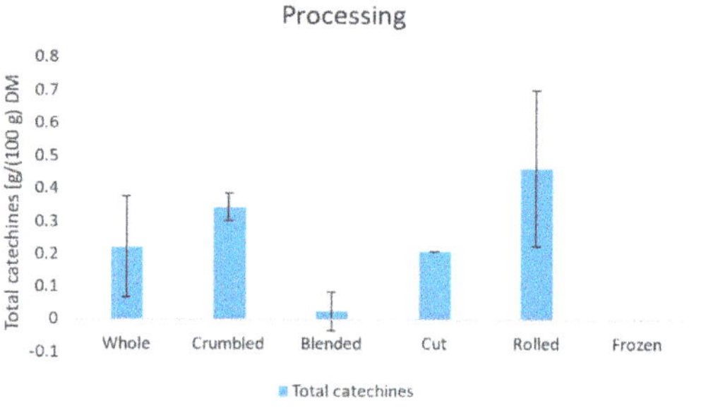

(b)

Figure 6. Influence of the drying method (**a**) and the processing method (**b**) on the total catechin content.

3.4.1. Caffeine

The caffeine content of the leaves varies between 0.37 g/100 g DM and 1.33 g/100 g DM. Here, the leaf type and the variety show significant influences on the caffeine content of the tea. Young leaves show the highest caffeine value (0.91 g/100 g DM) while yellow leaves show the lowest (0.44 g/100 g DM). The caffeine levels are approximately the same as those detected by Ratanamarno et al. [33]. The effect of caffeine reduction with leaf age has already been observed in some studies with different plants [34,35]. Song et al. [36] explained this effect mainly by the function of caffeine as a pesticide. Younger leaves of the plant must be more protected compared to old ones; therefore, the plant builds up higher concentrations in those leaves. Furthermore, the processing is a significant variable towards the caffeine content. Here, rolling (1.20 g/100 g DM) and freezing (1.00 g/100 g DM) of the leaf show the highest contents of caffeine. According to Astill et al. [37], the caffeine content decreases during the fermentation and drying stage. In case of freezing, it is possible that metabolic pathways are stopped, which result in less degradation during drying. Since the rolling process was only carried out on young leaves, further tests would be required to determine whether this influences the caffeine content. Furthermore, the low caffeine content of the blended samples may be due to the addition of water. Some of the caffeine may have been dissolved in the water during processing and then dripped off through the drying bed during drying. Theobromine, which is described in the metabolic pathway as a precursor of caffeine [38], could only be detected in small amounts in three samples.

3.4.2. Catechins

The results of the total catechin content show a clear influence of the blending process. All blended samples, with the exclusion of sample 6, have no catechins at all. According to scientific findings, this could be due to the oxidation of catechins by polyphenol oxidase to theaflavin [39]. This effect occurs during the fermentation of black tea [40]. The large surface area and added water of the mixed samples could be responsible for an enhanced enzymatic reaction. Furthermore, the drying parameters show an influence on the total catechin content. Air drying has a significantly higher average value (0.266 g/100 g DM) than oven drying (0.054 g/100 g DM) and a significantly lower value than roasting (0.479 g/100 g DM). Li et al. [41] investigated the correlation of temperature and duration of thermal processing on catechin content. Accordingly, the low content of the oven-dried sample can be attributed to the 4 h drying time. The roasted sample had a much shorter drying time (20 min), which resulted in the highest content.

The epigallocatechin gallate mentioned in the European Commission's novel food approval could not be detected in any sample. Since Ratanamarno et al. [33] have already detected epigallocatechin gallate in fresh coffee leaves, it can be concluded that it was degraded during processing, transport or storage. Turkmen et al. [42] observed an absence of epigallocatechin gallate in black tea, attributed to oxidation and fermentation processes [42]. This effect could also have occurred in the coffee leaf tea samples during the withering process.

3.5. Content of Total Polyphenols

The results of the photometric analysis of the total polyphenol content are shown in Table A3 in the Appendix A. Furthermore, the influence of the manufacturing methods is shown in Figure 7.

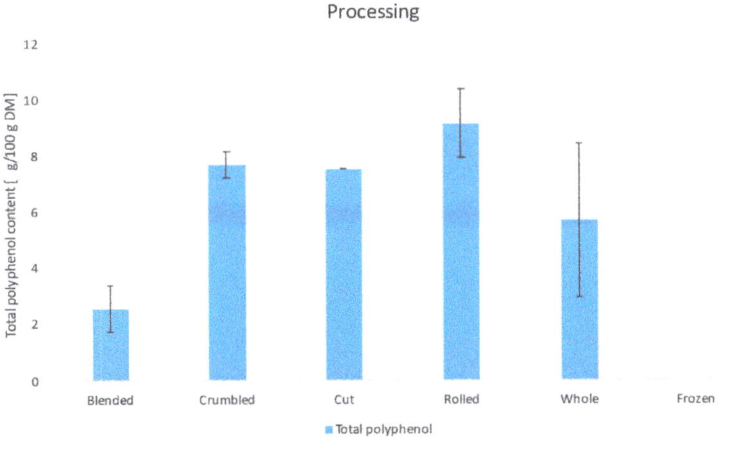

Figure 7. Total polyphenol content depending on process parameters.

The total polyphenol content varies between not detectable and 10.36 g/100 g DM and depends mainly on the processing. The blended samples show a significantly lower phenol content with 2.55 g/100 g DM compared to crumbling, cutting, rolling and whole leaves. Furthermore, no phenol could be detected in the frozen sample. The existing literature shows the influence of freezing on the polyphenol content. Oszmiański et al. [43] found a loss of up to 33.6% and Loncaric et al. [44] up to 48% after a freezing process. The low content in the blended samples can be explained by an increased oxidation process. Due to the large surface area of the blended leaves, the polyphenol oxidase can degrade the polyphenols faster compared to the other samples. Turkmen et al. [42] described a decrease in polyphenol content in fermented black tea compared to green tea. Here, the polyphenol oxidase has not been deactivated by a heat process, resulting in a loss that depends on the duration of the fermentation.

3.6. Content of Organic Acids and Trigonellin

The results of the NMR analysis for organic acids and trigonellin are shown in Table A4 in the Appendix A. Furthermore, the influence of the manufacturing methods is shown in Figures 8–11.

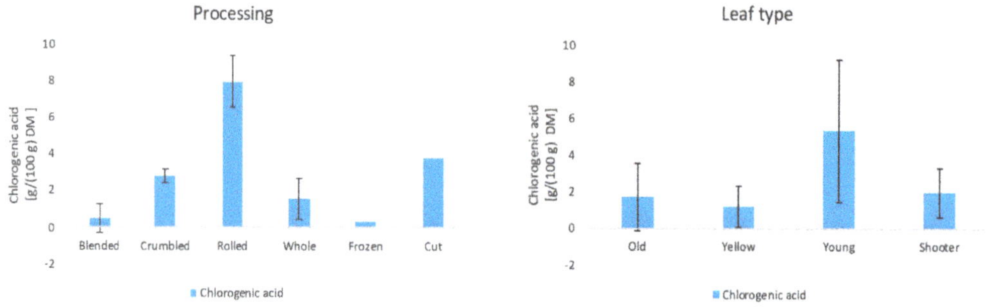

(a) (b)

Figure 8. (**a**) Influence of the processing method on the chlorogenic acid content. (**b**) Influence of the leaf type on the chlorogenic acid content.

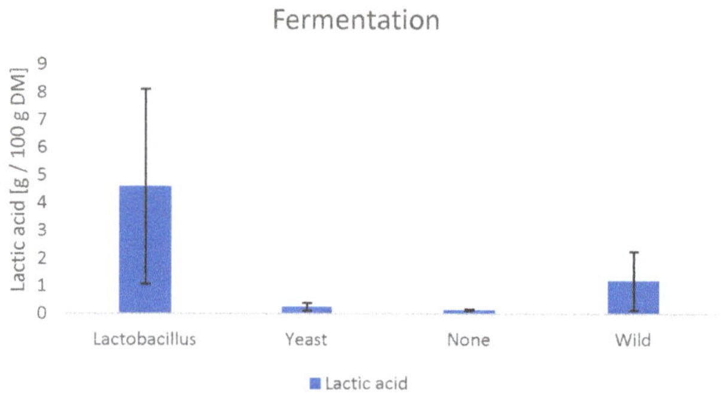

Figure 9. Influence of the fermentation on the lactic acid content.

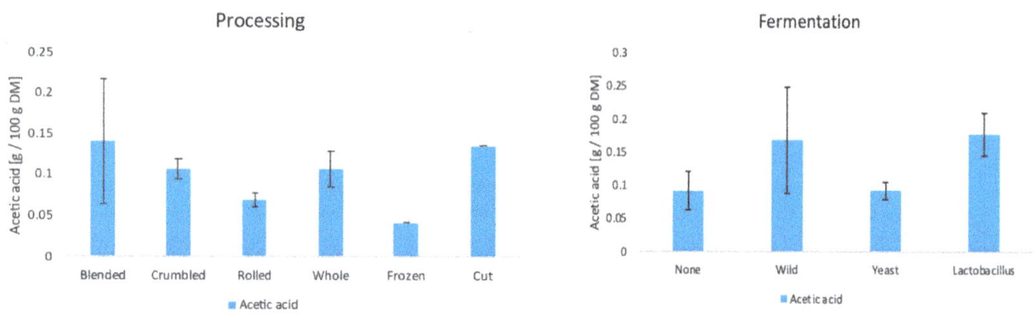

(a) (b)

Figure 10. (**a**) Influence of the processing method on the acetic acid content. (**b**) Influence of the fermentation on the acetic acid content.

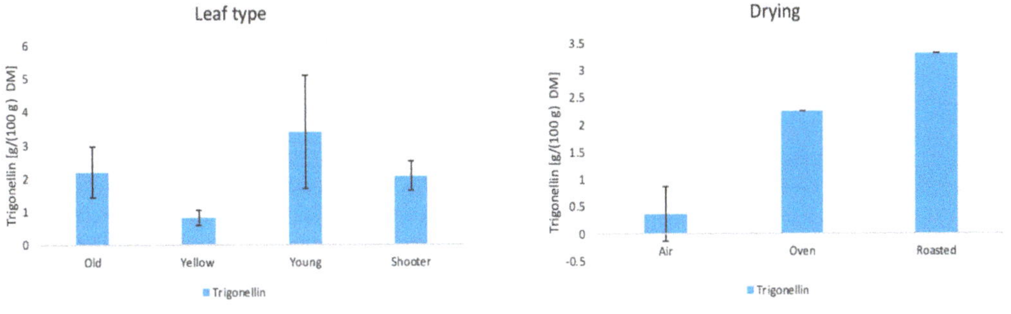

(a) (b)

Figure 11. (**a**) Influence of the leaf type on the trigonelline content. (**b**) Influence of the drying method on the trigonelline content.

3.6.1. Chlorogenic Acid

The content of chlorogenic acid (3-caffeoylquinic acid) varies between not detectable and 9.35 g/100 g DM. The significant influences here are the leaf type and the processing. The young leaves show the highest amount of chlorogenic acid with 5.33 g/100 g DM followed by shoots with 1.96 g/100 g DM, the old leaves with 1.71 g/100 g DM, and the yellow leaves with 1.21 g/100 g DM. As with caffeine, chlorogenic acid exerts a protective effect on the leaf through its antioxidant property. Therefore, here it is also present in an increased amount in the young leaves. These data also coincide with the analyses already carried out by Monteiro et al. [45]. Furthermore, it is shown that freezing (0.31 g/100 g DM) and blending (0.47 g/100 g DM) have a negative effect on the amount. The negative effect of freezing contradicts a study by Ścibisz et al. [46] where freezing had no effect.

3.6.2. Lactic Acid

Lactic acid varied from 0.11 g/100 g DM to 8.12 g/100 g DM in the samples. As expected, the samples fermented with the *Lactobacillus* showed the highest value. Furthermore, an increased amount is found in the wild fermented samples. This suggests that a certain percentage of lactic acid bacteria is present in the microbiota of the coffee leaf.

3.6.3. Acetic Acid

Acetic acid varied from 0.04 g/100 g DM to 0.29 g/100 g DM. Above all, the processing and the type of leaf influence the amount significantly. As with the other acids, freezing gives a significantly lower value (0.04 g/100 g DM) than the rest. Samples with larger surface area such as blended and cut samples show the highest values (0.14 g/100 g DM and 0.13 g/100 g DM). Samples fermented wild and with *Lactobacillus* have significantly higher values (0.17 g/100 g DM/0.18 g/100 g DM) than those fermented with yeast or not fermented (0.09 g/100 g DM). The yeast fermented samples do not differ from the non-fermented ones, indicating that this yeast strain produces few organic acids.

3.6.4. Trigonelline

Trigonelline is a substance, which is mainly found in the seeds of many plants. Quantities between 0.62 g/100 g DM and 4.87 g/100 g DM could be detected in the tea samples. The trigonelline content results in this study are up to four times higher than those found in a recent study by Monteiro et al. [45]. In the coffee beans, however, only 1–1.2 g/100 g DM are contained in the untreated state. The trigonelline content is influenced by the leaf type and the drying method. The higher content in young leaves coincides with the result of Monteiro et al. [45]. Furthermore, the roasting leads to a high trigonelline content. Zhu et al. [47] showed a similar relationship with hemp seeds where roasting had the highest influence on trigonelline levels.

3.7. Preliminary Sensory Evaluation

The following results of the sensory evaluation must be considered as being preliminary due to a restricted taste panel (n = 7).

3.7.1. Personal Acceptance

The results of the ranking test with the eight best rated teas are shown in Table 5. An evaluation of the results using the Friedmann test showed no statistical differences between the teas. Nevertheless, an improvement in flavour by yeast fermentation can be inferred by these results. The top 2 were both treated with the Anaferm yeast. For sample 369, however, a strong polarization could be detected within the panel. It was described as "the best" by two people and as "the most disliked" by two other people from the panel.

Table 5. Results of ranking test and descriptive terms for colour, odour (via orthonasal detection) and flavour (via retronasal detection) of the panel for the best 8 tea samples.

Test Number	Ranking	Colour	Odour	Flavour
687	1	sediment turbid red-brown	peach-like	peach-like
182	2	clear red	floral woody	sweet peach-like
369	3	clear yellow-brown	popcorn-like smoky roasty	popcorn-like roasty
147	4	very clear yellow-brown	chestnut flower-like floral	honey-like grassy
156	5	red clear	honey floral	honey-like acacia flower-like
234	6	clear amber	grassy rooibos-like	green bean-like vegetal
930	7	clear yellow-green	floral sweet	floral basil-like
743	8	turbid light orange	green bean-like	grassy green bean-like broccoli-like

3.7.2. Simple Descriptive Test

The results of the simple descriptive test are shown in Table 5. They show a wide range of flavours perceived by the panel. It can be recognised how the fermented samples 182, 687, 156 and 147 differ from the unfermented samples in the type of aromas. While fermentation tends to produce sweetish fruity notes, the unfermented samples tend to have green and vegetal aromas. Wang et al. [48] also found a correlation between fermentation and a loss of green flavour compounds and an increase of fruity flavours. There was also a wide difference in the various colours. Here, as shown in Figure 12, the fermented teas especially show a considerably darker colour compared to the rest. The exception is sample 147, which despite fermentation has the lightest colour of all the teas. The change of colour was also reported by Borah et al. [49]. In the study, the tea changed the colour from green to a darker copper red during the fermentation process.

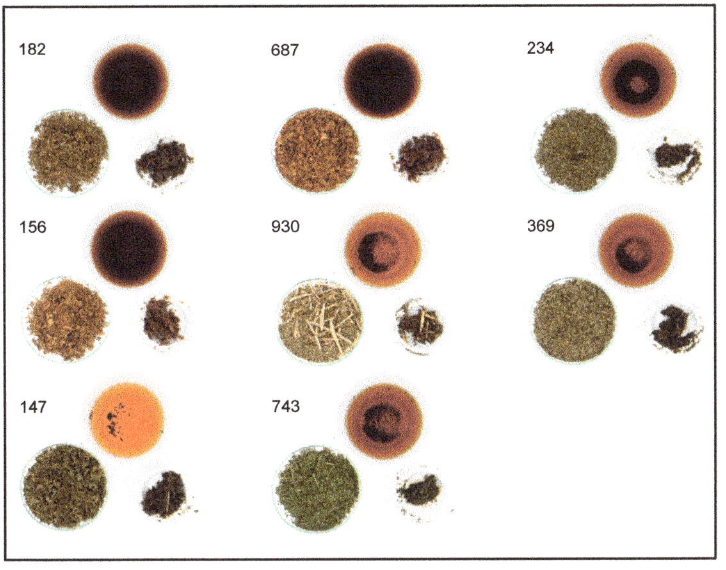

Figure 12. Brewed tea samples along with brewed and unbrewed tea leaves of the best eight teas.

3.7.3. Profile Test

The results of the profile test (Figure 13) show that the taste is mainly dominated by sweet and partly bitter impressions. Sweetness ranged from 2.8 to 4 in all samples, bitterness ranged from 1.5 to 2. Acidity was only detected with a value of 0.8 in sample 182 and saltiness could not be detected. Values of less than 0.5 were not considered in this study. This coincides with the description of Yuwono et al. [50], in which the authors describe the tea as sweetish, green and woody. The evaluation with the software programme Design Expert showed no significant influence of the different process parameters on sweetness, saltiness, bitterness, acidity, body and finish.

3.8. Aroma Analysis via DI-SBSE-GC-MS-O

3.8.1. Identification of Odour-Active Compounds

A total of 68 different olfactory impressions could be detected from the 8 samples (Table A5). Of these, 44 could be identified by mass spectrum, RI, and odour. Exemplary total ion chromatograms and mass spectra are shown in Figures A1 and A4. Of the 44 substances identified, 16 are aldehydes, 10 are ketones, 8 are alcohols, 3 are organic acids, 2 are pyrazines, 2 are ionenes, 1 is a terpene, 1 is an aromatic heterocyclic amine, and 1 is a fatty acid ester. A total of four substances were detected in all samples. These are (*E*,*E*)-3,5-octadien-2-one, 2,6-nonadienal, α-ionone and β-ionone. Pyrazines, which are typical roast aromas, could only be identified in the roasted sample 369. 4-Heptanal could only be detected in fermented samples and γ-dodecalactone just in yeast fermented samples.

The perceived odours could be described to a large extent as green and grassy (22 substances). Furthermore, some sweetish notes (11 substances) and notes in the area of melon or cucumber (11 substances) could be identified. In the area of fruity and citrus, 7 and 5 impressions, respectively, were perceived. Other attributes were roasty, herbal, honey, vanilla, aquarium, unique, nutty, forest and stable.

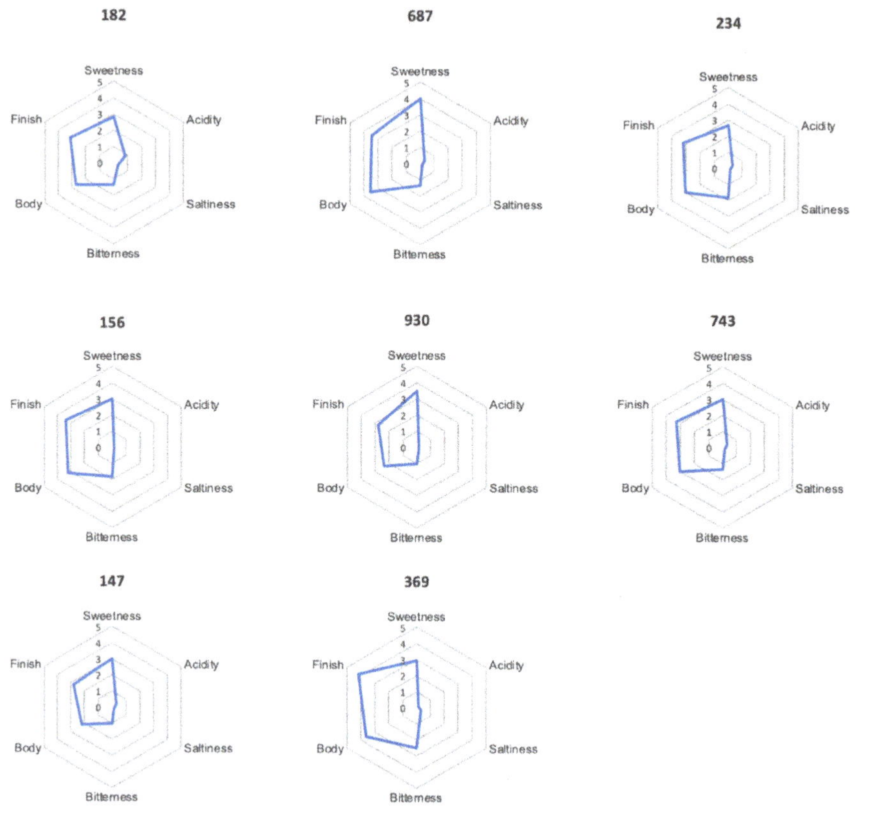

Figure 13. Graphical illustration of the results of the profile test.

3.8.2. Semi-Quantification and Odour Activity Values

The results of the semi-quantification and the calculated OAVs of each sample are shown in the Tables A6–A13 in the Appendix A. The associated OAVs can be found in Table A14 in the Appendix A. Additionally, the percentage of the OAVs in the total aroma for each sample is shown in Figures 14–21.

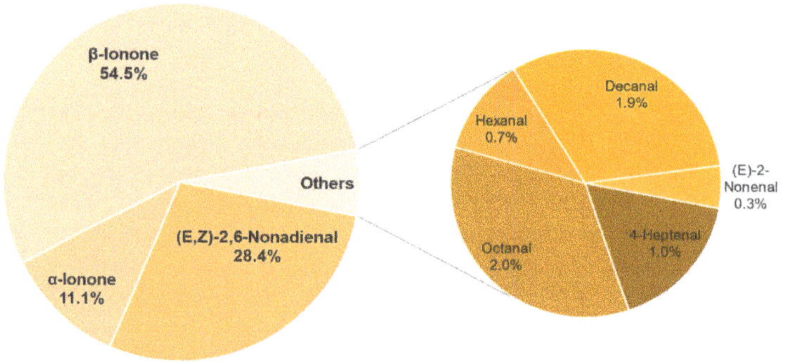

Figure 14. Percentage of OAV of each compound in the total aroma of sample 182.

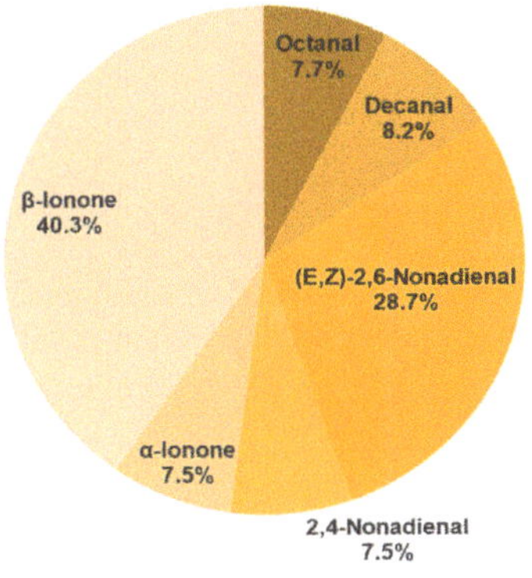

Figure 15. Percentage of OAV of each compound in the total aroma of sample 687.

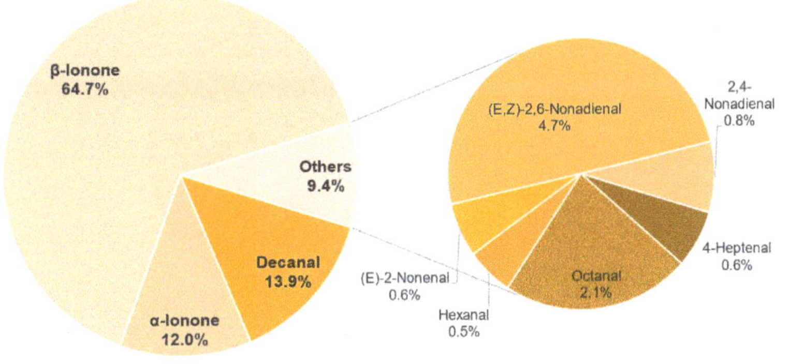

Figure 16. Percentage of OAV of each compound in the total aroma of sample 234.

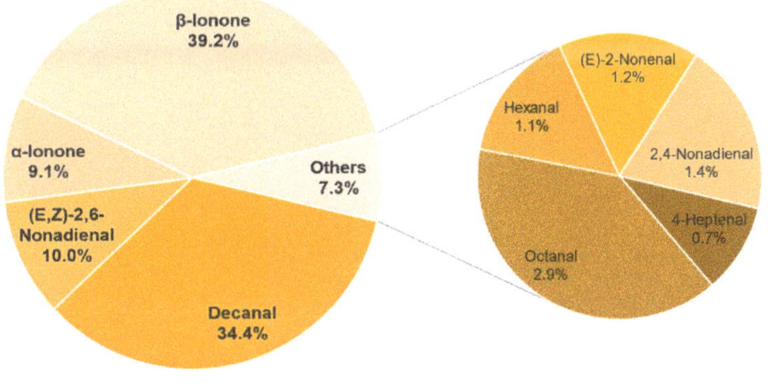

Figure 17. Percentage of OAV of each compound in the total aroma of sample 156.

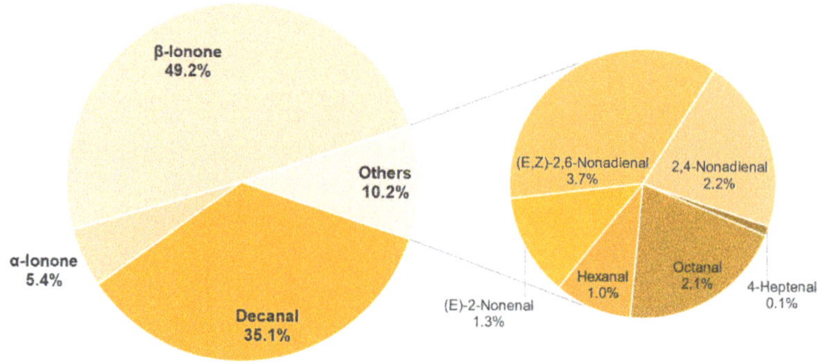

Figure 18. Percentage of OAV of each compound in the total aroma of sample 930.

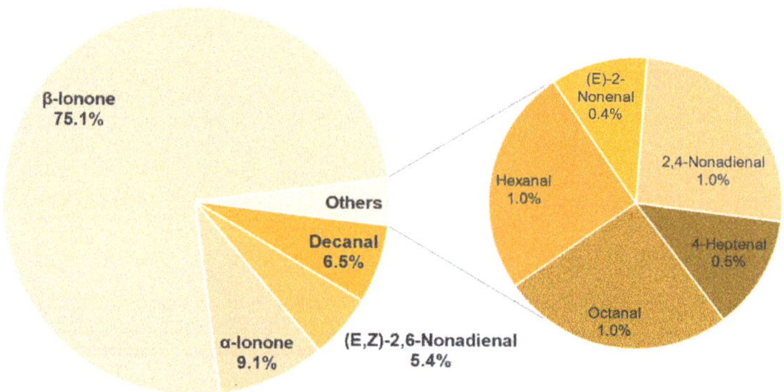

Figure 19. Percentage of OAV of each compound in the total aroma of sample 743.

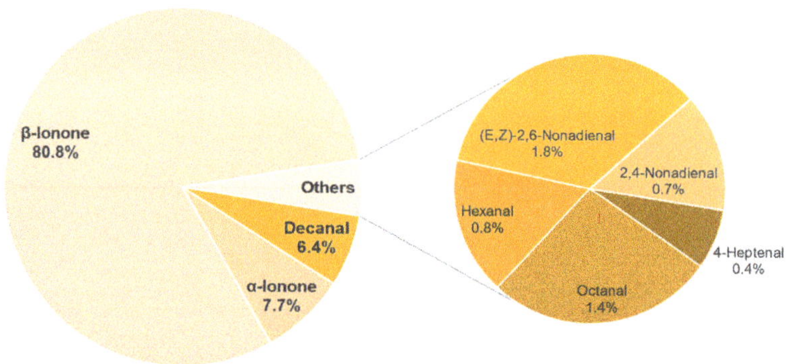

Figure 20. Percentage of OAV of each compound in the total aroma of sample 147.

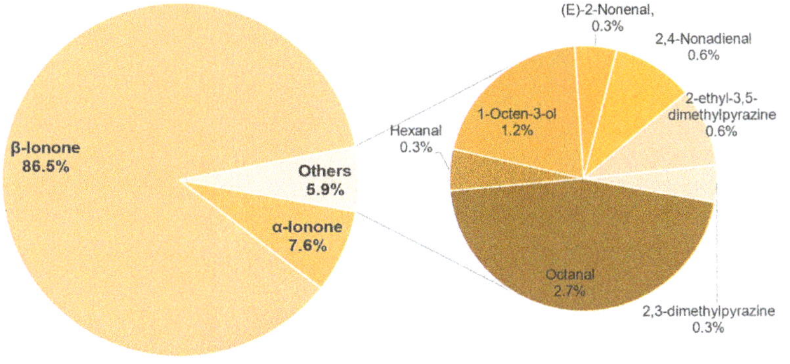

Figure 21. Percentage of OAV of each compound in the total aroma of sample 369.

Sample 182 showed a total of nine substances with an OAV greater than 1. Three of these substances, β-ionone (54.5%), (*E*,*Z*)-2,6-nonadienal (28.4%), and α-ionone (11.1%) accounting for 94.0% of the total aroma. Furthermore, the substances decanal, 4-heptenal, hexanal and (*E*)-2-nonenal are partly responsible for the aroma.

The smell of the tea was described in the previous tasting as floral and woody while the flavour was sweetish and peach via retronasal detection. The sweetness here can most likely be attributed to the β-ionone while the green and floral tones come from (*E*,*Z*)-2,6-nonadienal and α-ionone. Another role is probably played by hexanal (0.7%) which, according to Zhu and Xiao [51], is one of the key aroma compounds of peach. These, together with decanal (fruity) and octanal (citrus like) could have caused the perceived peach-like flavour.

Sample 687 shows six compounds with an OAV over 1. Here the main compounds are β-ionone (40.3%) and (*E*,*Z*)-2,6-nonadienal (28.7%). Furthermore, decanal (8.2%), octanal (7.7%) (*E*,*Z*)-2,4-nonadienal (7.5%) and α-ionone (7.5%) have a relatively large share. Overall, however, the sample showed significantly lower total concentrations than all other samples. This could be because the plant has extracted the substances from the yellow leaves before dropping them [52]. The resulting high percentage of fruity substances, such as (*E*,*Z*)-2,6-nonadienal and (*E*,*Z*)-2,4-nonadienal most likely leads to the detected fruity peach-like flavour of the tea.

A total of 9 aroma compounds with an OAV more than 1 were detected in sample 234. The main aroma substances are β-ionone (64.7%), decanal (13.9%) and α-ionone (12%). Furthermore, the substances (*E*,*Z*)-2,6-nonadienal, octanal, 2,4-nonadienal, 4–heptenal, (*E*)-2-nonenal and hexanal are partly responsible for the aroma. The strongly pronounced floral tones can therefore probably be attributed to decanal and α-ionone, together with (*E*,*Z*)-2,6-nonadienal (4.7%), hexanal (0.5%) and 4-heptanal (0.6%). The sensory tones of the green bean and the rooibos do not coincide here, however, with the aroma components of these two products described in the literature [53,54].

In sample 156, 9 aroma compounds could be identified with an OAV of more than 1. β-ionone (39.2%) and decanal (34.4%) form the main part of the aroma compounds. Furthermore, (*E*,*Z*)-2,6-nonadienal with 10% and α-ionone with 9.1% are main aroma compounds, while the substances octanal, 2,4-nonadienal, (*E*)-2-nonenal, hexanal and 4-heptenal account for the remaining 7.3%. Overall, the total concentrations of the various compounds are significantly higher compared to the comparable sample 687, which also consists of yellow leaves and was fermented. The difference here is in the type of fermentation, which in this sample was wild and uncontrolled. Microorganisms could have caused the increase in concentrations in this sample; however, this would have to be verified by further studies.

Similar to the previous sample, 9 aroma forming substances were detected in sample 930. Here, β-ionone (49.2%) and decanal (35.1%) make up the main component. Including α-ionone (5.4%), the substances (*E*,*Z*)-2,6-nonadienal, 2,4-nonadienal, octanal, (*E*)-2-nonenal, hexanal and 4-heptenal account for a total of 15.6% of the aroma. The aroma described as sweetish and floral is assumed to be composed mainly of the substances β-ionone, decanal, α-ionone, hexanal and 4-heptenal. Furthermore, no influence of the heat process (steaming) on the volatile substances can be identified for this sample.

Semi-quantification also identified 9 compounds of importance for the aroma in sample 743. β-Ionone and α-ionone are the main substances here with 75.1% and 9.1%. Furthermore, decanal (6.5%), (*E*,*Z*)-2,6-nonadienal (5.4%) and with 1% or less 2,4-nonadienal, octanal, hexanal, 4-heptenal,(*E*)-2-nonenal are odour-active. The aroma described as grassy and floral is assumed to be composed mainly of the substances, decanal, β-ionone, α-ionone, hexanal and 4-heptenal.

A total of 8 different substances with an OAV higher than 1 were detected in sample 147. The main flavouring agent is β-ionone with 80.8%. This is followed by α-ionone (7.7%), decanal (6.4%) and the others, (*E*,*Z*)-2,6-nonadienal, octanal, hexanal, 2,4-nonadienal, and 4-heptenal with 5.1%. In general, the sample had the highest concentration of β-ionone (6.49 µg/L, Table A12). This high level of β-ionone also seems to be confirmed in the odour perception, which would be described as chestnut blossom, and the flavour, which was determined by the panel to be honey.

In roasted sample 369, a total of 9 aroma-active compounds with an OAV above 1 could be detected. Here, β-ionone is the main aroma substance with 86.5%. Furthermore, 2,3-dimethylpyrazines (0.3%) and 2-ethyl-3,5-dimethylpyrazines (0.6%) with an OAV of 1 and 2 are found in the sample. However, the smoky and roasted flavour of this tea is strongly influenced by these components despite their minute contribution to the overall aroma. One reason for the strong perception of smoke is a natural protective instinct of humans against fire. The sensory cells perceive it very strongly in order to detect the danger of fire at an early stage [55].

In total, 9 main aroma compounds (Figure 22) were identified in coffee leaf tea via semi quantification and calculation of OAV.

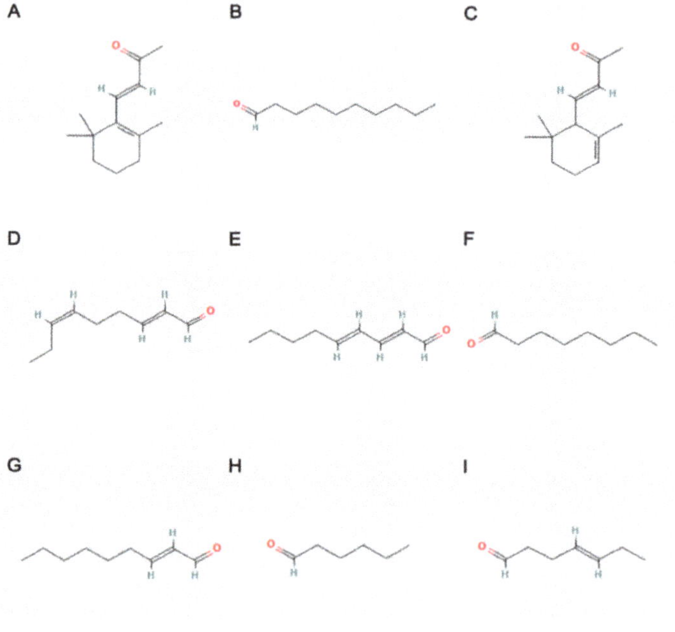

Figure 22. Structure of the main aroma compounds. (**A**): β-ionone, (**B**): decanal, (**C**): α-ionone, (**D**): (*E*,*Z*)-2,6-nonadienal, (**E**): 2,4-nonadienal, (**F**): octanal, (**G**): (*E*)-2 nonenal, (**H**): hexanal, (**I**): 4-heptenal [56].

Three of them, β-ionone, α-ionone and 2-ethyl-3,5-dimethylpyrazines are, according to Ho et al. [57], also main aroma substances of the tea plant. β-Ionone has the largest overall share in all samples and varies between 39.2% (156) and 86.5% (369). Furthermore, the percentage of decanal in two samples with a β-ionone value below 50% is very high, i.e., 34.4% (156) and 35.1% (930). α-Ionone which is also present in all samples varies from 5.4% (930) to 12% (234). Moreover, (*E*,*Z*)-2,6-nonadienal was found as a main aroma component in some samples. As in the case of decanal, the value here varies greatly between 1.8% (147) and 28.7% (687) depending on the sample. The citrusy note of (*E*)-2-nonenal could not be noted in any of the samples during tasting by the panel. Despite the large variation between the samples, it was not possible to correlate the differences with the parameters in this study. Further studies are required to provide comparable conditions between samples. Only the roasting process showed a large difference in these tests compared to the other samples, as the pyrazines also contribute a large proportion to the aroma.

The main disadvantage of the calculated OAV is that the interaction of individual aroma components is not considered. It is possible that individual aroma components, which were excluded because of an OAV below 1, may well contribute to the aroma together with other substances [58]. Additionally, the perceived intensity of an aroma is not proportional to the value of the OAV. A doubling of the OAV around the threshold can have a greater impact on the aroma than a doubling of an OAV very far above the threshold [59].

4. Conclusions

In this study, different ways of producing teas from coffee leaves directly on the farm were investigated. The samples differed both in popularity among consumers and in the chemical composition of the active ingredients and flavourings. In future, the data obtained in this study may help to adjust process parameters directly to consumer preferences and allow farmers to earn an extra income from this by-product. For this purpose, on-site experiments should be carried out to upscale the processes.

Young leaves showed a positive correlation on various plant protective ingredients such as caffeine content, chlorogenic acid and trigonelline. The variety played a role in essential oils and caffeine content in these experiments. Pacamara had an increased level of essential oil and a slightly lower level of caffeine compared to Bourbon. In the processing parameters, blending of the samples resulted in a strong decrease in caffeine, catechins, polyphenols and chlorogenic acid. In contrast, cell disruption processes such as rolling or crumbling led to increased values. Among the drying methods, roasting and oven drying had a positive effect on the essential oil and trigonelline content in the samples. Fermentation mainly affected the acidity of the samples. Increased levels of lactic and acetic acid were found here, particularly in wild fermented and *Lactobacillus* fermented samples.

The sensory analysis revealed that fermented teas in particular are ahead in terms of popularity with consumers. Green tones in particular are masked by more fruity notes (e.g., peach). Additionally, the roasting of the tea seems to lead to a polarizing product.

In the aroma analysis by gas chromatography, 68 aroma active compounds were detected by the ODP. By calculation of the OAV, 6–9 aroma compounds could be determined for each tea, which are the main components of the aroma profile. These are β-ionone (honey-like), decanal (citrus, floral), α-ionone (floral), (*E*,*Z*)-2,6-nonadienal (cucumber-like), 2,4-nonadienal (melon-like), octanal (fruity), (*E*)-2 nonenal (citrus), hexanal (grassy) and 4-heptenal (green). Additionally, the two substances 2,3-dimethylpyrazines and 2-ethyl-3,5-dimethylpyrazines were found in the roasted sample.

Author Contributions: Conceptualization, S.S. and D.W.L.; Data curation, M.C.S.; Formal analysis, M.C.S., P.B. and V.S.; Investigation, M.C.S., P.B. and V.S.; Methodology, M.R., A.Q.-B., J.R.-Z., S.S. and D.W.L.; Resources, S.S., D.W.L. and Y.Z.; Supervision, S.S., D.W.L. and Y.Z.; Validation, M.R. and D.W.L.; Visualization, M.C.S. and D.W.L.; Writing—original draft, M.C.S.; Writing—review and, M.C.S., M.R., P.B., V.S., A.Q.-B., J.M.B., J.R.-Z., S.S., D.W.L. and Y.Z. All authors have read and agreed to the published version of the manuscript.

Funding: This research received no external funding.

Institutional Review Board Statement: Ethical review and approval were waived for this study because the affiliated institutes do not require ethical clearance for sensory analysis of foods. Nevertheless, the authors considered guidelines for ethical and professional practices for the sensory analysis of foods. Potential adverse effects for the assessors were excluded as only small amounts of tea were taken into the mouth. All assessors had training in sensory analysis of foods, and regularly performed sensory testing of tea in line of their normal duties. All assessors have consented to participation in the study.

Informed Consent Statement: Not applicable because this is not a clinical study.

Data Availability Statement: Data is contained within the article or Appendix A.

Acknowledgments: The teams of the plant-based food and NMR laboratories at CVUA Karlsruhe are warmly thanked for excellent technical assistance.

Conflicts of Interest: S.S. is owner of and P.B. and M.C.S. are consultants for Coffee Consulate, Mannheim, Germany. Coffee Consulate is an independent training and research centre. A.Q.-B. is owner of Finca La Buena Esperanza, a coffee plantation in El Salvador. J.R.-Z. is managing director of Rubiacea Research and Development GmbH, Mannheim, Germany, which is providing consulting, research and development in the field of food and food technology with focus on coffee and coffee by-products. Coffee Consulate, Finca La Buena Esperanza and Rubiacea Research and Development GmbH are currently researching the potential of coffee by-products for production of tea and other foods. However, P.B., M.C.S., S.S., J.R.-Z. and A.Q.-B. report that there are no conflicts of interest related to the work under consideration and that the objectivity of the study was not affected. The other authors declare no conflict of interest.

Appendix A

Table A1. Results of the NIR analysis for water, essential oils and ash.

Sample Number	Water (g/100 g)	Essential Oil (mL/100 g)	Ash (g/100 g)
1	7.71 ± 0.15	1.08 ± 0.09	8.41 ± 0.19
2	9.68 ± 0.36	0.79 ± 0.07	10.20 ± 0.20
3	8.26 ± 0.13	0.46 ± 0.21	8.88 ± 0.99
4	10.07 ± 0.11	0.66 ± 0.11	9.85 ± 0.11
5	9.77 ± 0.24	0.52 ± 0.5	9.46 ± 1.01
6	7.86 ± 0.03	0.47 ± 0.06	7.94 ± 0.23
7	9.24 ± 0.23	0.76 ± 0.06	8.14 ± 0.08
8	5.99 ± 0.13	1.52 ± 0.08	9.98 ± 0.08
9	6.27 ± 0.11	1.28 ± 0.20	9.64 ± 0.10
10	7.11 ± 0.26	1.16 ± 0.07	9.00 ± 0.09
11	6.84 ± 0.14	1.20 ± 0.03	8.95 ± 0.10
12	8.12 ± 0.24	1.04 ± 0.29	9.29 ± 0.88
13	7.85 ± 0.22	1.31 ± 0.06	8.53 ± 0.16
14	8.35 ± 0.13	1.25 ± 0.16	8.52 ± 0.37
15	5.52 ± 0.26	0.00 ± 0.00	9.89 ± 0.14
16	17.42 ± 0.04	0.30 ± 0.01	3.22 ± 0.00
17	6.56 ± 0.12	0.02 ± 0.03	8.54 ± 0.09
18	6.75 ± 0.13	0.39 ± 0.02	7.82 ± 0.14
19	8.25 ± 0.30	0.00 ± 0.00	7.81 ± 0.05
20	5.39 ± 0.04	0.37 ± 0.31	9.00 ± 0.86
21	5.99 ± 0.1	0.01 ± 0.02	9.85 ± 0.92
22	6.58 ± 0.06	0.09 ± 0.04	8.62 ± 0.21
23	3.92 ± 0.06	1.48 ± 0.05	8.10 ± 0.16
24	6.58 ± 0.02	0.30 ± 0.26	9.55 ± 0.81

Table A2. Results of the HPLC analysis for caffeine and catechins.

Sample Number	Caffeine (g/100 g DM)	Epigallocatechin Gallate (g/100 g DM)	Total Catechins (g/100 g DM)	Theobromin (g/100 g DM)
1	0.740 ± 0.003	-	0.306 ± 0.006	-
2	0.565 ± 0.004	-	0.299 ± 0.002	0.021 ± 0.001
3	0.747 ± 0.010	-	0.390 ± 0.023	-
4	0.520 ± 0.001	-	0.290 ± 0.002	0.015 ± 0.001
5	0.699 ± 0.027	-	0.208 ± 0.090	-
6	0.693 ± 0.026	-	0.206 ± 0.089	-
7	0.713 ± 0.006	-	0.345 ± 0.004	-
8	0.669 ± 0.001	-	0.054 ± 0.003	-
9	0.396 ± 0.003	-	-	-
10	0.406 ± 0.003	-	-	-
11	0.444 ± 0.001	-	-	-
12	0.392 ± 0.003	-	-	-
13	0.370 ± 0.001	-	-	-
14	0.368 ± 0.004	-	-	-
15	0.380 ± 0.004	-	-	-
16	0.449 ± 0.001	-	-	-
17	0.462 ± 0.003	-	-	-
18	0.939 ± 0.596	-	0.221 ± 0.157	-
19	1.333 ± 0.019	-	0.699 ± 0.009	0.009 ± 0.003
20	0.381 ± 0.003	-	0.045 ± 0.007	-
21	0.761 ± 0.012	-	0.092 ± 0.008	-
22	1.000 ± 0.004	-	-	-
23	0.894 ± 0.008	-	0.479 ± 0.001	-
24	0.858 ± 0.003	-	0.091 ± 0.128	-

Table A3. Results of the photometric analysis of the total polyphenol content.

Sample Number	Total Polyphenol (g/100 g DM)
1	8.28 ± 0.33
2	9.34 ± 0.36
3	7.62 ± 0.26
4	8.31 ± 0.22
5	7.54 ± 0.67
6	3.73 ± 0.28
7	7.15 ± 0.18
8	4.76 ± 0.13
9	0.91 ± 0.12
10	1.68 ± 0.75
11	1.50 ± 0.81
12	2.35 ± 0.66
13	2.04 ± 0.82
14	2.18 ± 0.82
15	0.91 ± 0.08
16	1.08 ± 0.00
17	1.15 ± 0.00
18	7.91 ± 0.77
19	10.36 ± 0.25
20	2.73 ± 0.45
21	0.79 ± 0.11
22	-
23	5.93 ± 0.27
24	5.08 ± 0.19

Table A4. Results of the NMR analysis of the organic acid contents.

Sample	Chlorogenic Acid (g/100 g DM)	Acetic Acid (g/100 g DM)	Lactic Acid (g/100 g DM)	Trigonellin (g/100 g DM)
1	2.92	0.107	0.18	2.79
2	2.58	0.088	0.16	0.98
3	2.38	0.101	0.13	2.66
4	2.81	0.097	0.15	1.15
5	3.78	0.136	0.14	2.76
6	1.17	0.068	0.11	1.80
7	3.32	0.124	0.13	2.66
8	1.44	0.114	0.11	2.25
9	0.20	0.145	8.12	1.26
10	0.16	0.102	0.21	1.21
11	0.08	0.266	0.92	1.28
12	0.23	0.115	0.21	0.62
13	0.18	0.093	0.20	0.68
14	0.24	0.21	1.08	0.62
15	0.08	0.137	2.76	1.44
16	0.00	0.051	0.49	1.11
17	0.06	0.29	2.59	1.01
18	9.35	0.061	0.16	4.87
19	6.58	0.078	0.22	4.25
20	2.82	0.078	0.14	2.34
21	0.31	0.122	0.49	2.01
22	0.31	0.042	0.12	3.30
23	6.20	0.14	0.14	3.31
24	3.00	0.074	0.23	2.41

Table A5. Odour-active compounds of the eight tea samples sorted by their retention indices.

Number	Compound	RI	RI (lit/std)	Odour	Identification	182	687	234	156	930	743	147	369
1	Hexanal	1079	1079	green, grassy	RI,MS,O	X	X	X	X	X	X	X	X
2	4-Heptenal	1242	1239	green	RI,MS,O	X	X		X			X	
3	Octanal	1285	1286	fruity	RI,MS,O			X					
4	n.i.	1298	-	green	/					X			
5	1-Hydroxy-2-propanone [a]	1307	1305	sweetish	RI,O								X
6	2,3 Dimethylpyrazin	1347	1347	-	RI,MS								X
7	1-Hexanol [a]	1357	1357	sweetish green	RI,O						X		
8	n.i.	1370	-	green/cucumber	/								X
9	(*E*)-2-Octenal	1430	1430	cucumber	RI,MS,O		X	X	X				
10	(*E*)-4-Nonenal [a]	1435	1435	fruity	RI,O						X		
11	n.i.	1449	-	melon	/							X	
12	1-Octen-3-ol	1453	1452	forest	RI,MS,O			X					
13	n.i.	1465	-	fruity	/			X					
14	3-Ethyl-3,5-dimethyl-pyrazine [a]	1468	1466	roasty, coffee	RI,O								X
15	Decanal	1502	1497	citrus, floral	RI,MS,O	X	X	X	X	X	X	X	X
16	2-Ethyl-3,5-dimethylpyrazine	1512	1512	-	RI,MS								X
17	(*E*,*E*)-3,5-Octadien-2-one [a]	1519	1521	green, grassy	RI,O							X	
18	Benzaldehyde	1523	1522	sweetish fruity	RI,MS,O			X					
19	n.i.	1530	-	green, grassy	/								X
20	(*E*)-2-Nonenal	1536	1537	citrus like	RI,MS,O					X			
21	n.i.	1560	-	green, grassy	/							X	
22	n.i.	1562	-	green, grassy	/	X							
23	2,6-Nonadienal	1587	1586	cucumber	RI,MS,O	X	X	X	X	X	X	X	X
24	5H-5-Methyl-6,7-dihydrocyclopentapyrazine [a]	1629	1630	peanut	RI,O								X
25	(*E*,*E*)-2,4-Nonadienal [a]	1699	1698	melon,cucumber	RI,O						X		
26	α-Terpineol	1699	1699	citrus like	RI,MS,O					X		X	
27	n.i.	1700	-	citrus like	/			X	X				
28	2,4-Nonadienal	1702	1700	melon,cucumber	RI,MS,O	X							
29	2,3-Nonadienal [a]	1703	1703	melon,cucumber	RI,O	X							
30	4-Ethyl-benzaldehyde [a]	1709	1711	fruity	RI,O								X

Table A5. *Cont.*

Number	Compound	RI	RI (lit/std)	Odour	Identification	182	687	234	156	930	743	147	369
31	n.i.	1712	-	pineapple	/						X		
32	(*E*,*Z*)-2,6-Nonadien-1-ol [a]	1769	1770	green, cucumber	RI,O				X			X	X
33	3,7,11-Trimethyl-1-dodecanol [a]	1770		green	RI,O			X					
36	3,4-Dimethylbenzaldehyde [a]	1811	1790	sweetish, green	RI,O	X	X	X			X	X	
37	2,5-Dimethylbenzaldehyde [a]	1812	1812	sweetish, green	RI,O					X			
38	Hexanoic acid [a]	1844	1844	fatty	RI,O						X		
39	Dodecanoic acid, ethyl ester [a]	1846	1846	sweetish, floral	RI,O								X
40	α-Ionone	1857	1855	sweetish, floral	RI,MS,O	X	X	X	X	X	X	X	X
41	n.i.	1874	-	green	/					X			
42	Benzyl alcohol	1878	1877	fruity	RI,MS,O				X				
43	β-Ionone	1942	1943	honey, flower	RI,MS,O	X	X	X	X	X	X	X	X
44	1-Dodecanol	1969	1971	cocutnut	RI,MS,O				X				
45	n.i.	2005	-	green	/		X		X	X			
46	n.i.	2007	-	green	/	X							X
47	4-Methoxybenzaldehyde	2027	2025	vanilla	RI,MS,O		X		X			X	
48	n.i.	2034	-	cocutnut	RI,MS,O								X
49	n.i.	2078	-	green, cucumber	/				X	X			
50	n.i.	2081	-	green, cucumber	/		X						
51	1-Ethylundecylbenzene [a]	2092	2094	herbal	RI,O					X			
52	n.i.	2094	-	herbal	/						X	X	
53	n.i.	2096	-	herbal	/	X	X	X					
55	n.i.	2155	-	green, cucumber	/		X	X	X				
56	1-Methyldodecylbenzene [a]	2158		mint	RI,O					X			
57	Nonanoic acid	2168	2164	sweetish green	RI,MS,O		X	X					X
58	Tetrahydro-6-pentyl-2H-pyran-2-one [a]	2201	2201	cocotnut	RI,O								X
59	Isopropyl palmitate [a]	2242	2237	sweetish	RI,O	X							
60	n.i.	2251	-	citrus like	/						X		
61	Dihydro-5-pentyl-2(3H)-furanone [a]	2264	2266	cocutnut	RI,O					X			X
62	5-Heptyldihydro-2(3H)-furanone [a]	2264		roasty, coffee	RI,O				X				
63	γ-Dodecalactone	2381	2379	fruity	RI,MS,O	X	X						
64	n.i.	2397	-	aquarium	/							X	
65	(*Z*)-Dihydro-5(2-octenyl)-2(2H)-furanone [a]	2402	2390	green, sweetish	RI,O		X						
66	n.i.	2402	-	green, sweetish	/	X			X	X	X		
67	Indole	2449	2443	Nutty	RI,MS,O								X
68	n.i.	2471	-	Stable	/				X				

Abbreviations: RI, retention index; X, found in the sample; [a], provisionally identified; n.i.: not identified, MS: mass spectrometry, O: olfactometry.

Table A6. Results of semi-quantification and calculated OAVs for sample 182.

Compound	Concentration in μg/L	OAV
β-Ionone	3.44 ± 0.33	491
(*E*,*Z*)-2,6-Nonadienal	2.56 ± 0.36	256
α-Ionone	3.01 ± 0.59	100
Octanal	12.87 ± 0.93	18
Decanal	1.69 ± 0.15	17
4-Heptenal	7.15 ± 1.22	9
Hexanal	28.4 ± 4.19	6
(*E*)-2-Nonenal	0.22 ± 0.02	3
1-Dodecanol	0.33 ± 0.09	<1
3-Hexen-1-ol	28.91 ± 3.76	<1
α-Terpineol	0.76 ± 0.08	<1
1-Dodecanol	0.3 ± 0.03	<1
(*E*)-2-Octenal	1.78 ± 0.19	<1
Nonanoic acid	638.8 ± 36.55	<1
Benzyl alcohol	147.73 ± 26.28	<1
(*E*,*E*)-2,4-Heptadienal	101.17 ± 6.95	<1
γ-Dodecalactone	2.54 ± 0.43	<1

Table A7. Results of semi-quantification and calculated OAVs for sample 687.

Compound	Concentration in g/L	OAV
β-Ionone	0.93 ± 0.05	132
(*E*,*Z*)-2,6-Nonadienal	0.49 ± 0.06	49
α-Ionone	0.9 ± 0.04	30
Decanal	1.41 ± 0.16	14
Octanal	6.72 ± 1.05	10
2,4-Nonadienal	0.76 ± 0.13	8
Hexanal	23.46 ± 3.89	5
(*E*)-2-Nonenal	0.26 ± 0.01	3
4-Heptenal	2.21 ± 0.37	3
1-Dodecanol	0.38 ± 0.07	<1
3-Hexen-1-ol	41 ± 8.15	<1
α-Terpineol	0.82 ± 0.11	<1
1-Dodecanol	0.13 ± 0.01	<1
(*E*)-2-Octenal	0.39 ± 0.06	<1
Nonanoic acid	730.46 ± 13.3	<1
Benzyl alcohol	53.87 ± 7.15	<1
(*E*,*E*)-2,4-Heptadienal	25.07 ± 1.45	<1
γ-Dodecalactone	2 ± 0.36	<1
Benzaldehyde	2.71 ± 0.28	<1

Table A8. Results of semi-quantification and calculated OAVs for sample 234.

Compound	Concentration in µg/L	OAV
β-Ionone	1.75 ± 0.07	249
Decanal	5.37 ± 0.91	54
α-Ionone	1.39 ± 0.04	46
(*E*,*Z*)-2,6-Nonadienal	0.18 ± 0.03	18
Octanal	5.75 ± 0.71	8
2,4-Nonadienal	0.28 ± 0.03	3
4-Heptenal	1.97 ± 0.34	2
(*E*)-2-Nonenal	0.19 ± 0.02	2
Hexanal	9.28 ± 1.6	2
1-Dodecanol	0.24 ± 0.03	<1
1-Octen-3-ol	0.63 ± 0.1	<1
3-Hexen-1-ol	14.35 ± 2.77	<1
α-Terpineol	1.13 ± 0.06	<1
(*E*)-2-Octenal	0.1 ± 0.02	<1
Nonanoic acid	1199.03 ± 347.21	<1
Benzyl alcohol	33.6 ± 3.78	<1
γ-Dodecalactone	0.05 ± 0	<1
Benzaldehyde	5.66 ± 0.37	<1
Indole	3.32 ± 0.13	<1

Table A9. Results of semi-quantification and calculated OAVs for sample 156.

Compound	Concentration in µg/L	OAV
β-Ionone	2.14 ± 0.15	306
Decanal	26.83 ± 3.29	268
(*E*,*Z*)-2,6-Nonadienal	0.78 ± 0.21	78
α-Ionone	2.13 ± 0.06	71
Octanal	15.84 ± 2.13	23
2,4-Nonadienal	1.01 ± 0.01	11
(*E*)-2-Nonenal	0.74 ± 0.04	9
Hexanal	38.98 ± 1.25	9
4-Heptenal	4.65 ± 0.72	6
1-Dodecanol	0.73 ± 0.08	<1
3-Hexen-1-ol	40.6 ± 1.07	<1

Table A9. *Cont.*

Compound	Concentration in µg/L	OAV
α-Terpineol	1.46 ± 0.15	<1
(*E*)-2-Octenal	0.67 ± 0.08	<1
Nonanoic acid	1508.67 ± 310.48	<1
Benzyl alcohol	130.01 ± 14.89	<1
(*E*,*E*)-2,4-Heptadienal	57 ± 4.83	<1
γ-Dodecalactone	0.16 ± 0.03	<1
Benzaldehyde	6.5 ± 0.76	<1
Benzaldehyde, 4-methoxy-	1.83 ± 0.07	<1

Table A10. Results of semi-quantification and calculated OAVs for sample 930.

Compound	Concentration in µg/L	OAV
β-Ionone	2.95 ± 0.23	421
Decanal	30.06 ± 3.49	301
α-Ionone	1.39 ± 0.07	46
(*E*,*Z*)-2,6-Nonadienal	0.31 ± 0.03	31
2,4-Nonadienal	1.66 ± 0.06	18
Octanal	12.37 ± 1.9	18
(*E*)-2-Nonenal	0.86 ± 0.12	11
Hexanal	36.85 ± 3.52	8
4-Heptenal	0.81 ± 0.06	1
1-Dodecanol	0.96 ± 0.09	<1
3-Hexen-1-ol	57.73 ± 10.8	<1
α-Terpineol	6.01 ± 0.84	<1
(*E*)-2-Octenal	0.37 ± 0.06	<1
Nonanoic acid	1462.18 ± 228.82	<1
Benzyl alcohol	20.44 ± 0.05	<1
(*E*,*E*)-2,4-Heptadienal	21.27 ± 1.36	<1
γ-Dodecalactone	0.06 ± 0.01	<1
Benzaldehyde	3.56 ± 0.31	<1
Indole	1.9 ± 0.07	<1

Table A11. Results of semi-quantification and calculated OAVs for sample 743.

Compound	Concentration in µg/L	OAV
β-Ionone	3.79 ± 0.31	541
α-Ionone	1.96 ± 0.06	65
Decanal	4.7 ± 0.54	47
(*E*,*Z*)-2,6-Nonadienal	0.39 ± 0.04	39
2,4-Nonadienal	0.65 ± 0.06	7
Octanal	5.02 ± 0.19	7
Hexanal	31.71 ± 2.48	7
4-Heptenal	2.84 ± 0.29	4
(*E*)-2-Nonenal	0.24 ± 0.03	3
1-Dodecanol	0.66 ± 0.05	<1
1-Octen-3-ol	0.58 ± 0.11	<1
3-Hexen-1-ol	50.56 ± 6.39	<1
α-Terpineol	0.89 ± 0.08	<1
(*E*)-2-Octenal	0.19 ± 0.01	<1
Nonanoic acid	949.75 ± 121.15	<1
Benzyl alcohol	35.32 ± 3.23	<1
(*E*,*E*)-2,4-Heptadienal	24.74 ± 1.93	<1
γ-Dodecalactone	0.05 ± 0.01	<1
Benzaldehyde	6.07 ± 0.93	<1
Indole	3.14 ± 0.34	<1

Table A12. Results of semi-quantification and calculated OAVs for sample 147.

Compound	Concentration in μg/L	OAV
β-Ionone	6.49 ± 0.67	927
α-Ionone	2.66 ± 0.25	89
Decanal	7.38 ± 0.86	74
(E,Z)-2,6-Nonadienal	0.2 ± 0.03	20
Octanal	11 ± 0.89	16
Hexanal	42.82 ± 2.57	10
2,4-Nonadienal	0.74 ± 0.09	8
4-Heptenal	3.31 ± 0.07	4
1-Dodecanol	0.38 ± 0.04	<1
1-Octen-3-ol	0.3 ± 0.05	<1
3-Hexen-1-ol	44.86 ± 5	<1
α-Terpineol	1.7 ± 0.24	<1
(E)-2-Octenal	0.19 ± 0.02	<1
Nonanoic acid	1235.09 ± 11.07	<1
Benzyl alcohol	35.18 ± 2.8	<1
(E,E)-2,4-Heptadienal	11.5 ± 1.07	<1
γ-Dodecalactone	0.05 ± 0	<1
Benzaldehyde	5.84 ± 0.68	<1
Indole	1.61 ± 0.13	<1
Benzaldehyde, 4-methoxy-	0.9 ± 0.15	<1

Table A13. Results of semi-quantification and calculated OAVs for sample 369.

Compound	Concentration in μg/L	OAV
β-Ionone	2.57 ± 0.12	367
α-Ionone	0.97 ± 0.03	32
Octanal	7.97 ± 0.41	11
1-Octen-3-ol	5.07 ± 0.19	5
2,4-Nonadienal	0.22 ± 0.01	2
2-ethyl-3,5-dimethylpyrazine	0.1 ± 0.01	2
(E)-2-Nonenal	0.1 ± 0.01	1
Hexanal	5.66 ± 0.24	1
2,3-dimethylpyrazine	2.85 ± 0.11	1
1-Dodecanol	0.43 ± 0.02	<1
3-Hexen-1-ol	29.8 ± 1.1	<1
α-Terpineol	0.2 ± 0.02	<1
1-Dodecanol	0.29 ± 0.01	<1
(E)-2-Octenal	0.1 ± 0.01	<1
Nonanoic acid	951.9 ± 121.64	<1
Benzyl alcohol	47.66 ± 5.34	<1
(E,E)-2,4-Heptadienal	0.93 ± 0.05	<1
γ-Dodecalactone	0.06 ± 0.01	<1
Benzaldehyde	2.92 ± 0.2	<1
Indole	6.58 ± 0.43	<1
2,5-Dimethyl-3-(2-methylpropyl)-pyrazine	2.6 ± 0.03	<1

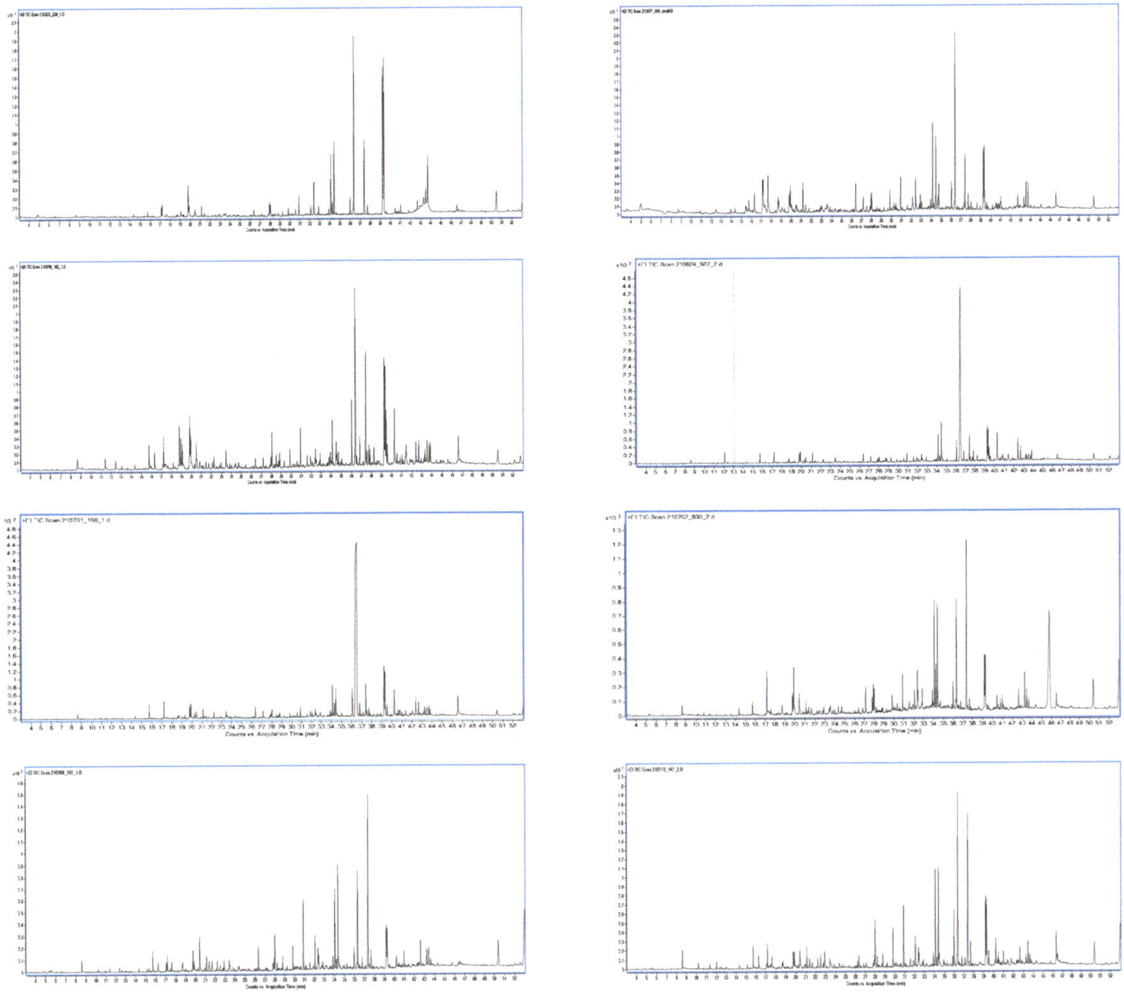

Figure A1. Total ion chromatograms of the tea samples.

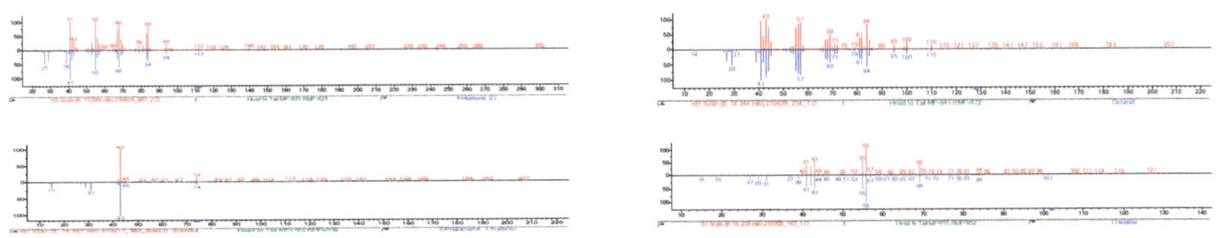

Figure A2. *Cont.*

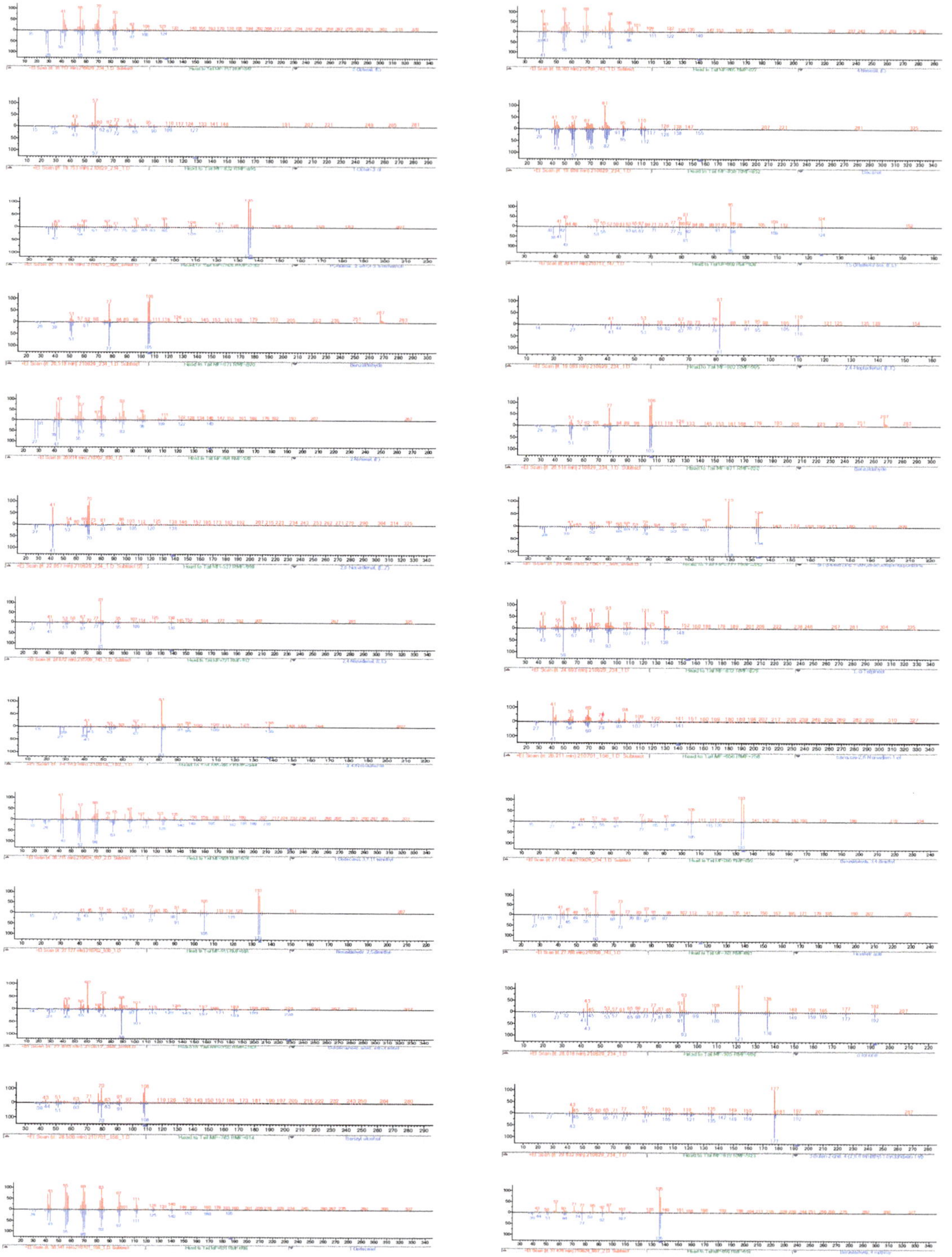

Figure A3. *Cont.*

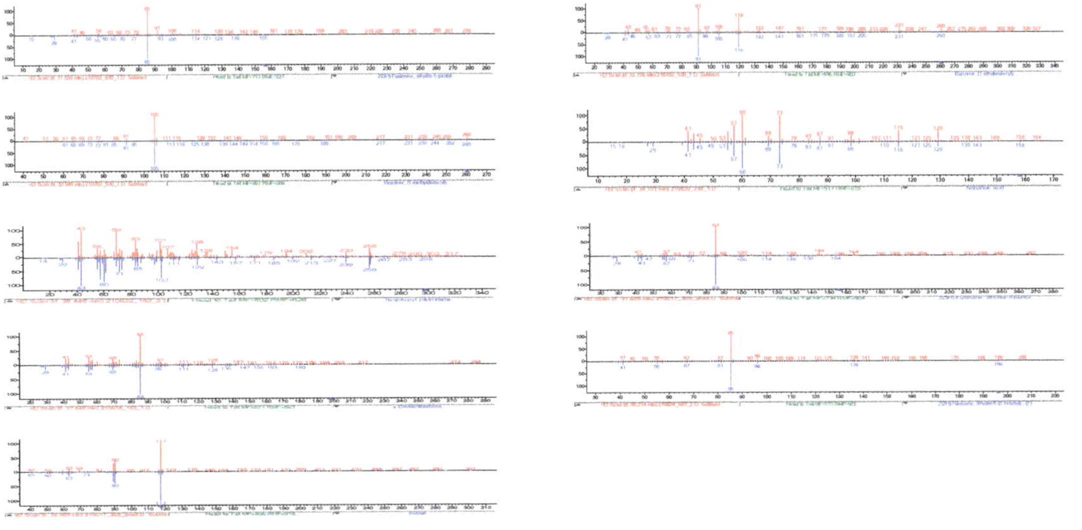

Figure A4. MS spectra of the identified substances (red beams) and the corresponding substance of the NIST Mass Spectral Library (blue beams).

Table A14. Chemical compounds and their associated odour thresholds.

Compound	Odour Threshold [ppb]	
α-Terpineol	330	Takeoka et al., 1990 [60]
1-Dodecanol	7.1	Pal et al., 2014 [61]
1-Octen-3-ol	1	Buttery et al., 1988 [62]
2-ethyl-3,5-dimethylpyrazine	0.04	Buttery and Ling, 1997 [63]
2-Nonenal, (*E*)-	0.08	Buttery et al., 1988 [62]
2-Octenal, (*E*)-	3	Guadagni et al., 1972 [64]
2,3-dimethylpyrazine	2.5	
2,4-Heptadienal, (*E*,*E*)-	778	
2,4-Nonadienal	0.09	Teranishi et al., 1974 [65]
2,5-Dimethyl-3-(2-methylpropyl)-pyrazine	800	
2,6-Nonadienal, (*E*,*Z*)-	0.01	Teranishi et al., 1974 [65]
3-Hexen-1-ol	70	Takeoka et al., 1990 [60]
4-Heptenal	0.8	
Benzaldehyde	350	Buttery et al., 1988 [62]
Benzaldehyde, 4-methoxy-	47	
Benzyl alcohol	10000	Buttery et al., 1988 [62]
Decanal	0.1	Guadagni et al., 1963 [66]
Hexanal	04.05.05	Buttery et al., 1988 [62]
Indole	140	Buttery et al., 1988 [62]
Nonanoic acid	3000	
Octanal	0.7	Buttery et al., 1988 [62]
α-Ionone	0.03	
β-Ionone	0.007	
γ-Dodecalactone	7	Engel et al., 1988 [67]

References

1. International Coffee Organisation (ICO). *Coffee Market Report*; International Coffee Organisation (ICO): London, UK, 2021.
2. Murthy, P.S.; Madhava Naidu, M. Sustainable management of coffee industry by-products and value addition—A review. *Resour. Conserv. Recycl.* **2012**, *66*, 45–58. [CrossRef]

3. European Comission. Commission Implementing Regulation (EU) 2020/917 of 1 July 2020 authorising the placing on the market of infusion from coffee leaves of *Coffea arabica* L. and/or *Coffea cnephora* Pierre Ex A. Froehner as a traditional food from a third country. *Off. J. Eur. Union* **2020**, *L209*, 10–13.
4. Lachenmeier, D.W.; Rajcic de Rezende, T.; Schwarz, S. An update on sustainable valorization of coffee by-products as novel foods within the European Union. *Biol. Life Sci. Forum* **2021**, *6*, 37. [CrossRef]
5. Wintgens, J.N. Coffee: Growing, processing, sustainable production. In *A Guidebook for Growers, Processors, Traders, and Researchers*; Wiley-VCH Verlag GmbH & Co. KGaA: Weinheim, Germany, 2004. [CrossRef]
6. Von Pröpper, L. *Der Kaffee- und Theetisch nebst Rezepten und Servierkarten*; Verlag von J. Engelhorn: Stuttgart, Germany, 1882.
7. Klingel, T.; Kremer, J.I.; Gottstein, V.; De Rezende, T.R.; Schwarz, S.; Lachenmeier, D.W. A review of coffee by-products including leaf, flower, cherry, husk, silver skin, and spent grounds as novel foods within the European Union. *Foods* **2020**, *9*, 665. [CrossRef]
8. Tritsch, N.; Steger, M.C.; Segatz, V.; Blumenthal, P.; Rigling, M.; Schwarz, S.; Zhang, Y.; Franke, H.; Lachenmeier, D.W. Risk assessment of caffeine and epigallocatechin gallate in coffee leaf tea. *Foods* **2022**, *11*, 263. [CrossRef]
9. Novita, R.; Kasim, A.; Anggraini, T.; Putra, D.P. Kahwa Daun: Traditional knowledge of a coffee leaf herbal tea from West Sumatera, Indonesia. *J. Ethn. Foods* **2018**, *5*, 286–291. [CrossRef]
10. Cramer, P.J.S. *A Review of Literature of Coffee Research in Indonesia*; SIC Editorial, Inter-American Institute of Agricultural Sciences: Turrialba, Costa Rica, 1957.
11. Lamorde, M.; Tabuti, J.R.S.; Obua, C.; Kukunda-Byobona, C.; Lanyero, H.; Byakika-Kibwika, P.; Bbosa, G.S.; Lubega, A.; Ogwal-Okeng, J.; Ryan, M.; et al. Medicinal plants used by traditional medicine practitioners for the treatment of HIV/AIDS and related conditions in Uganda. *J. Ethnopharmacol.* **2010**, *130*, 43–53. [CrossRef] [PubMed]
12. Ross, I.A. *Coffea arabica*. *Med. Plants World* **2005**, *3*, 155–195. [CrossRef]
13. Giday, M.; Asfaw, Z.; Woldu, Z. Ethnomedicinal study of plants used by Sheko ethnic group of Ethiopia. *J. Ethnopharmacol.* **2010**, *132*, 75–85. [CrossRef]
14. Martins, S.C.V.; Araújo, W.L.; Tohge, T.; Fernie, A.R.; DaMatta, F.M. In high-light-acclimated coffee plants the metabolic machinery is adjusted to avoid oxidative stress rather than to benefit from extra light enhancement in photosynthetic yield. *PLoS ONE* **2014**, *9*, e94862. [CrossRef]
15. Pompelli, M.F.; Martins, S.C.V.; Antunes, W.C.; Chaves, A.R.M.; DaMatta, F.M. Photosynthesis and photoprotection in coffee leaves is affected by nitrogen and light availabilities in winter conditions. *J. Plant Physiol.* **2010**, *167*, 1052–1060. [CrossRef] [PubMed]
16. Chen, X. A Review on coffee leaves: Phytochemicals, bioactivities and applications. *Crit. Rev. Food Sci. Nutr.* **2019**, *59*, 1008–1025. [CrossRef] [PubMed]
17. Chaves, A.R.M.; Ten-Caten, A.; Pinheiro, H.A.; Ribeiro, A.; Damatta, F.M. Seasonal changes in photoprotective mechanisms of leaves from shaded and unshaded field-grown coffee (*Coffea arabica* L.) trees. *Trees Struct. Funct.* **2008**, *22*, 351–361. [CrossRef]
18. The European Parliament and the Council of the European Union. Regulation (EU) 2015/2283 of the European Parliament and of the Council of 25 November 2015 on novel foods, amending Regulation (EU) No 1169/2011 of the European Parliament and of the Council and repealing Regulation (EC) No 258/97. *Off. J. Eur. Union* **2015**, *L327*, 1–22.
19. *DIN 10809:1988-08*; Analysis of Tea; Preparation of Liquor for Use in Sensory Tests. Beuth Verlag: Berlin, Germany, 1988. [CrossRef]
20. *ISO 1572:1980-08*; Tea; Preparation of Ground Sample of Known Dry Matter Content. International Organization for Standardization: Geneva, Switzerland, 1980.
21. Segatz, V.; Steger, M.C.; Blumenthal, P.; Gottstein, V.; Rigling, M.; Schwarz, S.; Zhang, Y.; Lachenmeier, D.W. Evaluation of analytical methods to determine regulatory compliance of coffee leaf tea. *Biol. Life Sci. Forum* **2021**, *6*, 45. [CrossRef]
22. Claassen, L.; Rinderknecht, M.; Porth, T.; Röhnisch, J.; Seren, H.Y.; Scharinger, A.; Gottstein, V.; Noack, D.; Schwarz, S.; Winkler, G.; et al. Cold brew coffee-pilot studies on definition, extraction, consumer preference, chemical characterization and microbiological hazards. *Foods* **2021**, *10*, 865. [CrossRef]
23. Rigling, M.; Yadav, M.; Yagishita, M.; Nedele, A.K.; Sun, J.; Zhang, Y. Biosynthesis of pleasant aroma by Enokitake (*Flammulina velutipes*) with a potential use in a novel tea drink. *LWT Food Sci. Technol.* **2021**, *140*, 110646. [CrossRef]
24. Weatherley, P.E. Studies in the water relations of the cotton plant: I. The field measurement of water deficits in leaves. *New Phytol.* **1950**, *49*, 81–97. [CrossRef]
25. Chabot, B.F.; Hicks, D.J. The ecology of leaf life spans. *Annu. Rev. Ecol. Syst.* **1982**, *13*, 229–259. [CrossRef]
26. Zerbini, S. The pruning of ornamental plants: Why topping is such a common practice. *Alberi Territ.* **2006**, *2*, 26–28.
27. Scherer, G.W. Theory of drying. *J. Am. Ceram. Soc.* **1990**, *73*, 3–14. [CrossRef]
28. Arslan, D.; Özcan, M.M. Evaluation of drying methods with respect to drying kinetics, mineral content and colour characteristics of rosemary leaves. *Energy Convers. Manag.* **2008**, *49*, 1258–1264. [CrossRef]
29. BMEL. *Leitsätze für Tee, Teeähnliche Erzeugnisse, deren Extrakte und Zubereitungen vom 2.12.1998*; Bundesministerium für Ernährung und Landwirtschaft: Berlin, Germany, 1998.
30. Hanaa, A.R.M.; Sallam, Y.I.; El-Leithy, A.S.; Aly, S.E. Lemongrass (*Cymbopogon citratus*) essential oil as affected by drying methods. *Ann. Agric. Sci.* **2012**, *57*, 113–116. [CrossRef]
31. Morsy, N.F. Quality Evaluation of The Essential Oil of Basil Plant Dried by Different Methods. Master's Thesis, Cairo University, Giza, Egypt, 2004.

32. Omidbaigi, R.; Sefidkon, F.; Kazemi, F. Influence of drying methods on the essential oil content and composition of Roman chamomile. *Flavour Fragr. J.* **2004**, *19*, 196–198. [CrossRef]
33. Ratanamarno, S.; Surbkar, S. Caffeine and catechins in fresh coffee leaf (*Coffea arabica*) and coffee leaf tea. *Maejo Int. J. Sci. Technol.* **2017**, *11*, 211–218.
34. Bowers, M.D.; Stamp, N.E. Effects of plant age, genotype, and herbivory on plantago performance and chemistry. *Ecology* **1993**, *74*, 1778–1791. [CrossRef]
35. Blum-Silva, C.H.; Chaves, V.C.; Schenkel, E.P.; Coelho, G.C.; Reginatto, F.H. The influence of leaf age on methylxanthines, total phenolic content, and free radical scavenging capacity of *Ilex paraguariensis* aqueous extracts. *Rev. Bras. Farmacogn.* **2015**, *25*, 1–6. [CrossRef]
36. Song, R.; Kelman, D.; Johns, K.L.; Wright, A.D. Correlation between leaf age, shade levels, and characteristic beneficial natural constituents of tea (*Camellia sinensis*) grown in Hawaii. *Food Chem.* **2012**, *133*, 707–714. [CrossRef]
37. Astill, C.; Birch, M.R.; Dacombe, C.; Humphrey, P.G.; Martin, P.T. Factors affecting the caffeine and polyphenol contents of black and green tea infusions. *J. Agric. Food Chem.* **2001**, *49*, 5340–5347. [CrossRef]
38. Ashihara, H.; Monteiro, A.M.; Gillies, F.M.; Crozier, A. Biosynthesis of caffeine in leaves of coffee. *Plant Physiol.* **1996**, *111*, 747–753. [CrossRef]
39. Friedman, M.; Levin, C.E.; Choi, S.H.; Lee, S.U.; Kozukue, N. Changes in the composition of raw tea leaves from the Korean yabukida plant during high-temperature processing to pan-fried Kamairi-Cha green tea. *J. Food Sci.* **2009**, *74*, C406–C412. [CrossRef] [PubMed]
40. Toschi, T.G.; Bordoni, A.; Hrelia, S.; Bendini, A.; Lercker, G.; Biagi, P.L. The protective role of different green tea extracts after oxidative damage is related to their catechin composition. *J. Agric. Food Chem.* **2000**, *48*, 3973–3978. [CrossRef] [PubMed]
41. Li, N.; Taylor, L.S.; Ferruzzi, M.G.; Mauer, L.J. Kinetic study of catechin stability: Effects of pH, concentration, and temperature. *J. Agric. Food Chem.* **2012**, *60*, 12531–12539. [CrossRef]
42. Turkmen, N.; Sarı, F.; Velioglu, Y.S. Factors affecting polyphenol content and composition of fresh and processed tea leaves. *Akad. Gıda* **2009**, *7*, 29–40.
43. Oszmiański, J.; Wojdyło, A.; Kolniak, J. Effect of L-ascorbic acid, sugar, pectin and freeze-thaw treatment on polyphenol content of frozen strawberries. *LWT Food Sci. Technol.* **2009**, *42*, 581–586. [CrossRef]
44. Loncaric, A.; Dugalic, K.; Mihaljevic, I.; Jakobek, L.; Pilizota, V. Effects of sugar addition on total polyphenol content and antioxidant activity of frozen and freeze-dried apple purée. *J. Agric. Food Chem.* **2014**, *62*, 1674–1682. [CrossRef]
45. Monteiro, Â.; Colomban, S.; Azinheira, H.G.; Guerra-Guimarães, L.; Silva, M.D.C.; Navarini, L.; Resmini, M. Dietary antioxidants in coffee leaves: Impact of botanical origin and maturity on chlorogenic acids and xanthones. *Antioxidants* **2020**, *9*, 6. [CrossRef]
46. Ścibisz, I.; Mitek, M. The changes of antioxidant properties in highbush blueberries (*Vaccinium corymbosum* L.) during freezing and long-term frozen storage. *Acta Sci. Pol. Technol. Aliment* **2007**, *6*, 75–82.
47. Zhu, X.; Wang, C.; Wang, X.; Cai, M.; Deng, S. Effects of different processing methods on the determination of trigonelline in fructus *Cannabis* decoction pieces. *J. Pharm. Res.* **2016**, *35*, 19–21.
48. Wang, L.F.; Lee, J.Y.; Chung, J.O.; Baik, J.H.; So, S.; Park, S.K. Discrimination of teas with different degrees of fermentation by SPME-GC Analysis of the characteristic volatile flavour compounds. *Food Chem.* **2008**, *109*, 196–206. [CrossRef]
49. Borah, S.; Bhuyan, M. Non-destructive testing of tea fermentation using image processing. *Insight Non-Destruct. Test. Cond. Monit.* **2003**, *45*, 55–58. [CrossRef]
50. Yuwono, S.S.; Fibrianto, K.; Wahibah, L.Y.; Wardhana, A.R. Sensory attributes profiling of dampit robusta coffee leaf tea (*Coffea canephora*). *Carpathian J. Food Sci. Technol.* **2019**, *11*, 165–176. [CrossRef]
51. Zhu, J.C.; Xiao, Z.B. Characterization of the key aroma compounds in peach by gas chromatography–olfactometry, quantitative measurements and sensory analysis. *Eur. Food Res. Technol.* **2019**, *245*, 129–141. [CrossRef]
52. Galmés, J.; Flexas, J.; Savé, R.; Medrano, H. Water relations and stomatal characteristics of Mediterranean plants with different growth forms and leaf habits: Responses to water stress and recovery. *Plant Soil* **2007**, *290*, 139–155. [CrossRef]
53. Kawakami, M.; Kobayashi, A.; Kator, K. Volatile constituents of rooibos tea (*Aspalathus linearis*) as affected by extraction process. *J. Agric. Food Chem.* **1993**, *41*, 633–636. [CrossRef]
54. Oomah, B.D.; Liang, L.S.Y.; Balasubramanian, P. Volatile compounds of dry beans (*Phaseolus vulgaris* L.). *Plant Foods Hum. Nutr.* **2007**, *62*, 177–183. [CrossRef] [PubMed]
55. Bryan, J.L.; DiNenno, P.J.; Drysdale, D.; Beyler, C.L. Behavioral response to fire and smoke. In *Handbook of Fire Protection Engineering*; DiNenno, P.J., Drysdale, D., Beyler, C.L., Eds.; The Society of Fire Protection Engineers (SFPE): Washington, DC, USA, 2002; pp. 315–341.
56. PubChem. National Library for Medicine. Available online: https://pubchem.ncbi.nlm.nih.gov/ (accessed on 10 March 2022).
57. Ho, C.T.; Zheng, X.; Li, S. Tea aroma formation. *Food Sci. Hum. Wellness* **2015**, *4*, 9–27. [CrossRef]
58. Paravisini, L.; Guichard, E. Interactions between aroma compounds and food matrix. In *Flavour: From Food to Perception*; Guichard, E., Salles, C., Morzel, M., Le Bon, A.-M., Eds.; John Wiley & Sons, Ltd.: Chichester, UK, 2016; pp. 208–234. [CrossRef]
59. Doty, R.L. Psychophysical measurement of odor perception in humans. In *The Human Sense of Smell*; Springer: Berlin/Heidelberg, Germany, 1991; pp. 95–134. [CrossRef]
60. Takeoka, G.R.; Flath, R.A.; Mon, T.R.; Teranishi, R.; Guentert, M. Volatile constituents of apricot (*Prunus armeniaca*). *J. Agric. Food Chem.* **1990**, *38*, 471–477. [CrossRef]

61. Pal, P.; Mitra, I.; Roy, K. QSPR modeling of odor threshold of aliphatic alcohols using extended topochemical atom (ETA) indices. *Croat. Chem. Acta* **2014**, *87*, 29–37. [CrossRef]
62. Buttery, R.G.; Turnbaugh, J.G.; Ling, L.C. Contribution of volatiles to rice aroma. *J. Agric. Food Chem.* **1988**, *36*, 1006–1009. [CrossRef]
63. Buttery, R.G.; Ling, L.C. 2-Ethyl-3,5-dimethylpyrazine and 2-ethyl-3,6-dimethylpyrazine: Odor thresholds in water solution. *LWT Food Sci. Technol.* **1997**, *30*, 109–110. [CrossRef]
64. Guadagni, D.G.; Buttery, R.G.; Turnbaugh, J.G. Odour thresholds and similarity ratings of some potato chip components. *J. Sci. Food Agric.* **1972**, *23*, 1435–1444. [CrossRef]
65. Teranishi, R.; Buttery, R.G.; Gaudagni, D.G. Odor quality and chemical structure in fruit and vegetable flavors. *Ann. N. Y. Acad. Sci.* **1974**, *237*, 209–216. [CrossRef] [PubMed]
66. Guadagni, D.G.; Buttery, R.G.; Okano, S. Odour Thresholds of some organic compounds associated with food flavours. *J. Sci. Food Agric.* **1963**, *14*, 761–765. [CrossRef]
67. Engel, K.H.; Flath, R.A.; Buttery, R.G.; Mon, T.R.; Teranishi, R.; Ramming, D.W. Investigation of volatile constituents in nectarines. 1. Analytical and sensory characterization of aroma components in some nectarine cultivars. *J. Agric. Food Chem.* **1988**, *36*, 549–553. [CrossRef]

foods

MDPI

Article

Characterization of Coffee Silver Skin as Potential Food-Safe Ingredient

Maria Martuscelli *, Luigi Esposito, Carla Daniela Di Mattia, Antonella Ricci and Dino Mastrocola

Faculty of Bioscience and Technology for Food, Agriculture and Environment, University of Teramo, Via R. Balzarini 1, 64100 Teramo, Italy; lesposito2@unite.it (L.E.); cdimattia@unite.it (C.D.D.M.); aricci@unite.it (A.R.); dmastrocola@unite.it (D.M.)
* Correspondence: mmartuscelli@unite.it; Tel.: +39-0861-266912

Abstract: By-products from the coffee industry are produced in large amounts each year. Among other wastes, coffee silver skin (CSS) is highly available and more stable due to its lower content of water. This research aimed to characterize coffee silver skin composition and evidence its potentiality for use as a food-safe ingredient in new formulations. Results showed an average total dietary fiber content of 50% but with a higher ratio for insoluble than soluble fiber. A high content of total phenolic compounds, chlorogenic acid, caffeine, and caffeic acid was found and correlated with the high measured antioxidant capacity. Moreover, minerals (e.g., calcium, magnesium, phosphorous, potassium, copper, iron, manganese) important for human wellbeing were found at a high level in CSS, while toxic minerals (e.g., nickel) were found at low levels. In conclusion, coffee silver skin could have an advantageous role for the recovery of valuable compounds and as a potential food-safe ingredient.

Keywords: coffee by-products; dietary fiber; color; minerals; bioactive compounds

1. Introduction

A real and effective solution to ensure access to safe and secure food is to limit food waste at the industry level and in home consumption [1]. Whatever the supply chain, a significant number of by-products, that are difficult to reuse and potentially responsible for environmental pollution, are produced [2,3]. Furthermore, the valorization and reuse of food by-products fulfills sustainable development goals, specifically the 12th (responsible consumption and production) and the 13th (climate action) goals, established by the Food and Agriculture Organization (FAO agenda 2030).

Coffee silver skin (CSS) is a thin layer that covers coffee seeds inside the coffee "cherries" of the genus *Coffea* of Rubiaceae plants (*Coffea arabica* and *Coffea canephora*). CSS is the unique by-product discarded after the roasting process; coffee production gives rise to other by-products, such as coffee pulp and coffee husks, before the roasting step and spent coffee grounds after coffee preparation [4]. Since coffee is one of the most consumed beverages worldwide and one of the most traded items globally, its industry generates a lot of residues [5,6]. CSS represents 4.2% (*w*/*w*) of seed total weight, with values of 1 ton of CSS for every 120 tons of roasted coffee [7]. Despite being a less abundant coffee by-product, the availability of CSS justifies the interest in its recovery and use in several applications in different sectors, not just for the food industry. CSS has been proposed as a material for renewable energy sources, as an electrode material, and as a biofuel source [7].

Due to its content of valuable compounds like fibers, sugars, and polyphenols, it would be interesting and useful to exploit it as an innovative and functional food ingredient [8]. Costa and co-authors reported that different cultivars (*C. arabica* and *C. canephora*) have comparable phenolic profiles and compositional traits, and they were successfully utilized as food antioxidants and fiber integrators [9].

Coffee's provenience greatly influences its composition; thus, it is difficult to exactly determine it in each variety cultivated. This substrate contains a lot of dietary fiber (up to 55%), including both insoluble (≈45%) and soluble (≈10%) fractions [10]. Cellulose is present at a high level (≈60%), which is helpful for the growth of many types of bacteria and molds to produce enzymes like α-amylase and fructooligosaccharides. According to Esquivel and Jiménez [11], CSS can be easily incorporated into preparations due to its low content in lipids and carbohydrates. Recently, some authors found that CSS is also a good source of relevant amino acids [12].

Some studies explored the possibility to use CSS as a food ingredient in several matrices, despite little information about levels of ochratoxin A and other contaminants being available. In bread, CSS alkalinization was proposed to increase water-holding capacity and the content of fibers [13]. More recently, CSS was added to yogurt [14], to biscuits as a source of dietary fibers and prebiotics [15], and was also blended with instant coffee for new coffee-based beverages [16]. Due to its content of melanoidins, high molecular weight compounds mainly constituting polysaccharides, proteins, and chlorogenic acid molecules [17], CSS can find application as a food dye. Moreover, melanoidin contributes to the antioxidant activity of CSS, showing anti-inflammatory effects on different organs [18].

For marketing within the European Union (EU), it is very important to define whether products, like CSS, need to obtain approval as an ingredient for novel food [4], with special reference to its legal status within the EU and potential options for producers to obtain approval according to the Novel Food Regulation (EU) No 2015/2283 and the two implementing regulations (EU) No. 2017/2468 and No. 2017/2469) [19,20].

In this scenario, the present research investigated the chemical and physical properties of CSS; it also found the volatile profile and mineral composition to enhance its global potentiality to be considered as an advantageous ingredient in newly formulated foods.

2. Materials and Methods

2.1. Coffee Silver Skin

Coffee silver skin (CSS) was received from the toasting plant Marcafè Torrefazione Adriatica s.p.a. (Giulianova, Italy). After 10 cycles of roasting (10 × 240 kg of roasted coffee), 3.3 kg of CSS were recovered. CSS used for this experiment was a blend of 5 arabica varieties (*C. arabica*) (India Arabica, India Cherry, Vietnam, India Mysore, and India Caracolito) and 5 robusta varieties (*C. canephora*) (India Parchment, Santos, Uganda CRV 18, Uganda CRV 17, Togo). CSS was ground at 10,200 rpm for 1 min (Bimby®, mod. TM 31, Wuppertal, Germany) until reaching a particle size of 250–125 μm.

Then, physico-chemical, physical, and chemical analyses were carried out. CSS was kept frozen at −20 °C until analysis.

2.2. Physico-Chemical, Color, and Compositional Analyses

Water activity values (a_w) were obtained with the Aqualab 4 TE kit (Court Pullman, WA, USA). Values of pH were taken with a pH meter (model 3510, Jenway, Stone, UK) in triplicate and expressed with their standard deviation.

Colorimetric parameters were obtained with a CR-5 colorimeter (spectrally based, Konica Minolta, Japan) with a D_{65} light source and observer 10°. Samples were measured in triplicate for a total of 9 measurements. To better define the final color observed, the saturation index (chroma, C^*) was calculated according to Formula (1):

$$C^* = (a^{*2} + b^{*2})^{1/2} \quad (1)$$

Proximate analysis of moisture, proteins, lipids, and ash was carried out with different protocols. Moisture, ash, and proteins were determined following the Association of Official Analytical Chemists procedure [21], as follows. Moisture content was determined by drying about 1 g of sample to a constant weight. Ash was determined by the calcination of 1 g of sample in a furnace at 525 ± 25 °C until a constant weight was reached.

Total nitrogen content (TN, % *w*/*w*) was determined using the standard Kjeldahl copper catalyst method and the protein content was calculated by multiplying TNx6.25.

Total lipids were measured using a modification of the chloroform to methanol procedure described by Folch et al. [22]. Total dietary fiber, insoluble fiber content, and soluble fiber content were determined with a Megazyme Integrated Total Dietary Fiber Assay Kit that was accepted as AOAC Method 2017.16 in 2018 and is the only method consistent with the CODEX Alimentarius definition of dietary fiber [21].

Briefly, 1 g dried food sample (duplicate) was subjected to sequential enzymatic digestion by heat-stable α-amylase, protease, and amyloglucosidase. Insoluble dietary fiber (IDF) was filtered, and then the residue was washed with warm distilled water. The combined solution of filtrate and water was precipitated with 4 volumes of 95% ethanol (EtOH) for soluble dietary fiber (SDF) determination. The precipitate was then filtered and dried. Both SDF and IDF residues were corrected for protein, ash, and blank, for the final calculation of SDF and IDF values.

2.2.1. Acrylamide Content

To determine acrylamide in CSS, an analysis was performed by the external laboratory Chelab s.r.l. (Accredia N° 0051L), by means of HPLC/MS (triple quadrupole) according to an internal procedure. The limit of quantification was 50 μg kg^{-1} (as CSS is a dried vegetable matrix); the limit of detectability was 15 μg kg^{-1} (as CSS is a dried vegetable matrix).

2.2.2. Water-Holding and Oil-Holding Capacity (WHC, OHC)

Water- and oil-holding capacities (WHC and OHC) were measured according to the procedure of Ballesteros et al. [23], with some modifications. Our results come from 6 independent measurements.

For WHC, 1 g of CSS was placed in 30 mL plastic tubes, with 10 mL of distilled water, then vortexed for 1 min and left in the dark for 24 h. After this time, samples were centrifuged for 20 min at 3000 rpm to separate the unabsorbed water from the CSS. The pellet obtained was weighed and the variation was expressed as grams (g) of absorbed water.

The same method was followed to determine the OHC by using 0.5 g of CSS and 5 g of sunflower oil.

2.3. *Polyphenol Extraction, Total Polyphenol Content (TPC), and Phenolic Profile Determination in CSS*

The extraction of the phenolic fractions from CSS samples was carried out by using different solvent mixtures: deionized water and methanolic aqueous solutions at two methanol concentrations (60:40 and 70:30 *v*:*v*). The procedure was as follows, 1 g of sample was added to 10 mL of the solvent; the mixture was vortexed for 1 min and then centrifuged (4000 rpm for 10 min). The supernatant was recovered and filtered through 0.45 μm nylon filters. For the extraction of the bound phenolic compounds, the insoluble residue from the extraction carried out with the 70% methanolic solution was recovered and re-extracted with 10 mL of a methanol:water:chloridric acid solution (70:29:1), vortexed for 1 min, stirred at room temperature for 2 h, and then centrifuged and filtered by using the same conditions as before. The extracted polyphenols were then stored at −32 °C until further analysis. These extracts were used for the evaluation of total polyphenol content and radical-scavenging activity. The analysis of the phenolic pattern was carried out on the 70% methanolic extract. The total polyphenol content (TPC) was determined following a modified procedure from Di Mattia et al. [24]. Water was added to 0.1 mL of sample extracts up to a volume of 5 mL; 500 μL of Folin–Ciocalteu reagent was then added. After 3 min, 1.5 mL of a 25% (*w*/*v*) Na_2CO_3 solution was added, followed by deionized water up to a 10 mL final volume. Solutions were maintained at room temperature away from light sources for 60 min. Absorbance was measured at 765 nm using a Perkin Elmer Lambda Bio 20 spectrophotometer. Gallic acid standard solutions

(Fluka, Buchs, CH, Switzerland) were used for calibration. Results were expressed as mg of gallic acid equivalents (GAE) per g of sample.

The identification and further quantification of phenolic acids were conducted by means of HPLC-DAD analysis (mod. HPLC: AGILENT series 1200 Agilent Technologies, Milan, Italy), in accordance with the method suggested by Horžić et al. [25]. A reverse-phase C18 column (5 μ 18 100A 250 × 4.6 mm, Phenomenex Kinetex, Bologna, Italy) was used. The solvent phases utilized were water and formic acid 5% (A), and methanol (B). The flux was fixed at 1 mL/min. Table 1 reports the gradient used for the analysis. Data for chlorogenic and caffeic acid were spectrophotometrically read at 280 nm while caffeine was read at 330 nm. The quantification was determined using calibration curves in a range from 6.25 to 100 mg/L.

Table 1. Time and gradient (B phase, %) used for the HPLC-DAD analysis of phenolic acids.

Time (min)	B Phase (%)
0	2
20	32
30	40
40	95
45	5
52	95
55	95

2.4. *Melanoidin Determination of CSS*

The melanoidin content was determined by spectrophotometric analysis by reading the absorption value at a wavelength of 345 nm [26] on the dialyzed fraction of the aqueous extract obtained as described in Section 2.3. Before melanoidin analysis, 5 mL of the aqueous extracts were dialyzed using cellulose dialysis tubing (MWCO: ~12,400 Da) (Sigma Aldrich Italy, Milan, Italy) placed in a glass vessel with 1 L of deionized water and maintained under stirring for 12 h. This procedure was repeated after refreshing the deionized water. The volume of the CSS aqueous extract which remained in the dialysis tube was diluted to a known volume with deionized water.

2.5. *Radical-Scavenging Activity (TEAC) of CSS*

The radical-scavenging activity was measured by the 2,2′-azino-nis (3-ethylbenzthiazoline-6-sulfuric acid (ABTS) radical cation discoloration assay [27]. The ABTS radical stock solution was prepared by dissolving the ABTS in water to a 7 mM concentration and by reacting this solution with 2.45 mM of potassium persulfate; the mixture was then left in the dark at room temperature for 12–16 h before use. The ABTS+• stock solution was diluted in water to an Abs of 0.70 ± 0.02 for the analysis and the reaction was started by the addition of 30 μL of CSS extract to 2.97 mL of ABTS+• radical solution.

The bleaching rate of ABTS+• in the presence of the sample was monitored at 25 °C at 734 nm using a Perkin Elmer (Boston, MA, USA) Lambda Bio 20 spectrophotometer, and the discoloration after 5 min was used for the calculation of the inhibition percentage.

Radical-scavenging activity was expressed as mol Trolox equivalents antioxidant capacity (TEAC-μmol of Trolox equivalents per g of sample) and calculated by the ratio of the correlation coefficient of the dose–response curve of the sample and the correlation coefficient of the dose–response curve of Trolox, the standard compound.

2.6. *Mineral Determination of CSS*

A microwave digestor (mod. Speedwave 2) (Berghof groups, Germany) coupled with inductively coupled plasma for the reading of the results (ICP-OES, mod. Avio200) (Perkin Elmer, Boston, MA, USA) were used, in accordance with UNI EN 13657:2002 [28].

Mineral determination followed the instructions of the manufacturer. Samples were weighed and added to 3 mL nitric acid and 3 mL oxygen peroxide left for 20 min at room temperature to start the digestion. After this, samples were submitted to the specific ramp in line with the Berghof manual procedure for coffee samples. Results are expressed as the average of triplicate analysis.

Once digested, samples were left at room temperature overnight. When cold, they were filtered with filter paper. A solution of 5% of nitric acid was used to clean vessels and to suspend the analytes obtained. This solution was brought to the final volume of 50 mL by adding nitric acid 5%.

The reading of samples was done by developing an internal method in ICP software (Syngistix) by selecting defined parameters and making a calibration curve. For this, nitric acid was used as a blank and washing solution. Certified standard solutions were used as standards containing all analytes traced at a concentration of 100 mg/L.

2.7. Volatile Compounds in CSS

To investigate the volatile compounds, profile GC-MS analysis was performed (Perkin Elmer Gas Chromatograph Clarus 580, Mass Spectrometer Perkin Elmer Clarus SQ 8 S Waltham, MA, USA). Approximately 2 g CSS was put in glassy vials of 20 mL capacity (Perkin Elmer), tightly closed, and stocked at −40 °C, assuring the highest headspace. For the GC-MS analysis, the method used was taken from Qi et al. [29] with some modifications

Briefly, volatiles from CSS were extracted with a headspace solid phase microextraction fiber (SPME 65 µm polydimethylsiloxane/divinylbenzene -PDMS/DVB-; Supelco, Bellofonte, USA) and collected for 30 min at 40 °C, then inserted into the GC injector and desorbed for 3 min at 250 °C. Volatile compounds were separated on a ZB-SemiVolatiles capillary GC column (30 m length × 0.25 mm internal diameter × 0.25 µm film thickness, Phenomenex, Torrance, CA, USA).

The oven temperature was maintained for 3 min at 40 °C, increased at 3 °C/min to 70 °C, then at 5 °C/min to 180 °C, then at 10 °C/min to 260 °C, and maintained for 5 min at 260 °C.

Helium was the carrier gas with a constant flow of 1mL/min.

The mass-selective detector was operated in electron impact mode (70 eV) and full scan mode (35–500 *m/z* range). The identification was performed using the National Institute of Standards and Technology mass spectral library (NIST Mass Spectral library, Search Program version 2.0, National Institute of Standards and Technology, U.S. Department of Commerce, Gaithersburg, MD, USA).

2.8. Statistical Analysis

All determinations were done in triplicate, except where indicated otherwise. Means and relative standard deviations were calculated. The statistical differences of the results on the content of phenolic compounds and on the antiradical activity between the soluble and bound fractions were determined by a *t*-test between means; the analyses were carried out by XLSTAT software (Addinsoft SARL, New York, NY, USA).

3. Results

3.1. pH, a_W, and Color of CSS

pH measurements gave values of 5.34 ± 0.018 (Table 2). No bibliographical references have been found to compare our results; however, other authors have reported pH levels of 4.9–5.6 for coffee beverages, and they considered these pH values suitable for the activity of biomolecules (e.g., polyphenols) [30]. As already mentioned, CSS has a coarse structure and dusty appearance which make it suitable for adding to food formulations and easily stored. Our values confirm this characteristic even if no reference is available to compare these values.

Table 2. Physico-chemical, color, and chemical characterization of CSS.

a_W	0.32 ± 0.004
pH	5.34 ± 0.02
L*	20.8 ± 2.62
a*	4.27 ± 0.18
b*	13.4 ± 0.80
C*	14.12 ± 0.79
Moisture (%)	2.78 ± 0.30
Ash (%)	6.79 ± 0.78
Proteins (%)	18.15 ± 2.17
Lipids (%)	2.31 ± 0.50
Fibers (%)	51.05 ± 0.35
Soluble Fibers (%)	7.86 ± 1.01
Insoluble Fibers (%)	48.85 ± 0.83
Acrylamide (μg kg^{-1})	720 ± 110
WHC (g water/g CSS)	4.89 ± 0.36
OHC (g oil/g CSS)	3.02 ± 0.11

Very low water activity values were found (0.319 ± 0.00), which is important evidence for the low availability of water for degradation reactions in CSS [31]; consequently, its stability depends on the environmental humidity conditions in which the CSS is stored, which must be such as to prevent water absorption and provide a prolonged shelf life to CSS.

Color, L*, a*, and b* parameters were like those reported in the literature [13]. The color mainly depends on the interaction of constitutive polyphenols and anthocyanins, which are involved in oxidation and polymerization reactions during the coffee roasting step [32]. Moreover, roasting processes cause degradation of proteins, and Maillard reactions occur, so melanoidins also contribute to the characteristic color of CSS.

3.2. Characteristics of CSS Composition

Table 2 reports data on the composition of CSS. As previously mentioned, CSS is a low-moisture ingredient due to the roasting step that also lowers the a_W value. Our results are close to data from Costa et al. [9] (4.76 ± 0.10 g/100 g). Gocmen et al. [33] substituted wheat flour with different concentrations of CSS (2.5, 5.0, and 7.5%) for cookie production, observing darker products. Similarly, Garcia De Serna et al. [15] used CSS as a substitute for sucrose in combination with stevia at concentrations of 1.33 and 3.33 g, obtaining darker biscuits. Finally, in sponge cakes, Ateş and Elmacı [34,35], used concentrations of 20, 25, and 30% as fat replacers, observing a decrease in L* and b* values of the crumb and an increase in a* value. They also added CSS as a substitution for flour [35], observing a higher index for crumb darkness and redness and an increase in fiber content, concluding that no strong difference could be observed with additions up to 30%.

Fiber content in by-products gave good results, confirming what is stated in the literature (Table 2). CSS has an average TDF content of 50% but a high ratio between IDF and SDF. Costa et al. [9] also found a TDF% content of 56.4 ± 0.70, IDF level of 49.1 ± 0.44, and SDF of 7.30 ± 0.07. Iriondo-De Hond et al. [10] found higher levels; their results were 67.7 ± 1.6 (TDF%) for arabica varieties (*C. arabica*), IDF content 64 ± 1.6, and SDF 3.7 ± 0.0. In comparison with these results, the study presented a major SDF content of 7.86%, although other studies gave SDF% values of around 8.8% for arabica varieties (*C. arabica*) [36]. As for other compounds, this variability depends on roasting conditions, geographical provenience, and the varieties analyzed. Despite these differences, the consumption of CSS can improve the daily intake of dietary fibers, known for their beneficial role in lowering the risk of cardiovascular diseases, obesity, type 2 diabetes, chronic constipation, and other unhealthy conditions [37,38]. Moreover, as highlighted by [39], the consumption of foods rich in dietary fibers are often good reservoirs of phenolic compounds which have antioxidant and anti-inflammatory effects. According to [40], the recommended dietary reference intake (DRI) daily allowance in men aged 19–50 years is 38 g/day and in women 25 g/day, and for men aged > 51 it is 31 g/day and in women aged > 51 it is 21 g/day, although the average consumption among Americans is about 15 g/day. The quantity of dietary fibers in vegetable foods varies a lot, and the ratio of IDF and SDF differs. Cereals and some fruits, vegetables, and nuts are considered good sources of dietary fibers. For wheat (*Triticum aestivum* L.), the common values are 11.6–17.0 g TDF, of which 10.2–14.7 g are IDF, and 1.4–2.3 g are SDF; for oats (*Avena sativa* L.), the values are 10.3 g TDF, of which 6.5 g are IDF, and 3.8 g are SDF [41]. In apples (*Malus domesticus*), these concentrations were found to be 4.4 g TDF, 3.3 g IDF, and 1.1 g SDF; oranges (*Citrus sinensis* L.) have 3.4 g TDF, 1.4 g SDF, and 2.0 g IDF [42].

A content of 720 $\pm$ 110 μg kg^{-1} of acrylamide was found. Recent limits established by the European Commission are 400 μg kg^{-1} for roast coffee, 850 μg kg^{-1} for instant (soluble) coffee, and 500–4000 μg kg^{-1} for coffee substitutes, reported in the Commission Regulation (EU) 2017/2158 of 20 November 2017 [43]. These values cannot be directly compared, as the use of CSS is not for coffee substitution. Instead, other acrylamide limit values must be considered to establish, depending on the final products, the amount of CSS as a functional component that can be added. The content we found was higher than 489 μg kg^{-1}, which was the value reported for coffee silver skin aqueous extracts [10]. However, the coffee silver skin analyzed in that study only came from the arabica variety (*C. arabica*) of coffee, while our sample included a contribution of 50% from the robusta variety (*C. canephora*), for which a higher content of acrylamide produced upon roasting has been described in the literature [44].

3.3. Water-Holding and Oil-Holding Capacity

Water- and oil-holding capacities are important indexes to express the capability of an ingredient to retain water or oil and give a formulation more moisture and softness, influencing the final texture and palatability of a product. CSS has already been tested in bakery products and bread, acting as a moisturizing ingredient and a fat replacer [13,15, 34].

In some of these studies, CSS was alkalizated to enhance its functional properties, but even when used raw, it had remarkable abilities as an adsorbent material.

Our results (Table 2) are comparable with those published by other authors [13,24] who have shown how the WHC is affected by the fiber content and type, which was mainly soluble, while differences for the OHC index can be imputable with the particle size of the investigated samples and the different oils used (corn oil vs. sunflower oil).

In their studies, Ateş and Elmacı [34,35] obtained higher results than us for OHC, (4.82 $\pm$ 0.23) but they are comparable, as they also used sunflower oil for the analysis. From their work, they determined that water-treated CSS can be a useful fat replacer in sponge cake and a flour replacer.

Previously, Behrouzian et al. [45] performed an optimization study for the utilization of CSS in food, establishing the most suitable conditions for an alkaline hydrogen peroxide treatment. With their attempts, they showed the effect of temperature and pH conditions on both WHC and OHC and also on cellulose extraction.

Belmiro et al. [46] suggested that matrices that are abundant in dietary fiber are affected by their molecular size, degree of branching, intermolecular aggregation, and water-binding sites and, consequently, their technological properties could be modified by physical or chemical treatments. This may induce a reduction in particle size and/or enhancement of particle porosity, and permeability and diffusion of water or oil into the fiber. These authors have shown that the WHC and OHC of coffee co-products are enhanced by higher pressures (100 MPa), by a minimum concentration of acetic anhydride (10%), or by enzymatic treatments (at least 5 U of cellulase for 30 min); in all tested methods, changes were attributed to the disruption or disordering of the dietary fiber structure and to the formation of aggregates. However, the same authors observed a significant negative effect of those technological treatments on the antioxidant status of coffee co-products [47].

3.4. Polyphenols, Melanoidin, and Radical-Scavenging Activity

The extraction of the phenolic fraction was preliminarily carried out by using deionized water and two solutions containing 60% and 70% (*v:v*) of methanol, respectively, in order to explore the extractive capability of solvent mixtures characterized by different polarities, widely used in the literature for vegetables [48].

Among the mixtures, the 70% methanol solution allowed the highest yield of the soluble fraction as evaluated by the content in total phenolic compounds (TPCs) and, for this reason, it was chosen for further investigations on the radical-scavenging activity as well as on the phenolic pattern. Moreover, as phenolic compounds can also occur in the bound form by either ester or ether bonds with sugars, alcohols, and other biomolecules from cell walls, the insoluble residue recovered from the first step was submitted to further extraction under acidified conditions. Çelik and Gocmen [49] have recently reviewed all the factors interacting with the availability and extraction of soluble and bound phenols, remarking on the dynamic balance of this system that can change according to the conditions of the medium. The results on the content of phenolic compounds and antiradical activity of the sample, along with the content in melanoidins, caffeine, and chlorogenic and caffeic acids, are reported in Table 3.

Table 3. Total polyphenol content (TPC), caffeic and chlorogenic acids, caffeine content, and radical-scavenging activity (TEAC) in CSS.

Phenolic Fraction	TPC (mg GAE 100 g^{-1})	TEAC (mmol Trolox eq 100 g^{-1})	Chlorogenic Acid (mg g^{-1})	Caffeic Acid (mg g^{-1})	Caffeine (mg g^{-1})
Soluble	578 ± 64	1.57 ± 0.04	0.39 ± 0.02	0.31 ± 0.01	17.45 ± 0.69
Bound	467 ± 29	1.11 ± 0.07	0.18 ± 0.01	0.21 ± 0.01	n.d.
Sign.	**	**	**	*	-

Legend: n.d.: not detectable; asterisks indicate significance at *: $p < 0.05$ and **: $p < 0.0$.

The TPC of the soluble fraction was significantly higher with respect to the bound fraction. Due to the differences in the solvent mixtures used, procedures adopted, and molecules used as reference compounds for the assays, the comparison of TPC results with others found in literature was very difficult; however, it can be generally affirmed that larger amounts were found with respect to Gemechu [8], whilst lower values were observed when the work of other authors was considered [36,49–52].

Besides the factors previously cited, such a discrepancy can also depend on the geographical provenience of CSS and the roasting process; however, a reduced content of these molecules can positively affect bitterness, astringency, and the cut grass smell sensations that are developed with cooking processes. As far as the radical-scavenging activity is concerned, the soluble and bound fractions differed significantly in the TEAC values, with the free fraction showing the highest value. This may be primarily related to the TPC values but maybe also a consequence of the different compositions in the patterns of phenolic compounds of the two extracts, with the soluble one being richer in molecules with higher antiradical activity with respect to the bound fraction. In general, the antiradical activity was comparable to results reported in the work by Borrelli et al. [53].

To gain a more in-depth understanding of the phenolic composition of CSS and to quantify caffein content, the extracts were submitted to HPLC analysis (Figure 1) and in Table 3 the contents of the most representative phenolic acids—chlorogenic and caffeic acids—are reported. Indeed, the major phenolic compounds in CSS are chlorogenic acids, which are esters of cinnamic acids with quinic acid; several isomers have been detected in CSS. However, 5-O-caffeoylquinic acid (5-CQA) was found to be the main one [54].

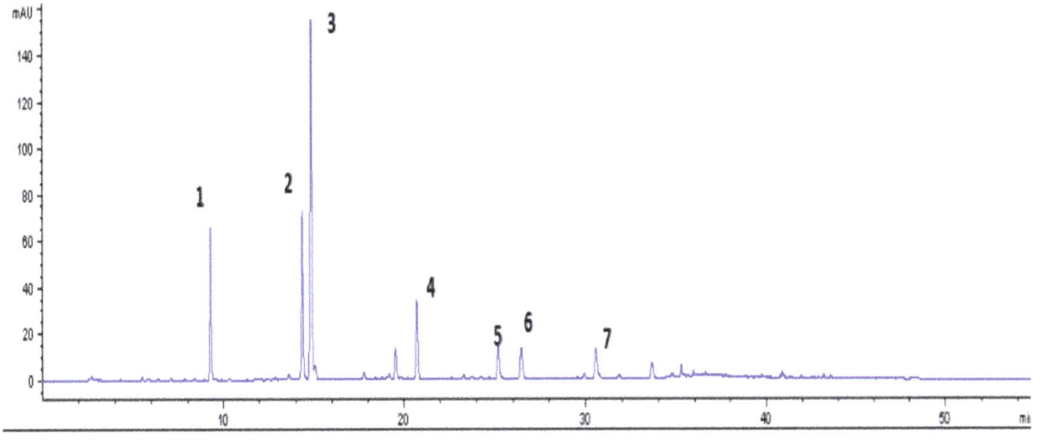

Figure 1. HPLC profile of coffee silver skin at 330 nm of phenolic acids in extract (MetOH:H_2O 70:30). Identified peaks are: caftaric acid (**1**), chlorogenic acid (**2**), caffeic acid (**3**), ferulic acid (**4**), chlorogenic acid isomers (**5**–**7**).

In the soluble and bound fractions, similar contents of caffeic and chlorogenic acids were found; generally, significantly higher contents were found in the soluble fraction. The amounts of chlorogenic acid found in the present work are generally in agreement with other works, in which contents of 0.12mg/g and 0.46 mg/g were found in CSS samples from India and Uganda, respectively [54].

The same HPLC method allowed the quantification of caffeine, which was found only in the soluble fraction at 17.45 ± 1.2 mg g^{-1}, a value which agrees with those found by Bessada et al. [2,54], as well as by Costa et al. [9]. Caffeine was not detected in the bound fraction, likely because, in the samples under investigation, it was not bound to other compounds and, under the experimental conditions used, it was thus possible to achieve a complete extraction with the first step.

Aqueous extracts were dialyzed and used to evaluate the melanoidin fraction; the absorbance at 345 nm had a value of 0.71 ± 0.01; Mésias and De Andrade [55] reported values of 4.5 g/100 g. Melanoidin content and the molecular weight are strictly dependent on the roasting process and temperature range at which coffee beans are roasted, as well as on the number of precursors available for the Maillard reaction.

CSS, and coffee by-products generally, mainly have water-soluble melanoidins, and so they were suggested as resources for the formulation of new products to impart antioxidant and antimicrobial functions when producing functional foods [54,55]. Tores de la Cruz et al. [30] suggest specific coffee compounds as being responsible for acidity, as well as chlorogenic acid, citric acid, quinic acid, trigonelline, caffeine, and arabinogalactan [30,32].

3.5. Minerals in CSS

Minerals are highly present in CSS [8]. As for fibers and bioactive molecules, minerals are important nutrients for human wellbeing, representing other valuable compounds to recover. We looked for those microelements that are useful for human health, such as calcium, potassium, and iron. We were able to trace out other minerals, such as phosphorous, magnesium, manganese, copper, sodium, zinc, nickel, chromium, and aluminum. The data obtained are shown in Table 4 and were compared with a few references available on mineral contents of coffee silver skin [9,51]. Costa et al. [9] reported CSS values of potassium (4977 ± 151 mg 100 g^{-1}), magnesium (2002 ± 72 mg 100 g^{-1}), calcium (584 ± 62 mg 100 g^{-1}), iron (41.8 ± 2.69 mg 100 g^{-1}), and sodium (5.32 ± 0.14 mg 100 g^{-1}). Ballesteros et al. [23] reported values of 470.60 ± 13.9 (mg/kg) for aluminum and 1.64 ± 0.34 (mg/kg) for nickel. This study confirms a high content of calcium, potassium, and phosphorus, and good values of iron, copper, and manganese in CSS. For aluminum, our results suggest lower contents with respect to cited data, as is the case for nickel.

Table 4. Mineral composition (mg Kg^{-1}) of CSS.

Minerals	(mg Kg^{-1})
Calcium	5465 ± 343
Magnesium	2227 ± 26
Phosphorus	1462 ± 27
Potassium	14,600 ± 387
Sodium	115.85 ± 2.31
Copper	72.15 ± 4.70
Chromium	5.55 ± 0.14
Iron	212.46 ± 6.56
Manganese	23.13 ± 2.23
Nickel	3.17 ± 0.23
Zinc	18.06 ± 1.56
Aluminum	223.67 ± 16.38

An important point is that CSS results to have large amounts of minerals and can be used as an integrator. Soetan et al. [56] remarked on how humans consume large amounts of minerals, with high needs for calcium and iron, used for bone renovation, the correct functioning of red blood cells, and cellular respiration. Magnesium, copper, selenium, zinc, iron, manganese, and molybdenum are important co-factors found in the structure of certain enzymes and are indispensable in numerous biochemical pathways. Conversely, some minerals are not involved in any biochemical pathway and can even be toxic or being used as a processing contamination index; aluminum and nickel are among them [56,57]. Minerals can have different health effects; however, their daily requirement levels are very low (as they are micronutrients), so a small amount of CSS in a newly formulated food could remarkably increase their intake. According to the World Health Organization (WHO), the daily calcium intake, for men and women from 19–50 years old, is fixed at 840 mg daily [58].

By looking at our data, it can be assumed that in 100 g of CSS, there is about 546.5 mg of calcium, so, keeping these values in mind, an addition of 1.5% of CSS to a final amount of 45 g of a new formulated food will give 16.38 mg of calcium (2% daily requirement). In some studies [59], the daily intake of calcium (for the same target population) is lower, tat 700 mg daily. Similarly, for potassium intake, the WHO [60] recommends 3510 mg daily for adults. Following the previous example, 3g of CSS (1.5% for a 45 g final product) provides 65.7 mg of potassium (1.91% of daily requirement). According to these data, potassium can be used as a replacement for salt (NaCl) in newly formulated foods.

The literature agrees that aluminum and nickel mainly come from fertilization, the high use of pesticides, and fungicides [56], which increase their content in the soil. Other causes are geographical and climatic conditions [57,61].

Aluminum and nickel are among the most abundant metals around the globe (they are normally contained in vegetable sources) and their concentration grows steeply due to the human activities of extraction and soil and water acidification [57,61,62]. Recently, the European Food Safety Agency reviewed the use of aluminum sulfates (E 520–523) and sodium aluminium phosphate (E 541) as food additives due to the higher intake in the human diet [63].

The EFSA statement [64] on the evaluation of aluminum oral bioavailability has shown drinking water values close to 0.3%. Often, the bioavailability of aluminum from food and beverages is lower, at about 0.1%.

However, the oral absorption of aluminum from food can vary at least 10-fold depending on the chemical forms present in the intestinal tract. The total burden of aluminum in healthy human subjects has been reported to be approximately 30–50 mg/kg bw. Considering a formulation containing 1.5% CSS, the aluminum intake would be 0.67g for each 100 g of food eaten. It was concluded that these additives are safe for the human diet, with a no observed adverse effect level (NOAEL) of 52 mg Al/kg body weight (bw) per day in rats. Limits for nickel were recently reviewed by the CONTAM group of EFSA [65] Although average intake from water does not represent a risk, some classes of the young population in Europe seem to be more exposed. The tolerable daily intake assumed to represent a risk had a lowest observed adverse effect level of 4.3 μg Ni/kg bw.

3.6. Volatile Organic Compounds (VOCs) in CSS

Volatile organic compounds (VOCs) in CSS, from this study, were grouped into seven classes (Table 5). The order of concentrations is nitrogen-containing compounds, pyrazines, aldehydes, alcohols, other compounds, hydrocarbons, and ketones. In coffee, many volatiles come from the roasting process of beans. Although there are few studies about VOCs from CSS, the typical volatiles found in coffee beans can be assumed for CSS.

Table 5. Qualitative profile (area %) of main volatile compounds (VOCs) from CSS.

VOCs from CSS	Area %
Aldehydes	
3-Furaldehyde	1.32
Benzaldehyde	2.51
Butanal, 2-methyl-	2.95
Butanal, 3-methyl-	3.19
Heptanal	1.51
Hexanal	8.83
Pentanal	0.81
Propanal, 2-methyl-	2.20
Alcohols	
1-Methylcyclopropanemethanol	3.00
1,3-Butanediol, (S)-	1.26
2-Furanmethanol	0.47
2-Furanmethanol, acetate	0.47
2-Nonen-1-ol	7.51
2-Octen-1-ol, (E)-	0.23
2-Octyn-1-ol	0.34
3-Hexen-1-ol, 2-ethyl-	0.43
Ketones	
2-Propanone, 1-(acetyloxy)-	1.11
Nitrogen-containing species	
(2-Aziridinylethyl)amine	1.30
3-Amino-2-oxazolidinone	1.98
Carane, 4,5-epoxy-, trans	1.58
Hydroxyurea	0.72
O-Methylisourea	0.43
Propanamide, 2-hydroxy-	0.32
Pyrazines	
1-Methyl-3-Phenylpiperazine	0.69
Pyrazine, methyl-	4.38
Pyrazine, 2-ethyl-6-methyl-	2.40
Pyrazine, 2-ethyl-3-methyl-	6.65
Pyrazine, 3-ethyl-2,5-dimethyl-	1.75
Pyrazole, 3,5-dimethyl-1-allyl-	0.78

Table 5. *Cont.*

VOCs from CSS	Area %
Hydrocarbons	
Ethylbenzene	1.08
Pentane, 1-(2-propenyloxy)-	1.44
Undecane,1,2-dibromo -2-methyl-	1.64
Other compounds	
3(2H)-Furanone,dihydro-2 methyl-	0.85
α-Pinene	0.27
Furan, 2-pentyl-	1.13
Methyl glyoxal	2.0
Pyrimidine, 4-methyl-	0.57
Pyrimidine, 2-methyl-	15.4

Recently, for the first time, Angeloni et al. [66] researched the volatiles of CSS. By HS-SPME-GC-MS, they found more than 80 active odorants, with many of them shown in the data reported here. Furthermore, by GC-O/FID and GC × GC-TOF analysis and comparison with coffee beans, they were able to identify some characteristic compounds, such as 1-octen-3-one, 2-phenylethanol, and γ-decalactone, not seen in this study.

Classes of compounds found here are also reported in other studies [67–69], although is very difficult to compare different samples because of the roasting parameters, provenience of coffee beans, their composition (also including the number of defective ones), the blending of cultivars chosen, and the selected processing method (wet vs. dry) [68]. Yang et al. [69] showed how the increase or decrease in thirty-seven typical compounds occurred during the roasting process, highly influencing the quality of the final coffee beans and used as quality markers. Specifically, five defects were found (light, scorched, dark, baked, and underdeveloped) and connected to elevated levels of indole, 4-ethyl-2-methoxy phenol, phenol, maltol, and 2,5-dimethylfuran, respectively. Besides these difficulties, Fisk and co-workers [70] listed fifteen "iconic" compounds which influence the coffee aroma. Among them, butanal, 2-methyl-butanal, 3-methyl-pyrazine, 2-ethyl-6-methyl-, pyrazine, and 3-ethyl-2,5-dimethyl were also present in CSS samples from this study (Table 5). A study on pyrolysis products (bio-oil) of coffee silver skin coupled with a two-dimensional (GC × GC/qMS) analysis [71] showed a great variety and complexity of compounds, as follows: N-compounds, phenols, unsaturated aliphatic hydrocarbons (8.28%), saturated hydrocarbons, aromatic hydrocarbons, ketones, esters, carboxylic acids, alcohols, and 10 compounds classified as "other". Pyrolysis allowed the authors to list the most abundant active compounds, find typical regions of elution, understand the compounds' provenience (i.e., phenols, coming from the degradation of lignin), and to adjust the pyrolysis conditions.

4. Conclusions

The present study provides data on the chemical and physical characteristics of coffee silver skin. Our findings show the advantage of the recovery of CSS that can help to lower the food waste generated by coffee production. More precisely, CSS can be considered a co-product of coffee beans and become a cheap but valuable integrator of fibers, bioactive molecules, and minerals such as calcium and potassium. Different proveniences of coffee beans and compositions of coffee blends after the roasting process can cause slight differences in mineral content and phenolic profile. As shown here, common features of the composition, fibers, minerals, and phenolic species can be seen. The aim for the characterization of CSS from this study is to encourage its use in different formulations, and the acrylamide content in each specific recipe should be evaluated for the maximum allowed CSS content.

This is a starting point of a deeper study to comprehend the best conditions for the use of this by-product and how to develop new formulations in which coffee silver skin could have a positive role in the control of quality and stability, in compliance with sustainable food production.

Author Contributions: Conceptualization, M.M., L.E., and C.D.D.M.; methodology, M.M., C.D.D.M., A.R., and L.E.; software, M.M. and L.E.; validation, M.M.; formal analysis, C.D.D.M., A.R., and L.E.; investigation, L.E. and M.M.; resources, M.M. and D.M.; data curation, M.M. and L.E.; writing—original draft preparation, M.M. and L.E.; writing—review and editing, M.M., L.E., A.R., C.D.D.M., and D.M.; visualization, M.M., L.E., C.D.D.M., A.R., and D.M.; supervision, M.M. All authors have read and agreed to the published version of the manuscript.

Funding: This research received no external funding.

Institutional Review Board Statement: Not applicable.

Informed Consent Statement: Not applicable.

Acknowledgments: The authors acknowledge Torrefazione Adriatica s.p.a., Giulianova (Teramo, Italy) for providing the coffee silver skin sample for this study.

Conflicts of Interest: The authors declare no conflict of interest.

References

1. Bahadur, K.C.; Dias, G.M.; Veeramani, A.; Swanton, C.J.; Fraser, D.; Steinke, D.; Lee, E.; Wittman, H.; Farber, J.M.; Dunfield, K.; et al. When too much isn't enough: Does current food production meet global nutritional needs? *PLoS ONE* **2018**, *13*, e0205683.
2. Bessada, S.M.F.; Alves, R.C.; Oliveira, M.B.P.P. Coffee Silverskin: A Review on Potential Cosmetic Applications. *Cosmetics* **2018**, *5*, 5. [CrossRef]
3. Balentić, J.P.; Ačkar, Đ.; Jokić, S.; Jozinović, A.; Babić, J.; Miličević, B.; Šubarić, D.; Pavlovic, N. Cocoa Shell: A By-Product with Great Potential for Wide Application. *Molecules* **2018**, *23*, 1404. [CrossRef]
4. Klingel, T.; Kremer, J.I.; Vera Gottstein, V.; Rajcic de Rezende, T.; Schwarz, S.; Lachenmeier, D.W. A Review of Coffee By-Products Including Leaf, Flower, Cherry, Husk, Silver Skin, and Spent Grounds as Novel Foods within the European Union. *Foods* **2020**, *9*, 655. [CrossRef] [PubMed]
5. Murthy, P.S.; Naidu, M.M. Sustainable management of coffee industry by-products and value addition—A review. *Resour. Conserv. Recycl.* **2012**, *66*, 45–48. [CrossRef]
6. Jannissen, B.; Huynh, T. Chemical composition and value-adding applications of coffee industry byproducts: A review. *Resour. Conserv. Recycl.* **2018**, *128*, 110–117. [CrossRef]
7. Alves, R.C.; Rodrigues, F.; Nunes, M.A.; Vinha, A.F.; Oliveira, M.B.P.P. State of the art in coffee processing by-products (chapter 1). In *Handbook of Coffee Processing By-Products*; Elsevier Inc.: Amsterdam, The Netherlands, 2017; pp. 1–26.
8. Gobena Gemechu, F. Embracing nutritional qualities, biological activities and technological properties of coffee byproducts in functional food formulation. *Trends Food Sci. Technol.* **2020**, *104*, 235–261. [CrossRef]
9. Costa, A.S.G.; Alves, R.C.; Vinha, A.F.; Costa, C.S.G.; Costa, M.A.; Nunes, A.A.; Almeida, A.; Santos-Silva Oliveira, M.B.P.P. Nutritional, chemical and antioxidant/pro-oxidant profiles of silverskin, a coffee roasting by-product. *Food Chem.* **2018**, *30*, 28–35. [CrossRef]
10. Iriondo-DeHond, A.; Garcia, N.A.; Fernandez-Gomez, B.; Guisantes-Batan, E.; Escobar, F.V.; Blanch, G.P.; San Andres, M.I.; Sanchez-Fortun, S.; del Castillo, M.D. Validation of coffee by-products as novel food ingredients. *Innov. Food Sci. Emerg. Technol.* **2019**, *51*, 194–204. [CrossRef]
11. Esquivel, P.; Jiménez, V.M. Functional properties of coffee and coffee by-products. *Food Res. Int.* **2012**, *46*, 488–495. [CrossRef]
12. Machado, S.; Costa, A.S.; Pimentel, F.; Oliveira, M.B.; Alves, R.C. A study on the protein fraction of coffee silverskin: Protein/non-protein nitrogen and free and total amino acid profiles. *Food Chem.* **2020**, *326*, 126940. [CrossRef]
13. Pourfarzad, A.; Mahdavian-Mehr, H.; Sedaghat, N. Coffee silverskin as a source of dietary fiber in bread-making: Optimization of chemical treatment using response surface methodology. *LWT-Food Sci. Technol.* **2013**, *50*, 599–606. [CrossRef]
14. Bertolino, M.; Barbosa-Pereira, L.; Ghirardello, D.; Botta, C.; Rolle, L.; Guglielmetti, A.; Borotto, C.; Dalla Vecchia, S.; Zeppa, G. Coffee silverskin as nutraceutical ingredient in yogurt: Its effect on functional properties and its bioaccessibility. *J. Sci. Food Agric.* **2019**, *99*, 4267–4275. [CrossRef] [PubMed]
15. García-Serna, E.; Martinez-Saez, N.; Mesias, M.; Morales, F.J.; del Castillo, M.D. Use of Coffee Silverskin and Stevia to Improve the Formulation of Biscuits. *Pol. J. Food Nutr. Sci.* **2014**, *64*, 243–251. [CrossRef]
16. Santos Ribeiro, V.; Leitão, A.E.; Cochicho Ramalho, J.; Cebola Lidon, F. Chemical characterization and antioxidant properties of a new coffee blend with cocoa, coffee silverskin and green coffee minimally processed. *Food Res. Int.* **2014**, *16*, 39–47. [CrossRef]
17. Martinez-Saez, N.; Ullate, M.; Martin-Cabrejas, M.A.; Martorell, P.; Genovés, S.; Ramon, D.; del Castillo, M.D. A novel antioxidant beverage for body weight control based on coffee silverskin. *Food Chem.* **2014**, *150*, 227–234. [CrossRef]

18. Walker, J.M.; Mennella, I.; Ferracane, R.; Tagliamonte, S.; Holik, A.K.; Hölz, K.; Somoza, M.M.; Somoza, V.; Fogliano, V.; Vitaglione, P. Melanoidins from coffee and bread differently influence energy intake: A randomized controlled trial of food intake and gut-brain axis response. *J. Funct. Food* **2020**, *72*, 104063. [CrossRef]
19. European Union. European Union Commission implementing regulation (EU) 2017/2468 of 20 December 2017 laying down administrative and scientific requirements concerning traditional foods from third countries in accordance with Regulation (EU) 2015/2283 of the European Parliament and of the Council on novel foods. *Off. J. Eur. Union* **2017**, *L351*, 55–63.
20. European Union. European Union Commission implementing regulation (EU) 2017/2469 of 20 December 2017 laying down administrative and scientific requirements for applications referred to in Article 10 of Regulation (EU) 2015/2283 of the European Parliament and of the Council on novel foods. *Off. J. Eur. Union* **2017**, *L351*, 64–71.
21. AOAC. *Official Methods of Analysis of AOAC International, Methods 2009.01, and 2011.25*, 19th ed.; AOAC International: Rockville, MD, USA, 2012.
22. Folch, J.; Lees, M.; Sloane-Stanley, G. A simple method for the isolation and purification of total lipids from animal tissues. *J. Biol. Chem.* **1957**, *226*, 497–509. [CrossRef]
23. Ballesteros, L.F.; Teixeira, J.A.; Mussatto, S.I. Chemical, functional, and structural properties of spent coffee grounds and coffee silverskin. *Food Bioprocess. Technol.* **2014**, *7*, 3493–3503. [CrossRef]
24. Di Mattia, C.D.; Martuscelli, M.; Sacchetti, G.; Scheirlinck, I.; Beheydt, B.; Mastrocola, D.; Pittia, P. Effect of Fermentation and Drying on Procyanidins, Antiradical Activity and Reducing Properties of Cocoa Beans. *Food Bioprocess. Technol.* **2013**, *6*, 3420–3432. [CrossRef]
25. Horžić, D.; Komes, D.; Belščak, A.; Kovačević Ganić, K.; Iveković, D.; Karlović, D. The composition of polyphenols and methylxanthines in teas and herbal infusions. *Food Chem.* **2009**, *115*, 441–448. [CrossRef]
26. Rivero-Perez, M.; Perez-Magarino, S.; Gonzalez-San, J. Role of melanoidins in sweet wines. *Anal. Chim. Acta* **2012**, *458*, 169–175. [CrossRef]
27. Delgado-Ospina, J.; Di Mattia, C.D.; Paparella, A.; Mastrocola, D.; Martuscelli, M.; Chaves-Lopez, C. Effect of Fermentation, Drying and Roasting on Biogenic Amines and Other Biocompounds in Colombian Criollo Cocoa Beans and Shells. *Foods* **2020**, *9*, 520. [CrossRef]
28. UNI EN 13657:2002. Characterization of waste—Digestion for subsequent determination of aqua regia soluble portion of elements.
29. Qi, J.; Wang, H.H.; Zhou, G.H.; Xu, X.L.; Li, X.; Bai, Y.; Yu, X. Evaluation of the taste-active and volatile compounds in stewed meat from the Chinese yellow-feather chicken breed. *Int. J. Food Prop.* **2017**, *20*, S2579–S2595. [CrossRef]
30. Tores de la Cruz, S.; Iriondo-DeHond, A.; Herrera, T.; Lopez-Tofiño, Y.; Galvez-Robleño, C.; Prodanov, M.; Velazquez-Escobar, F.; Abalo, R.; Del Castillo, M.D. An Assessment of the Bioactivity of Coffee Silverskin Melanoidins. *Foods* **2019**, *8*, 68. [CrossRef]
31. Mathlouthi, M. Water content, water activity, water structure and the stability of foodstuffs. *Food Control* **2001**, *12*, 409–417. [CrossRef]
32. Sacchetti, G.; Di Mattia, C.D.; Pittia, P.; Mastrocola, D. Effect of roasting degree, equivalent thermal effect and coffee type on the radical scavenging activity of coffee brews and their phenolic fraction. *J. Food Eng.* **2009**, *90*, 74–80. [CrossRef]
33. Gocmen, D.; Sahan, Y.; Yildiz, E.; Coskun, M.; Aroufai, I.A. Use of coffee silverskin to improve the functional properties of cookies. *J. Food Sci Technol.* **2019**. [CrossRef]
34. Ateş, G.; Elmacı, Y. Coffee silverskin as fat replacer in cake formulations and its effect on physical, chemical and sensory attributes of cakes. *LWT* **2018**, *90*, 519–525. [CrossRef]
35. Ateş, G.; Elmacı, Y. Physical, chemical and sensory characteristics of fiber-enriched cakes prepared with coffee silverskin as wheat flour substitution. *J. Food Meas. Charact.* **2019**. [CrossRef]
36. Narita, Y.; Inouye, K. Review on utilization and composition of coffee silverskin. *Food Res. Int.* **2014**, *61*, 16–22. [CrossRef]
37. Soliman, G.A. Dietary Fiber, Atherosclerosis, and Cardiovascular Disease. *Nutrients* **2019**, *11*, 1155. [CrossRef]
38. Dong, Y.; Chen, L.; Gutin, B.; Haidong, Z. Total, insoluble, and soluble dietary fiber intake and insulin resistance and blood pressure in adolescents. *Eur. J. Clin. Nutr.* **2019**, *73*, 1172–1178. [CrossRef]
39. González-Aguilar, G.A.; Blancas-Benítez, F.J.; Sáyago-Ayerdi, S.G. Polyphenols associated with dietary fibers in plant foods: Molecular interactions and bioaccessibility. *Curr. Opin. Food Sci.* **2017**, *13*, 84–88. [CrossRef]
40. Food and Nutrition Board. *Dietary Reference Intakes for Energy, Carbohydrates, Fiber, Protein, and Aminoacids*; National Academy of Sciences: Washington, DC, USA, 2002.
41. Prasadi, N.V.P.; Joyce, I.J. Dietary Fibre from Whole Grains and Their Benefits on Metabolic Health. *Nutrients* **2020**, *12*, 3045. [CrossRef]
42. Slavin, J.L.; Lloyd, B. Health Benefits of Fruits and Vegetables American Society for Nutrition. *Adv. Nutr.* **2012**, *3*, 506–516. [CrossRef]
43. Commission regulation (EU) 2017/2158 of 20 November 2017 establishing mitigation measures and benchmark levels for the reduction of the presence of acrylamide in food. *Off. J. Eur. Union* **2017**. Available online: http://data.europa.eu/eli/reg/2017/2158/oj (accessed on 12 May 2021).
44. Bagdonaite, K.; Derler, K.; Murkovic, M. Determination of acrylamide during roasting of coffee. *J. Agric. Food Chem.* **2008**, *56*, 6081–6086. [CrossRef]
45. Behrouzian, F.; Amini, A.M.; Alghooneh, A.; Razavi, S.M.A. Characterization of dietary fiber from coffee silverskin: An optimization study using response surface methodology. *Bioact. Carbohydr. Diet. Fibre* **2016**, *8*, 58–64. [CrossRef]

46. Belmiro, R.H.; Artigiani Lima Tribst, A.; Cristianini, M. Application of high pressure homogenization on gums. *J. Sci. Food Agric.* **2018**, *98*, 2060–2069. [CrossRef]
47. Belmiro, R.H.; de Carvalho Oliveira, L.; Geraldi, M.V.; Junior, M.R.; Cristianini, M. Modification of coffee coproducts by-products by dynamic high pressure, acetylation and hydrolysis by cellulase: A potential functional and sustainable food ingredient. *Innov. Food Sci. Emerg. Technol.* **2021**, *68*, 10268. [CrossRef]
48. Ignat, I.; Volf, I.; Popa, V.I. A critical review of methods for characterisation of polyphenolic compounds in fruits and vegetables. *Food Chem.* **2011**, *126*, 1821–1835. [CrossRef] [PubMed]
49. Çelik, E.E.; Gökmen, V. Interactions between free and bound antioxidants under different conditions in food systems. *Crit. Rev. Food Sci. Nutr.* **2021**, 1–17. [CrossRef] [PubMed]
50. Guglielmetti, A.; Fernandez-Gomez, B.; Zeppa, G.; Del Castillo, M.D. Nutritional quality, potential health promoting properties and sensory perception of an improved gluten-free bread formulation containing inulin, rice protein and bioactive compounds extracted from coffee byproducts. *Pol. J. Food Nutr. Sci.* **2019**, *69*, 157–166. [CrossRef]
51. Mesías, M.; Navarro, M.; Martínez-Saez, N.; Ullate, M.; del Castillo, M.D.; Morales, F.J. Antiglycative and carbonyl trapping properties of the water soluble fraction of coffee silverskin. *Food Res. Int.* **2014**, *62*, 1120–1126. [CrossRef]
52. Bresciani, L.; Calani, L.; Bruni, R.; Brighenti, F.; Del Rio, D. Phenolic composition, caffeine content and antioxidant capacity of coffee silverskin. *Food Res. Int.* **2014**, *61*, 196–201. [CrossRef]
53. Borrelli, R.C.; Esposito, F.; Napolitano, A.; Ritieni, A.; Fogliano, V. Characterization of a New Potential Functional Ingredient: Coffee Silverskin. *J. Agric. Food Chem.* **2004**, *52*, 1338–1343. [CrossRef]
54. Bessada, S.M.F.; Alves, R.C.; Costa, A.S.G.; Nunes, M.A.; Oliveira, M.B.P.P. Coffea canephora silverskin from different geographical origins: A comparative study. *Sci. Total Environ.* **2018**, *645*, 1021–1028. [CrossRef]
55. Mesías, M.; Delgado- Andrade, C. Melanoidins as a potential functional food ingredient. *Curr. Opin. Food Sci.* **2017**, *14*, 37–42. [CrossRef]
56. Soetan, K.O.; Olaiya, C.O.; Oyewole, O.E. The importance of mineral elements for humans, domestic animals and plants: A review. *Afr. J. Food Sci.* **2010**, *4*, 200–222.
57. Hardisson, A.; Revert, C.; Gonzales-Weler, D.; Rubio, C. Aluminium Exposure Through the Diet. *Food Sci. Nutr.* **2017**, *3*, 19.
58. FAO; WHO. *Human Vitamin and Mineral Requirements*; FAO: Rome, Italy; WHO: Geneva, Switzerland, 2001.
59. EFSA Panel on Dietetic Products, Nutrition and A (NDA). Scientific Opinion on principles for deriving and applying Dietary Reference Values. *EFSA J.* **2010**, *8*, 1–30.
60. World Health Organization. Guideline: Potassium Intake for Adults and Children. Geneva. 2009. Available online: http://www.who.int/nutrition/publications/guidelines/potassium_intake/en/ (accessed on 2 May 2021).
61. Sharma Ashimav, D. Low Nickel Diet in Dermatology. *Ind. J. Dermatol.* **2013**, *58*, 240–250. [CrossRef]
62. Vītola, V.; Ciproviča, I. The Effect of Cocoa Beans Heavy and Trace Elements on Safety and Stability of Confectionery Products. *Rural Sustain. Res.* **2016**, *35*, 19–23. [CrossRef]
63. European Food Safety Authority (EFSA). 2018. Available online: https://efsa.onlinelibrary.wiley.com/doi/full/10.2903/j.efsa.2018.5372 (accessed on 2 May 2021).
64. European Food Safety Authority (EFSA). On the Evaluation of a New Study Related to the Bioavailability of Aluminium in Food. *EFSA J.* **2011**, *9*, 2157. [CrossRef]
65. European Food Safety Authority (EFSA). Update of the risk assessment of Nickel in food and drinking water. *EFSA J.* **2020**, *18*, e06268. [CrossRef]
66. Angeloni, S.; Scortichini, S.; Fiorini, D.; Sagratini, G.; Vittori, S.; Neiens, S.D.; Steinhaus, M.; Zheljazkov, V.D.; Maggi, F.; Caprioli, G. Characterization of Odor-Active Compounds, Polyphenols, and Fatty Acids in Coffee Silverskin. *Molecules* **2020**, *25*, 2993. [CrossRef]
67. Caporaso, N.; Whitworth, M.B.; Cui, C.; Fisk, I.D. Variability of single bean coffee volatile compounds of Arabica and robusta roasted coffees analysed by SPME-GC-MS. *Food Res. Int.* **2018**, *108*, 628–640. [CrossRef]
68. De Toledo, P.A.B.; de Melo, M.M.R.; Pezza, H.R.; Toci, A.T.; Pezza, L.; Silva, C.M. Discriminant analysis for unveiling the origin of roasted coffee samples: A tool for quality control of coffee related products. *Food Control* **2017**, *73*, 164–174. [CrossRef]
69. Yang, N.; Liu, C.; Liu, X.; Kreuzfeldt Degn, T.; Munchow, M.; Fisk, I. Determination of volatile marker compounds of common coffee roast defects. *Food Chem.* **2016**, *15*, 206–214. [CrossRef]
70. Fisk, I.D.; Kettle, A.; Hofmeister, S.; Virdie, A.; Kenny, J.S. Discrimination of roast and ground coffee aroma. *Flavour* **2012**, *1*, 14. [CrossRef]
71. Dos Santos Polidoro, A.; Scapin, E.; Lazzari, E.; Nunes Silva, A.; Loreiro dos Santos, A.; Bastos Caramão, E.; Assis Jacques, R. Valorization of coffee silverskin industrial waste by pyrolysis: From optimization of bio-oil production to chemical characterization by GC × GC/qMS. *J. Anal. Appl. Pyrol.* **2018**, *129*, 43–52. [CrossRef]

foods

MDPI

Article

Cacao Pod Husk Flour as an Ingredient for Reformulating Frankfurters: Effects on Quality Properties

Johannes Delgado-Ospina [1,2,*], Maria Martuscelli [1], Carlos David Grande-Tovar [3], Raquel Lucas-González [4], Junior Bernardo Molina-Hernandez [1], Manuel Viuda-Martos [4], Juana Fernández-López [4], José Ángel Pérez-Álvarez [4,5,*] and Clemencia Chaves-López [1,4]

1 Faculty of Bioscience and Technology for Food, Agriculture and Environment, University of Teramo, Via R. Balzarini 1, 64100 Teramo, Italy; mmartuscelli@unite.it (M.M.); juniorbernardo.molinahernandez@studenti.unite.it (J.B.M.-H.); cchaveslopez@unite.it (C.C.-L.)

2 Grupo de Investigación Biotecnología, Facultad de Ingeniería, Universidad de San Buenaventura Cali, Carrera 122 # 6-65, Cali 76001, Colombia

3 Grupo de Investigación de Fotoquímica y Fotobiología, Universidad del Atlántico, Carrera 30 # 8-49, Puerto Colombia 081008, Colombia; carlosgrande@mail.uniatlantico.edu.co

4 IPOA Research Group, Centro de Investigación e Innovación Agroalimentaria y Agroambiental de la UMH (CIAGRO), Miguel Hernández University, Orihuela, CYTED-Healthy Meat. 119RT0568 "Productos Cárnicos más Saludables", 03312 Alicante, Spain; raquel.lucas@graduado.umh.es (R.L.-G.); mviuda@umh.es (M.V.-M.); j.fernandez@umh.es (J.F.-L.)

5 Faculty of Science, King Abdelaziz University, Jedda 21589, Saudi Arabia

* Correspondence: jdelgado1@usbcali.edu.co (J.D.-O.); ja.perez@goumh.umh.es (J.Á.P.-Á.)

Abstract: The cocoa pod husk is considered a source of dietary fiber with a high content of water-soluble pectins, bioactive compounds which should be viewed as a by-product with the potential to be incorporated into food. This study aimed to investigate the effect of adding different cocoa pod husk flour (CPHF) levels as a starch replacement for reformulating frankfurters. Results showed that the addition of 1.5 and 3.0% pod husk proportionally increased the frankfurter's fiber content by 0.49 ± 0.08 and 0.96 ± 0.19 g/100 g, which is acceptable for a product that does not contain fiber. Textural properties and sensory characteristics were affected when substituting the starch with CPHF, either totally or partially, although these samples had higher water content, hardness, and adhesiveness while springiness decreased. Non-adverse effects of nitrite on polyphenolic compounds content were evidenced in samples enriched with CPHF. The incorporation of CPHF did not significantly affect the color parameters ($\Delta E < 3$). Finally, the panelists indicated a sensation of the unsalted sausage, suggesting that CPHF may have natural mucoadhesion properties. In conclusion, in formulated meat products such as sausages, plant co-products such as cacao pod husks could be a valid new ingredient to improve technological parameters, functional characteristics, and stability.

Keywords: by-product; co-product; dietary fibers; functional foods; meat products; cacao pod husk flour

1. Introduction

The COVID-19 pandemic highlighted the need for healthier dietary patterns, encourage food producers to develop new products. The observatories that exist, globally and nationally, are alert to these changes to help companies re-direct their activities. That is why the pandemic will be an excellent opportunity to develop new products in the world of food, using research and development plus innovation strategies. These strategies are aimed at improving the health and wellness of consumers through products enriched with proteins of high biological value (chia, quinoa, amaranth) [1,2], vitamins, minerals, and, above all, dietary fiber, without forgetting that the environment must be respected [3]. In this context, extracts rich in dietary fiber derived from different parts of plants could be used as would-be functional ingredients due to their potential human health benefits that go far beyond supporting bowel regularity [4]. Moreover, the current consumer increasingly believes that food contributes to their health [5].

The reports of diseases that are directly related to low consumption of fiber in people's diet, including constipation, diarrhea, ulcerative colitis, diverticulosis, colorectal cancer, cardiovascular disease, and diabetes, are increasing alarmingly [6]. The low fiber consumption depends on eating habits in which foods with low fiber content or easily prepared processed foods to which the fiber has been removed as part of the process are favored. Increasing the incorporation of fiber in food also means looking for new sources of fiber to add to those products already in the market, which is why co-products are an excellent alternative in the search for this type of fiber [7].

Meat is a food group that is not usually a significant source of dietary fiber in the diet as vegetables. Meat products such as comminuted meat products (sausages, mortadella) and hams are no exception, although their formulation incorporates spices that provide small amounts of fiber. However, in ground meat products, there is the possibility of improving their formulation appropriately, managing to incorporate different attributes that allow them to become healthier foods. Several studies are related to meat product innovations. In those studies, several ingredients from spices, fruits, and vegetable by-products were added as bioactive compounds, improving the nutritional, physicochemical, textural, and sensory properties [8,9]. They have even been used as biopreservative additives [10].

Some studies have been carried out showing that some dietary fibers can be successfully incorporated into meat products, making an essential contribution of fiber in the diet and improving some food properties [11,12]. One of the benefits of fiber is that it increases satiety, so when incorporated into processed meat products, it could generate a decrease in the size of the portions, causing the net consumption of meat to decrease and, therefore, the greenhouse gas emissions from livestock. However, the incorporation of fiber is not easy. The incorporation of the fiber depends on the technological properties (concentration, particle size, water-holding capacity, and swelling capacity, among others), physical properties (color it contributes, texture), physicochemical properties (pH, oxidation–reduction potential, activity antioxidant, and prooxidant), chemical properties (composition, type of fiber, etc.), and sensory properties. Even industrial-scale production capacity must be assessed. Therefore, its addition requires extensive scientific and technological studies.

Some by-products, such as cacao pod husk, are important sources of dietary fiber that can be used in processed foods [13]. According to the International Cocoa Organization (ICCO) [14], in the 2019–2020 period, 4,697,000 tons of cocoa were produced globally, representing only 15% of the fruit. The remainder, more than 22,000,000 tons, is represented by pod husk (70 to 75% of the fruit).

Cocoa pod husk (CPH) is considered as a source of bioactive compounds [3,15], antioxidant dietary fiber with a high content (up to 6%) of minerals (K, P, Ca, and Mg) [16], high content of water-soluble pectins [17], and a source of hydrolase enzymes [18]. Its techno-functional properties were recently evaluated, and the high water-holding capacity suggests that it can be used as a good substitute for replacement of emulsifiers or water retainers, incorporating the fiber and antioxidant properties into new food products at the same time [3]. On the other hand, the use of by-products such as cacao pod husk flour generates benefits for the environment. Cacao pod husk is usually left to degrade on the cocoa plantation, causing inoculum of fungal diseases in cacao crops such as black pod rot [3]. In addition, it has resulted in silting, blockage of water drainage systems, and flooding of rivers and other waterbodies.

The ingestion of soluble pectins present multiple benefits. They can increase satiety and decrease caloric intake and adiposity during a high-fat diet, such as when there is consumption of sausages. The objective of the present work was to study the effect of CPH as a partial or total substitute for starch in the reformulation of frankfurters.

2. Materials and Methods

2.1. Cacao Pod Husk Flour (CPHF)

Fresh cocoa fruits were collected from a Criollo cacao processing farm in Colombia and taken to the laboratory where they were opened for extracting their grains and mucilage. Cocoa pods husks (CPH) were cut into 1 cm wide strips and immediately placed in boiling water for 10 min to inactivate the enzymes, then cooled and taken to undergo a freeze-drying process in a Liophilizer FreeZone 4.5 L (Labconco Corporation, Kansas, USA). Freeze-dried CPH were ground in a Hammer Mill M20 (IKA, Staufen, Germany), passed through a sieve (Dp < 200 μm) to obtain the flour (hereinafter CPHF), and stored at 4 °C until being used in the analyses. The chemical, physical–chemical, and techno-functional properties of CS were investigated in our previous study [3].

2.2. Frankfurters Elaboration Process and Treatments

Frankfurters were prepared following a traditional formulation described in [19]. The ingredients were pork shoulder lean meat (70%) and pork backfat (30%). The following percentages of ingredients refer to meat: 15% water (ice form *w*/*w*), 2.5% sodium chloride (*w*/*w*), 1.5% casein, 500 mg/kg sodium ascorbate, 300 mg/kg sodium tripolyphosphate, 150 mg/kg sodium nitrite, 0.5% (*w*/*w*) liquid smoke, 0.2% (*w*/*w*) white pepper, 0.05% (*w*/*w*) nutmeg, and 0.05% (*w*/*w*) coriander powder. For the treatments, the concentrations of CPHF and potato starch (PS) were varied (Table 1).

Table 1. Treatments in the frankfurter elaboration.

Treatment	CPHF (%)	PS (%)
Control	0	3.0
CPHF 1.5 *	1.5	0
CPHF 3.0	3.0	0
CPHF-PS 1.5	1.5	1.5

CPHF: Cacao pod husk flour. PS: Potato starch. * In this treatment, no other ingredient was added to complete the 3.0% generalized in the other treatments.

The frankfurters were prepared at the IPOA Research Group pilot plant at the Miguel Hernández University, Orihuela Spain. Briefly, pork lean meat and pork backfat previously cooled below 12 °C were ground in a cutter (1094-Homogeneizer, Tekator, Höganäs, Sweden) and mixed according to the treatments with the rest of the ingredients for 2 min. After homogenization, the emulsion formed was filled using an EM-12 piston stuffer (Mainca, Granollers, Barcelona, Spain) in cellulose wraps 20 mm in diameter (Fibran, Girona, Spain). The 20 cm long frankfurters were cooked in a water bath (80 °C) until reaching an internal temperature of 72 °C which was verified using a thermocouple probe (Omega Engineering, Inc., Stamford, CT, USA) placed in the geometric center of a sample. The frankfurters were immediately cooled in a blast chiller for 5 min, peeled, vacuum-packed, and stored at 4 °C until used in the analyses [19].

2.3. Proximate Composition

The proximate composition was determined in CPHF and frankfurters according to the following AOAC methods for meat: lipid (AOAC 991.36), protein (AOAC 981.10), moisture (AOAC 925.45), ash (AOAC 923.03), and general method for total dietary fiber in foods (TDF) (AOAC 985.29) [20], and the carbohydrate content was determined by calculating the percent remaining after all the other components have been measured (%carbohydrates = 100 − %moisture − %protein − %lipid − %TDF − %ash).

2.4. Physicochemical Analysis

The pH of CPHF was measured with an electrode probe connected to a pH meter (GLP 21, Crison, Barcelona, Spain) diluting the CPHF in distilled water (1:10), and the pH of frankfurters was measured for direct penetration into frankfurter at three different points using an electrode for penetration measurement. The water activity (a_w) for CPHF and Frankfurters was measured at 25 °C using a water activity analyzer (Novasina TH-500, Pfaeffikon, Zürich, Switzerland).

2.5. Residual Nitrite Level

Residual nitrite level was determined according to standards ISO/DIS 2918.26 [21] with modifications. In brief, ten grams of the previously homogenized samples were placed in an Erlenmeyer flask with 100 mL of water and incubated at 100 °C for 15 min with constant stirring. The samples were cooled in an ice bath, then 2.0 mL of Carrez I reagent and 2.0 mL Carrez II reagent (Merck, Darmstadt, Hesse, Germany) were added, it was left standing for 30 min, its volume was made up to 200 mL in a volumetric flask, and it was filtered through Whatman qualitative filter paper. In a test tube, 5.0 mL of the solution and 5.0 mL of the color reagent (alpha-naphthylamine chloride 0.03% w/v in acetic acid 20% v/v) were mixed, left to rest for 20 min, and absorbance at 520 nm determined using a spectrophotometer. The calibration curve was prepared from successive dilutions of a sodium nitrite standard, adding the color reagent in the same way and proportion as the samples. The results were compared against a sodium nitrite calibration curve and expressed as μg $NaNO_2$/g sample.

2.6. Measurement of Lipid Oxidation: Thiobarbituric Acid Index (TBARS)

The extent of lipid oxidation was determined by measuring the TBARS-reacting substances in frankfurters. Values were expressed as mg of malondialdehyde (MDA)/kg of a sample using the procedure described in [22]. In brief, 2.0 g of sample was homogenized with 16 mL of 10% trichloroacetic acid (TCA) for 15 min. The sample was placed at rest for 30 min in an ice bath. The sample homogenized was filtered through Whatman qualitative filter paper (grade 1) into 25 mL Erlenmeyer flasks. Filtered solution (2 mL) was mixed with 2 mL of 0.5% thiobarbituric acid (TBA) in distilled water in capped test tubes. Tubes were incubated in boiling water for 35 min. The absorbance was determined at 532 nm against a blank containing 2 mL of 10% TCA and 2 mL of 0.5% TBA solution. The calibration curve was prepared from successive dilutions of a malonaldehyde standard (prepared by acid hydrolysis of 1,1,3,3-tetramethoxypropane), adding the TBA reagent in the same way and proportion as the samples.

2.7. Color, Reflectance Spectra and Reflectance Ratios

Colorimetric analysis was performed using a CM-700d Spectrophotometer (Konica Minolta, Osaka, Japan). The CIELAB color coordinates (L*, a*, and b*), chroma C* (Equation (1)), hue angle h_{ab} (Equation (2)), color difference ΔE* (Equation (3)), and reflectance spectra (360 nm–740 nm) were determined with the following settings: illuminant D65, observer 10°. Guidelines for meat color evaluation were applied [23] and the method of [24] was followed. For the CPHF, it was deposited in a rectangular cell of 20 mm path length (CM-A132) and the measurement was carried out on both sides of the cell, while the frankfurters were cut into several 2 cm cross sections for the analysis. The reported values correspond to the average of four measurements. The reflectance ratios R630/R580, R650/R570, and R560/R500 were calculated from the reflectance values obtained for the wavelength defined for each.

$$C^* = \sqrt{a^{*2} + b^{*2}} \tag{1}$$

$$h_{ab} = \arctan\frac{b^*}{a^*} \tag{2}$$

$$\Delta E^* = \sqrt{(\Delta L^*)^2 + (\Delta a^*)^2 + (\Delta b^*)^2} \tag{3}$$

2.8. Textural Properties

Texture profile analysis was performed out on frankfurters according to the methodology used in [19]. A Texture Analyzer TA-XT2i (Stable Micro Systems, Surrey, England) was used. Frankfurter cross-sections (15 mm long) were subjected to a two-cycle compression to 75% deformation of their original height with an activation force of 5 g, speed of 5 mm/s, and 5 s retrieval between cycles. The following attributes were obtained: hardness (N), adhesiveness, springiness (mm), cohesiveness (Nmm), gumminess, chewiness (Nmm), and resilience. Four determinations were made per sample.

2.9. Sensory Evaluation

The frankfurters were cooked, and were tested after storage at 4 °C (Fridge, Siemens BSH, Zaragoza, Spain) for two days to evaluate color, taste (saltiness), texture (compactness, greasiness, juiciness, and hardness), flavor, and general acceptability using 25 sensory assessors untrained in the sensory analysis of frankfurters (12 males and 13 females) aged 18–55 years, according to protocols described in [19]. In brief, four pieces of 2.0 cm width (one from each treatment) were cut from the frankfurters and served with unsalted crackers and mineral water to clean the palate between samples (all served at room temperature). This analysis was performed under white fluorescent lights in individual booths, with the room temperature kept at 25 °C. The hedonic scale consisted of 9 levels (1—dislike extremely and 9—like extremely).

2.10. Statistical Analysis

Three independent experiments were made (10 kg of sausages for each experiment were elaborated), three replications of each factor and level were made, and three repeats were analyzed for each sample. The analysis of variance (ANOVA) was performed to determine the statistical significance ($p < 0.05$) of adding different concentrations of CPHF and PS: control (3.0% PS); CPHF 1.5 (1.5% CPHF); CPHF 3.0 (3.0% CPHF); and CPHF-PS 1.5 (1.5% PS and 1.5% CPHF). The Tukey HDS test ($p \leq 0.05$) was used for means comparisons. Data are presented as mean values ± standard deviation (SD). The Statgraphics Centurion XVI program (Statgraphics Technologies, Virginia, USA) was used for these statistical analyses.

3. Results and Discussion

3.1. Proximate Composition of Frankfurters

The compositional analysis of CPHF was previously reported in [3] and is presented in Table 2 to understand the frankfurter results. The technological properties depend significantly on the material's chemical composition, mainly due to the chemical interactions between proteins, lipids, and fiber with water and oil molecules. In CPH, the composition reported by different authors varies widely. However, the protein and lipid values are within the reported ranges in the literature [15]. The soluble fiber content, which corresponds to 30% of the total dietary fiber, is mainly attributed to the presence of pectins [25].

As expected, the addition of CPHF increased the frankfurter's fiber content ($p < 0.05$). Thus, the contribution from the fiber to the nutritional content of the sausage is essential. Similarly, a significant increase in protein content was found, consistent with the content provided by CPHF. A significant increase was found concerning the control, related to the high water-holding capacity (WHC) of the CPHF (28.90 ± 1.82 g/g) found by us in a previous study [3], which increases the hydration of the sausage. Similar results are found with the addition of chia seed flour to sausages [19].

The lipids showed a significant decrease ($p < 0.01$) concerning the control. This decrease is related to the low oil-holding capacity (OHC) of the CPHF (1.99 ± 0.02 g/g) found by us in a previous study [3], which slightly decreases the ability to stabilize the emulsion, which can be quickly released during cooking. This trend can be something positive from the point of view of having foods with lower fat content as long as it does not affect the other sensory and texture characteristics of the product. Proximate composition data expressed on a dry matter is shown in Table S1.

Table 2. Proximate composition (g/100 g sample), TBA value, residual nitrite level, pH, and a_w of cacao pod husk and frankfurters.

	CPHF	Frankfurters				
		Control	CPHF 1.5	CPHF 3.0	CPHF-PS 1.5	Sig.
Protein (g/100 g)	4.85 ± 0.01	12.82 ± 0.39 a	13.76 ± 0.31 ab	14.32 ± 0.61 b	13.89 ± 0.19 b	*
Lipid (g/100 g)	0.77 ± 0.15	24.43 ± 0.11 c	20.72 ± 0.46 b	16.09 ± 0.10 a	20.01 ± 1.35 b	**
Total dietary fiber (g/100 g)	37.4 ± 1.0	0.06 ± 0.17 a	0.49 ± 0.08 b	0.96 ± 0.19 c	0.52 ± 0.09 b	*
Carbohydrates (g/100 g)	49.7 ± 1.0	6.4 ± 0.4 c	4.8 ± 0.4 b	5.8 ± 0.6 c	3.1 ± 1.3 a	*
Moisture (g/100 g)	1.08 ± 0.20	53.29 ± 0.41 a	57.43 ± 0.06 b	60.02 ± 0.21 c	59.84 ± 0.21 c	**
Ash (g/100 g)	7.3 ± 0.1	2.91 ± 0.03 b	2.75 ± 0.18 ab	2.80 ± 0.01 b	2.57 ± 0.09 a	*
pH	5.47 ± 0.03	6.30 ± 0.03 b	6.36 ± 0.04 b	6.38 ± 0.06 b	6.09 ± 0.06 a	**
a_w	0.317 ± 0.015	0.941 ± 0.020	0.946 ± 0.010	0.949 ± 0.008	0.954 ± 0.009	n.s.
Residual nitrite level (mg $NaNO_2$)/kg		47.51 ± 1.82 b	41.18 ± 1.21 a	47.70 ± 0.23 b	36.46 ± 4.62 a	*
TBA * (mg MDA/kg product)		0.18 ± 0.03 a	0.25 ± 0.04 b	0.37 ± 0.09 c	0.28 ± 0.07 bc	*

* TBA, thiobarbituric acid value; MDA, malonaldehyde. Results are expressed as means of three samples ± standard deviations. Different letters in the same row indicate significant differences (LSD test, $p < 0.05$). Asterisks indicate significance at * $p < 0.05$; ** $p < 0.01$; n.s. not significant. CPHF Dp < 200 µm.

3.2. Physicochemical Analysis. pH and a_w

The addition of the CPHF to the frankfurters did show significant differences ($p < 0.01$) in the pH concerning the control, although the pH of the CPHF is slightly lower. A decrease in the treatment that combines starch and CPHF (CPHF-PS 1.5), due to starch–protein interactions being favored, leaves the CPHF groups without interaction, which causes the carboxyl groups to be exposed, showing a decrease in the pH of the sausage. All sausages showed a similar a_w value ($p > 0.05$), results that are slightly lower than those reported for other sausages (0.984 a_w) [19]. Water activity, pH, and temperature are the main parameters that directly impact the growth of microorganisms. Decreasing the water activity in the food limits the growth of microorganisms. In the case of meat products, the growth of bacteria generally occurs due to the high a_w, and it is not recommended in terms of the sensory properties to take the product to an intermediate humidity (a_w 0.6–0.85). Therefore, it is recommended to lower the pH of the product and maintain a low temperature during storage.

3.3. Residual Nitrite Level

The use of sodium nitrite in pork processing mainly has important effects on the development of color, acts as an antioxidant additive, prevents warmed-over flavor, and ensures microbiological safety; therefore, the level of residual nitrite is vital to prevent *Clostridium botulinum* spore development, which causes botulism, although decreasing its daily consumption is essential to alleviating the potential risk of carcinogenic, teratogenic, and mutagenic nitroso compounds [26].

Some previous studies have reported that the level of added nitrite in processed meat products decreases over time because nitrite is reduced to nitric oxide (NO), which reacts with myoglobin to form nitric oxide myoglobin [27]. The decrease in nitrite content also depends on factors such as pH, initial nitrite concentration, processing and storage temperatures, the meat/water ratio, and the presence of reducers [28].

In CPHF, residual nitrites levels were not detected. The inclusion of CPHF in the sausage does not alter the residual nitrite level, as is usual when vegetable co-products are used. The addition of vegetables by-products or co-products can incorporate nitrites into meat products since nitrates are naturally found in some parts of plants, especially in green leaves [29]. It has been reported that meat products made with by-products derived from quinoa provide some considerable amount of nitrates due to their presence in seeds, which could be used in place of nitrate/nitrite addition in the formulation [1].

The content found in the treatments is higher than 30 mg/kg, which is the lower limit to avoid toxinogenesis under storage conditions, and without exceeding the upper limit of 150 mg/kg [30]. The higher content was found in the control and the CPHF 3.0 treatment, while CPHF 1.5 and CPHF-PS 1.5 presented statistically lower values (Table 2). Although it has been reported that nitrite can react with bioactive compounds, especially with polyphenolic compounds [31], and that CPHF contains large amounts of bioactive compounds [3], no evidence of this was observed in the treatments.

3.4. Lipid Oxidation. Thiobarbituric Acid Index (TBARS)

The TBARS values are presented in Table 2. Frankfurters containing CPHF showed a statistically significant increase in TBARS ($p < 0.05$). The sample with the highest co-product content (CPHF 3.0) presented an increase of 105% concerning the control. However, all samples' TBARS values were lower than the acceptance limit of TBARS for rancidity (0.5 mg MDA/kg) [32]. In this study, the sausages' TBARS values were slightly higher than those mentioned by Fernández [19], in Frankfurter-type sausages added with chia co-product.

Many factors can influence the degree of oxidation of meat and meat products. However, the rate and extent of oxidation can be retarded, reduced, or prevented by applying natural antioxidants [33]. Although not conclusive, the results obtained contrast with the antioxidant capacity of CPHF (ABTS (139.9 μmol TE g^{-1}) and DPPH (132.8 μmol TE g^{-1})) reported in [3], and more in-depth studies should be done in this regard.

In some cases, the antioxidant effect of several types of polyphenol-rich dietary fibers is reversible in animal-based foods. This effect is concentration-dependent; in low concentrations it acts as an antioxidant, and when the concentration increases, the fiber acts as a prooxidant. Moreover, this effect could be the reason that higher CPHF concentrations showed higher TBA values.

3.5. Color

The hydrated CPHF showed a significant decrease in the color parameters (L*, a*, C*, h_{ab}) concerning the dry CPHF. Only the b* value increased. The dry and hydrated CPHF has a pale-yellow color, so the frankfurter color may slightly change when added. Color data are reported in Table 3.

Table 3. Color parameters of cocoa pods husk and frankfurters.

	L*	a*	b*	C*	h_{ab}	ΔE*
Cocoa pod husk						
Dry	78.92 ± 1.24	4.02 ± 0.35	25.23 ± 0.46	25.55 ± 0.50	80.96 ± 0.59	
Hydrated	53.25 ± 0.68	4.56 ± 0.69	22.95 ± 1.63	23.40 ± 1.74	78.81 ± 0.87	
Frankfurters						
Control	69.32 ± 0.69	6.42 ± 0.15 ab	10.19 ± 0.16 a	12.05 ± 0.17 a	57.79 ± 0.58 a	-
CPHF 1.5	70.63 ± 0.83	6.23 ± 0.37 ab	11.25 ± 0.18 b	12.87 ± 0.25 b	61.05 ± 1.45 b	1.82 ± 0.59 ab
CPHF 3.0	69.25 ± 0.86	6.53 ± 0.19 b	12.31 ± 0.41 c	13.94 ± 0.39 c	62.05 ± 0.99 b	2.28 ± 0.35 a
CPHF-PS 1.5	69.74 ± 0.14	6.04 ± 0.08 a	11.46 ± 0.15 b	12.96 ± 0.17 b	62.22 ± 0.08 b	1.40 ± 0.08 b
Significance	n.s.	*	*	*	**	*

L*, lightness; a*, red/green coordinate; b*, yellow/blue coordinate; C*, chroma; H*, hue; color differences ΔE*. Results are expressed as means of four samples ± standard deviations. Different letters in the same column (only frankfurters) indicate significant differences (LSD test, $p < 0.05$). Asterisks indicate significance at * $p < 0.05$; ** $p < 0.01$; n.s. not significant. CPHF Dp < 200 μm.

The addition of CPHF to the frankfurters did not show significant changes in L* values, indicating homogeneous incorporation within the meat emulsion in concordance with [34]. On the other hand, the a* and b* values are useful for identifying the evolution of a meat product since they both decrease during oxidation, being the best indicators of metmyoglobin or nitrosyl hemochrome changes during oxidation [35,36]. The CPHF addition did not cause a significant variation in the a* values, while there was an increase in the b* values. In the frankfurter samples, the measurement of the parameters a* and b* suggests that the changes are related to the CPHF color's contribution and not to oxidative processes. Although the TBA analysis showed slight increases with the addition of CPHF, this oxidative process did not directly affect the color, or the concentration of oxidizing substances was not high enough to affect the color, since it would be "limited" by the antioxidant compounds present in the product.

The increase in CPHF in the frankfurters produced an increase in the C* and h_{ab} values so that the samples showed a greater intensity of color and the tone changed from red to yellow. This indicates no myoglobin oxidation to metmyoglobin, which occurs when the chroma value decreases and h_{ab} remains constant [36], reinforcing the argument that the color changes are related to the color contribution of the CPHF.

There were no significant differences in the ΔE* values between the treatments, and these presented values below 2.3. Most reports indicate that values > 3 correspond to changes perceptible by the observer [19]; in this sense, the addition of CPHF produced values below the threshold, which implies that this addition, although not having established values for meat products such as sausages, is not perceptible to the human eye.

In general, the changes presented in the different color parameters are not perceptible, so the CPHF incorporation does not affect frankfurter presentation. Uniformity and color stability are important parameters for the visual acceptance of meat products [23].

3.5.1. Reflectance Spectra

In terms of color, the relationships between dietary fibers, meat matrix, and water are crucial to understanding the color evolution during processing storage and shelf-life. Reflectance spectra are the meat product fingerprint, so it is very important for its evaluation and reflectance ratio.

Figure 1 presents the reflectance spectrum (360–740 nm) obtained from frankfurters and CPHF. A pronounced increase in dry CPHF was observed above 500 nm, indicating a contribution from photosynthetic pigments (green, yellow, and orange pigments) and phenolic compounds. In hydrated CPHF, the spectrum decreased ($p < 0.05$). Hydration can occur by absorption, a process in which water is absorbed into the matrix and the reflection is reduced (diffuse reflectance), which was observed in hydrated CPHF. It can also occur by adsorption, a process in which solid particles with water on the surface reflect more light (specular reflectance) [3].

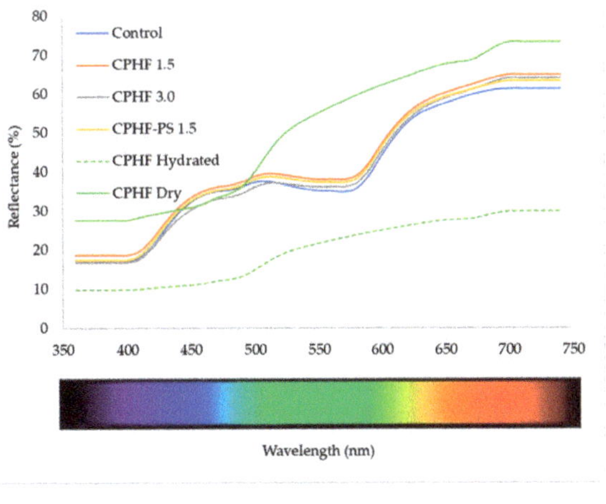

Figure 1. Reflectance spectra (360–740 nm) of cacao pod husk flour (CPHF) (dry and hydrated) and the frankfurters with the addition of CPHF. CPHF Dp < 200 μm.

In the frankfurters, a relative maximum was observed in the blue region (~500 nm), a greater maximum in the red region (~650 nm), and a minimum in the central area corresponding to the green region (~580 nm). The spectrum corresponds to a characteristic spectrum of a cured meat product with which it is possible to calculate the different relationships of wavelengths (ratios), which allow evaluating the changes of the states of the hemopigments during the meat product elaboration process [37].

The oxygenation process also induces a decrease in reflectance in the blue region and increases the red region, changing the meat's color from purplish red to bright red. In all the treatments, a non-significant increase in reflectance was observed in the entire spectrum compared to the control. Only CPHF 3.0 presented a decrease in reflectance in the blue region (430–500 nm), indicating the beginning of oxygenation of the meat without presenting another distinctive sign of the presence of metmyoglobin.

3.5.2. Reflectance Ratios R630/R580, R650/R570, and R560/R500

Table 4 shows the reflectance ratios, and all treatments present statistically significant differences with respect to the control ($p < 0.05$) but not between CPHF treatments.

Table 4. Reflectance ratios R630/R580, R650/R570, and R560/R500 in frankfurter samples.

Treatments	R630/R580	R650/R570	R560/R500
Control	1.53 ± 0.01 b	1.64 ± 0.02 b	0.94 ± 0.00 b
CPHF 1.5	1.47 ± 0.03 a	1.58 ± 0.04 a	0.98 ± 0.01 a
CPHF 3.0	1.48 ± 0.02 a	1.61 ± 0.02 a	1.00 ± 0.00 a
CPHF-PS 1.5	1.47 ± 0.01 a	1.58 ± 0.01 a	0.98 ± 0.00 a
Significance	*	*	*

Results are expressed as means ± standard deviations. Different letters in the same column indicate significant differences (LSD test, $p < 0.05$). Asterisks indicate significance at * $p < 0.05$; n.s. not significant. CPHF Dp < 200 μm.

The myoglobin and metmyoglobin that have not reacted with nitrite are adequately quantified using the reflectance ratio R630/R580 (pigment transformation index), a ratio of 1.0 would indicate essentially 100% metmyoglobin. During oxidation, this ratio decreases, causing a loss of vibrancy and a brownish appearance [38]. In our reflectance data, this relation for the control is 1.53 ± 0.01 (Table 4). There was a maximum reduction of 0.06 units (CPHF 1.5 and CPHF-PS 1.5), which corresponds to a 1.5% increase in metmyoglobin, and according to Hernandez et al. [36], when 0.2 units of this relationship are reduced, there is a 5% increase in metmyoglobin. A value below the threshold for rejection by consumers (20%).

In the treatments, the values found for the ratio R650/R570 (red color stability index) indicate that the addition of CPHF tends to moderately decrease the cured color, increasing metmyoglobin formation. The red color stability index determines the intensity of the cured color by estimating the relative proportion of reduced (Fe^{2+} pigment) and oxidized (Fe^{3+} pigment) pigments present on the sample's surface [37]. The values found close to 1.6 indicate a moderate fading towards the metmyoglobin formation.

The reflectance ratio R560/R500 (nitrosation index) determines the relative proportion of native Fe^{2+} pigments and Fe^{2+} pigments stabilized by surface nitrosation, representing the curing process's efficiency [37]. The statistically higher ratio R560/R500 values found in the treatments concerning the control indicate a lower conversion of myoglobin and oxymyoglobin into nitrosyl myoglobin because of CPHF.

Contrary to what is shown by the color parameters a* and b*, the three indices show that the treatments tend to increase the formation of metmyoglobin (gray) and decrease myoglobin and oxymyoglobin's conversion into nitrosyl myoglobin (pink) in the frankfurters. In other words, there are minor oxidative processes associated with the incorporation of CPHF, without appreciable changes in color and without jeopardizing the integrity of the product.

3.6. TPA Analysis

The TPA analysis performed on frankfurters (Table 5) showed that in most of the parameters, there was significant variation ($p < 0.05$). At this point, the texture parameters impact the initial consumer perception. Hardness increased (81.82 to 94.63 and 86.30 N) when the starch was replaced with the addition of CPHF 1.5% and CPHF 3.0%, respectively ($p < 0.01$) in accordance with [39], who found that the addition of insoluble fiber to sausages increased their hardness, attributed to the ability of some fibers to promote or strengthen connections among the matrix components [40,41]. This textural behavior could be explained by the technological properties of CPHF [3], whose emulsifying ability is probably sufficient to provide stability of the emulsion and gelling properties, helping to maintain the protein network and the continuous phase in the meat emulsion [42]. However, it was noted that when more CPHF is added (3.0%), the hardness decreases. This can be related to its high water-holding capacity (WHC 28.90 ± 1.82 g/g), with more moisture retained in the sausage (Table 2), which causes a decrease in hardness. By replacing only 50% of the starch (CPHF-PS 1.5 treatment), it was noted that the hardness increased to a similar value to the CPHF 3.0 treatment, even with similar water content. This indicates a good integration between fiber and starch within the meat matrix.

Table 5. Texture profile analysis (TPA) parameters of frankfurters.

Parameter	Control	CPHF 1.5	CPHF 3.0	CPHF-PS 1.5	Sign.
Hardness (N)	81.82 ± 8.29 a	94.63 ± 15.07 b	86.30 ± 8.15 ab	85.60 ± 8.16 ab	**
Adhesiveness	0.20 ± 0.04 a	1.07 ± 0.12 c	0.72 ± 0.24 b	0.62 ± 0.15 b	*
Springiness (mm)	0.29 ± 0.02 b	0.24 ± 0.02 a	0.27 ± 0.01 b	0.28 ± 0.02 b	*
Cohesiveness	0.76 ± 0.01 b	0.75 ± 0.04 b	0.69 ± 0.11 a	0.79 ± 0.04 b	*
Gumminess (N)	61.83 ± 6.09 ab	70.90 ± 8.07 b	58.90 ± 4.94 a	65.60 ± 3.54 b	*
Chewiness (N mm)	17.86 ± 2.11	17.05 ± 2.90	15.98 ± 1.60	17.58 ± 1.81	n.s.
Resilience	0.42 ± 0.01	0.43 ± 0.03	0.37 ± 0.07	0.39 ± 0.08	n.s.

Legend: Results are expressed as means ± standard deviations. Different letters in the same row indicate significant differences (LSD test, $p < 0.05$). Asterisks indicate significance at * $p < 0.05$; ** $p < 0.01$; n.s. not significant. CPHF Dp < 200 μm.

Adhesiveness showed a significant increase ($p < 0.05$) with the addition of CPHF, presenting the same tendency for hardness. While springiness presented a trend contrary to hardness, when hardness increased, springiness decreased in the samples. Although the product increases its hardness, it can be easily broken, which can promote chewing. Cohesiveness showed a tendency to decrease with the addition of CPHF (only significant for CPHF 3.0), possibly explained by the greater swelling capacity of CPHF (SWC 14.72 ± 1.13 mL/g), which may decrease the cohesiveness of the sausage. However, when CPHF was mixed with starch, the cohesiveness of CPHF-PS 1.5 increased above the control (not significant). Chewiness and resilience remained constant.

From the point of view of the textural properties, when substituting the starch with CPHF totally or partially, the hardness and the adhesiveness increase are the two most important parameters to consider since they increase significantly.

3.7. Sensory Evaluation

In general, the samples were positively evaluated. On a scale of 1 to 9, the scores were between 6.0 and 7.9. It was observed in all aspects that the best-evaluated sample was always the control (Figure 2).

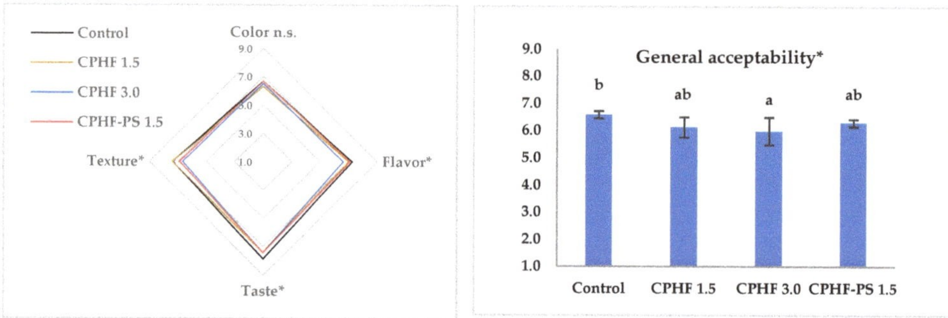

Figure 2. Spider diagram and column diagram of sensory evaluation of frankfurters samples with added cacao pod husk (control, CPHF 1.5%, CPHF 3.0%, and CPHF-PS 1.5%). Mean score of duplicate analysis (n = 21). Letters different indicate significant differences (LSD test, $p < 0.05$). Asterisks indicate significance at * $p < 0.05$; n.s. not significant.

Regarding color, the lowest scores ($p > 0.05$) were for CPHF 1.5 and CPHF 3.0, which are the ones with the most significant color difference, $\Delta E = 1.82 \pm 0.59$ and 2.28 ± 0.35 respectively. That is, the panelists were able to detect these color differences. This fact contrasts with the report that in color differences below $\Delta E = 3$, where the observer does not notice the color difference [43], suggesting that the color differences should be evaluated for each particular product and cannot be taken as a standard. In our case, the CPHF-PS 1.5 sample with a $\Delta E = 1.40 \pm 0.08$ presented the same score as the control.

Regarding texture, the lowest scores ($p < 0.05$) were in the CPHF 3.0 and CPHF-PS 1.5 samples. The questions about juiciness and hardness were the questions with the lowest score. It is possible that although these samples had higher water content, the water is bound and is not quickly released during chewing and generates a sensation of dryness in the mouth. Additionally, as already mentioned, addition of insoluble fiber to the sausage increases its hardness [39].

For the taste, the panelists gave lower scores ($p < 0.05$) to the samples with CPHF. For the question about the salty taste, the panelists indicated a sensation of unsalted sausage. The addition of CPHF in the sausage can hide the taste of salt, and the polysaccharides in CPHF may have natural mucoadhesion properties. This phenomenon is widely exploited in the pharmaceutical industry to prolong drug residence. It has been investigated to retain taste or aroma molecules in the oral cavity, and other uses [44]. This characteristic is not desirable in food since there is currently a tendency to reduce salt consumption without sacrificing taste [45].

For flavor and general acceptability, the sample with the lowest score was CPHF 3.0, although none of the treatments was rejected and all were judged acceptable (> 5) by the panelists. The panelists felt the difference due to fiber addition, which is in agreement with the results of studies that incorporate vegetable fibers in meat products [46].

4. Conclusions

This work demonstrated that CPHF up to 3.0% resulted in being a good substitute for starch for reformulating frankfurters, showing good emulsion stability. In particular, CPHF significantly affects the rheological and composition of the cooked frankfurters with the increase of the total dietary fiber content. TPA analyses showed CPHF fiber contributed to increasing hardness and adhesiveness and decreased cohesiveness in overall samples added with 3% CPHF. Panelists detected clear differences concerning the control frankfurters sausages during the sensory evaluation of the products. Although all the products were considered acceptable, the descriptor texture and flavor were determinant for the panelist to attribute the best score to the samples added with 1.5% CPHF.

Supplementary Materials: The following are available online at https://www.mdpi.com/article/10.3390/foods10061243/s1, Table S1: Proximate composition (g/100 g dry sample) of frankfurters.

Author Contributions: Conceptualization, J.D.-O., J.Á.P.-Á., M.V.-M., J.F.-L., M.M., and C.C.-L.; methodology, J.D.-O. R.L.-G., M.M., C.D.G.-T., J.B.M.-H., and M.V.-M.; validation, J.D.-O., M.V.-M. and J.Á.P.-Á.; formal analysis, J.D.-O. and J.B.M.-H.; investigation, J.D.-O., J.Á.P.-Á., J.F.-L., M.M., and C.C.-L.; resources, J.D.-O., J.Á.P.-Á., and C.C.-L.; data curation, J.D.-O.; writing—original draft preparation, J.D.-O.; writing—review and editing, J.D.-O., J.Á.P.-Á., M.V.-M., M.M., C.D.G.-T., and C.C.-L.; visualization, J.F.-L., M.M., and M.V.-M.; supervision, J.Á.P.-Á. and C.C.-L.; project administration, J.D.-O.; funding acquisition, J.D.-O., and C.D.G.-T. All authors have read and agreed to the published version of the manuscript.

Funding: This work was supported by Colciencias, Patrimonio Autónomo Fondo Nacional de Financiamiento para la Ciencia, la Tecnología y la Innovación Francisco José de Caldas. (C. 808-2018. Agreement 240-2019. Number 123280864259).

Data Availability Statement: The data presented in this study are available on request from the corresponding author.

Acknowledgments: To the Asociación de Productores de Cacao del Centro del Valle (Tuluá, Colombia). For facilitating the cacao samples for the study.

Conflicts of Interest: The authors declare no conflict of interest.

References

1. Fernández-López, J.; Viuda-Martos, M.; Pérez-Alvarez, J.A. Quinoa and chia products as ingredients for healthier processed meat products: Technological strategies for their application and effects on the final product. *Curr. Opin. Food Sci.* **2021**, *40*, 26–32. [CrossRef]
2. Fernández-López, J.; Viuda-Martos, M.; Sayas-Barberá, M.E.; de Vera Navarro-Rodríguez, C.; Lucas-González, R.; Roldán-Verdú, A.; Botella-Martínez, C.; Pérez-Alvarez, J.A. Chia, Quinoa, and Their Coproducts as Potential Antioxidants for the Meat Industry. *Plants* **2020**, *9*, 1359. [CrossRef]
3. Delgado-Ospina, J.; Lucas-González, R.; Viuda-Martos, M.; Fernández-López, J.; Pérez-Álvarez, J.Á.; Martuscelli, M.; Chaves-López, C. Bioactive compounds and techno-functional properties of high-fiber co-products of the cacao agro-industrial chain. *Heliyon* **2021**, *7*, e06799. [CrossRef]
4. Viuda-Martos, M.; López-Marcos, M.C.; Fernández-López, J.; Sendra, E.; López-Vargas, J.H.; Pérez-Álvarez, J.A. Role of Fiber in Cardiovascular Diseases: A Review. *Compr. Rev. Food Sci. Food Saf.* **2010**, *9*, 240–258. [CrossRef]
5. Pérez-Álvarez, J.Á.; Botella-Martínez, C.M.; de Vera Navarro-Rodríguez, C.; Sayas-Barberá, E.; Viuda-Martos, M.; Fernández-López, J.; Sánchez-Zapata, E. A Preliminary Study on the Incorporation of Quinoa Flour in Organic Pumpkin Creams: Effect on the Physicochemical Properties. *Proceedigs* **2021**, *70*, 71.
6. Lim, S.S.; Vos, T.; Flaxman, A.D.; Danaei, G.; Shibuya, K.; Adair-Rohani, H.; Amann, M.; Anderson, H.R.; Andrews, K.G.; Aryee, M.; et al. A comparative risk assessment of burden of disease and injury attributable to 67 risk factors and risk factor clusters in 21 regions, 1990-2010: A systematic analysis for the Global Burden of Disease Study 2010. *Lancet* **2012**, *380*, 2224–2260. [CrossRef]
7. Föste, M.; Verheyen, C.; Jekle, M.; Becker, T. Fibres of milling and fruit processing by-products in gluten-free bread making: A review of hydration properties, dough formation and quality-improving strategies. *Food Chem.* **2020**, *306*, 125451. [CrossRef] [PubMed]
8. Ribeiro, J.S.; Santos, M.J.M.C.; Silva, L.K.R.; Pereira, L.C.L.; Santos, I.A.; da Silva Lannes, S.C.; da Silva, M.V. Natural antioxidants used in meat products: A brief review. *Meat Sci.* **2019**, *148*, 181–188. [CrossRef]
9. Das, A.K.; Nanda, P.K.; Madane, P.; Biswas, S.; Das, A.; Zhang, W.; Lorenzo, J.M. A comprehensive review on antioxidant dietary fibre enriched meat-based functional foods. *Trends Food Sci. Technol.* **2020**, *99*, 323–336. [CrossRef]
10. Chaves-López, C.; Serio, A.; Mazzarrino, G.; Martuscelli, M.; Scarpone, E.; Paparella, A. Control of household mycoflora in fermented sausages using phenolic fractions from olive mill wastewaters. *Int. J. Food Microbiol.* **2015**, *207*, 49–56. [CrossRef]
11. Longato, E.; Meineri, G.; Peiretti, P.G.; Gai, F.; Viuda-Martos, M.; Pérez-Álvarez, J.Á.; Amarowicz, R.; Fernández-López, J. Effects of hazelnut skin addition on the cooking, antioxidant and sensory properties of chicken burgers. *J. Food Sci. Technol.* **2019**, *57*, 3329–3336. [CrossRef] [PubMed]
12. Mehta, N.; Ahlawat, S.S.; Sharma, D.P.; Dabur, R.S. Novel trends in development of dietary fiber rich meat products—A critical review. *J. Food Sci. Technol.* **2015**, *52*, 633–647. [CrossRef]
13. Lu, F.; Rodriguez-Garcia, J.; Van Damme, I.; Westwood, N.J.; Shaw, L.; Robinson, J.S.; Warren, G.; Chatzifragkou, A.; McQueen Mason, S.; Gomez, L.; et al. Valorisation strategies for cocoa pod husk and its fractions. *Curr. Opin. Green Sustain. Chem.* **2018**, *14*, 80–88. [CrossRef]
14. International Cocoa Organization ICCO. *Quarterly Bulletin of Cocoa Statistics*; No. 4, Cocoa Year 2019/20; International Cocoa Organization ICCO: Abidjan, Côte d'Ivoire, 2020; Volume XLVI.

15. Campos-Vega, R.; Nieto-Figueroa, K.H.; Oomah, B.D. Cocoa (*Theobroma cacao* L.) pod husk: Renewable source of bioactive compounds. *Trends Food Sci. Technol.* **2018**, *81*, 172–184. [CrossRef]
16. Yapo, B.M.; Besson, V.; Koubala, B.B.; Koffi, K.L. Adding Value to Cacao Pod Husks as a Potential Antioxidant-Dietary Fiber Source. *Am. J. Food Nutr.* **2013**, *1*, 38–46. [CrossRef]
17. Vriesmann, L.C.; de Mello Castanho Amboni, R.D.; De Oliveira Petkowicz, C.L. Cacao pod husks (Theobroma cacao L.): Composition and hot-water-soluble pectins. *Ind. Crop. Prod.* **2011**, *34*, 1173–1181. [CrossRef]
18. Yusof, F.; Khanahmadi, S.; Amid, A.; Mahmod, S.S. Cocoa pod husk, a new source of hydrolase enzymes for preparation of cross-linked enzyme aggregate. *Springerplus* **2016**, *5*, 1–18. [CrossRef] [PubMed]
19. Fernández-López, J.; Lucas-González, R.; Viuda-Martos, M.; Sayas-Barberá, E.; Navarro, C.; Haros, C.M.; Pérez-Álvarez, J.A. Chia (Salvia hispanica L.) products as ingredients for reformulating frankfurters: Effects on quality properties and shelf-life. *Meat Sci.* **2019**, *156*, 139–145. [CrossRef]
20. Horwitz, W. *Official Methods of Analysis of AOAC International*, 17th ed.; AOAC International: Gaithersburg, MD, USA, 2000; ISBN 0935584544.
21. International Organization for Standardization. *Meat and Meat Products: Determination of Nitrite Content*; International Organization for Standardization: Genevè, Switzerland, 1975.
22. Rosmini, M.R.; Perlo, F.; Pérez-Alvarez, J.A.; Pagán-Moreno, M.J.; Gago-Gago, A.; López-Santoveña, F.; Aranda-Catalá, V. TBA test by an extractive method applied to "Paté. " *Meat Sci.* **1996**, *42*, 103–110. [CrossRef]
23. AMSA. *Meat Color Measurement Guidelines*; Association, A.M.S., Ed.; American Meat Science Association: Champaign, IL, USA, 2012; ISBN 8005172672.
24. Sánchez-Zapata, E.; Fuentes-Zaragoza, E.; de Vera Navarro-Rodríguez, C.; Sayas, E.; Sendra, E.; Fernández-López, J.; Pérez-Alvarez, J.A. Effects of tuna pâté thickness and background on CIEL*a*b* color parameters and reflectance spectra. *Food Control* **2011**, *22*, 1226–1232. [CrossRef]
25. Muñoz-Almagro, N.; Valadez-Carmona, L.; Mendiola, J.A.; Ibáñez, E.; Villamiel, M. Structural characterisation of pectin obtained from cacao pod husk. Comparison of conventional and subcritical water extraction. *Carbohydr. Polym.* **2019**, *217*, 69–78. [CrossRef]
26. Viuda-Martos, M.; Fernández-López, J.; Sayas-Barbera, E.; Sendra, E.; Navarro, C.; Pérez-Álvarez, J.A. Citrus Co-Products as Technological Strategy to Reduce Residual Nitrite Content in Meat Products. *J. Food Sci.* **2009**, *74*, R93–R100. [CrossRef] [PubMed]
27. Pérez-Rodríguez, M.L.; Bosch-Bosch, N.; Garciá-Mata, M. Monitoring nitrite and nitrate residues in frankfurters during processing and storage. *Meat Sci.* **1996**, *44*, 65–73. [CrossRef]
28. Merino, L.; Darnerud, P.; Toldrá, F.; Ilbäck, N.-G. Time-dependent depletion of nitrite in pork-beef and chicken meat products affects nitrite intake estimation. *Food Addit. Contam. Part A* **2016**, *33*, 186–192. [CrossRef]
29. Martín León, V.; Luzardo, O.P. Evaluation of nitrate contents in regulated and non-regulated leafy vegetables of high consumption in the Canary Islands, Spain: Risk assessment. *Food Chem. Toxicol.* **2020**, *146*, 111812. [CrossRef]
30. Lebrun, S.; Van Nieuwenhuysen, T.; Crèvecoeur, S.; Vanleyssem, R.; Thimister, J.; Denayer, S.; Jeuge, S.; Daube, G.; Clinquart, A.; Fremaux, B. Influence of reduced levels or suppression of sodium nitrite on the outgrowth and toxinogenesis of psychrotrophic Clostridium botulinum Group II type B in cooked ham. *Int. J. Food Microbiol.* **2020**, *334*, 108853. [CrossRef]
31. Garrote, G.; Cruz, J.M.; Moure, A.; Domínguez, H.; Parajó, J.C. Antioxidant activity of byproducts from the hydrolytic processing of selected lignocellulosic materials. *Trends Food Sci. Technol.* **2004**, *15*, 191–200. [CrossRef]
32. Sheard, P.R.; Enser, M.; Wood, J.D.; Nute, G.R.; Gill, B.P.; Richardson, R.I. Shelf life and quality of pork and pork products with raised n-3 PUFA. *Meat Sci.* **2000**, *55*, 213–221. [CrossRef]
33. Falowo, A.B.; Fayemi, P.O.; Muchenje, V. Natural antioxidants against lipid–protein oxidative deterioration in meat and meat products: A review. *Food Res. Int.* **2014**, *64*, 171–181. [CrossRef] [PubMed]
34. Hernández, B.; Sáenz, C.; Alberdi, C.; Diñeiro, J.M. CIELAB color coordinates versus relative proportions of myoglobin redox forms in the description of fresh meat appearance. *J. Food Sci. Technol.* **2016**, *53*, 4159–4167. [CrossRef] [PubMed]
35. Shimokomaki, M.; Youssef Youssef, E.; Terra, N. Curing. In *Encyclopedia of Food Sciences and Nutrition*; Caballero, B., Finglas, P., Toldra, F., Eds.; Academic Press: Cambridge, MA, USA, 2003; pp. 1702–18708. ISBN 978-0-12-227055-0.
36. Hernández Salueña, B.; Sáenz Gamasa, C.; Diñeiro Rubial, J.M.; Alberdi Odriozola, C. CIELAB color paths during meat shelf life. *Meat Sci.* **2019**, *157*, 107889. [CrossRef] [PubMed]
37. Martín-Sánchez, A.M.; Sánchez-Zapata, E.; Viuda-Martos, M.; Sendra, E.; Sayas-Barberá, E.; Fernández-López, J.; Pérez-Álvarez, J.Á.; Chaves-López, C. Influencia de la adición de sorbato sobre los cocientes de reflectancia R650/R570, R560/R500 y R630/R580 en productos cárnicos crudo-curados. *Opt. Pura y Apl.* **2010**, *43*, 185–191.
38. Strange, E.D.; Benedict, R.C.; Gugger, R.E.; Metzger, V.G.; Swift, C.E. Simplified Methodology for Measuring Meat Color. *J. Food Sci.* **1974**, *39*, 988–992. [CrossRef]
39. Sánchez-Zapata, E.; Zunino, V.; Pérez-Alvarez, J.A.; Fernández-López, J. Effect of tiger nut fibre addition on the quality and safety of a dry-cured pork sausage ("Chorizo") during the dry-curing process. *Meat Sci.* **2013**, *95*, 562–568. [CrossRef] [PubMed]
40. Cruz, A.G.; Cadena, R.S.; Walter, E.H.M.; Mortazavian, A.M.; Granato, D.; Faria, J.A.F.; Bolini, H.M.A. Sensory analysis: Relevance for prebiotic, probiotic, and synbiotic product development. *Compr. Rev. Food Sci. Food Saf.* **2010**, *9*, 358–373. [CrossRef]
41. Biswas, A.K.; Kumar, V.; Bhosle, S.; Sahoo, J.; Chatli, M.K. Dietary fibers as functional ingredients in meat products and their role in human health. *Int. J. Livest. Prod.* **2011**, *2*, 45–54. [CrossRef]

42. Ahmad, S.S.; Khalid, M.; Younis, K. Interaction study of dietary fibers (pectin and cellulose) with meat proteins using bioinformatics analysis: An In-Silico study. *LWT* **2020**, *119*, 108889. [CrossRef]
43. Martínez, J.A.; Melgosa, M.; Pérez, M.M.; Hita, E.; Negueruela, A.I. Note. Visual and Instrumental Color Evaluation in Red Wines. *Food Sci. Technol. Int.* **2001**, *7*, 439–444. [CrossRef]
44. Cook, S.L.; Woods, S.; Methven, L.; Parker, J.K.; Khutoryanskiy, V. Mucoadhesive polysaccharides modulate sodium retention, release and taste perception. *Food Chem.* **2018**, *240*, 482–489. [CrossRef]
45. Sun, C.; Zhou, X.; Hu, Z.; Lu, W.; Zhao, Y.; Fang, Y. Food and salt structure design for salt reducing. *Innov. Food Sci. Emerg. Technol.* **2020**, *67*, 102570. [CrossRef]
46. Pintado, T.; Herrero, A.M.; Jiménez-Colmenero, F.; Ruiz-Capillas, C. Strategies for incorporation of chia (Salvia hispanica L.) in frankfurters as a health-promoting ingredient. *Meat Sci.* **2016**, *114*, 75–84. [CrossRef] [PubMed]

foods

MDPI

Review

Risk Assessment of Caffeine and Epigallocatechin Gallate in Coffee Leaf Tea

Nadine Tritsch [1,2], Marc C. Steger [3,4], Valerie Segatz [2,5], Patrik Blumenthal [3], Marina Rigling [4], Steffen Schwarz [3], Yanyan Zhang [4], Heike Franke [1] and Dirk W. Lachenmeier [2,*]

1 Postgraduate Study of Toxicology and Environmental Toxicology, Rudolf-Boehm-Institut für Pharmakologie und Toxikologie, Universität Leipzig, Härtelstraße 16-18, 04107 Leipzig, Germany; nt60rawo@studserv.uni-leipzig.de (N.T.); Heike.Franke@medizin.uni-leipzig.de (H.F.)
2 Chemisches und Veterinäruntersuchungsamt (CVUA) Karlsruhe, Weissenburger Strasse 3, 76187 Karlsruhe, Germany; Valerie.Segatz@stud.hs-coburg.de
3 Coffee Consulate, Hans-Thoma-Strasse 20, 68163 Mannheim, Germany; marcsteger2@googlemail.com (M.C.S.); patrik.blumenthal@live.de (P.B.); schwarz@coffee-consulate.com (S.S.)
4 Department of Flavor Chemistry, University of Hohenheim, Fruwirthstrasse 12, 70599 Stuttgart, Germany; marina.rigling@uni-hohenheim.de (M.R.); yanyan.zhang@uni-hohenheim.de (Y.Z.)
5 Hochschule für Angewandte Wissenschaften Coburg, Friedrich-Streib-Strasse 2, 96450 Coburg, Germany
* Correspondence: Lachenmeier@web.de; Tel.: +49-721-926-5434

Abstract: Coffee leaf tea is prepared as an infusion of dried leaves of *Coffea* spp. in hot water. It is a traditional beverage in some coffee-producing countries and has been authorized in 2020 within the European Union (EU) according to its novel food regulation. This article reviews current knowledge on the safety of coffee leaf tea. From the various ingredients contained in coffee leaves, only two were highlighted as possibly hazardous to human health, namely, caffeine and epigallocatechin gallate (EGCG), with maximum limits implemented in EU legislation, which is why this article focuses on these two substances. While the caffeine content is comparable to that of roasted coffee beans and subject to strong fluctuations in relation to the age of the leaves, climate, coffee species, and variety, a maximum of 1–3 cups per day may be recommended. The EGCG content is typically absent or below the intake of 800 mg/day classified as hepatotoxic by the European Food Safety Authority (EFSA), so this compound is suggested as toxicologically uncritical. Depending on selection and processing (age of the leaves, drying, fermentation, roasting, etc.), coffee leaf tea may exhibit a wide variety of flavors, and its full potential is currently almost unexplored. As a coffee by-product, it is certainly interesting to increase the income of coffee farmers. Our review has shown that coffee leaf tea is not assumed to exhibit risks for the consumer, apart from the well-known risk of caffeine inherent to all coffee-related beverages. This conclusion is corroborated by the history of its safe use in several countries around the world.

Keywords: coffee leaf tea; novel food; coffee by-products; *Coffea arabica*; risk assessment; caffeine; epigallocatechin gallate

1. Introduction

Coffee consumption has been known worldwide for decades. As early as 1450, pilgrims brought coffee from Ethiopia to Mecca and during the 17th century to Europe [1,2]. However, the so-called coffee by-products, such as coffee flowers, leaves, pulp, husk, parchment, and silver skin, are rarely known outside of the countries of manufacture [3,4].

The *Coffea* genus consists of several species and approximately 6000 varieties [5]. The most important coffee species are *C. arabica* and *C. canephora*.

C. arabica accounts for around 60–75% of global production and grows in plantations at high altitudes of 600–2000 m, whereas *C. canephora* accounts for the remaining proportion and grows in plantations from 300–800 m [5,6]. For completeness, the species *C. liberica* and *C. excelsa* are still cultivated to a limited extent [5,7].

Coffea spp. needs a consistently warm and at the same time humid climate without extremes. Therefore, the plants grow in tropical and subtropical regions between the 25th latitude north and south along the equator [5], the so-called "coffee belt" [8,9]. The main producing countries include Brazil, Mexico, Guatemala, Colombia, Honduras, Vietnam, Indonesia, India, Ethiopia, and Uganda [5].

As mentioned by Haller et al. [10], *Camellia sinensis* tea is the most widely consumed drink in the world directly after water. It is obvious that making tea from coffee leaves may offer economic potential in the coffee-growing regions [11]. Figures 1 and 2 show examples of coffee leaves on coffee plants.

Figure 1. Coffee leaves (*C. arabica*) ((**A–C**): strong, dark brown, new shoots).

Figure 2. Fresh shoots on branches with flowers (**A,B**) and fruits (**C**) (*C. arabica;* newly sprouted leaves are darker than older ones).

According to Rimbach et al. [5], around half of German consumers drink tea every day. That means that every year around 30 L of tea (250 g of tea leaves/year) is consumed, although there are large regional differences. In contrast, coffee consumption is higher. Finland, for example, has the highest coffee consumption per capita. Every resident drinks an average of five cups of coffee a day, while in Germany, the average consumption is four cups a day or 6.7 kg of coffee beans per year [5]. According to the International Coffee Organization, exports of all forms of coffee amounted to almost 119 million bags (60 kg bags) for the coffee year 2020/21 (October 2020–August 2021) [12].

Food and beverages can be marketed on the European Union (EU) market without authorization if there was a significant human consumption within the EU before 15 May 1997 (see EU's guidance on "human consumption to a significant degree" [13]). If this was not the case, a particular food can still be authorized according to the novel food regulation, which also provides a separate pathway for notification of traditional foods from third countries [14]. As described by Chen et al. [15], coffee leaf tea is a well-known traditional beverage in coffee plant-growing countries, for example, "copi daon" in Indonesia, "giser" in Yemen or "Kahwa dau" in West Sumatra, Indonesia [15,16]. From the authors' own travel experience, we can say that coffee leaf tea is less known among tourists and still rarely offered. For illustration, Figure 3 shows some examples of brewed coffee leaf tea, whereas Figure 4 shows some examples of dried and chopped coffee leaves for tea infusions. Samples (Figures 3 and 4) were prepared using large tasting bowls (Figure 5) according to International Organization for Standardization (ISO) standard ISO 3103 [17].

(**A**) (**B**) (**C**)

Figure 3. Examples of brewed coffee leaf tea infusions (water extraction)—different processing shows different colors. Preparation: 2 g of dried coffee leaves/100 mL water ((**A**) = Bourbon, old leaves, air/sun dried, wild fermentation; (**B**) = Pacamara, old leaves, whole leaves, oven-dried, no fermentation; (**C**) = Bourbon, old leaves, whole leaves, roasted, no fermentation).

(**A**) (**B**) (**C**)

Figure 4. Examples of dried coffee leaf tea ((**A**) = Pacamara, yellow leaves, mixed, air/sun dried, fermentation by yeast; (**B**) = Bourbon, whole water shoots, chopped, 1-h steam distilled, air/sun dried; (**C**) = Bourbon, old leaves, mixed, air/sun dried, wild fermentation).

The company AM Breweries (Denmark) submitted a notification request [18] to the EU commission according to Article 14 "Notification of a traditional food from a third country" (EU regulation 2015/2283) [14]. Before granting authorization, the European Commission can ask the European Food Safety Authority (EFSA) for a scientific risk assessment [19]. In 2020, the EFSA published its 'Technical Report on the notification of infusion from coffee leaves . . . as a traditional food from a third country' [20]. Subsequently, an implementing regulation was published [21] that allows coffee leaf tea to be placed on the EU market (positive list in the annex [21,22]) as a traditional food from a third country, in accordance with the novel food regulation (regulation 2015/2283) [14]. Since the applicant AM Breweries [18] did not provide evidence of the use of coffee leaves as an ingredient in other beverages, only the infusion of coffee leaves was approved as a novel food [21,22]. Among other factors, such as microbiological criteria, critical values have been established for the substances caffeine and (−)-epigallocatechin-3-gallate (EGCG) [22].

Figure 5. Spent coffee leaves after the preparation of a coffee leaf tea infusion.

2. Materials and Methods

For this article, electronic searches of the literature were conducted, including the databases PubMed (National Library of Medicine, Bethesda, MD, USA) and Google Scholar (Google LLC, Mountain View, CA, USA). A wide range of search terms was used, including coffee, coffee leaf tea, caffeine, and EGCG. In addition, standard works of literature were used concerning knowledge of the product. Furthermore, databases such as the US Department of Agriculture (USDA) database for the flavonoid content of selected foods were searched for terms such as caffeine and EGCG, as well as the EFSA homepage for the terms coffee leaf tea, caffeine, and EGCG.

In addition to the data from the literature, our own data of the *C. arabica* varieties (Bourbon and Pacamara) of El Salvador from a recent Master's thesis were considered [23]. The 14502-1 test regulations that were used to determine the total polyphenol content in tea and ISO 14502-2, used to determine the catechin content in green tea, were used [24].

3. Composition of Coffee Leaf Tea

3.1. Ingredients

Caffeine and EGCG are naturally found in coffee leaves [15,25]. As described by Zheng et al. [26], caffeine is a purine alkaloid found mainly in young leaves and shoots including buds and cotyledons, whereas caffeine is not detected in roots, aged cotyledons or older brown parts of shoots [26]. The results of the aforementioned publication suggest that caffeine accumulation is specific to the above-ground parts (leaves, cotyledons, and shoots) of the seedlings and that biosynthesis is performed in young tissues. The authors describe caffeine synthesis as a defense mechanism of the (soft) parts of plants against predators [26]. Frischknecht et al. [27] found that the formation of purine alkaloids is costly for the plant [27].

Plant-based phenols such as EGCG are products of the plant's secondary metabolism [6]. In contrast to black tea, green tea is neither fermented nor oxidized. Due to that, polyphenols such as epicatechin (EC), epicatechin-3-gallate (ECG), epigallocatechin (EGC), and epigallocatechin gallate (EGCG) can make up about 59% of the catechin content [10]. The proportion of phenolic compounds in fresh, young tea leaves of the tea plant is high [5]. Rimbach et al. [5] mention that four-fifths of these compounds are made up of catechins (flavan-3-ols) and gallocatechins (flavan-3,4-diols), which belong to flavanols [5]. According to Frede et al. [6], fresh tea shoots contain several flavanols (catechins), for example, epigallocatechin gallate (9–13%), epicatechin gallate (3–6%), and epigallocatechin (3–6%), as well as epicatechin (1–3%) and others (catechin, gallocatechin, 1–2%) [6].

The infusion of coffee leaves is a beverage rich in several bioactive compounds (e.g., chlorogenic acids, xanthones [28], trigonelline, adenine-7-glucosyl, quercetin, and mangiferin [29,30], not only polyphenols (catechins) and caffeine described above.

Basically, the EFSA has assessed the ingredients and focused on caffeine and EGCG. All other compounds did not pose any significant risk. This article therefore focuses on these two compounds, knowing that there are other ingredients in coffee leaf tea.

3.2. Analytical Results

In a recent project conducted with coffee leaves from El Salvador, various coffee leaf teas were analyzed, among other factors, for their caffeine and EGCG content [23]. The samples examined differed in terms of the age of the leaves used, drying, fermentation, and other processing conditions (see examples in Table 1). A total of 24 different samples were produced.

Table 1. Some examples of coffee leaf tea samples prepared in El Salvador.

Picture	Processing		
	***Coffea arabica* Variety**	**Leaves**	**Drying, Fermentation**
	Pacamara	old leaves, mixed	air/sun dried, fermentation by yeast (*Saccharomyces cerevisiae,* var. *bayanus*)
	Pacamara	yellow leaves, mixed	air/sun dried, fermentation by yeast (*Saccharomyces cerevisiae,* var. *bayanus*)
	Pacamara	old leaves, whole leaves	Oven-dried, no fermentation
	Pacamara	yellow leaves, mixed	air/sun dried, wild fermentation
	Bourbon	whole water shoots, chopped	1 h steam distilled, air/sun dried

Table 1. *Cont.*

Picture	Processing		
	Coffea arabica Variety	Leaves	Drying, Fermentation
	Bourbon	old leaves, whole leaves	air/sun dried, no fermentation
	Bourbon	old leaves, mixed	air/sun dried, wild fermentation
	Bourbon	old leaves, whole leaves	roasted, no fermentation

According to the results of the study (Table 2), the samples contained 3.4 g/kg DW (dry weight) to 23.9 g/kg DW of caffeine, and catechins from not detectable to 6.5 g/kg DW (sum of all analyzed catechins, only epicatechin and epicatechin gallate were present) [23,24]. EGCG was not detected in the samples [23,24]. The moisture content of the samples ranged from 3.9 to 16.1% [23,24]. Figure 6 shows some dried coffee leaves.

Table 2. Analytical results for fresh and dried coffee leaves.

	Fresh Coffee Leaves (mg/g) [25]	Dried Coffee Leaves (mg/g DW) [23,24]
Caffeine	1.8–3.2	3.4–23.9
Epigallocatechin gallate	5.5–16.4	n.d.
Epicatechin gallate	0.26–0.48	n.d.–0.6
Epicatechin	0.27–0.40	n.d.–6.5
Catechins, sum	0.05–0.18	n.d.–6.5

n.d., not detectable.

(a) (b) (c) (d)

Figure 6. Air-dried coffee leaves, *C. arabica*. (**a**) Coffee leaves that had fallen from the bush themselves, old leaves. (**b**) Coffee leaves, fresh shoots, only brown leaves. (**c**) Coffee leaves, fresh shoots, only the green leaves. (**d**) Coffee leaves, fresh shoots, mixture of leaves from branches/shoots.

Ratanamarno et al. [25] found that the amount of these substances depends on the age of the leaf [25]. They recorded higher amounts in young leaves than in full-grown leaves. Fibrinato et al. [31] came to the same conclusion [31]. Acidri et al. [32] showed that total phytochemicals, for example, catechins, in coffee leaves decreased with leaf age [32]. However, Steger [23] could not confirm this for the dried coffee leaf teas examined, as no significant correlation with the age of the coffee leaves was observed (Figure 7).

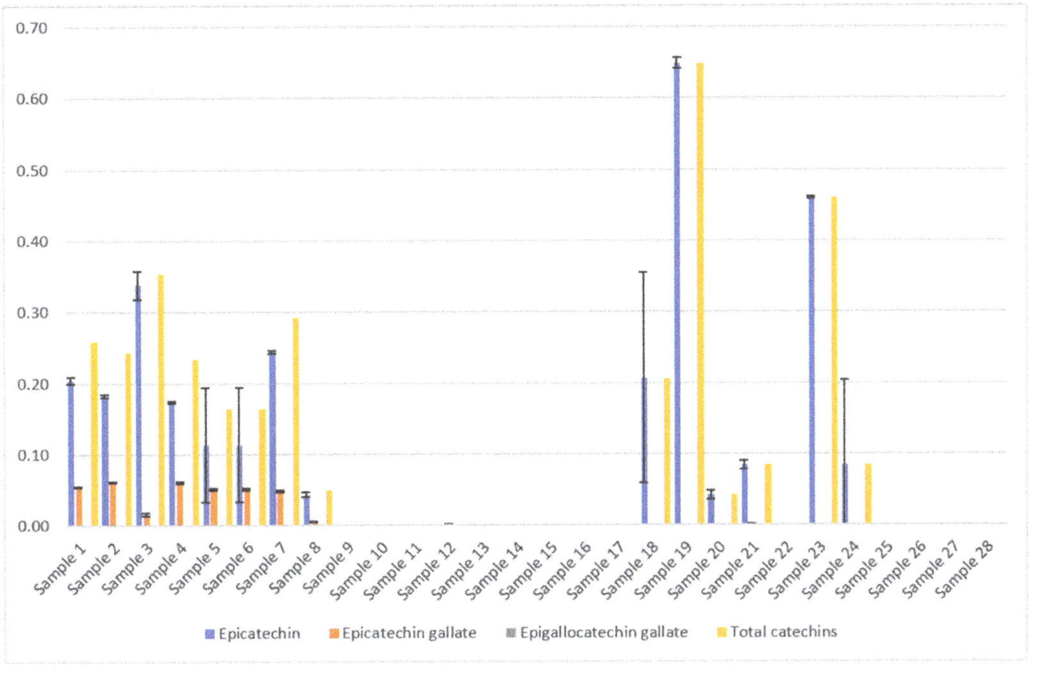

Figure 7. Results of the coffee leaf tea samples examined for the catechin content [23,24], analyzed by HPLC. The mean value in g/100 g DW is given from three determinations in each case. The total catechin content depends on the drying method and processing. Epigallocatechin gallate could not be detected in the samples examined. The error bars show the standard deviation of the three measured values. The total catechin contents were calculated as the sum of the mean values of all catechins (only epicatechin and epicatechin gallate were detected). Samples 1–24 picked in El Salvador from two different plantations, *C. arabica* [23]. Samples 25–28 are Indian *C. canephora*. [Sample 1: Pacamara old whole air Palma, Sample 2: Pacamara yellow whole air Palma, Sample 3: Pacamara old crumbled air Palma, Sample 4: Pacamara yellow crumbled air Palma, Sample 5: Pacamara old cutted air Palma, Sample 6: Pacamara old blended air Palma, Sample 7: Pacamara old crumbled air Palma Yeast fermented, Sample 8: Pacamara old whole oven Palma, Sample 9: Pacamara old blended air Palma Lacto., Sample 10: Pacamara old blended air Palma Yeast, Sample 11: Pacamara old blended air Palma Wild, Sample 12: Pacamara yellow blended air Palma Wild, Sample 13: Pacamara yellow blended air Palma Yeast, Sample 14: Pacamara yellow blended air Palma Lacto., Sample 15: Bourbon water shoot whole blended air Quintanilla wild, Sample 16: Bourbon water shoot/old blended air Quintanilla Yeast, Sample 17: Bourbon water shoot/old blended air Quintanilla Wild, Sample 18: Bourbon water shoot/young green tea air Quintanilla, Sample 19: Bourbon water shoot/young black tea air Quintanilla Wild, Sample 20: Bourbon water shoot/whole blended/steamed air Quintanilla, Sample 21: Bourbon water shoot/old whole air Quintanilla wild, Sample 22: Bourbon water shoot/old whole/frozen air Quintanilla, Sample 23: Water shoot old whole roasted Quintanilla, Sample 24: Water shoot whole air Quintanilla, Sample 25: India prepackage, Sample 26: India bulkware, Sample 27: India CxR, Sample 28: India old Paradenia. Yeast = *Saccharomyces cerevisiae, var. bayanus*, lacto. = *Lactobacillus plantarum*].

3.2.1. Catechins

Samples 1–24 *C. arabica*: The results of the total catechin content show a clear influence of the blending process (Figure 8). Except for one sample, all blended samples have no catechins at all (Figure 7). According to scientific research this could be due to the oxidation of catechins by polyphenol oxidase to theaflavins or similar oligomeric structures [33]. This effect also occurs during black tea fermentation [34]. The large surface area and added water of the mixed samples could be responsible for an enhanced enzymatic reaction. Furthermore, the drying parameters show an influence on the total catechin content. Air drying has a significantly higher average value (0.266 g/100 g DW) than oven drying (0.054 g/100 g DW) and a significantly lower value than roasting (0.479 g/100 g DW) [23] (Figure 8).

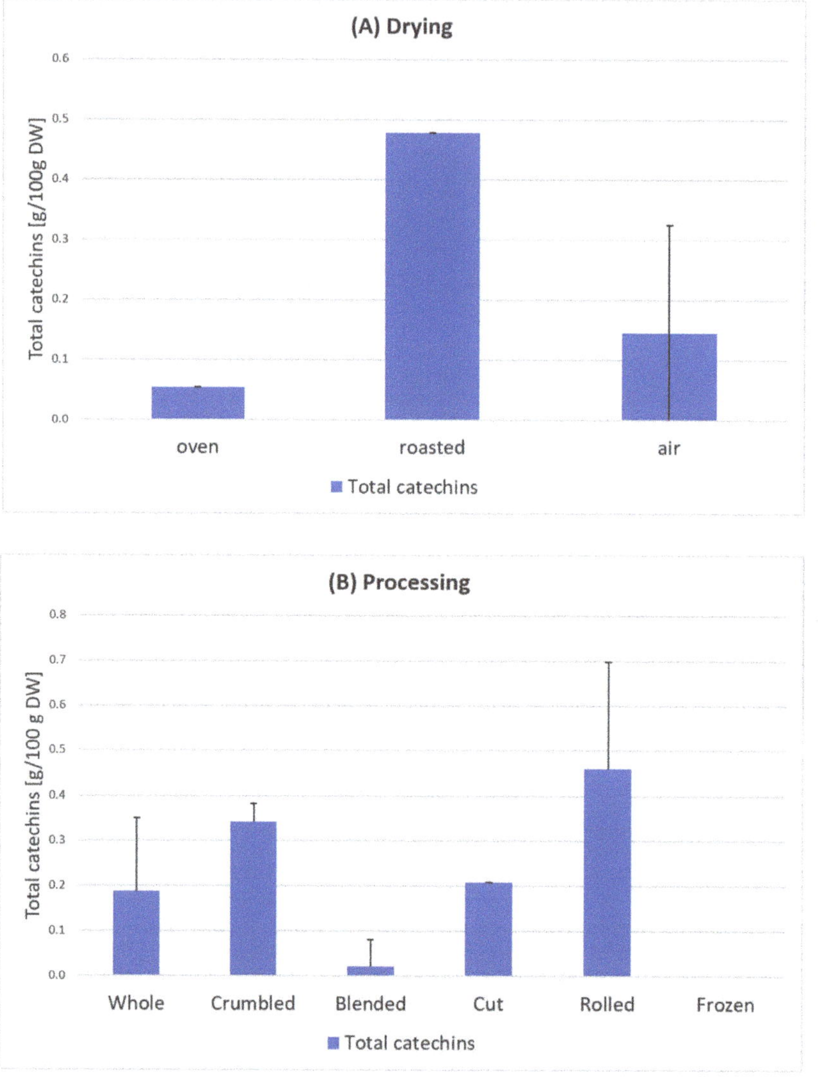

Figure 8. Total catechin content in leaves depending on the drying method (**A**) and processing (**B**), based on samples of *C. arabica* from El Salvador [23]. Description of the samples, see Figure 7. Drying: Oven-dried sample 8; roasted samples 23; air-dried samples 1–7; 9–22, 24. Processing: Whole samples 1–2, 8, 21–24, Crumbled samples 3–4, 7; Blended samples 6, 9–17, 20; Cut sample 5; Rolled sample 19; Frozen sample 22. The results show considerable variances depending on the individual processing and drying steps (sample 1–24).

Krol et al. [35] report that the polyphenol content decreases as the coffee beans are roasted. This is because polyphenolic components are thermolabile and easily disintegrate at temperatures above 80 °C. This process occurs due to the thermal instability during roasting [35]. Our tests on coffee leaves could not confirm this. The highest catechin contents were measured here in the roasted samples [23]. Further research will have to show whether there is a difference between coffee beans and coffee leaves in this regard.

Li et al. [36] investigated the correlation between the temperature and duration of the thermal treatment with the catechin content. In accordance with this, the low content of the oven-dried sample can be attributed to the 4-h drying time. The roasted samples had a much shorter drying time (20 min), resulting in the highest content (Figure 8). The EGCG mentioned in the European Commission's approval [22] could not be detected in any sample. Turkmen et al. [37] stated that the absence of EGCG in black tea was due to the oxidation and fermentation processes. This effect could have also occurred in the coffee leaf tea samples during the withering process.

Samples 25–28, *C. canephora*: Samples of *C. canephora* do not contain catechins, according to Figure 7. No data on this result could be found during scientific research. Our four samples represent only a small spot check. Therefore, this result should be verified by examining a larger number of samples.

3.2.2. Caffeine

Samples 1–24 *C. arabica*: The caffeine content of the leaves varies between 0.37 g/100 g of DW and 1.33 g/100 g DW. Here, the type of leaf and variety show significant influences on the caffeine content of the tea. Young leaves show the highest caffeine value (0.91 g/100 g DW) while yellow leaves show the lowest (0.44 g/100 g DW). Caffeine levels for *C. arabica* are approximately the same as those detected by Ratanamarno et al. [25].

The effect of caffeine reduction with leaf age has already been observed in some studies with different plants, e.g., *Ilex paraguariensis* and *Plantago lanceolata* [38,39]. Song et al. [40] explained this effect mainly by the function of caffeine as a pesticide. Younger leaves of the plant must be more protected compared to older ones; therefore, the plant builds higher concentrations in these leaves. Furthermore, processing is a significant variable toward caffeine content. Here, rolling (1.2 g/100 g DW) and freezing (1.00 g/100 g DW) of the leaf show the highest caffeine content. According to Astill et al. [41] the caffeine content decreases during the fermentation and drying stage. In the case of freezing, it is possible that metabolic pathways are stopped, resulting in less degradation during drying. Since the rolling process was conducted on young leaves, further tests would be needed to determine whether this influences the caffeine content. Furthermore, the low caffeine content of the blended samples may be due to the addition of water. Some of the caffeine may have been dissolved in the water during processing and then dripped off through the drying bed during drying. The caffeine content in the infusion prescribed in the approval of the EU Commission [22] can be exceeded depending on the dose of the coffee leaves per water.

Samples 25–28, *C. canephora*: The *C. canephora* samples have the highest levels of caffeine according to Figure 9. Rodriges et al. [42] and Perrois et al. [43] also showed that *C. canephora* has a higher caffeine content than *C. arabica*. Miedaner confirmed this observation [44].

Figure 10 shows the caffeine content depending on the leaf type, variety, and processing based on our samples (*C. arabica*) from El Salvador (coffee plantation: Finca la Quintanilla and Finca la Palma).

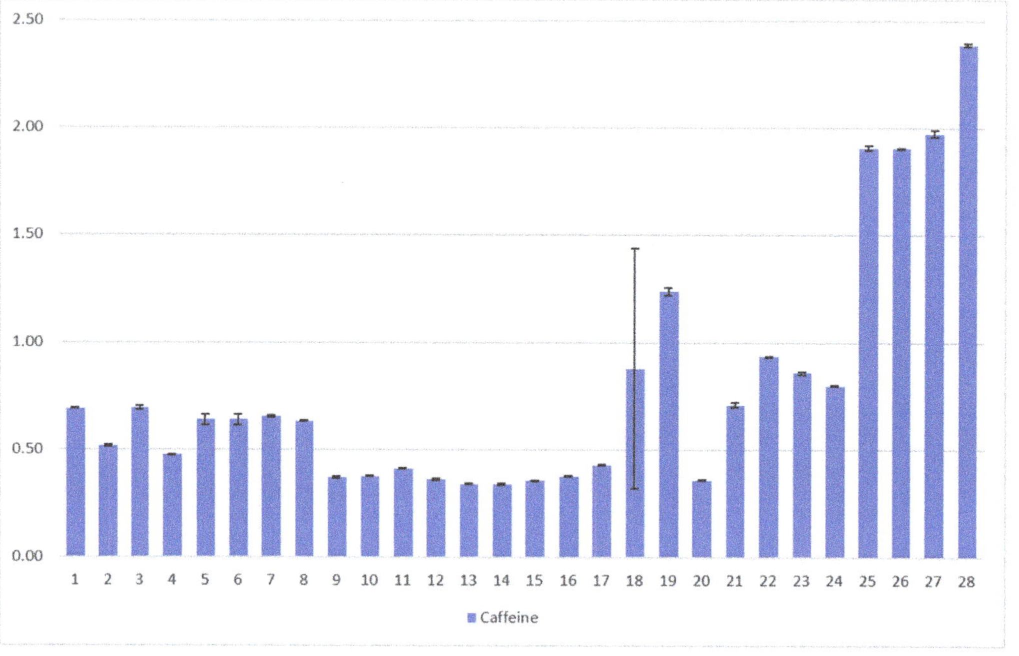

Figure 9. Results of the coffee leaf tea samples examined for the caffeine content [23,24], analyzed by HPLC. The mean value in g/100 g DW is given from three determinations in each case. The error bars show the standard deviation of the three measured values. The first 24 samples were collected in El Salvador from two different plantations and were dried and fermented in different ways, *C. arabica* [23]. The last four samples are Indian *C. canephora*. [Description of the samples, see Figure 7].

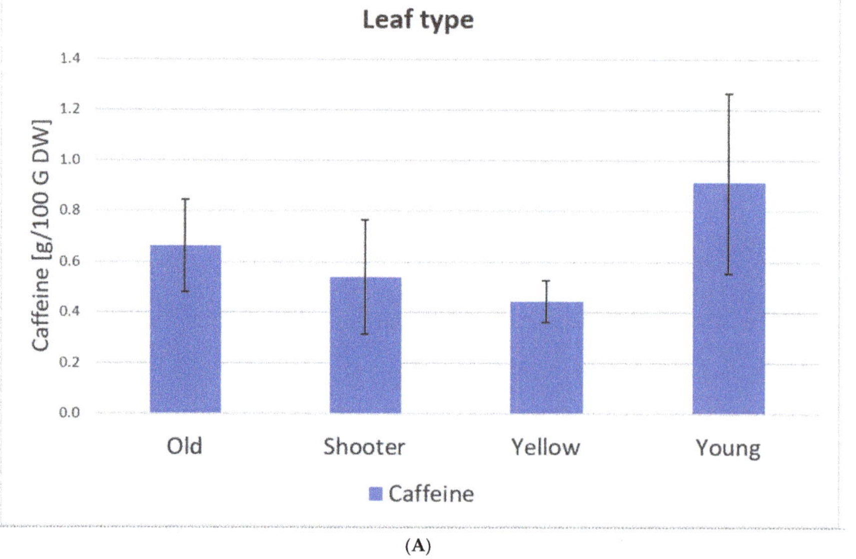

(A)

Figure 10. *Cont.*

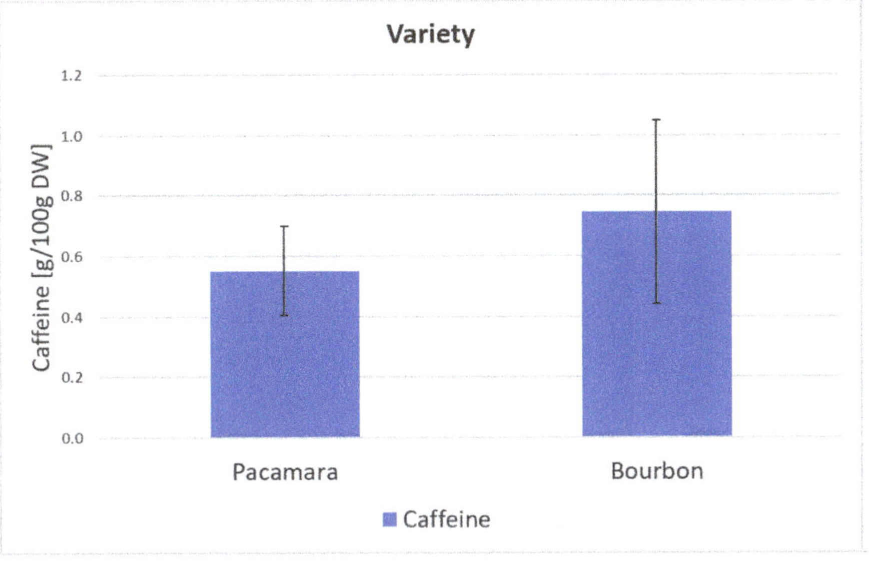

(B)

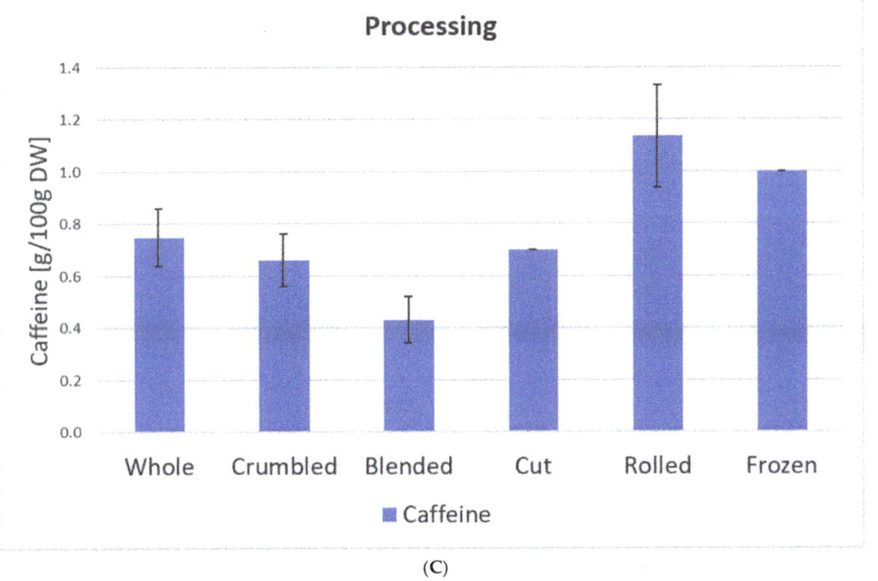

(C)

Figure 10. Caffeine content depending on leaf type (**A**), variety (**B**), and processing (**C**) based on the samples from El Salvador [23], *C. arabica*. Description of the samples, see Figure 7. (**A**): Old samples 1, 3, 5–11, 16, 21–23; Water shoot samples 15, 20, 24; Yellow samples 2, 4, 12–14; 17–19. (**B**): Pacamara samples 1–14; Bourbon samples 15–24. (**C**): Whole samples 1, 2, 8, 21, 23–24: Crumbled samples 3, 4, 7; Blended samples 6, 9–17, 20; Cut sample 5; Rolled samples 18–19; Frozen samples 22.

The higher caffeine content of *C. canephora* coffee leaves (Figure 8) is plausible in correspondence with the higher caffeine content in *C. canephora* beans [44,45].

4. Exposure Assessment

4.1. National Consumption Study

The National Consumption Study II (NVS II, Nationale Verzehrsstudie II) [46,47] of the Max Rubner Institute (MRI) served as a data basis concerning the consumption of tea infusions in adolescents and adults (Table 3). NVS II is still the most up-to-date representative study on the food and beverage consumption of the German population. In this study, around 20,000 people between the ages of 14 and 80 were asked about their consumption habits. It was conducted across Germany between 2005/2006.

Table 3. NVS (National Consumption Study) results [46].

	Average Male (n = 6160)		**Average Female (n = 7593)**	
[g/day]	Arithmetic Average	Confidence Interval	Arithmetic Average	Confidence Interval
Coffee and tea (black/green)	538	527/550	506	497/515
Herbal tea/fruit tea	133	125/141	251	241/261

This study includes consumption information for individual days (24-h method, [46]). Therefore, this study is suitable for exposure estimates for both acute and chronic risks. Individual body weights of the participants were used as a basis for intake estimates [46]. Table 3 shows the average results for men and women. For simplicity, details on the individual age groups are not given here but can be found in the study.

The daily value for fluid intake is at least 1.5 L/day according to NVS [46]. The amount of nonalcoholic beverages consumed hardly differs between men (2.351 g/day) and women (2.285 g/day) [47]. About one-third of this amount is covered by coffee and (black/green) tea (men: 571 g/day or 24%; women: 506 g/day or 22%), while the consumption of herbal tea/fruit tea is much lower. The proportion of herbal tea and fruit tea is more than twice as high in women (318 g/day or 14%) than in men (149 g/day or 6%) [47].

For further considerations, the total value for coffee and tea is used since coffee leaf tea may replace either coffee or tea in the daily diet in a worst-case scenario.

The term "tea" refers to infusing plant parts in hot water. However, according to the definition in the German Food Book, the term tea may be used only for products made from the tea plant (*Camellia sinensis*). Other products made from leaves, fruits or flowers (e.g., mint, rosehip, hibiscus) are grouped under the term tea-like products (sometimes also referred to as herbal tea or tisane). Therefore, coffee leaf tea is under legal consideration a tea-like product [48]. Nevertheless, the exposure of coffee leaf tea should better be assessed using the data from the caffeine-containing coffee/tea category than the caffeine-free herbal tea category.

In the approval of coffee leaf tea under the novel food regulation, maximum levels have been established for some substances that must be met to sell the product [22]. These include caffeine, chlorogenic acid (5-CQA), and EGCG. Table 4 lists the limit values according to the Novel Food Regulation [22]. Two of the substances (caffeine and EGCG) are discussed in more detail below.

Table 4. Maximum permitted levels of caffeine, 5-CQA, and EGCG as approved by the European Commission [22].

Substance	Maximum Amount
Caffeine	<80 mg/L
5-CQA	<100 mg/L
EGCG	<700 mg/L

4.2. Caffeine Exposure

The taste of Arabica coffee is particularly aromatic, soft, and mild. In contrast, *C. canephora* coffee of most commercial qualities is perceived to be less delicate and develops a higher caffeine content [44]. In addition to many ingredients such as flavorings and aromas, the coffee beans of both varieties contain caffeine. In small amounts, this primarily affects the central nervous system and increases the capacity for mental absorption and memory while at the same time the resulting fatigue is reduced [44]. Therefore, it is not surprising that coffee has an image as a stimulant (central nervous system stimulant) [44,45]. However, consuming it in excessive quantities can cause anxiety, insomnia as well as nausea, vomiting, diarrhea, and gastrointestinal upset [44,45]. Table 5 lists some examples of beverages concerning the caffeine content. Average values of caffeine in selected foods are shown in Table 6.

Table 5. Average values for selected foods [49].

Beverage	Caffeine Amounts [a]
Filter coffee	90 mg/200 mL
Energy drink	80 mg/250 mL
Black tea	45 mg/200 mL
Coke	35 mg/330 mL
Dark chocolate	25 mg/50 g
Iced-tea	17 mg/330 mL

[a] It should be noted that these are plant-based, natural raw materials and the caffeine content can therefore vary [49].

Table 6. Caffeine amount/cup [25].

Beverage	Caffeine Amounts
Coffee [a]	74 mg/237 mL
Tea (*Camellia sinensis*)	15 mg/237 mL
Coffee leaf tea [a]	12 mg/250 mL

[a] Without specifying the variety, probably *C. arabica*, as according to Miedaner, around two-thirds of the coffee sold worldwide is made from Arabica beans [44].

The oral LD_{50} of caffeine in rats is 192 mg/kg, and the iv LD_{50} in rats is 105 mg/kg [45]. An acute fatal overdose of caffeine in humans is approximately 10–14 g (equivalent to 150–200 mg/kg of body weight) [45].

An estimated 90% of adults in the United States consume caffeine daily, which means that caffeine intake is almost universal. The average daily intake is 200 mg caffeine/day, which according to Table 5 corresponds to approximately two cups of filter coffee. There are no known effects on the liver. Epidemiological studies suggest that regular coffee consumption has a modest protective effect against the progression of chronic liver disease and the development of liver cancer [45,50].

High levels of caffeine were observed in energy drinks, in particular, see Table 5 [49]. Several reports of liver damage related to the use of caffeine-rich energy drinks have been published, but they are, as described in the source, uninformative as they are not well documented, and the causality is questionable [45]. Therefore, caffeine is unlikely to cause liver damage, but the various high-caffeinated energy drinks widely available can possibly cause liver damage if used in excess [45]. The likelihood score for caffeine in coffee is rated E, which means that it is an unlikely cause of clinically apparent liver injury [45].

According to the WHO assessment of the carcinogenicity of caffeine in humans and also in laboratory animals, there is insufficient evidence that caffeine is carcinogenic. Caffeine was therefore classified as group 3, i.e., not classifiable, concerning its carcinogenicity for humans [51]. Intoxication with caffeine is also included in the World Health Organization's International Classification of Diseases (ICD-10) [45]. Caffeine levels of 57 mg/kg intravenously or oral doses of 18–50 g are lethal in human adults [45].

Zhang et al. [52] showed the preventive benefits of coffee and tea or their combination with respect to stroke and dementia. Two to three cups of coffee/day or three to five cups of tea/day or their combined consumption of four to six cups/day was shown to have the lowest hazard ratio (HR) for stroke and dementia.

As mentioned above, *C. arabica* coffees normally have a lower caffeine content (0.9 and 1.4 g/100 g DW) compared to *C. canephora* varieties (1.5 and 2.6 g/100 g DW) [5,6]. According to Rimbach [5], the caffeine content in tea is 1–4% of the dry weight on average.

Analysis of coffee leaf tea made of *C. arabica* var. Pacamara and Bourbon showed amounts between 0.34 and 1.24 g/100 g or 0.4–1.3% of the dry weight [23,24] (Table 7).

Here, the type and variety of leaves had significant influences on the caffeine content of the tea (Figure 10). Young leaves had the highest caffeine value (0.91 g/100 g DW) while yellow leaves had the lowest (0.44 g/100 g DW) [23].

The *C. canephora* samples showed amounts between 1.91 and 2.39 g/100 g or 2.1–2.6% of the dry weight. Fibrianto et al. reported that "coffee leaves contain caffeine at 21.9 g/kg of dry weight" [31]. There is 87.6 mg caffeine/cup of coffee leaf tea. This roughly corresponds to the content of coffee (Table 6). As shown in Table 2, the caffeine content of coffee leaf tea has a wide range since it is a natural raw material.

The EFSA published a scientific opinion on the safety of caffeine [53] in 2015. As mentioned by EFSA "caffeine intakes of no concern derived for acute caffeine consumption by adults (3 mg/kg bw per day) may serve as a basis to derive single doses of caffeine and daily caffeine intakes of no concern for these population subgroups". According to the Ministry of Food and Agriculture (BMEL), Germany, and the EFSA, up to 200 mg of caffeine as a single dose or 400 mg spread over a day is supposed to be virtually free of risk for healthy adults [49,53]. Table 7 lists some examples of beverages concerning the caffeine content and dosage.

Table 7. Caffeine Content.

<table>
<tr><th>Product</th><th>Portion Size</th><th>Dosage per Drink</th><th>Amount of Liquid [mL]</th><th>Caffeine Content [%/DW]</th></tr>
<tr><td colspan="5">Coffee</td></tr>
<tr><td>Coffee (C. canephora, roasted) [54]</td><td>1 cup</td><td>4 g</td><td>125–150</td><td>1.5–3.0</td></tr>
<tr><td>Coffee (C. arabica, roasted) [54]</td><td>1 cup</td><td>4 g</td><td>125–150</td><td>1.2–1.5</td></tr>
<tr><td colspan="5">Tea</td></tr>
<tr><td>Black tea [54]</td><td>1 cup</td><td>2 tea spoons</td><td>300</td><td>1.3–4.0</td></tr>
<tr><td>Green tea [54]</td><td>1 cup</td><td>2–3 heaped tea spoons</td><td>300</td><td>0.1–5.9</td></tr>
<tr><td>Coffee leaf tea (C. arabica)</td><td rowspan="2">1 cup</td><td rowspan="2">5 g</td><td rowspan="2">250 (1 tea cup)</td><td>0.4–1.3</td></tr>
<tr><td>Coffee leaf tea (C. canephora)</td><td>2.1–2.6</td></tr>
<tr><td>Rooibos-tea [54]</td><td colspan="4">No caffeine</td></tr>
</table>

For tea infusions, it is mentioned that significantly less caffeine dissolves in water depending on the preparation [54]. According to Frede et al. [6], the transfer of substances into the drink depends on the size of the leaf and brewing time and the water temperature. For a 2-min infusion (tea bag), Frede et al. [6] mentioned that nearly 100% of caffeine and around 50% of catechins pass into the drink while other sources [54] mention that not all caffeine dissolves in the water during the preparation of the infusion.

The caffeine content is not only influenced by the *Coffea* spp. used but also by brewing temperature and time. The recommended optimal brewing conditions as reported by Hariyadi et al. [55] are shown in Table 8. As already described above, the maximum caffeine content is <80 mg/L [22]. To not exceed this value, several scenarios were calculated in Table 9.

Table 8. Recommended optimal processing point [55].

	Canephora Coffee Leaf Tea	*Liberica* Coffee Leaf Tea
Brewing conditions: temperature [°C]	93.4	91.7
Brewing conditions: duration [min]	4.80	4.84
caffeine level [mg/100 mL]	74.9	72.5

Table 9. Maximum daily caffeine intake based on coffee leaf tea.

Product	Caffeine Content [g/100 g] [23,24]	Virtually Safe Intake (Coffee Leaf Tea) [a]
Coffee leaf tea (*C. arabica*)	Min. 0.34 Max. 1.24	Min. caffeine content: 1.18 L Max. caffeine content: 0.32 L
Coffee leaf tea (*C. canephora*)	Min. 1.91 Max. 2.39	Min. caffeine content: 0.21 L Max. caffeine content: 0.17 L

[a] The following assumptions were made: Complete (100%) extraction of caffeine. Weight of water [g] = volume of water [mL]. To prepare the coffee leaf tea, 2 g of dried coffee leaves/100 mL of infusion is required (which corresponds to typical manufacturer's information). Calculation: Caffeine content [g/100 g]/50 = caffeine content in 2 g of dried coffee leaves [1]. The caffeine content given in g/100 g of tea, calculated at the 2 g of tea used [1]. Caffeine content in 2 g of dried coffee leaves with a maximum caffeine content of 0.8 g/100 g [2]. How often the caffeine content of 2 g of coffee leaf tea falls within the safe value of 0.8 g of caffeine [2]. Finally, the value is recalculated from 100 mL to 1 L [3].

The maximum daily caffeine intake possible with coffee leaf tea consumption depends on its content in the plant. With *C. canephora*, the recommended amounts are much lower than with *C. arabica*. The *C. canephora* samples (Samples 25–28) have the highest levels of caffeine, according to Figure 8. Rodrigues et al. [42] and Perrois et al. [43] also showed that *C. canephora* has a higher caffeine content than *C. arabica*. Miedaner confirmed this observation [44]. The calculations were performed as an example of the minimum and maximum caffeine content of the measured coffee leaf teas measured.

C. arabica: With the minimum caffeine content, up to five cups (200 mL) can be consumed, while with the maximum caffeine content, only one cup would be advisable.

C. canephora: With the minimum caffeine content, one cup (200 mL) can be consumed, while with the maximum caffeine content, half a cup would be advisable.

The mathematical values represent the worst-case considerations. Today, three to five cups of coffee a day are considered safe for healthy people [44]. Naturally, the assessment also depends on the individual dosage, with higher dosages such as 80 g coffee leaves per L of water often being preferable for taste reasons.

Information on the number of coffee/tea cups also fits well with the protective effects concerning stroke and dementia [52]. The caffeine content depends on the type of plant used and the processing of the coffee leaves. Therefore, the daily intake considered safe in terms of the caffeine content is between half a cup and five cups of coffee leaf tea (Table 9).

Since caffeine levels correspond largely to those of coffee, but the limit value of caffeine in coffee leaf tea is comparably low [22], the national consumption study (NVS) [46,47] should also be placed into focus. According to the NVS, the intake of herbal teas is not high. Values with health risk due to caffeine are not reached with normal tea consumption.

The price of coffee leaf tea is currently high compared to other tea products, so it is probably not consumed daily.

4.3. Exposure to (−)-Epigallocatechin-3-Gallate (EGCG)

According to Haller et al. [10], polyphenols are the most common antioxidants. Naturally, they are available in fruits, vegetables, grains, legumes, chocolate, and beverages such as tea, coffee or wine [10]. According to the USDA database, EGCG is found in many everyday foods, such as dairy products, apples or bananas [56].

In 2018, EFSA was asked to provide a scientific opinion on green tea catechins [57]. Evaluation of the ingredient EGCG, as the main representative of catechins, can also be applied to the tea made from coffee leaves. The EFSA panel considered the possible association between the consumption of EGCG, the most relevant catechin in green tea, and hepatotoxicity [57]. According to EFSA, EGCG is present in unconjugated form in plasma after oral ingestion and it is the most cytotoxic catechin (compared to EGC and ECG) in primary rat hepatocytes. In this assessment, EFSA concluded that the amount of catechins absorbed through *Camellia sinensis* tea infusions is safe. However, it should be noted that on rare occasions, liver damage has occurred after ingesting green tea (more than 800 mg of EGCG/day). The reason given is likely an idiosyncratic reaction [57]. The panel states that 'hepatotoxicity is the adverse effect listed in the compendium, which also states that the doses causing hepatotoxicity are not indicated in human case reports' [57]. Additionally, there are also positive effects of EGCG, such as reduced risk of type 2 diabetes and its cardiovascular complications [58], prevention of metabolic syndrome [59], improved endothelial function and insulin sensitivity, reduced blood pressure, and protection against myocardial injury [60]. The combination of EGCG (blood pressure lowering) and caffeine (stimulating, blood pressure-increasing effect) in coffee leaf tea could be an interesting area of research.

The daily intake of tea is shown in Table 3. According to the table, men consume 538 g/day and women 506 g/day of green tea on average every day. The quantities of herbal tea are much lower (Table 3). According to EFSA, an adult consumes an average of 90–300 mg of EGCG/day. When consumed in large amounts, the value can reach 866 mg [57]. Similar to coffee leaf tea, the chemical composition of green tea varies depending on the type of plant, environment, season, age of the leaves, and thus also the EGCG content, which is in the range of 1600 to 20,320 mg/100 g of dried leaves (13 times). Regarding green tea infusions, the EGCG content varied over more than an 88-fold range (2.3–203 mg/100 g of infusion) [57].

By evaluating studies, the EFSA concluded that there is no evidence of hepatotoxicity below 800 mg of EGCG/day for up to 12 months [57]. EFSA concludes that catechins from (green) tea infusions and similar beverages are generally safe.

According to Figure 7, catechins could only be detected in *C. arabica*, i.e., not in *C. canephora* coffee leaf tea [23,24]. Since the *C. canephora* leaf teas examined were only a small sample, further research is necessary.

The EGCG content, which is potentially liver damaging (above 800 mg of EGCG/day), does not relate to tea drinks but to extracts [57]. Therefore, from a toxicological perspective, the caffeine content is the focus of attention. However, a value of more than 800 mg of EGCG/day is not reached with normal daily coffee leaf tea consumption.

5. Conclusions

In contrast to black and green tea infusions, which are consumed as traditional foods, coffee leaf tea contains higher levels of caffeine, with the content roughly corresponding to that of coffee beans. The amount of caffeine in coffee leaf tea depends on the processing, the plant (species, variety), the harvest season, the age of the leaves, and the beverage preparation (dosage, temperature, etc.). Due to the caffeine content, up to three cups of coffee leaf tea, drunk throughout the day, are considered safe. Since coffee leaf tea is a niche item in Europe, it is unlikely that this amount will be exceeded.

The positive effects of caffeine (see Section 4.2) and EGCG (Section 4.3) are less obvious due to the (still) low consumption of coffee leaf tea and should be examined more closely. No EGCG could be detected in our coffee leaf tea samples examined. The results of the analytical data (all species and varieties) for caffeine and EGCG are below the EFSA Safety Thresholds and can be considered safe with moderate consumption in terms of toxicity or side effects.

According to Miedaner [44], it takes around nine months before coffee fruits can be harvested. In the press, it has been repeatedly reported that the current coffee prices are well below the production costs. Since it is a plant with evergreen leaves, its leaves can be harvested all year long and therefore represent an additional source of income for coffee farmers, as coffee beans and leaves are from the same plant [11].

Finally, it should be noted that EGCG and caffeine in coffee leaf tea occur in a specific matrix rich in polyphenols. This results in differences regarding bioavailability and effects compared with the isolated substances investigated in most studies.

Author Contributions: Conceptualization, N.T. and D.W.L.; methodology, N.T.; formal analysis, N.T.; investigation, N.T., M.C.S., V.S. and P.B.; writing—original draft preparation, N.T.; writing—review and editing, M.C.S., V.S., P.B., M.R., S.S., Y.Z., H.F. and D.W.L.; visualization, N.T.; supervision, S.S., Y.Z., H.F. and D.W.L.; project administration, D.W.L. All authors have read and agreed to the published version of the manuscript.

Funding: This research received no external funding.

Institutional Review Board Statement: Not applicable.

Informed Consent Statement: Not applicable.

Data Availability Statement: No new data were created or analyzed in this study. Data sharing is not applicable to this article.

Acknowledgments: This manuscript was published as part of the Postgraduate Study of "Toxicology and Environmental Protection" at the University of Leipzig, Germany.

Conflicts of Interest: S.S. is the owner of and P.B. is a consultant for Coffee Consulate, Mannheim, Germany. Coffee Consulate is an independent training and research center. Coffee Consulate is currently researching the potential of coffee by-products. However, S.S. and P.B. report that there is no conflict of interest related to the work under consideration. The other authors declare that they have no conflict of interest.

References

1. Baltes, W.; Matissek, R. *Lebensmittelchemie*, 7th ed.; Springer: Berlin/Heidelberg, Germany, 2011.
2. Belitz, H.-D.; Grosch, W.; Schieberle, P. *Lehrbuch der Lebensmittelchemie*, 6th ed.; Springer: Berlin/Heidelberg, Germany, 2008.
3. Klingel, T.; Kremer, J.I.; Gottstein, V.; Rajcic de Rezende, T.; Schwarz, S.; Lachenmeier, D.W. A Review of Coffee By-Products Including Leaf, Flower, Cherry, Husk, Silver Skin, and Spent Grounds as Novel Foods within the European Union. *Foods* **2020**, *9*, 665. [CrossRef]
4. Lachenmeier, D.W.; Schwarz, S.; Rieke-Zapp, J.; Cantergiani, E.; Rawel, H.; Martín-Cabrejas, M.A.; Martuscelli, M.; Gottstein, V.; Angeloni, S. Coffee By-Products as Sustainable Novel Foods: Report of the 2nd International Electronic Conference on Foods—"Future Foods and Food Technologies for a Sustainable World". *Foods* **2022**, *11*, 3. [CrossRef] [PubMed]
5. Rimbach, G.; Nagursky, J.; Erbersdobler, H. F *Lebensmittel-Warenkunde für Einsteiger*, 2nd ed.; Springer: Berlin, Germany, 2015.
6. Frede, W. *Taschenbuch für Lebensmittelchemiker: Lebensmittel—Bedarfsgegenstände—Kosmetika—Futtermittel*; Springer: Berlin/Heidelberg, Germany, 2006.
7. Lachenmeier, D.W.; Schwarz, S. Digested Civet Coffee Beans (Kopi Luwak)-An Unfortunate Trend in Specialty Coffee Caused by Mislabeling of *Coffea liberica*? *Foods* **2021**, *10*, 1329. [CrossRef]
8. Rösemann, U. Coffea—Für viele müde Menschen die erste Pflanze des Tages. *Biol. Unserer Zeit* **2017**, *47*, 336–337. [CrossRef]
9. Bibium. What Is the Coffee Belt. Available online: https://bibium.co.uk/bean-belt-infographic/ (accessed on 30 October 2021).
10. Haller, D.; Rimbach, G.; Grune, T. *Biofunktionalität der Lebensmittelinhaltsstoffe*; Springer: Berlin/Heidelberg, Germany, 2013.
11. Lachenmeier, D.W.; Rajcic de, R.T.; Schwarz, S. An update on sustainable valorization of coffee by-products as novel foods within the European Union. *Biol. Life Sci. Forum* **2022**, *10*. [CrossRef]
12. International Coffee Organization. Coffee Market Report—September 2021. Available online: https://www.ico.org/documents/cy2020-21/cmr-0921-e.pdf (accessed on 26 October 2021).
13. European Union. Human Consumption to a Significant Degree; Information and Guidance Document. Available online: https://ec.europa.eu/food/system/files_de?file=2016-10/novel-food_guidance_human-consumption_en.pdf (accessed on 6 June 2021).

14. European Union. European Union Regulation (EU) 2015/2283 of the European parliament and of the council of 25 November 2015 on novel foods, amending Regulation (EU) No 1169/2011 of the European Parliament and of the Council and repealing Regulation (EC) No 258/97 of the European Parliament and of the Council and Commission Regulation (EC) No 1852/2001. *Off. J. Eur. Union* **2015**, *L327*, 1–22.
15. Chen, X. A review on coffee leaves: Phytochemicals, bioactivities and applications. *Crit. Rev. Food Sci. Nutr.* **2019**, *59*, 1008–1025. [CrossRef]
16. Novita, R.; Kasim, A.; Anggraini, T.; Putra, D.P. Kahwa daun: Traditional knowledge of a coffee leaf herbal tea from West Sumatera, Indonesia. *J. Ethnic Foods* **2018**, *5*, 286–291.
17. DIN 10809 Tee Bereitung Eines Aufgusses für Sensorische Prüfungen, Normenausschuß Lebensmittel und Landwirtschaftliche Produkte (NAL) im DIN Deutsches Institut für Normung e.V. Available online: https://www.din.de/de/mitwirken/normenausschuesse/nal (accessed on 4 November 2021).
18. AM Breweries IVS, Raffinaderivej 8, 1. Floor, DK-2300 Amager, Denmark Herbal Infusion Made from Coffee Leaves. Available online: https://ec.europa.eu/food/sites/food/files/safety/docs/novel-food_sum_ongoing-not_2018-0740.pdf (accessed on 30 May 2021).
19. EFSA. Novel Food. Available online: https://www.efsa.europa.eu/en/topics/topic/novel-food (accessed on 6 June 2021).
20. European Food Safety Authority (EFSA). Technical Report on the notification of infusion from coffee leaves (*Coffea arabica* L. and/or *Coffea canephora* Pierre ex A. Froehner) as a traditional food from a third country pursuant to Article 14 of Regulation (EU) 2015/2283. *EFSA Supp. Publ.* **2020**, *17*, 1783E. [CrossRef]
21. European Union. European Union implementing regulation (EU) 2017/2470 of 20 December 2017 establishing the Union list of novel foods in accordance with Regulation (EU) 2015/2283 of the European Parliament and of the Council on novel foods. *Off. J. Eur. Union* **2017**, *L351*, 72–201.
22. European Commission. Commission Implementing Regulation (EU) 2020/917 of 1 July 2020 authorising the placing on the market of infusion from coffee leaves of *Coffea arabica* L. and/or Coffea canephora Pierre ex A. Froehner as a traditional food from a third country under Regulation (EU) 2015/2283 of the European Parliament and of the Council and amending Implementing Regulation (EU) 2017/2470. *Off. J. Eur. Union* **2020**, *L209*, 10–13.
23. Steger, M.C. Coffee Leaf Tea from El Salvador: On-Site Production Considering Influences of Processing on Chemical Composition. Master's Thesis, University of Hohenheim, Hohenheim, Germany, 2021.
24. Segatz, V.; Steger, M.; Blumenthal, P.; Gottstein, V.; Rigling, M.; Schwarz, S.; Zhang, Y.; Lachenmeier, D.W. Evaluation of analytical methods to determine regulatory compliance of coffee leaf tea. *Biol. Life Sci. Forum* **2022**, *10*. [CrossRef]
25. Ratanamarno, S.; Surbkar, S. Caffeine and catechins in fresh coffee leaf (*Coffea arabica*) and coffee leaf tea. *Maejo Int. J. Sci. Technol.* **2017**, *11*, 211–218.
26. Zheng, X.; Ashihara, H. Distribution, biosynthesis and function of purine and pyridine alkaloids in *Coffea arabica* seedlings. *Plant Sci.* **2004**, *166*, 807–813. [CrossRef]
27. Frischknecht, P.M.; Ulmer-Dufek, J.; Baumann, T.W. Purine alkaloid formation in buds and developing leaflets of *Coffea arabica*: Expression of an optimal defence strategy? *Phytochemistry* **1986**, *25*, 613–616. [CrossRef]
28. Monteiro, Â.; Colomban, S.; Azinheira, H.G.; Guerra-Guimarães, L.; Do Céu Silva, M.; Navarini, L.; Resmini, M. Dietary Antioxidants in Coffee Leaves: Impact of Botanical Origin and Maturity on Chlorogenic Acids and Xanthones. *Antioxidants* **2020**, *9*, 6. [CrossRef] [PubMed]
29. Chen, X.; Mu, K.; Kitts, D.D. Characterization of phytochemical mixtures with inflammatory modulation potential from coffee leaves processed by green and black tea processing methods. *Food Chem.* **2019**, *271*, 248–258. [CrossRef]
30. Chen, X.-M.; Ma, Z.; Kitts, D.D. Effects of processing method and age of leaves on phytochemical profiles and bioactivity of coffee leaves. *Food Chem.* **2018**, *249*, 143–153. [CrossRef]
31. Fibrianto, K.; Wardhana, A.R.; Wahibah, L.Y.; Wulandari, E.S. The Influence of Leaf Age, Oxidizing Pre-Treatment and Serving Temperature on Sensory Characteristics of Ampelgading Robusta Coffee Leaves Tea. *J. Apl. Teknol. Pangan* **2019**, *8*, 100. [CrossRef]
32. Acidri, R.; Sawai, Y.; Sugimoto, Y.; Handa, T.; Sasagawa, D.; Masunaga, T.; Yamamoto, S.; Nishihara, E. Phytochemical Profile and Antioxidant Capacity of Coffee Plant Organs Compared to Green and Roasted Coffee Beans. *Antioxidants* **2020**, *9*, 93. [CrossRef]
33. Friedman, M.; Levin, C.E.; Choi, S.-H.; Lee, S.-U.; Kozukue, N. Changes in the composition of raw tea leaves from the Korean Yabukida plant during high-temperature processing to pan-fried Kamairi-cha green tea. *J. Food Sci.* **2009**, *74*, C406–C412. [CrossRef] [PubMed]
34. Toschi, T.G.; Cardenia, V.; Bonaga, G.; Mandrioli, M.; Rodriguez-Estrada, M.T. Coffee silverskin: Characterization, possible uses, and safety aspects. *J. Agric. Food Chem.* **2014**, *62*, 10836–10844. [CrossRef]
35. Król, K.; Gantner, M.; Tatarak, A.; Hallmann, E. The content of polyphenols in coffee beans as roasting, origin and storage effect. *Eur. Food Res. Technol.* **2020**, *246*, 33–39. [CrossRef]
36. Li, N.; Taylor, L.S.; Ferruzzi, M.G.; Mauer, L.J. Kinetic study of catechin stability: Effects of pH, concentration, and temperature. *J. Agric. Food Chem.* **2012**, *60*, 12531–12539. [CrossRef]
37. Turkmen, N.; Ferda, S.A.R.I.; Velioglu, Y.S. Factors Affecting Polyphenol Content and Composition of Fresh and Processed Tea Leaves. *Akad. Gıda* **2009**, *7*, 29–40.
38. Bowers, M.D.; Stamp, N.E. Effects of Plant Age, Genotype and Herbivory on Plantago Performance and Chemistry. *Ecology* **1993**, *74*, 1778–1791. [CrossRef]

39. Blum-Silva, C.H.; Chaves, V.C.; Schenkel, E.P.; Coelho, G.C.; Reginatto, F.H. The influence of leaf age on methylxanthines, total phenolic content, and free radical scavenging capacity of Ilex paraguariensis aqueous extracts. *Rev. Bras. Farmacogn.* **2015**, *25*, 1–6. [CrossRef]
40. Song, R.; Kelman, D.; Johns, K.L.; Wright, A.D. Correlation between leaf age, shade levels, and characteristic beneficial natural constituents of tea (*Camellia sinensis*) grown in Hawaii. *Food Chem.* **2012**, *133*, 707–714. [CrossRef]
41. Astill, C.; Birch, M.R.; Dacombe, C.; Humphrey, P.G.; Martin, P.T. Factors affecting the caffeine and polyphenol contents of black and green tea infusions. *J. Agric. Food Chem.* **2001**, *49*, 5340–5347. [CrossRef]
42. Rodrigues, C.I.; Marta, L.; Maia, R.; Miranda, M.; Ribeirinho, M.; Máguas, C. Application of solid-phase extraction to brewed coffee caffeine and organic acid determination by UV/HPLC. *J. Food Compos. Anal.* **2007**, *20*, 440–448. [CrossRef]
43. Perrois, C.; Strickler, S.R.; Mathieu, G.; Lepelley, M.; Bedon, L.; Michaux, S.; Husson, J.; Mueller, L.; Privat, I. Differential regulation of caffeine metabolism in *Coffea arabica* (*Arabica*) and *Coffea canephora* (*Robusta*). *Planta* **2015**, *241*, 179–191. [CrossRef]
44. Miedaner, T. *Genusspflanzen*, 1st ed.; Springer: Berlin/Heidelberg, Germany, 2018.
45. National Center for Biotechnology Information. PubChem Compound Summary for CID 2519, Caffeine. Available online: https://pubchem.ncbi.nlm.nih.gov/compound/Caffeine (accessed on 30 October 2021).
46. Max Rubner-Institut. *Nationale Verzehrsstudie II Lebensmittelverzehr und Nährstoffzufuhr auf Basis von 24h-Recalls*; Max Rubner-Institut: Karlsruhe, Germany, 2013.
47. Max Rubner-Institut; Nationale Verzehrsstudie II. *Ernährung*; Max Rubner-Institut: Karlsruhe, Germany, 2008; Volume 2, pp. 77–81.
48. BMEL. *Leitsätze für Tee, teeähnliche Erzeugnisse, Deren Extrakte und Zubereitungen vom 2.12.1998 (BAnz. Nr. 66a vom 9.4.1999, GMBl. Nr. 11 S.228 vom 26.4.1999), Zuletzt Geändert am 5.3.2013 (BAnz. AT vom 12.12.2013 B6, GMBl. Nr. 63 S.1265 vom 17.12.2013)*; BMEL: Berlin, Germany, 2013.
49. Bundesministerium für Ernährung und Landwirtschaft. Koffein—Checken Was Sicher Ist. Available online: https://www.bmel.de/SharedDocs/Downloads/DE/Broschueren/Koffein.pdf?__blob=publicationFile&v=3 (accessed on 27 June 2021).
50. Loomis, D.; Guyton, K.Z.; Grosse, Y.; Lauby-Secretan, B.; El Ghissassi, F.; Bouvard, V.; Benbrahim-Tallaa, L.; Guha, N.; Mattock, H.; Straif, K. Carcinogenicity of drinking coffee, mate, and very hot beverages. *Lancet Oncol.* **2016**, *17*, 877–878. [CrossRef]
51. IARC Working Group on the Evaluation of Carcinogenic Risks to Humans; International Agency for Research on Cancer; World Health Organization. *Coffee, Tea, Mate, Methylxanthines and Methylglyoxal*; WHO: Lyon, France, 1991.
52. Zhang, Y.; Yang, H.; Li, S.; Li, W.-D.; Wang, Y. Consumption of coffee and tea and risk of developing stroke, dementia, and poststroke dementia: A cohort study in the UK Biobank. *PLoS Med.* **2021**, *18*, e1003830. [CrossRef]
53. EFSA. Panel on Dietetic Products, Nutrition and Allergies Scientific Opinion on the Safety of Caffeine. *EFSA J.* **2015**, *13*, 4254.
54. Dr. Schweikart Verlag. Grüner Tee & Gesundheit. Available online: https://www.gruenertee.com/koffeingehalt-tee-kaffee/ (accessed on 7 June 2021).
55. Hariyadi, D.M.; Tedja, C.A.; Zubaidah, E.; Yuwono, S.S.; Fibrianto, K. Optimization of brewing time and temperature for caffeine and tannin levels in Dampit coffee leaf tea of Robusta (*Coffea canephora*) and Liberica (*Coffea liberica*). *Potravinarstvo* **2020**, *14*, 58–68. [CrossRef]
56. Haytowitz, D.B. *USDA Database for the Flavonoid Content of Selected Foods, Release 3.1*; USDA: Washington, DC, USA, 2014.
57. EFSA. Panel on Food Additives and Nutrient Sources added to Food Scientific opinion on the safety of green tea catechins. *EFSA J.* **2018**, *16*, e05239.
58. Keske, M.A.; Ng, H.L.H.; Premilovac, D.; Rattigan, S.; Kim, J.-A.; Munir, K.; Yang, P.; Quon, M.J. Vascular and metabolic actions of the green tea polyphenol epigallocatechin gallate. *CMC* **2015**, *22*, 59–69. [CrossRef]
59. Legeay, S.; Rodier, M.; Fillon, L.; Faure, S.; Clere, N. Epigallocatechin Gallate: A Review of Its Beneficial Properties to Prevent Metabolic Syndrome. *Nutrients* **2015**, *7*, 5443–5468. [CrossRef]
60. Potenza, M.A.; Marasciulo, F.L.; Tarquinio, M.; Tiravanti, E.; Colantuono, G.; Federici, A.; Kim, J.-A.; Quon, M.J.; Montagnani, M. EGCG, a green tea polyphenol, improves endothelial function and insulin sensitivity, reduces blood pressure, and protects against myocardial I/R injury in SHR. American journal of physiology. *Endocrinol. Metab.* **2007**, *292*, E1378–E1387. [CrossRef] [PubMed]

MDPI AG
Grosspeteranlage 5
4052 Basel
Switzerland
Tel.: +41 61 683 77 34

Foods Editorial Office
E-mail: foods@mdpi.com
www.mdpi.com/journal/foods

www.ingramcontent.com/pod-product-compliance
Lightning Source LLC
Chambersburg PA
CBHW061834210926
51850CB00006B/885